Selected Titles in This Series

72 **Gerald Teschl,** Jacobi operators and completely integrable nonlinear lattices, 2000

71 **Lajos Pukánszky,** Characters of connected Lie groups, 1999

70 **Carmen Chicone and Yuri Latushkin,** Evolution semigroups in dynamical systems and differential equations, 1999

69 **C. T. C. Wall (A. A. Ranicki, Editor),** Surgery on compact manifolds, second edition, 1999

68 **David A. Cox and Sheldon Katz,** Mirror symmetry and algebraic geometry, 1999

67 **A. Borel and N. Wallach,** Continuous cohomology, discrete subgroups, and representations of reductive groups, second edition, 2000

66 **Yu. Ilyashenko and Weigu Li,** Nonlocal bifurcations, 1999

65 **Carl Faith,** Rings and things and a fine array of twentieth century associative algebra, 1999

64 **Rene A. Carmona and Boris Rozovskii, Editors,** Stochastic partial differential equations: Six perspectives, 1999

63 **Mark Hovey,** Model categories, 1999

62 **Vladimir I. Bogachev,** Gaussian measures, 1998

61 **W. Norrie Everitt and Lawrence Markus,** Boundary value problems and symplectic algebra for ordinary differential and quasi-differential operators, 1999

60 **Iain Raeburn and Dana P. Williams,** Morita equivalence and continuous-trace C^*-algebras, 1998

59 **Paul Howard and Jean E. Rubin,** Consequences of the axiom of choice, 1998

58 **Pavel I. Etingof, Igor B. Frenkel, and Alexander A. Kirillov, Jr.,** Lectures on representation theory and Knizhnik-Zamolodchikov equations, 1998

57 **Marc Levine,** Mixed motives, 1998

56 **Leonid I. Korogodski and Yan S. Soibelman,** Algebras of functions on quantum groups: Part I, 1998

55 **J. Scott Carter and Masahico Saito,** Knotted surfaces and their diagrams, 1998

54 **Casper Goffman, Togo Nishiura, and Daniel Waterman,** Homeomorphisms in analysis, 1997

53 **Andreas Kriegl and Peter W. Michor,** The convenient setting of global analysis, 1997

52 **V. A. Kozlov, V. G. Maz'ya, and J. Rossmann,** Elliptic boundary value problems in domains with point singularities, 1997

51 **Jan Malý and William P. Ziemer,** Fine regularity of solutions of elliptic partial differential equations, 1997

50 **Jon Aaronson,** An introduction to infinite ergodic theory, 1997

49 **R. E. Showalter,** Monotone operators in Banach space and nonlinear partial differential equations, 1997

48 **Paul-Jean Cahen and Jean-Luc Chabert,** Integer-valued polynomials, 1997

47 **A. D. Elmendorf, I. Kriz, M. A. Mandell, and J. P. May (with an appendix by M. Cole),** Rings, modules, and algebras in stable homotopy theory, 1997

46 **Stephen Lipscomb,** Symmetric inverse semigroups, 1996

45 **George M. Bergman and Adam O. Hausknecht,** Cogroups and co-rings in categories of associative rings, 1996

44 **J. Amorós, M. Burger, K. Corlette, D. Kotschick, and D. Toledo,** Fundamental groups of compact Kähler manifolds, 1996

43 **James E. Humphreys,** Conjugacy classes in semisimple algebraic groups, 1995

42 **Ralph Freese, Jaroslav Ježek, and J. B. Nation,** Free lattices, 1995

41 **Hal L. Smith,** Monotone dynamical systems: an introduction to the theory of competitive and cooperative systems, 1995

(Continued in the back of this publication)

i

Jacobi Operators and Completely Integrable Nonlinear Lattices

Mathematical Surveys and Monographs

Volume 72

Jacobi Operators and Completely Integrable Nonlinear Lattices

Gerald Teschl

American Mathematical Society

Editorial Board

Georgia M. Benkart
Peter Landweber

Michael Loss
Tudor Stefan Ratiu, Chair

1991 *Mathematics Subject Classification.* Primary 39Axx, 47B39, 58F07.

ABSTRACT. This book is intended to serve both as an introduction and a reference to spectral and inverse spectral theory of Jacobi operators (i.e., second order symmetric difference operators) and applications of these theories to the Toda and Kac-van Moerbeke hierarchy.

Starting from second order difference equations we move on to self-adjoint operators and develop discrete Weyl-Titchmarsh-Kodaira theory, covering all classical aspects like Weyl m-functions, spectral functions, the moment problem, inverse spectral theory, and uniqueness results. Next, we investigate some more advanced topics like locating the essential, absolutely continuous, and discrete spectrum, subordinacy, oscillation theory, trace formulas, random operators, almost periodic operators, (quasi-)periodic operators, scattering theory, and spectral deformations.

Then, the Lax approach is used to introduce the Toda hierarchy and its modified counterpart, the Kac-van Moerbeke hierarchy. Uniqueness and existence theorems for the initial value problem, solutions in terms of Riemann theta functions, the inverse scattering transform, Bäcklund transformations, and soliton solutions are discussed.

Corrections and complements to this book are available from:

http://www.mat.univie.ac.at/~gerald/ftp/book-jac/

Mathematica® is a registered trademark of Wolfram Research Inc.

Library of Congress Cataloging-in-Publication Data

Teschl, Gerald, 1970–
  Jacobi operators and complete integrable nonlinear lattices / Gerald Teschl.
    p. cm. — (Mathematical surveys and monographs, ISSN 0076-5376 ; V. 72)
  Includes bibliographical references and index.
  ISBN 0-8218-1940-2
  1. Jacobi operators. 2. Differentiable dynamical systems. I. Title. II. Series: Mathematical surveys and monographs ; no. 72.
QA329.2.T47 1999
515′.7242–dc21
                                                                                     99-39165
                                                                                     CIP

Copying and reprinting. Individual readers of this publication, and nonprofit libraries acting for them, are permitted to make fair use of the material, such as to copy a chapter for use in teaching or research. Permission is granted to quote brief passages from this publication in reviews, provided the customary acknowledgment of the source is given.

Republication, systematic copying, or multiple reproduction of any material in this publication is permitted only under license from the American Mathematical Society. Requests for such permission should be addressed to the Assistant to the Publisher, American Mathematical Society, P. O. Box 6248, Providence, Rhode Island 02940-6248. Requests can also be made by e-mail to reprint-permission@ams.org.

© 2000 by the American Mathematical Society. All rights reserved.
The American Mathematical Society retains all rights
except those granted to the United States Government.
Printed in the United States of America.

∞ The paper used in this book is acid-free and falls within the guidelines
established to ensure permanence and durability.
Visit the AMS home page at URL: http://www.ams.org/

10 9 8 7 6 5 4 3 2 1   05 04 03 02 01 00

To Susanne

Contents

Preface   xiii

Part 1. Jacobi Operators

Chapter 1.   Jacobi operators   3
   1.1.   General properties   3
   1.2.   Jacobi operators   13
   1.3.   A simple example   19
   1.4.   General second order difference expressions   20
   1.5.   The infinite harmonic crystal in one dimension   22

Chapter 2.   Foundations of spectral theory for Jacobi operators   27
   2.1.   Weyl m-functions   27
   2.2.   Properties of solutions   29
   2.3.   Positive solutions   33
   2.4.   Weyl circles   37
   2.5.   Canonical forms of Jacobi operators and the moment problem   40
   2.6.   Some remarks on unbounded operators   47
   2.7.   Inverse spectral theory   53

Chapter 3.   Qualitative theory of spectra   57
   3.1.   Locating the spectrum and spectral multiplicities   57
   3.2.   Locating the essential spectrum   60
   3.3.   Locating the absolutely continuous spectrum   63
   3.4.   A bound on the number of eigenvalues   70

Chapter 4.   Oscillation theory   73
   4.1.   Prüfer variables and Sturm's separation theorem   73

4.2.	Classical oscillation theory	78
4.3.	Renormalized oscillation theory	81

Chapter 5. Random Jacobi operators — 87

5.1.	Random Jacobi operators	87
5.2.	The Lyapunov exponent and the density of states	91
5.3.	Almost periodic Jacobi operators	100

Chapter 6. Trace formulas — 105

6.1.	Asymptotic expansions	105
6.2.	General trace formulas and xi functions	109

Chapter 7. Jacobi operators with periodic coefficients — 115

7.1.	Floquet theory	115
7.2.	Connections with the spectra of finite Jacobi operators	119
7.3.	Polynomial identities	123
7.4.	Two examples: period one and two	124
7.5.	Perturbations of periodic operators	126

Chapter 8. Reflectionless Jacobi operators — 133

8.1.	Spectral analysis and trace formulas	133
8.2.	Isospectral operators	140
8.3.	The finite-gap case	142
8.4.	Further spectral interpretation	150

Chapter 9. Quasi-periodic Jacobi operators and Riemann theta functions — 153

9.1.	Riemann surfaces	153
9.2.	Solutions in terms of theta functions	155
9.3.	The elliptic case, genus one	162
9.4.	Some illustrations of the Riemann-Roch theorem	165

Chapter 10. Scattering theory — 167

10.1.	Transformation operators	167
10.2.	The scattering matrix	171
10.3.	The Gel'fand-Levitan-Marchenko equations	175
10.4.	Inverse scattering theory	180

Chapter 11. Spectral deformations – Commutation methods — 185

11.1.	Commuting first order difference expressions	185
11.2.	The single commutation method	187
11.3.	Iteration of the single commutation method	191
11.4.	Application of the single commutation method	194
11.5.	A formal second commutation	196

11.6.	The double commutation method	198
11.7.	Iteration of the double commutation method	204
11.8.	The Dirichlet deformation method	207

Notes on literature   215

Part 2. Completely Integrable Nonlinear Lattices

Chapter 12. The Toda system   221
12.1.	The Toda lattice	221
12.2.	Lax pairs, the Toda hierarchy, and hyperelliptic curves	224
12.3.	Stationary solutions	231
12.4.	Time evolution of associated quantities	234

Chapter 13. The initial value problem for the Toda system   239
13.1.	Finite-gap solutions of the Toda hierarchy	239
13.2.	Quasi-periodic finite-gap solutions and the time-dependent Baker-Akhiezer function	245
13.3.	A simple example – continued	248
13.4.	Inverse scattering transform	249
13.5.	Some additions in case of the Toda lattice	252
13.6.	The elliptic case – continued	254

Chapter 14. The Kac-van Moerbeke system   255
14.1.	The Kac-van Moerbeke hierarchy and its relation to the Toda hierarchy	255
14.2.	Kac and van Moerbeke's original equations	261
14.3.	Spectral theory for supersymmetric Dirac-type difference operators	262
14.4.	Associated solutions	263
14.5.	N-soliton solutions on arbitrary background	265

Notes on literature   271

Appendix A. Compact Riemann surfaces – a review   273
A.1.	Basic notation	273
A.2.	Abelian differentials	275
A.3.	Divisors and the Riemann-Roch theorem	277
A.4.	Jacobian variety and Abel's map	281
A.5.	Riemann's theta function	284
A.6.	The zeros of the Riemann theta function	286
A.7.	Hyperelliptic Riemann surfaces	289

Appendix B. Herglotz functions   297

Appendix C. Jacobi Difference Equations with `Mathematica`   307

C.1.	The package *DiffEqs* and first order difference equations	307
C.2.	The package *JacDEqs* and Jacobi difference equations	311
C.3.	Simple properties of Jacobi difference equations	313
C.4.	Orthogonal Polynomials	317
C.5.	Recursions	319
C.6.	Commutation methods	320
C.7.	Toda lattice	326
C.8.	Kac-van Moerbeke lattice	328

Bibliography	333
Glossary of notations	343
Index	349

Preface

The name of the game

Jacobi operators appear in a variety of applications. They can be viewed as the discrete analogue of Sturm-Liouville operators and their investigation has many similarities with Sturm-Liouville theory. Spectral and inverse spectral theory for Jacobi operators plays a fundamental role in the investigation of completely integrable nonlinear lattices, in particular the Toda lattice and its modified counterpart, the Kac-van Moerbeke lattice.

Why I have written this book

Whereas numerous books about Sturm-Liouville operators have been written, only few on Jacobi operators exist. In particular, there is currently no monograph available which covers all basic topics (like spectral and inverse spectral theory, scattering theory, oscillation theory and positive solutions, (quasi-)periodic operators, spectral deformations, etc.) typically found in textbooks on Sturm-Liouville operators.

In the case of the Toda lattice a textbook by M. Toda [**230**] exists, but none of the recent advances in the theory of nonlinear lattices are covered there.

Audience and prerequisites

As audience I had researchers in mind. This book can be used to get acquainted with selected topics as well as to look up specific results. Nevertheless, no previous knowledge on difference equations is assumed and all results are derived in a self-contained manner. Hence the present book is accessible to graduate students as well. Previous experience with Sturm-Liouville operators might be helpful but is not necessary. Still, a solid working knowledge from other branches of mathematics is needed. In particular, I have assumed that the reader is familiar with the theory of (linear) self-adjoint operators in Hilbert spaces which can be found in (e.g.)

[**192**] or [**241**]. This theory is heavily used in the first part. In addition, the reader might have to review material from complex analysis (see Appendix A and B) and differential equations on Banach manifolds (second part only) if (s)he feels (s)he does not have the necessary background. However, this knowledge is mainly needed for understanding proofs rather than the results themselves.

The style of this book

The style of this monograph is strongly influenced by my personal bias. I have striven to present an intuitive approach to each subject and to include the simplest possible proof for every result. Most proofs are rather sketchy in character, so that the main idea becomes clear instead of being drowned by technicalities. Nevertheless, I have always tried to include enough information for the reader to fill in the remaining details (her)himself if desired. To help researchers, using this monograph as a reference, to quickly spot the result they are looking for, most information is found in display style formulas.

The entire treatment is supposed to be mathematically rigorous. I have tried to prove *every* statement I make and, in particular, these little obvious things, which turn out less obvious once one tries to prove them. In this respect I had Marchenko's monograph on Sturm-Liouville operators [**167**] and the one by Weidmann [**241**] on functional analysis in mind.

Literature

The first two chapters are of an introductory nature and collect some well-known "folklore", the successive more advanced chapters are a synthesis of results from research papers. In most cases I have rearranged the material, streamlined proofs, and added further facts which are not published elsewhere. All results appear without special attribution to who first obtained them but there is a section entitled "Notes on literature" in each part which contains references to the literature plus hints for additional reading. The bibliography is selective and far from being complete. It contains mainly references I (am aware of and which I) have actually used in the process of writing this book.

Terminology and notation

For the most part, the terminology used agrees with generally accepted usage. Whenever possible, I have tried to preserve original notation. Unfortunately I had to break with this policy at various points, since I have given higher priority to a consistent (and self-explaining) notation throughout the entire monograph. A glossary of notation can be found towards the end.

Contents

For convenience of the reader, I have split the material into two parts; one on Jacobi operators and one on completely integrable lattices. In particular, the second part is to a wide extent independent of the first one and anybody interested only in completely integrable lattices can move directly to the second part (after browsing Chapter 1 to get acquainted with the notation).

Part I

Chapter 1 gives an introduction to the theory of second order difference equations and bounded Jacobi operators. All basic notations and properties are presented here. In addition, this chapter provides several easy but extremely helpful gadgets. We investigate the case of constant coefficients and, as a motivation for the reader, the infinite harmonic crystal in one dimension is discussed.

Chapter 2 establishes the pillars of spectral and inverse spectral theory for Jacobi operators. Here we develop what is known as discrete Weyl-Titchmarsh-Kodaira theory. Basic things like eigenfunction expansions, connections with the moment problem, and important properties of solutions of the Jacobi equation are shown in this chapter.

Chapter 3 considers qualitative theory of spectra. It is shown how the essential, absolutely continuous, and point spectrum of specific Jacobi operators can be located in some cases. The connection between existence of α-subordinate solutions and α-continuity of spectral measures is discussed. In addition, we investigate under which conditions the number of discrete eigenvalues is finite.

Chapter 4 covers discrete Sturm-Liouville theory. Both classical oscillation and renormalized oscillation theory are developed.

Chapter 5 gives an introduction to the theory of random Jacobi operators. Since there are monographs (e.g., [**40**]) devoted entirely to this topic, only basic results on the spectra and some applications to almost periodic operators are presented.

Chapter 6 deals with trace formulas and asymptotic expansions which play a fundamental role in inverse spectral theory. In some sense this can be viewed as an application of Krein's spectral shift theory to Jacobi operators. In particular, the tools developed here will lead to a powerful reconstruction procedure from spectral data for reflectionless (e.g., periodic) operators in Chapter 8.

Chapter 7 considers the special class of operators with periodic coefficients. This class is of particular interest in the one-dimensional crystal model and several profound results are obtained using Floquet theory. In addition, the case of impurities in one-dimensional crystals (i.e., perturbation of periodic operators) is studied.

Chapter 8 again considers a special class of Jacobi operators, namely reflectionless ones, which exhibit an algebraic structure similar to periodic operators. Moreover, this class will show up again in Chapter 10 as the stationary solutions of the Toda equations.

Chapter 9 shows how reflectionless operators with no eigenvalues (which turn out to be associated with quasi-periodic coefficients) can be expressed in terms of Riemann theta functions. These results will be used in Chapter 13 to compute explicit formulas for solutions of the Toda equations in terms of Riemann theta functions.

Chapter 10 provides a comprehensive treatment of (inverse) scattering theory for Jacobi operators with constant background. All important objects like reflection/transmission coefficients, Jost solutions and Gel'fand-Levitan-Marchenko equations are considered. Again this applies to impurities in one-dimensional crystals. Furthermore, this chapter forms the main ingredient of the inverse scattering transform for the Toda equations.

Chapter 11 tries to deform the spectra of Jacobi operators in certain ways. We compute transformations which are isospectral and such which insert a finite number of eigenvalues. The standard transformations like single, double, or Dirichlet commutation methods are developed. These transformations can be used as powerful tools in inverse spectral theory and they allow us to compute new solutions from old solutions of the Toda and Kac-van Moerbeke equations in Chapter 14.

Part II

Chapter 12 is the first chapter on integrable lattices and introduces the Toda system as hierarchy of evolution equations associated with the Jacobi operator via the standard Lax approach. Moreover, the basic (global) existence and uniqueness theorem for solutions of the initial value problem is proven. Finally, the stationary hierarchy is investigated and the Burchnall-Chaundy polynomial computed.

Chapter 13 studies various aspects of the initial value problem. Explicit formulas in case of reflectionless (e.g., (quasi-)periodic) initial conditions are given in terms of polynomials and Riemann theta functions. Moreover, the inverse scattering transform is established.

The final Chapter 14 introduces the Kac van-Moerbeke hierarchy as modified counterpart of the Toda hierarchy. Again the Lax approach is used to establish the basic (global) existence and uniqueness theorem for solutions of the initial value problem. Finally, its connection with the Toda hierarchy via a Miura-type transformation is studied and used to compute N-soliton solutions on arbitrary background.

Appendix

Appendix A reviews the theory of Riemann surfaces as needed in this monograph. While most of us will know Riemann surfaces from a basic course on complex analysis or algebraic geometry, this will be mainly from an abstract viewpoint like in [**86**] or [**129**], respectively. Here we will need a more "computational" approach and I hope that the reader can extract this knowledge from Appendix A.

Appendix B compiles some relevant results from the theory of Herglotz functions and Borel measures. Since not everybody is familiar with them, they are included for easy reference.

Appendix C shows how a program for symbolic computation, Mathematica®, can be used to do some of the computations encountered during the main bulk. While I don't believe that programs for symbolic computations are an indispensable tool for doing research on Jacobi operators (or completely integrable lattices), they are at least useful for checking formulas. Further information and Mathematica® notebooks can be found at

> http://www.mat.univie.ac.at/~gerald/ftp/book-jac/

respectively

> ftp://ftp.mat.univie.ac.at/pub/teschl/book-jac/

Acknowledgments

This book has greatly profited from collaborations and discussions with W. Bulla, F. Gesztesy, H. Holden, M. Krishna, and B. Simon. In addition, many people generously devoted considerable time and effort to reading earlier versions of the manuscript and making many corrections. In particular, I wish to thank D. Damanik, H. Hanßmann, A. von der Heyden, R. Killip, T. Sørensen, S. Timischl, K. Unterkofler, and H. Widom. Next, I am happy to express my gratitude to P. Deift, J. Geronimo, and E. Lieb for helpful suggestions and advise. I also like to thank the staff at the American Mathematical Society for the fast and professional production of this book.

Partly supported by the Austrian Science Fund (http://www.fwf.ac.at/) under Grant No. P12864-MAT.

Finally, no book is free of errors. So if you find one, or if you have comments or suggestions, please let me know. I will make all corrections and complements available at the URL above.

Gerald Teschl

Vienna, Austria
May, 1999

Gerald Teschl
Institut für Mathematik
Universität Wien
Strudlhofgasse 4
1090 Wien, Austria

E-mail address: gerald@mat.univie.ac.at
URL: http://www.mat.univie.ac.at/~gerald/

Part 1

Jacobi Operators

Chapter 1

Jacobi operators

This chapter introduces to the theory of second order difference equations and Jacobi operators. All the basic notation and properties are presented here. In addition, it provides several easy but extremely helpful gadgets. We investigate the case of constant coefficients and, as an application, discuss the infinite harmonic crystal in one dimension.

1.1. General properties

The issue of this section is mainly to fix notation and to establish all for us relevant properties of symmetric three-term recurrence relations in a self-contained manner.

We start with some preliminary notation. For $I \subseteq \mathbb{Z}$ and M a set we denote by $\ell(I, M)$ the set of M-valued sequences $(f(n))_{n \in I}$. Following common usage we will frequently identify the sequence $f = f(.) = (f(n))_{n \in I}$ with $f(n)$ whenever it is clear that n is the index (I being understood). We will only deal with the cases $M = \mathbb{R}, \mathbb{R}^2, \mathbb{C},$ and $\mathbb{C}^2$. Since most of the time we will have $M = \mathbb{C}$, we omit M in this case, that is, $\ell(I) = \ell(I, \mathbb{C})$. For $N_1, N_2 \in \mathbb{Z}$ we abbreviate $\ell(N_1, N_2) = \ell(\{n \in \mathbb{Z} | N_1 < n < N_2\})$, $\ell(N_1, \infty) = \ell(\{n \in \mathbb{Z} | N_1 < n\})$, and $\ell(-\infty, N_2) = \ell(\{n \in \mathbb{Z} | n < N_2\})$ (sometimes we will also write $\ell(N_2, -\infty)$ instead of $\ell(-\infty, N_2)$ for notational convenience). If M is a Banach space with norm $|.|$, we define

$$\ell^p(I, M) = \{f \in \ell(I, M) | \sum_{n \in I} |f(n)|^p < \infty\}, \quad 1 \leq p < \infty,$$

(1.1) $\quad \ell^\infty(I, M) = \{f \in \ell(I, M) | \sup_{n \in I} |f(n)| < \infty\}.$

Introducing the following norms

(1.2) $\quad \|f\|_p = \left(\sum_{n \in I} |f(n)|^p \right)^{1/p}, \quad 1 \leq p < \infty, \quad \|f\|_\infty = \sup_{n \in I} |f(n)|,$

makes $\ell^p(I, M)$, $1 \leq p \leq \infty$, a Banach space as well.

Furthermore, $\ell_0(I, M)$ denotes the set of sequences with only finitely many values being nonzero and $\ell_\pm^p(\mathbb{Z}, M)$ denotes the set of sequences in $\ell(\mathbb{Z}, M)$ which are ℓ^p near $\pm\infty$, respectively (i.e., sequences whose restriction to $\ell(\pm\mathbb{N}, M)$ belongs to $\ell^p(\pm\mathbb{N}, M)$. Here $\mathbb{N}$ denotes the set of positive integers). Note that, according to our definition, we have

$$(1.3) \qquad \ell_0(I, M) \subseteq \ell^p(I, M) \subseteq \ell^q(I, M) \subseteq \ell^\infty(I, M), \quad p < q,$$

with equality holding if and only if I is finite (assuming $\dim M > 0$).

In addition, if M is a (separable) Hilbert space with scalar product $\langle.,..\rangle_M$, then the same is true for $\ell^2(I, M)$ with scalar product and norm defined by

$$\langle f, g \rangle = \sum_{n \in I} \langle f(n), g(n) \rangle_M,$$

$$(1.4) \qquad \|f\| = \|f\|_2 = \sqrt{\langle f, f \rangle}, \quad f, g \in \ell^2(I, M).$$

For what follows we will choose $I = \mathbb{Z}$ for simplicity. However, straightforward modifications can be made to accommodate the general case $I \subset \mathbb{Z}$.

During most of our investigations we will be concerned with difference expressions, that is, endomorphisms of $\ell(\mathbb{Z})$;

$$(1.5) \qquad \begin{array}{rcl} R: \ell(\mathbb{Z}) & \to & \ell(\mathbb{Z}) \\ f & \mapsto & Rf \end{array}$$

(we reserve the name difference operator for difference expressions defined on a subset of $\ell^2(\mathbb{Z})$). Any difference expression R is uniquely determined by its corresponding matrix representation

$$(1.6) \qquad \big(R(m, n)\big)_{m,n \in \mathbb{Z}}, \quad R(m, n) = (R\delta_n)(m) = \langle \delta_m, R\delta_n \rangle,$$

where

$$(1.7) \qquad \delta_n(m) = \delta_{m,n} = \begin{cases} 0 & n \neq m \\ 1 & n = m \end{cases}$$

is the canonical basis of $\ell(\mathbb{Z})$. The **order** of R is the smallest nonnegative integer $N = N_+ + N_-$ such that $R(m, n) = 0$ for all m, n with $n - m > N_+$ and $m - n > N_-$. If no such number exists, the order is infinite.

We call R **symmetric** (resp. **skew-symmetric**) if $R(m, n) = R(n, m)$ (resp. $R(m, n) = -R(n, m)$).

Maybe the simplest examples for a difference expression are the **shift expressions**

$$(1.8) \qquad (S^\pm f)(n) = f(n \pm 1).$$

They are of particular importance due to the fact that their powers form a basis for the space of all difference expressions (viewed as a module over the ring $\ell(\mathbb{Z})$). Indeed, we have

$$(1.9) \qquad R = \sum_{k \in \mathbb{Z}} R(., . + k)(S^+)^k, \quad (S^\pm)^{-1} = S^\mp.$$

Here $R(., . + k)$ denotes the multiplication expression with the sequence $(R(n, n + k))_{n \in \mathbb{Z}}$, that is, $R(., . + k) : f(n) \mapsto R(n, n + k)f(n)$. In order to simplify notation we agree to use the short cuts

$$(1.10) \qquad f^\pm = S^\pm f, \quad f^{++} = S^+ S^+ f, \quad \text{etc.,}$$

1.1. General properties

whenever convenient. In connection with the difference expression (1.5) we also define the diagonal, upper, and lower triangular parts of R as follows

(1.11) $$[R]_0 = R(.,.), \quad [R]_{\pm} = \sum_{k \in \mathbb{N}} R(.,. \pm k)(S^{\pm})^k,$$

implying $R = [R]_+ + [R]_0 + [R]_-$.

Having these preparations out of the way, we are ready to start our investigation of second order symmetric difference expressions. To set the stage, let $a, b \in \ell(\mathbb{Z}, \mathbb{R})$ be two real-valued sequences satisfying

(1.12) $$a(n) \in \mathbb{R} \setminus \{0\}, \quad b(n) \in \mathbb{R}, \quad n \in \mathbb{Z},$$

and introduce the corresponding **second order, symmetric difference expression**

(1.13) $$\begin{aligned} \tau : \ell(\mathbb{Z}) &\to \ell(\mathbb{Z}) \\ f(n) &\mapsto a(n)f(n+1) + a(n-1)f(n-1) + b(n)f(n) \end{aligned}.$$

It is associated with the tridiagonal matrix

(1.14) $$\begin{pmatrix} \ddots & \ddots & \ddots & & & \\ a(n-2) & b(n-1) & a(n-1) & & & \\ & a(n-1) & b(n) & a(n) & & \\ & & a(n) & b(n+1) & a(n+1) & \\ & & & \ddots & \ddots & \ddots \end{pmatrix}$$

and will be our main object for the rest of this section and the tools derived here – even though simple – will be indispensable for us.

Before going any further, I want to point out that there is a close connection between second order, symmetric difference expressions and second order, symmetric differential expressions. This connection becomes more apparent if we use the difference expressions

(1.15) $$(\partial f)(n) = f(n+1) - f(n), \quad (\partial^* f)(n) = f(n-1) - f(n),$$

(note that ∂, ∂^* are formally adjoint) to rewrite τ in the following way

(1.16) $$\begin{aligned} (\tau f)(n) &= -(\partial^* a \partial f)(n) + (a(n-1) + a(n) + b(n))f(n) \\ &= -(\partial a^- \partial^* f)(n) + (a(n-1) + a(n) + b(n))f(n). \end{aligned}$$

This form resembles very much the Sturm-Liouville differential expression, well-known in the theory of ordinary differential equations.

In fact, the reader will soon realize that there are a whole lot more similarities between differentials, integrals and their discrete counterparts differences and sums. Two of these similarities are the product rules

(1.17) $$\begin{aligned} (\partial fg)(n) &= f(n)(\partial g)(n) + g(n+1)(\partial f)(n), \\ (\partial^* fg)(n) &= f(n)(\partial^* g)(n) + g(n-1)(\partial^* f)(n) \end{aligned}$$

and the **summation by parts** formula (also known as **Abel transform**)

(1.18) $$\sum_{j=m}^{n} g(j)(\partial f)(j) = g(n)f(n+1) - g(m-1)f(m) + \sum_{j=m}^{n} (\partial^* g)(j)f(j).$$

Both are readily verified. Nevertheless, let me remark that ∂, ∂^* are no derivations since they do not satisfy Leibnitz rule. This very often makes the discrete case different (and sometimes also harder) from the continuous one. In particular, many calculations become much messier and formulas longer.

There is much more to say about relations for the difference expressions (1.15) analogous to the ones for differentiation. We refer the reader to, for instance, [4], [**87**], or [**147**] and return to (1.13).

Associated with τ is the eigenvalue problem $\tau u = zu$. The appropriate setting for this eigenvalue problem is the Hilbert space $\ell^2(\mathbb{Z})$. However, before we can pursue the investigation of the eigenvalue problem in $\ell^2(\mathbb{Z})$, we need to consider the **Jacobi difference equation**

$$(1.19) \qquad \tau u = z\, u, \qquad u \in \ell(\mathbb{Z}),\, z \in \mathbb{C}.$$

Using $a(n) \neq 0$ we see that a solution u is uniquely determined by the values $u(n_0)$ and $u(n_0 + 1)$ at two consecutive points n_0, $n_0 + 1$ (you have to work much harder to obtain the corresponding result for differential equations). It follows, that there are exactly two linearly independent solutions.

Combining (1.16) and the summation by parts formula yields **Green's formula**

$$(1.20) \qquad \sum_{j=m}^{n} \Big(f(\tau g) - (\tau f)g\Big)(j) = W_n(f, g) - W_{m-1}(f, g)$$

for $f, g \in \ell(\mathbb{Z})$, where we have introduced the (modified) **Wronskian**

$$(1.21) \qquad W_n(f, g) = a(n)\Big(f(n)g(n+1) - g(n)f(n+1)\Big).$$

Green's formula will be the key to self-adjointness of the operator associated with τ in the Hilbert space $\ell^2(\mathbb{Z})$ (cf. Theorem 1.5) and the Wronskian is much more than a suitable abbreviation as we will show next.

Evaluating (1.20) in the special case where f and g both solve (1.19) (with the same parameter z) shows that the Wronskian is constant (i.e., does not depend on n) in this case. (The index n will be omitted in this case.) Moreover, it is nonzero if and only if f and g are linearly independent.

Since the (linear) space of solutions is two dimensional (as observed above) we can pick two linearly independent solutions c, s of (1.19) and write any solution u of (1.19) as a linear combination of these two solutions

$$(1.22) \qquad u(n) = \frac{W(u, s)}{W(c, s)} c(n) - \frac{W(u, c)}{W(c, s)} s(n).$$

For this purpose it is convenient to introduce the following **fundamental solutions** $c, s \in \ell(\mathbb{Z})$

$$(1.23) \qquad \tau c(z, ., n_0) = z\, c(z, ., n_0), \qquad \tau s(z, ., n_0) = z\, s(z, ., n_0),$$

fulfilling the **initial conditions**

$$(1.24) \qquad \begin{aligned} c(z, n_0, n_0) &= 1, & c(z, n_0 + 1, n_0) &= 0, \\ s(z, n_0, n_0) &= 0, & s(z, n_0 + 1, n_0) &= 1. \end{aligned}$$

Most of the time the base point n_0 will be unessential and we will choose $n_0 = 0$ for simplicity. In particular, we agree to omit n_0 whenever it is 0, that is,

$$(1.25) \qquad c(z, n) = c(z, n, 0), \qquad s(z, n) = s(z, n, 0).$$

Since the Wronskian of $c(z,.,n_0)$ and $s(z,.,n_0)$ does not depend on n we can evaluate it at n_0

(1.26) $$W(c(z,.,n_0), s(z,.,n_0)) = a(n_0)$$

and consequently equation (1.22) simplifies to

(1.27) $$u(n) = u(n_0)c(z,n,n_0) + u(n_0+1)s(z,n,n_0).$$

Sometimes a lot of things get more transparent if (1.19) is regarded from the viewpoint of dynamical systems. If we introduce $\underline{u} = (u, u^+) \in \ell(\mathbb{Z}, \mathbb{C}^2)$, then (1.19) is equivalent to

(1.28) $$\underline{u}(n+1) = U(z, n+1)\underline{u}(n), \qquad \underline{u}(n-1) = U(z,n)^{-1}\underline{u}(n),$$

where $U(z,.)$ is given by

(1.29) $$U(z,n) = \frac{1}{a(n)} \begin{pmatrix} 0 & a(n) \\ -a(n-1) & z - b(n) \end{pmatrix},$$
$$U^{-1}(z,n) = \frac{1}{a(n-1)} \begin{pmatrix} z - b(n) & -a(n) \\ a(n-1) & 0 \end{pmatrix}.$$

The matrix $U(z,n)$ is often referred to as **transfer matrix**. The corresponding (non-autonomous) flow on $\ell(\mathbb{Z}, \mathbb{C}^2)$ is given by the **fundamental matrix**

(1.30) $$\Phi(z, n, n_0) = \begin{pmatrix} c(z, n, n_0) & s(z, n, n_0) \\ c(z, n+1, n_0) & s(z, n+1, n_0) \end{pmatrix}$$
$$= \begin{cases} U(z, n) \cdots U(z, n_0 + 1) & n > n_0 \\ \mathbb{1} & n = n_0 \\ U^{-1}(z, n+1) \cdots U^{-1}(z, n_0) & n < n_0 \end{cases}.$$

More explicitly, equation (1.27) is now equivalent to

(1.31) $$\begin{pmatrix} u(n) \\ u(n+1) \end{pmatrix} = \Phi(z, n, n_0) \begin{pmatrix} u(n_0) \\ u(n_0+1) \end{pmatrix}.$$

Using (1.31) we learn that $\Phi(z, n, n_0)$ satisfies the usual group law

(1.32) $$\Phi(z, n, n_0) = \Phi(z, n, n_1)\Phi(z, n_1, n_0), \qquad \Phi(z, n_0, n_0) = \mathbb{1}$$

and constancy of the Wronskian (1.26) implies

(1.33) $$\det \Phi(z, n, n_0) = \frac{a(n_0)}{a(n)}.$$

Let us use $\Phi(z, n) = \Phi(z, n, 0)$ and define the upper, lower **Lyapunov exponents**

(1.34) $$\overline{\gamma}^{\pm}(z) = \limsup_{n \to \pm\infty} \frac{1}{|n|} \ln \|\Phi(z, n, n_0)\| = \limsup_{n \to \pm\infty} \frac{1}{|n|} \ln \|\Phi(z, n)\|,$$
$$\underline{\gamma}^{\pm}(z) = \liminf_{n \to \pm\infty} \frac{1}{|n|} \ln \|\Phi(z, n, n_0)\| = \liminf_{n \to \pm\infty} \frac{1}{|n|} \ln \|\Phi(z, n)\|.$$

Here

(1.35) $$\|\Phi\| = \sup_{u \in \mathbb{C}^2 \setminus \{0\}} \frac{\|\Phi u\|_{\mathbb{C}^2}}{\|u\|_{\mathbb{C}^2}}$$

denotes the operator norm of Φ. By virtue of (use (1.32))

(1.36) $$\|\Phi(z, n_0)\|^{-1} \|\Phi(z, n)\| \leq \|\Phi(z, n, n_0)\| \leq \|\Phi(z, n_0)^{-1}\| \|\Phi(z, n)\|$$

the definition of $\overline{\gamma}^\pm(z)$, $\underline{\gamma}^\pm(z)$ is indeed independent of n_0. Moreover, $\underline{\gamma}^\pm(z) \geq 0$ if $a(n)$ is bounded. In fact, since $\underline{\gamma}^\pm(z) < 0$ would imply $\lim_{j\to\pm\infty} \|\Phi(z, n_j, n_0)\| = 0$ for some subsequence n_j contradicting (1.33).

If $\underline{\gamma}^\pm(z) = \overline{\gamma}^\pm(z)$ we will omit the bars. A number $\lambda \in \mathbb{R}$ is said to be hyperbolic at $\pm\infty$ if $\underline{\gamma}^\pm(\lambda) = \overline{\gamma}^\pm(\lambda) > 0$, respectively. The set of all hyperbolic numbers is denoted by $\mathrm{Hyp}_\pm(\Phi)$. For $\lambda \in \mathrm{Hyp}_\pm(\Phi)$ one has existence of corresponding **stable** and **unstable manifolds** $V^\pm(\lambda)$.

Lemma 1.1. *Suppose that $|a(n)|$ does not grow or decrease exponentially and that $|b(n)|$ does not grow exponentially, that is,*

$$(1.37) \qquad \lim_{n\to\pm\infty} \frac{1}{|n|} \ln|a(n)| = 0, \qquad \lim_{n\to\pm\infty} \frac{1}{|n|} \ln(1+|b(n)|) = 0.$$

If $\lambda \in \mathrm{Hyp}_\pm(\Phi)$, then there exist one-dimensional linear subspaces $V^\pm(\lambda) \subseteq \mathbb{R}^2$ such that

$$\underline{v} \in V^\pm(\lambda) \Leftrightarrow \lim_{n\to\pm\infty} \frac{1}{|n|} \ln \|\Phi(\lambda, n)\underline{v}\| = -\gamma^\pm(\lambda),$$

$$(1.38) \qquad \underline{v} \notin V^\pm(\lambda) \Leftrightarrow \lim_{n\to\pm\infty} \frac{1}{|n|} \ln \|\Phi(\lambda, n)\underline{v}\| = \gamma^\pm(\lambda),$$

respectively.

Proof. Set

$$(1.39) \qquad A(n) = \begin{pmatrix} 1 & 0 \\ 0 & a(n) \end{pmatrix}$$

and abbreviate

$$\tilde{U}(z,n) = A(n)U(z,n)A(n-1)^{-1} = \frac{1}{a(n-1)}\begin{pmatrix} 0 & 1 \\ -a(n-1)^2 & z-b(n) \end{pmatrix},$$

$$(1.40) \qquad \tilde{\Phi}(z,n) = A(n)\Phi(z,n)A(0)^{-1}.$$

Then (1.28) translates into

$$(1.41) \qquad \underline{\tilde{u}}(n+1) = \tilde{U}(z, n+1)\underline{\tilde{u}}(n), \qquad \underline{\tilde{u}}(n-1) = \tilde{U}(z,n)^{-1}\underline{\tilde{u}}(n),$$

where $\underline{\tilde{u}} = A\underline{u} = (u, au^+)$, and we have

$$(1.42) \qquad \det \tilde{U}(z,n) = 1 \quad \text{and} \quad \lim_{n\to\infty} \frac{1}{|n|} \ln \|\tilde{U}(z,n)\| = 0$$

due to our assumption (1.37). Moreover,

$$(1.43) \qquad \frac{\min(1, a(n))}{\max(1, a(0))} \leq \frac{\|\tilde{\Phi}(z,n)\|}{\|\Phi(z,n)\|} \leq \frac{\max(1, a(n))}{\min(1, a(0))}$$

and hence $\lim_{n\to\pm\infty} |n|^{-1} \ln \|\tilde{\Phi}(z,n)\| = \lim_{n\to\pm\infty} |n|^{-1} \ln \|\Phi(z,n)\|$ whenever one of the limits exists. The same is true for the limits of $|n|^{-1} \ln \|\tilde{\Phi}(z,n)v\|$ and $|n|^{-1} \ln \|\Phi(z,n)v\|$. Hence it suffices to prove the result for matrices $\tilde{\Phi}$ satisfying (1.42). But this is precisely the (deterministic) multiplicative ergodic theorem of Osceledec (see [**201**]). $\square$

1.1. General properties

Observe that by looking at the Wronskian of two solutions $u \in V^\pm(\lambda)$, $v \notin V^\pm(\lambda)$ it is not hard to see that the lemma becomes false if $a(n)$ is exponentially decreasing.

For later use observe that

$$(1.44) \qquad \Phi(z,n,m)^{-1} = \frac{a(n)}{a(m)} J\, \Phi(z,n,m)^\top J^{-1}, \qquad J = \begin{pmatrix} 0 & -1 \\ 1 & 0 \end{pmatrix},$$

(where $\Phi^\top$ denotes the transposed matrix of Φ) and hence

$$(1.45) \qquad |a(m)|\|\Phi(z,n,m)^{-1}\| = |a(n)|\|\Phi(z,n,m)\|.$$

We will exploit this notation later in this monograph but for the moment we return to our original point of view.

The equation

$$(1.46) \qquad (\tau - z)f = g$$

for fixed $z \in \mathbb{C}$, $g \in \ell(\mathbb{Z})$, is referred to as **inhomogeneous Jacobi equation**. Its solution can can be completely reduced to the solution of the corresponding **homogeneous Jacobi equation** (1.19) as follows. Introduce

$$(1.47) \qquad K(z,n,m) = \frac{s(z,n,m)}{a(m)}.$$

Then the sequence

$$(1.48) \qquad f(n) = f_0\, c(z,n,n_0) + f_1 s(z,n,n_0) + \sum_{m=n_0+1}^{n}{}^* K(z,n,m) g(m),$$

where

$$(1.49) \qquad \sum_{j=n_0}^{n-1}{}^* f(j) = \begin{cases} \sum_{j=n_0}^{n-1} f(j) & \text{for } n > n_0 \\ 0 & \text{for } n = n_0 \\ -\sum_{j=n}^{n_0-1} f(j) & \text{for } n < n_0 \end{cases},$$

satisfies (1.46) and the initial conditions $f(n_0) = f_0$, $f(n_0+1) = f_1$ as can be checked directly. The summation kernel $K(z,n,m)$ has the following properties: $K(z,n,n) = 0$, $K(z,n+1,n) = a(n)^{-1}$, $K(z,n,m) = -K(z,m,n)$, and

$$(1.50) \qquad K(z,n,m) = \frac{u(z,m)v(z,n) - u(z,n)v(z,m)}{W(u(z),v(z))}$$

for any pair $u(z)$, $v(z)$ of linearly independent solutions of $\tau u = zu$.

Another useful result is the **variation of constants** formula. It says that if one solution u of (1.19) with $u(n) \neq 0$ for all $n \in \mathbb{Z}$ is known, then a second (linearly independent, $W(u,v) = 1$) solution of (1.19) is given by

$$(1.51) \qquad v(n) = u(n) \sum_{j=n_0}^{n-1}{}^* \frac{1}{a(j)u(j)u(j+1)}.$$

It can be verified directly as well.

Sometimes transformations can help to simplify a problem. The following two are of particular interest to us. If u fulfills (1.19) and $u(n) \neq 0$, then the sequence $\phi(n) = u(n+1)/u(n)$ satisfies the (discrete) **Riccati equation**

$$(1.52) \qquad a(n)\phi(n) + \frac{a(n-1)}{\phi(n-1)} = z - b(n).$$

Conversely, if ϕ fulfills (1.52), then the sequence

$$(1.53) \qquad u(n) = \prod_{j=n_0}^{n-1}{}^* \phi(j) = \begin{cases} \prod_{j=n_0}^{n-1} \phi(j) & \text{for } n > n_0 \\ 1 & \text{for } n = n_0 \\ \prod_{j=n}^{n_0-1} \phi(j)^{-1} & \text{for } n < n_0 \end{cases},$$

fulfills (1.19) and is normalized such that $u(n_0) = 1$. In addition, we remark that the sequence $\phi(n)$ might be written as finite continued fraction,

$$(1.54) \quad a(n)\phi(n) = z - b(n) - \cfrac{a(n-1)^2}{z - b(n-1) - \cfrac{\ddots}{\quad - \cfrac{a(n_0+1)^2}{z - b(n_0+1) - \cfrac{a(n_0)}{\phi(n_0)}}}}$$

for $n > n_0$ and

$$(1.55) \qquad a(n)\phi(n) = \cfrac{a(n)^2}{z - b(n+1) - \cfrac{\ddots}{\quad - \cfrac{a(n_0-1)^2}{z - b(n_0) - a(n_0)\phi(n_0)}}}$$

for $n < n_0$.

If $\tilde{a}$ is a sequence with $\tilde{a}(n) \neq 0$ and u fulfills (1.19), then the sequence

$$(1.56) \qquad \tilde{u}(n) = u(n) \prod_{j=n_0}^{n-1}{}^* \tilde{a}(j),$$

fulfills

$$(1.57) \qquad \frac{a(n)}{\tilde{a}(n)}\tilde{u}(n+1) + a(n-1)\tilde{a}(n-1)\tilde{u}(n-1) + b(n)\tilde{u}(n) = z\,\tilde{u}(n).$$

Especially, taking $\tilde{a}(n) = \text{sgn}(a(n))$ (resp. $\tilde{a}(n) = -\text{sgn}(a(n)))$, we see that it is no restriction to assume $a(n) > 0$ (resp. $a(n) < 0$) (compare also Lemma 1.6 below).

We conclude this section with a detailed investigation of the fundamental solutions $c(z, n, n_0)$ and $s(z, n, n_0)$. To begin with, we note (use induction) that both $c(z, n \pm k, n)$ and $s(z, n \pm k, n)$, $k \geq 0$, are polynomials of degree at most k with respect to z. Hence we may set

$$(1.58) \qquad s(z, n \pm k, n) = \sum_{j=0}^{k} s_{j,\pm k}(n) z^j, \quad c(z, n \pm k, n) = \sum_{j=0}^{k} c_{j,\pm k}(n) z^j.$$

1.1. General properties

Using the coefficients $s_{j,\pm k}(n)$ and $c_{j,\pm k}(n)$ we can derive a neat expansion for arbitrary difference expressions. By (1.9) it suffices to consider $(S^\pm)^k$.

Lemma 1.2. *Any difference expression R of order at most $2k+1$ can be expressed as*

$$(1.59) \qquad R = \sum_{j=0}^{k} \left(c_j + s_j S^+ \right) \tau^j, \qquad c_j, s_j \in \ell(\mathbb{Z}),\ k \in \mathbb{N}_0,$$

with $c_j = s_j = 0$ if and only if $R = 0$. In other words, the set $\{\tau^j, S^+ \tau^j\}_{j \in \mathbb{N}_0}$ forms a basis for the space of all difference expressions.

We have

$$(1.60) \qquad (S^\pm)^k = \sum_{j=0}^{k} \left(c_{j,\pm k} + s_{j,\pm k} S^+ \right) \tau^j,$$

where $s_{j,\pm k}(n)$ and $c_{j,\pm k}(n)$ are defined in (1.58).

Proof. We first prove (1.59) by induction on k. The case $k=0$ is trivial. Since the matrix element $\tau^k(n, n\pm k) = \prod_{j=0}^{k-1} a(n \pm j - \binom{0}{1}) \neq 0$ is nonzero we can choose $s_k(n) = R(n, n+k+1)/\tau^k(n-1, n+k-1)$, $c_k(n) = R(n, n-k)/\tau^k(n, n-k)$ and apply the induction hypothesis to $R - (c_k - s_k S^+)\tau^k$. This proves (1.59). The rest is immediate from

$$(1.61) \qquad (R(s(z,.,n)))(n) = \sum_{j=0}^{k} s_j(n) z^j, \qquad (R(c(z,.,n)))(n) = \sum_{j=0}^{k} c_j(n) z^j.$$

$\square$

As a consequence of (1.61) we note

Corollary 1.3. *Suppose R is a difference expression of order k. Then $R = 0$ if and only if $R|_{\mathrm{Ker}(\tau-z)} = 0$ for $k+1$ values of $z \in \mathbb{C}$. (Here $R|_{\mathrm{Ker}(\tau-z)} = 0$ says that $Ru = 0$ for any solution u of $\tau u = zu$.)*

Next, $\Phi(z, n_0, n_1) = \Phi(z, n_1, n_0)^{-1}$ provides the useful relations

$$(1.62) \qquad \begin{pmatrix} c(z, n_0, n_1) & s(z, n_0, n_1) \\ c(z, n_0+1, n_1) & s(z, n_0+1, n_1) \end{pmatrix} = \frac{a(n_1)}{a(n_0)} \begin{pmatrix} s(z, n_1+1, n_0) & -s(z, n_1, n_0) \\ -c(z, n_1+1, n_0) & c(z, n_1, n_0) \end{pmatrix}$$

and a straightforward calculation (using (1.27)) yields

$$s(z, n, n_0+1) = -\frac{a(n_0+1)}{a(n_0)} c(z, n, n_0),$$

$$s(z, n, n_0-1) = c(z, n, n_0) + \frac{z - b(n_0)}{a(n_0)} s(z, n, n_0),$$

$$c(z, n, n_0+1) = \frac{z - b(n_0+1)}{a(n_0)} c(z, n, n_0) + s(z, n, n_0),$$

$$(1.63) \qquad c(z, n, n_0-1) = -\frac{a(n_0-1)}{a(n_0)} s(z, n, n_0).$$

Our next task will be expansions of $c(z, n, n_0)$, $s(z, n, n_0)$ for large z. Let J_{n_1,n_2} be the **Jacobi matrix**

$$(1.64) \quad J_{n_1,n_2} = \begin{pmatrix} b(n_1+1) & a(n_1+1) & & & \\ a(n_1+1) & b(n_1+2) & \ddots & & \\ & \ddots & \ddots & \ddots & \\ & & \ddots & b(n_2-2) & a(n_2-2) \\ & & & a(n_2-2) & b(n_2-1) \end{pmatrix}.$$

Then we have the following expansion for $s(z, n, n_0)$, $n > n_0$,

$$(1.65) \quad s(z, n, n_0) = \frac{\det(z - J_{n_0,n})}{\prod_{j=n_0+1}^{n-1} a(j)} = \frac{z^k - \sum_{j=1}^{k} p_{n_0,n}(j) z^{k-j}}{\prod_{j=1}^{k} a(n_0+j)},$$

where $k = n - n_0 - 1 \geq 0$ and

$$(1.66) \quad p_{n_0,n}(j) = \frac{\operatorname{tr}(J_{n_0,n}^j) - \sum_{\ell=1}^{j-1} p_{n_0,n}(\ell) \operatorname{tr}(J_{n_0,n}^{j-\ell})}{j}, \quad 1 \leq j \leq k.$$

To verify the first equation, use that if z is a zero of $s(., n, n_0)$, then $(s(z, n_0+1, n_0), \ldots, s(z, n-1, n_0))$ is an eigenvector of (1.64) corresponding to the eigenvalue z. Since the converse statement is also true, the polynomials (in z) $s(z, n, n_0)$ and $\det(z - J_{n_0,n})$ only differ by a constant which can be deduced from (1.30). The second is a well-known property of characteristic polynomials (cf., e.g., [**91**]).

The first few traces are given by

$$\operatorname{tr}(J_{n_0,n_0+k+1}) = \sum_{j=n_0+1}^{n_0+k} b(j),$$

$$\operatorname{tr}(J_{n_0,n_0+k+1}^2) = \sum_{j=n_0+1}^{n_0+k} b(j)^2 + 2 \sum_{j=n_0+1}^{n_0+k-1} a(j)^2,$$

$$\operatorname{tr}(J_{n_0,n_0+k+1}^3) = \sum_{j=n_0+1}^{n_0+k} b(j)^3 + 3 \sum_{j=n_0+1}^{n_0+k-1} a(j)^2 (b(j) + b(j+1)),$$

$$\operatorname{tr}(J_{n_0,n_0+k+1}^4) = \sum_{j=n_0+1}^{n_0+k} b(j)^4 - 4 \sum_{j=n_0+1}^{n_0+k-1} a(j)^2 \Big(b(j)^2 + b(j+1)b(j)$$

$$(1.67) \qquad\qquad + b(j+1)^2 + \frac{a(j)^2}{2}\Big) + 4 \sum_{j=n_0+1}^{n_0+k-2} a(j+1)^2.$$

1.2. Jacobi operators

An explicit calculation yields for $n > n_0 + 1$

$$c(z,n,n_0) = \frac{-a(n_0)z^{n-n_0-2}}{\prod_{j=n_0+1}^{n-1} a(j)} \left(1 - \frac{1}{z}\sum_{j=n_0+2}^{n-1} b(j) + O(\frac{1}{z^2})\right),$$

(1.68) $$s(z,n,n_0) = \frac{z^{n-n_0-1}}{\prod_{j=n_0+1}^{n-1} a(j)} \left(1 - \frac{1}{z}\sum_{j=n_0+1}^{n-1} b(j) + O(\frac{1}{z^2})\right),$$

and (using (1.63) and (1.62)) for $n < n_0$

$$c(z,n,n_0) = \frac{z^{n_0-n}}{\prod_{j=n}^{n_0-1} a(j)} \left(1 - \frac{1}{z}\sum_{j=n+1}^{n_0} b(j) + O(\frac{1}{z^2})\right),$$

(1.69) $$s(z,n,n_0) = \frac{-a(n_0)z^{n_0-n-1}}{\prod_{j=n}^{n_0-1} a(j)} \left(1 - \frac{1}{z}\sum_{j=n+1}^{n_0-1} b(j) + O(\frac{1}{z^2})\right).$$

1.2. Jacobi operators

In this section we scrutinize the eigenvalue problem associated with (1.19) in the Hilbert space $\ell^2(\mathbb{Z})$.

Recall that the scalar product and norm is given by

(1.70) $$\langle f, g \rangle = \sum_{n \in \mathbb{Z}} \overline{f(n)} g(n), \quad \|f\| = \sqrt{\langle f, f \rangle}, \quad f, g \in \ell^2(\mathbb{Z}),$$

where the bar denotes complex conjugation.

For simplicity we assume from now on (and for the rest of this monograph) that a, b are bounded sequences.

Hypothesis H. 1.4. Suppose

(1.71) $$a, b \in \ell^\infty(\mathbb{Z}, \mathbb{R}), \quad a(n) \neq 0.$$

Associated with a, b is the **Jacobi operator**

(1.72) $$\begin{aligned} H: \ell^2(\mathbb{Z}) &\to \ell^2(\mathbb{Z}) \\ f &\mapsto \tau f \end{aligned},$$

whose basic properties are summarized in our first theorem.

Theorem 1.5. *Assume (H.1.4). Then H is a bounded self-adjoint operator. Moreover, a,b bounded is equivalent to H bounded since we have $\|a\|_\infty \leq \|H\|$, $\|b\|_\infty \leq \|H\|$ and*

(1.73) $$\|H\| \leq 2\|a\|_\infty + \|b\|_\infty,$$

where $\|H\|$ denotes the operator norm of H.

Proof. The fact that $\lim_{n\to\pm\infty} W_n(f,g) = 0$, $f, g \in \ell^2(\mathbb{Z})$, together with Green's formula (1.20) shows that H is self-adjoint, that is,

(1.74) $$\langle f, Hg \rangle = \langle Hf, g \rangle, \qquad f, g \in \ell^2(\mathbb{Z}).$$

For the rest consider $a(n)^2 + a(n-1)^2 + b(n)^2 = \|H\delta_n\|^2 \leq \|H\|^2$ and

(1.75) $$|\langle f, Hf \rangle| \leq (2\|a\|_\infty + \|b\|_\infty)\|f\|^2.$$

□

Before we pursue our investigation of Jacobi operators H, let us have a closer look at Hypothesis (H.1.4).

The previous theorem shows that the boundedness of H is due to the boundedness of a and b. This restriction on a, b is by no means necessary. However, it significantly simplifies the functional analysis involved and is satisfied in most cases of practical interest. You can find out how to avoid this restriction in Section 2.6.

The assumption $a(n) \neq 0$ is also not really necessary. In fact, we have not even used it in the proof of Theorem 1.5. If $a(n_0) = 0$, this implies that H can be decomposed into the direct sum $H_{n_0+1,-} \oplus H_{n_0,+}$ on $\ell^2(-\infty, n_0+1) \oplus \ell^2(n_0, \infty)$ (cf. (1.90) for notation). Nevertheless I want to emphasize that $a(n) \neq 0$ was crucial in the previous section and is connected with the existence of (precisely) two linearly independent solutions, which again is related to the fact that the spectrum of H has multiplicity at most two (cf. Section 2.5).

Hence the analysis of H in the case $a(n_0) = 0$ can be reduced to the analysis of restrictions of H which will be covered later in this section. In addition, the following lemma shows that the case $a(n) \neq 0$ can be reduced to the case $a(n) > 0$ or $a(n) < 0$.

Lemma 1.6. *Assume (H.1.4) and pick $\varepsilon \in \ell(\mathbb{Z}, \{-1, +1\})$. Introduce a_ε, b_ε by*

(1.76) $$a_\varepsilon(n) = \varepsilon(n)a(n), \quad b_\varepsilon(n) = b(n), \quad n \in \mathbb{Z},$$

and the unitary involution U_ε by

(1.77) $$U_\varepsilon = U_\varepsilon^{-1} = \Big(\prod_{j=0}^{n-1}{}^{*} \varepsilon(j)\, \delta_{m,n} \Big)_{m,n \in \mathbb{Z}}.$$

Let H be a Jacobi operator associated with the difference expression (1.13). Then H_ε defined as

(1.78) $$H_\varepsilon = U_\varepsilon H U_\varepsilon^{-1}$$

is associated with the difference expression

(1.79) $$(\tau_\varepsilon f)(n) = a_\varepsilon(n)f(n+1) + a_\varepsilon(n-1)f(n-1) + b_\varepsilon(n)f(n)$$

and is unitarily equivalent to H.

Proof. Straightforward. □

The next transformation is equally useful and will be referred to as **reflection** at n_0. It shows how information obtained near one endpoint, say $+\infty$, can be transformed into information near the other, $-\infty$.

1.2. Jacobi operators

Lemma 1.7. *Fix $n_0 \in \mathbb{Z}$ and consider the unitary involution*

(1.80) $$(U_R f)(n) = (U_R^{-1} f)(n) = f(2n_0 - n)$$

or equivalently $(U_R f)(n_0 + k) = f(n_0 - k)$. *Then the operator*

(1.81) $$H_R = U_R H U_R^{-1},$$

is associated with the sequences

(1.82) $$a_R(n_0 - k - 1) = a(n_0 + k), \quad b_R(n_0 - k) = b(n_0 + k), \quad k \in \mathbb{Z},$$

or equivalently $a_R(n) = a(2n_0 - n - 1)$, $b_R(n) = b(2n_0 - n)$.

Proof. Again straightforward. □

Associated with U_R are the two orthogonal projections

(1.83) $$P_R^\pm = \frac{1}{2}(\mathbb{1} \pm U_R), \qquad P_R^- + P_R^+ = \mathbb{1}, \; P_R^- P_R^+ = P_R^+ P_R^- = 0$$

and a corresponding splitting of H into two parts $H = H_R^+ \oplus H_R^-$, where

(1.84) $$H_R^\pm = P_R^\pm H P_R^\pm + P_R^\mp H P_R^\mp = \frac{1}{2}(H \pm H_R).$$

The symmetric part H_R^+ (resp. antisymmetric part H_R^-) commutes (resp. anticommutes) with U_R, that is, $[U_R, H_R^+] = U_R H_R^+ - H_R^+ U_R = 0$ (resp. $\{U_R, H_R^-\} = U_R H_R^- + H_R^- U_R = 0$). If $H = H_R^-$ we enter the realm of supersymmetric quantum mechanics (cf., e.g., [**225**] and Section 14.3).

After these two transformations we will say a little more about the spectrum $\sigma(H)$ of H. More precisely, we will estimate the location of $\sigma(H)$.

Lemma 1.8. *Let*

(1.85) $$c_\pm(n) = b(n) \pm \big(|a(n)| + |a(n-1)|\big).$$

Then we have

(1.86) $$\sigma(H) \subseteq [\inf_{n \in \mathbb{Z}} c_-(n), \sup_{n \in \mathbb{Z}} c_+(n)].$$

Proof. We will first show that H is semi-bounded from above by $\sup c_+$. From (1.16) we infer

(1.87) $$\langle f, Hf \rangle = \sum_{n \in \mathbb{Z}} \Big(-a(n)|f(n+1) - f(n)|^2 + \big(a(n-1) + a(n) + b(n)\big)|f(n)|^2 \Big).$$

By Lemma 1.6 we can first choose $a(n) > 0$ to obtain

(1.88) $$\langle f, Hf \rangle \leq \sup_{n \in \mathbb{Z}} c_+(n) \|f\|^2.$$

Similarly, choosing $a(n) < 0$ we see that H is semibounded from below by $\inf c_-$,

(1.89) $$\langle f, Hf \rangle \geq \inf_{n \in \mathbb{Z}} c_-(n) \|f\|^2,$$

completing the proof. □

We remark that these bounds are optimal in the sense that equality is attained for (e.g.) $a(n) = 1/2, b(n) = 0$ (cf. Section 1.3).

We will not only consider H but also restrictions of H; partly because they are of interest on their own, partly because their investigation gives information about H.

To begin with, we define the following restrictions $H_{\pm,n_0}$ of H to the subspaces $\ell^2(n_0, \pm\infty)$,

$$H_{+,n_0} f(n) = \begin{cases} a(n_0+1)f(n_0+2) + b(n_0+1)f(n_0+1), & n = n_0+1 \\ (\tau f)(n), & n > n_0+1 \end{cases},$$

(1.90) $H_{-,n_0} f(n) = \begin{cases} a(n_0-2)f(n_0-2) + b(n_0-1)f(n_0-1), & n = n_0-1 \\ (\tau f)(n), & n < n_0-1 \end{cases}.$

In addition, we also define for $\beta \in \mathbb{R} \cup \{\infty\}$

$$H^0_{+,n_0} = H_{+,n_0+1}, \quad H^\beta_{+,n_0} = H_{+,n_0} - a(n_0)\beta^{-1}\langle \delta_{n_0+1}, \cdot \rangle \delta_{n_0+1}, \quad \beta \neq 0,$$

(1.91) $H^\infty_{-,n_0} = H_{-,n_0}, \quad H^\beta_{-,n_0} = H_{-,n_0+1} - a(n_0)\beta \langle \delta_{n_0}, \cdot \rangle \delta_{n_0}, \quad \beta \neq \infty.$

All operators $H^\beta_{\pm,n_0}$ are bounded and self-adjoint.

Last, we define the following finite restriction H_{n_1,n_2} to the subspaces $\ell^2(n_1, n_2)$

(1.92) $H_{n_1,n_2} f(n) = \begin{cases} a(n_1+1)f(n_1+2) + b(n_1+1)f(n_1+1), & n = n_1+1 \\ (\tau f)(n), & n_1+1 < n < n_2-1 \\ a(n_2-2)f(n_2-2) + b(n_2-1)f(n_2-1), & n = n_2-1 \end{cases}.$

The operator H_{n_1,n_2} is clearly associated with the Jacobi matrix J_{n_1,n_2} (cf. (1.64)). Moreover, we set $H^{\infty,\infty}_{n_1,n_2} = H_{n_1,n_2}$, $H^{0,\beta_2}_{n_1,n_2} = H^{\infty,\beta_2}_{n_1+1,n_2}$, and

$$H^{\beta_1,\beta_2}_{n_1,n_2} = H^{\infty,\beta_2}_{n_1,n_2} - a(n_1)\beta_1^{-1}\langle \delta_{n_1+1}, \cdot \rangle \delta_{n_1+1}, \quad \beta_1 \neq 0,$$

(1.93) $H^{\beta_1,\beta_2}_{n_1,n_2} = H^{\beta_1,\infty}_{n_1,n_2+1} - a(n_2)\beta_2 \langle \delta_{n_2}, \cdot \rangle \delta_{n_2}, \quad \beta_2 \neq \infty.$

Remark 1.9. H^β_{+,n_0} can be associated with the following domain

(1.94) $\mathfrak{D}(H^\beta_{+,n_0}) = \{f \in \ell^2(n_0, \infty) | \cos(\alpha)f(n_0) + \sin(\alpha)f(n_0+1) = 0\},$

$\beta = \cot(\alpha) \neq 0$, if one agrees that only points with $n > n_0$ are of significance and that the last point is only added as a dummy variable so that one does not have to specify an extra expression for $(\tau f)(n_0+1)$. In particular, the case $\beta = \infty$ (i.e., corresponding to the boundary condition $f(n_0) = 0$) will be referred to as **Dirichlet boundary condition** at n_0. Analogously for H^β_{-,n_0}, $H^{\beta_1,\beta_2}_{n_1,n_2}$.

One of the most important objects in spectral theory is the **resolvent** $(H-z)^{-1}$, $z \in \rho(H)$, of H. Here $\rho(H) = \mathbb{C} \backslash \sigma(H)$ denotes the resolvent set of H. The matrix elements of $(H-z)^{-1}$ are called **Green function**

(1.95) $\quad G(z,n,m) = \langle \delta_n, (H-z)^{-1} \delta_m \rangle, \quad z \in \rho(H).$

Clearly,

(1.96) $\quad (\tau - z)G(z, \cdot, m) = \delta_m(\cdot), \quad G(z,m,n) = G(z,n,m)$

and

(1.97) $\quad ((H-z)^{-1}f)(n) = \sum_{m \in \mathbb{Z}} G(z,n,m) f(m), \quad f \in \ell^2(\mathbb{Z}), z \in \rho(H).$

1.2. Jacobi operators

We will derive an explicit formula for $G(z, n, m)$ in a moment. Before that we need to construct solutions $u_\pm(z)$ of (1.19) being square summable near $\pm\infty$.

Set

(1.98) $\qquad u(z,.) = (H - z)^{-1}\delta_0(.) = G(z,.,0), \qquad z \in \rho(H).$

By construction u fulfills (1.19) only for $n > 0$ and $n < 0$. But if we take $u(z, -2)$, $u(z, -1)$ as initial condition we can obtain a solution $u_-(z, n)$ of $\tau u = zu$ on the whole of $\ell(\mathbb{Z})$ which coincides with $u(z, n)$ for $n < 0$. Hence $u_-(z)$ satisfies $u_-(z) \in \ell_-^2(\mathbb{Z})$ as desired. A solution $u_+(z) \in \ell_+^2(\mathbb{Z})$ is constructed similarly.

As anticipated, these solutions allow us to write down the Green function in a somewhat more explicit way

(1.99) $\qquad G(z, n, m) = \dfrac{1}{W(u_-(z), u_+(z))} \begin{cases} u_+(z, n)u_-(z, m) & \text{for } m \leq n \\ u_+(z, m)u_-(z, n) & \text{for } n \leq m \end{cases},$

$z \in \rho(H)$. Indeed, since the right hand side of (1.99) satisfies (1.96) and is square summable with respect to n, it must be the Green function of H.

For later use we also introduce the convenient abbreviations

$$g(z, n) = G(z, n, n) = \frac{u_+(z, n)u_-(z, n)}{W(u_-(z), u_+(z))},$$

$$h(z, n) = 2a(n)G(z, n, n+1) - 1$$

(1.100) $\qquad = \dfrac{a(n)(u_+(z, n)u_-(z, n+1) + u_+(z, n)u_-(z, n+1))}{W(u_-(z), u_+(z))}.$

Note that for $n \leq m$ we have

$$G(z, n, m) = g(z, n_0)c(z, n, n_0)c(z, m, n_0)$$
$$+ g(z, n_0 + 1)s(z, n, n_0)s(z, m, n_0)$$
$$+ h(z, n_0)\frac{c(z, n, n_0)s(z, m, n_0) + c(z, m, n_0)s(z, n, n_0)}{2a(n_0)}$$

(1.101) $\qquad - \dfrac{c(z, n, n_0)s(z, m, n_0) - c(z, m, n_0)s(z, n, n_0)}{2a(n_0)}.$

Similar results hold for the restrictions: Let

(1.102) $\qquad s_\beta(z, n, n_0) = \sin(\alpha)c(z, n, n_0) - \cos(\alpha)s(z, n, n_0)$

with $\beta = \cot(\alpha)$ (i.e., the sequence $s_\beta(z, n, n_0)$ fulfills the boundary condition $\cos(\alpha)s_\beta(z, n_0, n_0) + \sin(\alpha)s_\beta(z, n_0 + 1, n_0) = 0$). Then we obtain for the resolvent of $H_{\pm,n_0}^\beta$

(1.103) $\qquad ((H_{\pm,n_0}^\beta - z)^{-1}u)(n) = \sum_{m \gtrless n_0} G_{\pm,n_0}^\beta(z, m, n)u(m), \quad z \in \rho(H_{\pm,n_0}^\beta),$

where

(1.104) $\qquad G_{\pm,n_0}^\beta(z, m, n) = \dfrac{\pm 1}{W(s_\beta(z), u_\pm(z))} \begin{cases} s_\beta(z, n, n_0)u_\pm(z, m) & \text{for } m \gtrless n \\ s_\beta(z, m, n_0)u_\pm(z, n) & \text{for } n \gtrless m \end{cases}$

(use $(H_{\pm,n_0}^\beta - z)^{-1}$ to show the existence of $u_\pm(z,.)$ for $z \in \rho(H_{\pm,n_0}^\beta)$).

Remark 1.10. The solutions being square summable near $\pm\infty$ (resp. satisfying the boundary condition $\cos(\alpha)f(n_0)+\sin(\alpha)f(n_0+1) = 0$) are unique up to constant multiples since the Wronskian of two such solutions vanishes (evaluate it at $\pm\infty$ (resp. n_0)). This implies that the point spectrum of H, $H_{\pm,n_0}^{\beta}$ is always simple.

In addition to $H_{\pm,n_0}^{\beta}$ we will be interested in the following direct sums of these operators

(1.105) $$H_{n_0}^{\beta} = H_{-,n_0}^{\beta} \oplus H_{+,n_0}^{\beta},$$

in the Hilbert space $\{f \in \ell^2(\mathbb{Z}) | \cos(\alpha)f(n_0) + \sin(\alpha)f(n_0 + 1) = 0\}$. The reason why $H_{n_0}^{\beta}$ is of interest to us follows from the close spectral relation to H as can be seen from their resolvents (resp. Green functions)

$$G_{n_0}^{\infty}(z, n, m) = G(z, n, m) - \frac{G(z, n, n_0)G(z, n_0, m)}{G(z, n_0, n_0)},$$

$$G_{n_0}^{\beta}(z, n, m) = G(z, n, m) - \gamma^{\beta}(z, n_0)^{-1}\bigl(G(z, n, n_0 + 1) + \beta G(z, n, n_0)\bigr)$$
(1.106) $$\times \bigl(G(z, n_0 + 1, m) + \beta G(z, n_0, m)\bigr), \quad \beta \in \mathbb{R},$$

where

$$\gamma^{\beta}(z, n) = \frac{(u_+(z, n+1) + \beta u_+(z, n))(u_-(z, n+1) + \beta u_-(z, n))}{W(u_-(z), u_+(z))}$$

(1.107) $$= g(z, n+1) + \frac{\beta}{a(n)} h(z, n) + \beta^2 g(z, n).$$

Remark 1.11. The operator $H_{n_0}^{\beta}$ is equivalently given by

(1.108) $$H_{n_0}^{\beta} = (\mathbb{1} - P_{n_0}^{\beta}) H (\mathbb{1} - P_{n_0}^{\beta})$$

in the Hilbert space $(\mathbb{1} - P_{n_0}^{\beta})\ell^2(\mathbb{Z}) = \{f \in \ell^2(\mathbb{Z}) | \langle \delta_{n_0}^{\beta}, f \rangle = 0\}$, where $P_{n_0}^{\beta}$ denotes the orthogonal projection onto the one-dimensional subspace spanned by $\delta_{n_0}^{\beta} = \cos(\alpha)\delta_{n_0} + \sin(\alpha)\delta_{n_0+1}$, $\beta = \cot(\alpha)$, $\alpha \in [0, \pi)$ in $\ell^2(\mathbb{Z})$.

Finally, we derive some interesting difference equations for $g(z, n)$ to be used in Section 6.1.

Lemma 1.12. Let u, v be two solutions of (1.19). Then $g(n) = u(n)v(n)$ satisfies

(1.109) $$\frac{(a^+)^2 g^{++} - a^2 g}{z - b^+} + \frac{a^2 g^+ - (a^-)^2 g^-}{z - b} = (z - b^+)g^+ - (z - b)g,$$

and

(1.110) $$\Bigl(a^2 g^+ - (a^-)^2 g^- + (z - b)^2 g\Bigr)^2 = (z - b)^2 \Bigl(W(u, v)^2 + 4a^2 g g^+\Bigr).$$

Proof. First we calculate (using (1.19))

(1.111) $$a^2 g^+ - (a^-)^2 g^- = -(z - b)^2 g + a(z - b)(uv^+ + u^+ v).$$

Adding $(z - b)^2 g$ and taking squares yields the second equation. Dividing both sides by $z - b$ and adding the equations corresponding to n and $n + 1$ yields the first. $\square$

Remark 1.13. There exists a similar equation for $\gamma^\beta(z,n)$. Since it is quite complicated, it seems less useful. Set $\gamma^\beta(n) = (u(n+1) + \beta u(n))(v(n+1) + \beta v(n))$, then we have

$$
\begin{aligned}
&\left((a^+ A^-)^2 (\gamma^\beta)^+ - (a^- A)^2 (\gamma^\beta)^- + B^2 \gamma^\beta\right)^2 \\
&\quad = (A^- B)^2 \left((\frac{A}{a} W(u,v))^2 + 4(a^+)^2 \gamma^\beta (\gamma^\beta)^+\right),
\end{aligned}
\tag{1.112}
$$

with

$$
\begin{aligned}
A &= a + \beta(z - b^+) + \beta^2 a^+, \\
B &= a^-(z - b^+) + \beta((z - b^+)(z - b) + a^+ a^- - a^2) \\
&\quad + \beta^2 a^+ (z - b).
\end{aligned}
\tag{1.113}
$$

It can be verified by a long and tedious (but straightforward) calculation.

1.3. A simple example

We have been talking about Jacobi operators for quite some time now, but we have not seen a single example yet. Well, here is one, the free Jacobi operator H_0 associated with constant sequences $a(n) = a$, $b(n) = b$. The transformation $z \to 2az + b$ reduces this problem to the one with $a_0(n) = 1/2$, $b_0(n) = 0$. Thus we will consider the equation

$$
\frac{1}{2}\Big(u(n+1) + u(n-1)\Big) = z u(n). \tag{1.114}
$$

Without restriction we choose $n_0 = 0$ throughout this section (note that we have $s(z, n, n_0) = s(z, n - n_0)$, etc.) and omit n_0 in all formulas. By inspection (try the ansatz $u(n) = k^n$) $u_\pm(z,.)$ are given by

$$
u_\pm(z, n) = (z \pm R_2^{1/2}(z))^n, \tag{1.115}
$$

where $R_2^{1/2}(z) = -\sqrt{z-1}\sqrt{z+1}$. Here and in the sequel $\sqrt{.}$ always denotes the standard branch of the square root, that is,

$$
\sqrt{z} = |\sqrt{z}|\exp(i \arg(z)/2), \quad \arg(z) \in (-\pi, \pi], \quad z \in \mathbb{C}. \tag{1.116}
$$

Since $W(u_-, u_+) = R_2^{1/2}(z)$ we need a second solution for $z^2 = 1$, which is given by $s(\pm 1, n) = (\pm 1)^{n+1} n$. For the fundamental solutions we obtain

$$
\begin{aligned}
s(z, n) &= \frac{(z + R_2^{1/2}(z))^n - (z - R_2^{1/2}(z))^n}{2 R_2^{1/2}(z)}, \\
c(z, n) &= \frac{s(z, n-1)}{s(z, -1)} = -s(z, n-1).
\end{aligned}
\tag{1.117}
$$

Notice that $s(-z, n) = (-1)^{n+1} s(z, n)$. For $n > 0$ we have the following expansion

$$
\begin{aligned}
s(z, n) &= \sum_{j=0}^{[n/2]} \binom{n}{2j+1} (z^2 - 1)^j z^{n-2j-1} \\
&= \sum_{k=0}^{[n/2]} \left((-1)^k \sum_{j=k}^{[n/2]} \binom{n}{2j+1}\binom{j}{k}\right) z^{n-2k-1},
\end{aligned}
\tag{1.118}
$$

where $[\![x]\!] = \sup\{n \in \mathbb{Z} \,|\, n < x\}$. It is easily seen that we have $\|H_0\| = 1$ and further that

(1.119) $$\sigma(H_0) = [-1, 1].$$

For example, use unitarity of the Fourier transform

(1.120) $$\begin{aligned} U : \ell^2(\mathbb{Z}) &\to L^2(-\pi, \pi) \\ u(n) &\mapsto \sum_{n \in \mathbb{Z}} u(n) e^{inx} \end{aligned}.$$

which maps H_0 to the multiplication operator by $\cos(x)$.

The Green function of H_0 explicitly reads ($z \in \mathbb{C}\backslash[-1,1]$)

(1.121) $$G_0(z, m, n) = \frac{(z + R_2^{1/2}(z))^{|m-n|}}{R_2^{1/2}(z)}.$$

In particular, we have

(1.122) $$\begin{aligned} g_0(z, n) &= \frac{1}{R_2^{1/2}(z)} = -2 \sum_{j=0}^{\infty} \binom{2j}{j} \frac{1}{(2z)^{2j+1}} \\ h_0(z, n) &= \frac{z}{R_2^{1/2}(z)} = -\sum_{j=0}^{\infty} \binom{2j}{j} \frac{1}{(2z)^{2j}}. \end{aligned}$$

Note that it is sometimes convenient to set $k = z + R_2^{1/2}(z)$ (and conversely $z = \frac{1}{2}(k + k^{-1})$), or $k = \exp(i\kappa)$ (and conversely $z = \cos(\kappa)$) implying

(1.123) $$u_\pm(z, n) = k^{\pm n} = e^{\pm i\kappa n}.$$

The map $z \mapsto k = z + R_2^{1/2}(z)$ is a holomorphic mapping from the set $\Pi_+ \simeq (\mathbb{C} \cup \{\infty\})\backslash[-1,1]$ to the unit disk $\{z \in \mathbb{C}\,|\,|z| < 1\}$. In addition, viewed as a map on the Riemann surface of $R_2^{1/2}(z)$, it provides an explicit isomorphism between the Riemann surface of $R_2^{1/2}(z)$ and the Riemann sphere $\mathbb{C} \cup \{\infty\}$.

1.4. General second order difference expressions

We consider the difference expression

(1.124) $$\hat{\tau} f(n) = \frac{1}{w(n)} \Big(f(n+1) + f(n-1) + d(n)f(n) \Big),$$

where $w(n) > 0$, $d(n) \in \mathbb{R}$, and $(w(n)w(n+1))^{-1}$, $w(n)^{-1}d(n)$ are bounded sequences. It gives rise to an operator $\hat{H}$, called **Helmholtz operator**, in the weighted Hilbert space $\ell^2(\mathbb{Z}; w)$ with scalar product

(1.125) $$\langle f, g \rangle = \sum_{n \in \mathbb{Z}} w(n) \overline{f(n)} g(n), \qquad f, g \in \ell^2(\mathbb{Z}; w).$$

Green's formula (1.20) holds with little modifications and $\hat{H}$ is easily seen to be bounded and self-adjoint. There is an interesting connection between Jacobi and Helmholtz operators stated in the next theorem.

1.4. General second order difference expressions

Theorem 1.14. *Let H be the Jacobi operator associated with the sequences $a(n) > 0$, $b(n)$ and let $\hat{H}$ be the Helmholtz operator associated with the sequences $w(n) > 0$, $d(n)$. If we relate these sequences by*

$$(1.126) \quad w(2m) = w(0) \prod_{j=0}^{m-1}{}^* \frac{a(2j)^2}{a(2j+1)^2}, \quad d(n) = w(n)b(n),$$

$$w(2m+1) = \frac{1}{a(2m)^2 w(2m)}$$

respectively

$$(1.127) \quad a(n) = \frac{1}{\sqrt{w(n)w(n+1)}}, \quad b(n) = \frac{d(n)}{w(n)},$$

then the operators H and $\hat{H}$ are unitarily equivalent, that is, $H = U\hat{H}U^{-1}$, where U is the unitary transformation

$$(1.128) \quad \begin{array}{rcl} U: \ell^2(\mathbb{Z};w) & \to & \ell^2(\mathbb{Z}) \\ u(n) & \mapsto & \sqrt{w(n)}u(n) \end{array}.$$

Proof. Straightforward. □

Remark 1.15. (i). The most general three-term recurrence relation

$$(1.129) \quad \tilde{\tau} f(n) = \tilde{a}(n)f(n+1) + \tilde{b}(n)f(n) + \tilde{c}(n)f(n-1),$$

with $\tilde{a}(n)\tilde{c}(n+1) > 0$, can be transformed to a Jacobi recurrence relation as follows. First we render $\tilde{\tau}$ symmetric,

$$(1.130) \quad \tilde{\tau} f(n) = \frac{1}{w(n)}\Big(c(n)f(n+1) + c(n-1)f(n-1) + d(n)f(n)\Big),$$

where

$$w(n) = \prod_{j=n_0}^{n-1}{}^* \frac{\tilde{a}(j)}{\tilde{c}(j+1)},$$

$$(1.131) \quad c(n) = w(n)\tilde{a}(n) = w(n+1)\tilde{c}(n+1), \quad d(n) = w(n)\tilde{b}(n).$$

Let $\tilde{H}$ be the self-adjoint operator associated with $\tilde{\tau}$ in $\ell^2(\mathbb{Z};w)$. Then the unitary operator

$$(1.132) \quad \begin{array}{rcl} U: \ell^2(\mathbb{Z};w) & \to & \ell^2(\mathbb{Z}) \\ u(n) & \mapsto & \sqrt{w(n)}u(n) \end{array}$$

transforms $\tilde{H}$ into a Jacobi operator $H = U\tilde{H}U^{-1}$ in $\ell^2(\mathbb{Z})$ associated with the sequences

$$a(n) = \frac{c(n)}{\sqrt{w(n)w(n+1)}} = \mathrm{sgn}(\tilde{a}(n))\sqrt{\tilde{a}(n)\tilde{c}(n+1)},$$

$$(1.133) \quad b(n) = \frac{d(n)}{w(n)} = \tilde{b}(n).$$

In addition, the Wronskians are related by

$$c(n)\Big(f(n)g(n+1) - f(n+1)g(n)\Big) =$$
(1.134) $$a(n)\Big((Uf)(n)(Ug)(n+1) - (Uf)(n+1)(Ug)(n)\Big).$$

(ii). Let $c(n) > 0$ be given. Defining

(1.135)
$$w(2m) = \prod_{j=0}^{m-1}{}^* \left(\frac{a(2j)c(2j+1)}{c(2j)a(2j+1)}\right)^2, \quad d(n) = w(n)b(n),$$
$$w(2m+1) = \frac{c(2m)^2}{a(2m)^2 w(2m)}$$

the transformation U maps H to an operator $\tilde{H} = UHU^{-1}$ associated with the difference expression (1.130).

1.5. The infinite harmonic crystal in one dimension

Finally, I want to say something about how Jacobi operators arise in applications. Despite the variety of possible applications of difference equations we will focus on only one model from solid state physics: the infinite harmonic crystal in one dimension. Hence we consider a linear chain of particles with harmonic nearest neighbor interaction. If $x(n,t)$ denotes the deviation of the n-th particle from its equilibrium position, the equations of motion read

(1.136)
$$m(n)\frac{d^2}{dt^2}x(n,t) = -k(n)\big(x(n+1,t) - x(n,t)\big) - k(n-1)\big(x(n-1,t) - x(n,t)\big)$$
$$= (\partial k \partial^* x)(n,t),$$

where $m(n) > 0$ is the mass of the n-th particle and $k(n)$ is the force constant between the n-th and $(n+1)$-th particle.

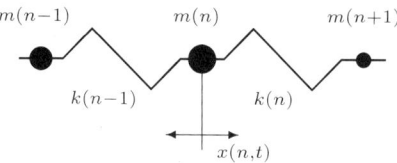

This model is only valid as long as the relative displacement is not too large (i.e., at least smaller than the distance of the particles in the equilibrium position). Moreover, from a physical viewpoint it is natural to assume $k, m, m^{-1} \in \ell^\infty(\mathbb{Z}, \mathbb{R})$ and $k(n) \neq 0$. Introducing conjugate coordinates

(1.137) $$p(n,t) = m(n)\frac{d}{dt}x(n,t), \quad q(n,t) = x(n,t),$$

the system (1.136) can be written as Hamiltonian system with Hamiltonian given by

(1.138) $$\mathcal{H}(p,q) = \sum_{n \in \mathbb{Z}} \left(\frac{p(n)^2}{2m(n)} + \frac{k(n)}{2}\big(q(n+1) - q(n)\big)^2\right).$$

1.5. The infinite harmonic crystal in one dimension

Since the total energy of the system is supposed to be finite, a natural phase space for this system is $(p,q) \in \ell^2(\mathbb{Z}, \mathbb{R}^2)$ with symplectic form

$$(1.139) \qquad \omega\big((p_1,q_1),(p_2,q_2)\big) = \sum_{n\in\mathbb{Z}} \Big(p_1(n)q_2(n) - p_2(n)q_1(n)\Big).$$

Using the symplectic transform

$$(1.140) \qquad (p,q) \to (v,u) = \Big(\frac{p}{\sqrt{m}}, \sqrt{m}\,q\Big)$$

we get a new Hamiltonian

$$(1.141) \qquad \tilde{\mathcal{H}}(v,u) = \frac{1}{2}\sum_{n\in\mathbb{Z}} \Big(v(n)^2 + 2a(n)u(n)u(n+1) + b(n)u(n)^2\Big),$$

where

$$(1.142) \qquad a(n) = \frac{-k(n)}{\sqrt{m(n)m(n+1)}}, \quad b(n) = \frac{k(n)+k(n-1)}{m(n)}.$$

The corresponding equations of evolution read

$$(1.143) \qquad \begin{aligned} \frac{d}{dt}u(n,t) &= \frac{\partial \tilde{\mathcal{H}}(v,u)}{\partial v(n,t)} = v(n,t), \\ \frac{d}{dt}v(n,t) &= -\frac{\partial \tilde{\mathcal{H}}(v,u)}{\partial u(n,t)} = -Hu(n,t), \end{aligned}$$

where H is our usual Jacobi operator associated with the sequences (1.142). Equivalently we have

$$(1.144) \qquad \frac{d^2}{dt^2}u(n,t) = -Hu(n,t).$$

Since this system is linear, standard theory implies

$$(1.145) \qquad \begin{aligned} u(n,t) &= \cos(t\sqrt{H})u(n,0) + \frac{\sin(t\sqrt{H})}{\sqrt{H}}v(n,0), \\ v(n,t) &= \cos(t\sqrt{H})v(n,0) - \frac{\sin(t\sqrt{H})}{\sqrt{H}}Hu(n,0), \end{aligned}$$

where

$$(1.146) \qquad \cos(t\sqrt{H}) = \sum_{\ell=0}^{\infty} \frac{(-1)^\ell t^{2\ell}}{(2\ell)!}H^\ell, \quad \frac{\sin(t\sqrt{H})}{\sqrt{H}} = \sum_{\ell=0}^{\infty} \frac{(-1)^\ell t^{2\ell+1}}{(2\ell+1)!}H^\ell.$$

In particular, introducing

$$(1.147) \qquad \underline{S}(\lambda,n) = \big(c(\lambda,n), s(\lambda,n)\big)$$

and expanding the initial conditions in terms of eigenfunctions (cf. Section 2.5, equation (2.133))

$$(1.148) \qquad \underline{u}(\lambda) = \sum_{n\in\mathbb{Z}} u(n,0)\underline{S}(\lambda,n), \quad \underline{v}(\lambda) = \sum_{n\in\mathbb{Z}} v(n,0)\underline{S}(\lambda,n)$$

we infer

$$u(n,t) = \int_{\sigma(H)} \left(\underline{u}(\lambda)\cos(t\sqrt{\lambda}) + \underline{v}(\lambda)\frac{\sin(t\sqrt{\lambda})}{\sqrt{\lambda}}\right)\underline{S}(\lambda,n)d\rho(\lambda),$$

(1.149) $$v(n,t) = \int_{\sigma(H)} \left(\underline{v}(\lambda)\cos(t\sqrt{\lambda}) - \underline{u}(\lambda)\frac{\sin(t\sqrt{\lambda})}{\sqrt{\lambda}}\lambda\right)\underline{S}(\lambda,n)d\rho(\lambda),$$

where $\rho(\lambda)$ is the spectral matrix of H.

This shows that in order to understand the dynamics of (1.144) one needs to understand the spectrum of H.

For example, in the case where $H \geq 0$ (or if $\underline{u}(\lambda), \underline{u}(\lambda) = 0$ for $\lambda \leq 0$) we clearly infer

(1.150)
$$\|u(t)\| \leq \|u(0)\| + t\|v(0)\|,$$
$$\|v(t)\| \leq \|v(0)\| + \|\sqrt{H}v(0)\|,$$

since $|\cos(t\sqrt{\lambda})| \leq 1$, $|\frac{\sin(t\sqrt{\lambda})}{t\sqrt{\lambda}}| \leq 1$ for $t \in \mathbb{R}$, $\lambda \geq 0$.

Given this setting two questions naturally arise. Firstly, given $m(n)$, $k(n)$ what can be said about the characteristic frequencies of the crystal? Spectral theory for H deals with this problem. Secondly, given the set of characteristic frequencies (i.e., the spectrum of H), is it possible to reconstruct $m(n)$, $k(n)$? This question is equivalent to inverse spectral theory for H once we establish how to reconstruct k, m from a, b. This will be done next.

Note that $u(n) = \sqrt{m(n)} > 0$ solves $\tau u = 0$. Hence, if we assume $k > 0$, then $H \geq 0$ by Corollary 11.2. In particular,

(1.151) $$\sigma(H) \subseteq [0, 2(\|k\|_\infty \|m^{-1}\|_\infty)]$$

and 0 is in the essential spectrum of H by Lemma 3.8. Moreover, since

(1.152) $$\sum_{n \in \pm \mathbb{N}} \frac{1}{-a(n)u(n)u(n+1)} = \sum_{n \in \pm \mathbb{N}} \frac{1}{k(n)} = \infty$$

(recall $k \in \ell^\infty(\mathbb{Z})$), H is critical (cf. Section 2.3). Thus we can recover k, m via

(1.153) $$k(n) = -m(0)a(n)u(n)u(n+1), \quad m(n) = m(0)u(n)^2,$$

where $u(n)$ is the unique positive solution of $\tau u = 0$ satisfying $u(0) = \sqrt{m(0)}$. That m, k can only be recovered up to a constant multiple (i.e., $m(0)$) is not surprising due to the corresponding scaling invariance of (1.136). If the positive solution of $\tau u = 0$ is not unique, we get a second free parameter. However, note that for each choice the corresponding Helmholtz operator $\hat{H} = m^{-1}\partial k \partial^*$ is unitary equivalent to H (and so are $\hat{H}_\pm$ and $H_\pm$).

From a physical viewpoint the case of a crystal with N atoms in the base cell is of particular interest, that is, k and m are assumed to be periodic with period N. In this case, the same is true for a, b and we are led to the study of periodic Jacobi operators. The next step is to consider impurities in this crystal. If such impurities appear only local, we are led to scattering theory with periodic background (to be dealt with in Chapter 7). If they are randomly distributed over the crystal we are led to random operators (to be dealt with in Chapter 5).

1.5. The infinite harmonic crystal in one dimension

But for the moment let us say a little more about the simplest case where m and k are both constants (i.e., $N = 1$). After the transformation $t \mapsto \sqrt{\frac{m}{2k}}$ we can even assume $m = k = 1$, that is,

$$\text{(1.154)} \qquad \frac{d^2}{dt^2} x(n,t) = -\frac{1}{2}\bigl(x(n+1,t) + x(n-1,t)\bigr) + x(n,t).$$

The so-called plane wave solutions are given by

$$\text{(1.155)} \qquad u^{\pm}(n,t) = e^{i(\kappa n \pm \nu(\kappa)t)},$$

where the wavelength κ^{-1} and the frequency ν are connected by the **dispersion relation**

$$\text{(1.156)} \qquad \nu(\kappa) = \sqrt{1 - \cos(\kappa)} = \sqrt{2}\sin(\frac{\kappa}{2}).$$

Since $u^{\pm}(n,t)$ is only meaningful for $n \in \mathbb{Z}$, we can restrict κ to values in the first Brillouin zone, that is, $\kappa \in (-\pi, \pi]$.

These solutions correspond to infinite total energy of the crystal. However, one can remedy this problem by taking (continuous) superpositions of these plane waves. Introducing

$$\text{(1.157)} \qquad u(\kappa) = \sum_{n \in \mathbb{Z}} u(n,0) e^{i\kappa n}, \quad v(\kappa) = \sum_{n \in \mathbb{Z}} v(n,0) e^{i\kappa n}$$

we obtain

$$\text{(1.158)} \qquad u(n,t) = \frac{1}{2\pi} \int_{-\pi}^{\pi} \Bigl(u(\kappa) \cos(\nu(\kappa)t) + v(\kappa) \frac{\sin(\nu(\kappa)t)}{\nu(\kappa)}\Bigr) e^{-i\kappa n} d\kappa.$$

Or, equivalently,

$$\text{(1.159)} \qquad u(n,t) = \sum_{m \in \mathbb{Z}} c_{n-m}(t) u(m,0) + s_{n-m}(t) v(m,0),$$

where

$$c_n(t) = \frac{1}{2\pi} \int_{-\pi}^{\pi} \cos(\nu(\kappa)t) e^{i\kappa n} d\kappa = J_{2|n|}(\sqrt{2}t),$$

$$s_n(t) = \frac{1}{2\pi} \int_{-\pi}^{\pi} \frac{\sin(\nu(\kappa)t)}{\nu(\kappa)} e^{i\kappa n} d\kappa = \int_0^t c_n(s) ds$$

$$\text{(1.160)} \qquad = \frac{t^{2|n|+1}}{2^{|n|}(|n|+1)!} {}_1F_2\bigl(\frac{2|n|+1}{2}; (\frac{2|n|+3}{2}, 2|n|+1); -\frac{t^2}{2}\bigr).$$

Here $J_n(x)$, ${}_pF_q(\underline{u}; \underline{v}; x)$ denote the Bessel and generalized hypergeometric functions, respectively. From this form one can deduce that a localized wave (say compactly supported at $t = 0$) will spread as t increases (cf. [**194**], Corollary to Theorem XI.14). This phenomenon is due to the fact that different plane waves travel with different speed and is called dispersion.

You might want to observe the following fact: if the coupling constant $k(n_0)$ between the n_0-th and $(n_0 + 1)$-th particle is zero (i.e., no interaction between these particles), then the chain separates into two parts (cf. the discussion after Theorem 1.5 and note that $k(n_0) = 0$ implies $a(n_0) = 0$ for the corresponding Jacobi operator).

We will encounter chains of particles again in Section 12.1. However, there we will consider a certain nonlinear interaction.

Chapter 2

Foundations of spectral theory for Jacobi operators

The theory presented in this chapter is the discrete analog of what is known as Weyl-Titchmarsh-Kodaira theory for Sturm-Liouville operators. The discrete version has the advantage of being less technical and more transparent.

Again, the present chapter is of fundamental importance and the tools developed here are the pillars of spectral theory for Jacobi operators.

2.1. Weyl m-functions

In this section we will introduce and investigate Weyl m-functions. Rather than the classical approach of Weyl (cf. Section 2.4) we advocate a different one which is more natural in the discrete case.

As in the previous chapter, $u_\pm(z,.)$ denote the solutions of (1.19) in $\ell(\mathbb{Z})$ which are square summable near $\pm\infty$, respectively.

We start by defining the **Weyl m-functions**

$$(2.1) \quad m_\pm(z, n_0) = \langle \delta_{n_0 \pm 1}, (H_{\pm, n_0} - z)^{-1} \delta_{n_0 \pm 1} \rangle = G_{\pm, n_0}(z, n_0 \pm 1, n_0 \pm 1).$$

By virtue of (1.104) we also have the more explicit form

$$(2.2) \quad m_+(z, n_0) = -\frac{u_+(z, n_0 + 1)}{a(n_0) u_+(z, n_0)}, \quad m_-(z, n_0) = -\frac{u_-(z, n_0 - 1)}{a(n_0 - 1) u_-(z, n_0)}.$$

The base point n_0 is of no importance in what follows and we will only consider $m_\pm(z) = m_\pm(z, 0)$ most of the time. Moreover, all results for $m_-(z)$ can be obtained from the corresponding results for $m_+(z)$ using reflection (cf. Lemma 1.7).

The definition (2.1) implies that the function $m_\pm(z)$ is holomorphic in $\mathbb{C} \backslash \sigma(H_\pm)$ and that it satisfies

$$(2.3) \quad m_\pm(\bar{z}) = \overline{m_\pm(z)}, \quad |m_\pm(z)| \leq \|(H_\pm - z)^{-1}\| \leq \frac{1}{|\mathrm{Im}(z)|}.$$

Moreover, $m_\pm(z)$ is a Herglotz function (i.e., it maps the upper half plane into itself, cf. Appendix B). In fact, this is a simple consequence of the first resolvent

identity
$$\text{Im}(m_\pm(z)) = \text{Im}(z)\langle \delta_{\pm 1}, (H_\pm - \bar{z})^{-1}(H_\pm - z)^{-1}\delta_{\pm 1}\rangle$$
(2.4)
$$= \text{Im}(z)\|(H_\pm - z)^{-1}\delta_{\pm 1}\|^2.$$

Hence by Theorem B.2, $m_\pm(z)$ has the following representation

(2.5)
$$m_\pm(z) = \int_\mathbb{R} \frac{d\rho_\pm(\lambda)}{\lambda - z}, \quad z \in \mathbb{C}\backslash\mathbb{R},$$

where $\rho_\pm = \int_{(-\infty,\lambda]} d\rho_\pm$ is a nondecreasing bounded function which is given by Stieltjes inversion formula (cf. Theorem B.2)

(2.6)
$$\rho_\pm(\lambda) = \frac{1}{\pi} \lim_{\delta \downarrow 0} \lim_{\varepsilon \downarrow 0} \int_{-\infty}^{\lambda + \delta} \text{Im}(m_\pm(x + i\varepsilon))dx.$$

Here we have normalized $\rho_\pm$ such that it is right continuous and obeys $\rho_\pm(\lambda) = 0$, $\lambda < \sigma(H_\pm)$.

Let $P_\Lambda(H_\pm)$, $\Lambda \subseteq \mathbb{R}$, denote the family of spectral projections corresponding to $H_\pm$ (spectral resolution of the identity). Then $d\rho_\pm$ can be identified using the spectral theorem,

(2.7) $$m_\pm(z) = \langle \delta_{\pm 1}, (H_\pm - z)^{-1}\delta_{\pm 1}\rangle = \int_\mathbb{R} \frac{d\langle \delta_{\pm 1}, P_{(-\infty,\lambda]}(H_\pm)\delta_{\pm 1}\rangle}{\lambda - z}.$$

Thus we see that $d\rho_\pm = d\langle \delta_{\pm 1}, P_{(-\infty,\lambda]}(H_\pm)\delta_{\pm 1}\rangle$ is the spectral measure of $H_\pm$ associated to the sequence $\delta_{\pm 1}$.

Remark 2.1. (i). Clearly, similar considerations hold for arbitrary expectations of resolvents of self-adjoint operators (cf. Lemma 6.1).
(ii). Let me note at this point that the fact which makes discrete life so much easier than continuous life is, that δ_1 is an element of our Hilbert space. In contradistinction, the continuous analog of δ_n, the delta distribution δ_x, is not a square integrable function. (However, if one considers non-negative Sturm-Liouville operators H, then δ_x lies in the scale of spaces associated to H. See [206] for further details.)

It follows in addition that all moments $m_{\pm,\ell}$ of $d\rho_\pm$ are finite and given by

(2.8)
$$m_{\pm,\ell} = \int_\mathbb{R} \lambda^\ell d\rho_\pm(\lambda) = \langle \delta_{\pm 1}, (H_+)^\ell \delta_{\pm 1}\rangle.$$

Moreover, there is a close connection between the so-called moment problem (i.e., determining $d\rho_\pm$ from all its moments $m_{\pm,\ell}$) and the reconstruction of $H_\pm$ from $d\rho_\pm$. Indeed, since $m_{\pm,0} = 1$, $m_{\pm,1} = b(\pm 1)$, $m_{\pm,2} = a(\pm 1 - \genfrac{}{}{0pt}{}{0}{1})^2 + b(\pm 1)^2$, etc., we infer

(2.9) $$b(\pm 1) = m_{\pm,1}, \quad a(\pm 1 - \genfrac{}{}{0pt}{}{0}{1})^2 = m_{\pm,2} - (m_{\pm,1})^2, \quad \text{etc.}.$$

We will consider this topic in Section 2.5.

You might have noticed, that $m_\pm(z, n)$ has (up to the factor $-a(n - \genfrac{}{}{0pt}{}{0}{1})$) the same structure as the function $\phi(n)$ used when deriving the Riccati equation (1.52).

Comparison with the formulas for $\phi(n)$ shows that

$$(2.10) \quad u_\pm(z,n) = u_\pm(z,n_0) \prod_{j=n_0}^{n-1}{}^* (-a(j-{}_1^0)m_\pm(z,j))^{\pm 1}$$

and that $m_+(z,n)$ satisfies the following recurrence relation

$$(2.11) \quad a(n-{}_1^0)^2 m_\pm(z,n) + \frac{1}{m_\pm(z,n\mp 1)} + z - b(n) = 0.$$

The functions $m_\pm(z,n)$ are the correct objects from the viewpoint of spectral theory. However, when it comes to calculations, the following pair of Weyl m-functions

$$(2.12) \quad \tilde{m}_\pm(z,n) = \mp\frac{u_\pm(z,n+1)}{a(n)u_\pm(z,n)}, \quad \tilde{m}_\pm(z) = \tilde{m}_\pm(z,0)$$

is often more convenient than the original one. The connection is given by

$$(2.13) \quad m_+(z,n) = \tilde{m}_+(z,n), \quad m_-(z,n) = \frac{a(n)^2 \tilde{m}_-(z,n) - z + b(n)}{a(n-1)^2}$$

(note also $m_-(z,n+1)^{-1} = -a(n)^2 \tilde{m}_-(z,n)$) and the corresponding spectral measures (for $n=0$) are related by

$$(2.14) \quad d\rho_+ = d\tilde{\rho}_+, \quad d\rho_- = \frac{a(0)^2}{a(-1)^2} d\tilde{\rho}_-.$$

You might want to note that $\tilde{m}_-(z)$ does not tend to 0 as $\text{Im}(z) \to \infty$ since the linear part is present in its Herglotz representation

$$(2.15) \quad \tilde{m}_-(z) = \frac{z - b(n)}{a(0)^2} + \int_\mathbb{R} \frac{d\tilde{\rho}_-(\lambda)}{\lambda - z}.$$

Finally, we introduce the Weyl m-functions $m_\pm^\beta(z,n)$ associated with $H_{\pm,n}^\beta$. They are defined analogously to $m_\pm(z,n)$. Moreover, the definition of $H_{+,n}^\beta$ in terms of $H_{+,n}$ suggests that $m_+^\beta(z,n)$, $\beta \neq 0$, can be expressed in terms of $m_+(z,n)$. Using (1.91) and the second resolvent identity we obtain

$$(2.16) \quad m_+^\beta(z,n) = \langle \delta_{n+1}, (H_{+,n}^\beta - z)^{-1} \delta_{n+1}\rangle = \frac{\beta m_+(z,n)}{\beta - a(n)m_+(z,n)}.$$

Similarly, the Weyl m-functions $m_-^\beta(z,n)$ associated with $H_{-,n}^\beta$, $\beta \neq \infty$, can be expressed in terms of $m_-(z,n+1)$,

$$(2.17) \quad m_-^\beta(z,n) = \langle \delta_n, (H_{-,n}^\beta - z)^{-1} \delta_n \rangle = \frac{m_-(z,n+1)}{1 - \beta a(n) m_-(z,n+1)}.$$

2.2. Properties of solutions

The aim of the present section is to establish some fundamental properties of special solutions of (1.19) which will be indispensable later on.

As an application of Weyl m-functions we first derive some additional properties of the solutions $u_\pm(z,.)$. By (1.27) we have

$$(2.18) \quad u_\pm(z,n) = a(0)u_\pm(z,0)\Big(a(0)^{-1}c(z,n) \mp \tilde{m}_\pm(z)s(z,n)\Big),$$

where the constant (with respect to n) $a(0)u_\pm(z,0)$ is at our disposal. If we choose $a(0)u_\pm(z,0) = 1$ (bearing in mind that $c(z,n)$, $s(z,n)$ are polynomials with respect to z), we infer for instance (using (2.4))

$$\overline{u_\pm(z,n)} = u_\pm(\bar{z},n). \tag{2.19}$$

Moreover, $u_\pm(z,n)$ are holomorphic with respect to $z \in \mathbb{C}\backslash\sigma(H_\pm)$. But we can do even better. If μ is an isolated eigenvalue of $H_\pm$, then $\tilde{m}_\pm(z)$ has a simple pole at $z = \mu$ (since it is Herglotz; see also (2.36)). By choosing $a(0)u_\pm(z,0) = (z-\mu)$ we can remove the singularity at $z = \mu$. In summary,

Lemma 2.2. *The solution $u_\pm(z,n)$ of (1.19) which is square summable near $\pm\infty$ exist for $z \in \mathbb{C}\backslash\sigma_{ess}(H_\pm)$, respectively. If we choose*

$$u_\pm(z,n) = \frac{c(z,n)}{a(0)} \mp \tilde{m}_\pm(z)s(z,n), \tag{2.20}$$

then $u_\pm(z,n)$ is holomorphic for $z \in \mathbb{C}\backslash\sigma(H_\pm)$, $u_\pm(z,.) \not\equiv 0$, and $\overline{u_\pm(z,.)} = u_\pm(\bar{z},.)$. In addition, we can include a finite number of isolated eigenvalues in the domain of holomorphy of $u_\pm(z,n)$.

Moreover, the sums

$$\sum_{j=n}^{\infty} u_+(z_1,j)u_+(z_2,j), \quad \sum_{j=-\infty}^{n} u_-(z_1,j)u_-(z_2,j) \tag{2.21}$$

are holomorphic with respect to z_1 (resp. z_2) provided $u_\pm(z_1,.)$ (resp. $u_\pm(z_2,.)$) are.

Proof. Only the last assertion has not been shown yet. It suffices to prove the claim for one n, say $n = 1$. Without restriction we suppose $u_+(z,0) = -a(0)^{-1}$ (if $u_+(z,0) = 0$, shift everything by one). Then $u_+(z,n) = (H_+ - z)^{-1}\delta_1(n)$, $n \geq 1$, and hence

$$\sum_{j=1}^{\infty} u_+(z_1,j)u_+(z_2,j) = \langle\delta_1, (H_+ - z_1)^{-1}(H_+ - z_2)^{-1}\delta_1\rangle_{\ell^2(\mathbb{N})}. \tag{2.22}$$

The remaining case is similar. □

Remark 2.3. If $(\lambda_0,\lambda_1) \subset \rho(H)$, we can define $u_\pm(\lambda,n)$, $\lambda \in [\lambda_0,\lambda_1]$. Indeed, by (2.20) it suffices to prove that $m_\pm(\lambda)$ tends to a limit (in $\mathbb{R} \cup \{\infty\}$) as $\lambda \downarrow \lambda_0$ or $\lambda \uparrow \lambda_1$. This follows from monotonicity of $m_\pm(\lambda)$,

$$m'_\pm(\lambda) = -\langle\delta_{\pm 1}, (H_\pm - \lambda)^{-2}\delta_{\pm 1}\rangle < 0, \quad \lambda \in (\lambda_0,\lambda_1) \tag{2.23}$$

(compare also equation (B.18)). Here the prime denotes the derivative with respect to λ. However, $u_\pm(\lambda_{0,1},n)$ might not be square summable near $\pm\infty$ in general.

In addition, the discrete eigenvalues of H_+^β are the zeros of $\frac{\beta}{a(0)} - m_+(\lambda)$ (see (2.16)) and hence decreasing as a function of β. Similarly, the discrete eigenvalues of H_-^β are the zeros of $\frac{1}{a(0)\beta} - m_-(\lambda)$ (see (2.17)) and hence increasing as a function of β.

2.2. Properties of solutions

Let $u(z)$ be a solution of (1.19) with $z \in \mathbb{C}\backslash\mathbb{R}$. If we choose $f = u(z)$, $g = \overline{u(z)}$ in (1.20), we obtain

$$[u(z)]_n = [u(z)]_{m-1} - \sum_{j=m}^{n}{}^{*} |u(z,j)|^2, \tag{2.24}$$

where $[.]_n$ denotes the **Weyl bracket**,

$$[u(z)]_n = \frac{W_n(u(z), \overline{u(z)})}{2\mathrm{i} \mathrm{Im}(z)} = a(n) \frac{\mathrm{Im}(u(z,n)\overline{u(z,n+1)})}{\mathrm{Im}(z)}. \tag{2.25}$$

Especially for $u_\pm(z,n)$ as in Lemma 2.2 we get

$$[u_\pm(z)]_n = \begin{cases} \sum_{j=n+1}^{\infty} |u_+(z,j)|^2 \\ -\sum_{j=-\infty}^{n} |u_-(z,j)|^2 \end{cases} \tag{2.26}$$

and for $s(z,n)$

$$[s(z)]_n = \begin{cases} -\sum_{j=0}^{n} |s(z,j)|^2, & n \geq 0 \\ \sum_{j=n+1}^{0} |s(z,j)|^2, & n < 0 \end{cases}. \tag{2.27}$$

Moreover, let $u(z,n), v(z,n)$ be solutions of (1.19). If $u(z,n)$ is differentiable with respect to z, we obtain

$$(\tau - z)u'(z,n) = u(z,n) \tag{2.28}$$

(here the prime denotes the derivative with respect to z) and using Green's formula (1.20) we infer

$$\sum_{j=m}^{n} v(z,j)u(z,j) = W_n(v(z), u'(z)) - W_{m-1}(v(z), u'(z)). \tag{2.29}$$

Even more interesting is the following result.

Lemma 2.4. *Let $u_\pm(z,n)$ be solutions of (1.19) as in Lemma 2.2. Then we have*

$$W_n(u_\pm(z), \frac{d}{dz} u_\pm(z)) = \begin{cases} -\sum_{j=n+1}^{\infty} u_+(z,j)^2 \\ \sum_{j=-\infty}^{n} u_-(z,j)^2 \end{cases}. \tag{2.30}$$

Proof. Green's formula (1.20) implies

$$W_n(u_+(z), u_+(\tilde{z})) = (z - \tilde{z}) \sum_{j=n+1}^{\infty} u_+(z,j) u_+(\tilde{z}, j) \tag{2.31}$$

and furthermore,

$$W_n(u_+(z), u'_+(z)) = \lim_{\tilde{z} \to z} W_n\left(u_+(z), \frac{u_+(z) - u_+(\tilde{z})}{z - \tilde{z}}\right)$$

$$= \sum_{j=n+1}^{\infty} u_+(z,j)^2. \tag{2.32}$$

As a first application of this lemma, let us investigate the isolated poles of $G(z,n,m)$. If $z = \lambda_0$ is such an isolated pole (i.e., an isolated eigenvalue of H), then $W(u_-(\lambda_0), u_+(\lambda_0)) = 0$ and hence $u_\pm(\lambda_0, n)$ differ only by a (nonzero) constant multiple. Moreover,

$$\frac{d}{dz}W(u_-(z), u_+(z))\Big|_{z=\lambda_0} = W_n(u_-(\lambda_0), u'_+(\lambda_0)) + W_n(u'_-(\lambda_0), u_+(\lambda_0))$$

$$(2.33) \qquad = -\sum_{j\in\mathbb{Z}} u_-(\lambda_0, j) u_+(\lambda_0, j)$$

by the lemma and hence

$$(2.34) \qquad G(z,n,m) = -\frac{P(\lambda_0, n, m)}{z - \lambda_0} + O(z - \lambda_0)^0,$$

where

$$(2.35) \qquad P(\lambda_0, n, m) = \frac{u_\pm(\lambda_0, n) u_\pm(\lambda_0, m)}{\sum_{j\in\mathbb{Z}} u_\pm(\lambda_0, j)^2}.$$

Similarly, for $H_\pm$ we obtain

$$(2.36) \qquad \lim_{z\to\mu}(z - \mu)G_\pm(z,n,m) = -\frac{s(\mu, n) s(\mu, m)}{\sum_{j\in\pm\mathbb{N}} s(\mu, j)^2}.$$

Thus the poles of the kernel of the resolvent at isolated eigenvalues are simple. Moreover, the negative residue equals the kernel of the projector onto the corresponding eigenspace.

As a second application we show monotonicity of $G(z, n, n)$ with respect to z in a spectral gap. Differentiating (1.99), a straightforward calculation shows

$$(2.37) \quad G'(z,n,n) = \frac{u_+(z,n)^2 W_n(u_-(z), \dot{u}_-(z)) - u_-(z,n)^2 W_n(u_+(z), \dot{u}_+(z))}{W(u_-(z), u_+(z))^2},$$

which is positive for $z \in \mathbb{R}\backslash\sigma(H)$ by Lemma 2.4. The same result also follows from

$$G'(z,n,m) = \frac{d}{dz}\langle\delta_n, (H-z)^{-1}\delta_m\rangle = \langle\delta_n, (H-z)^{-2}\delta_m\rangle$$

$$(2.38) \qquad = \sum_{j\in\mathbb{Z}} G(z,n,j) G(z,j,m).$$

Finally, let us investigate the qualitative behavior of the solutions $u_\pm(z)$ more closely.

Lemma 2.5. *Suppose $z \in \mathbb{C}\backslash\sigma_{ess}(H_\pm)$. Then we can find $C_\pm, \gamma^\pm > 0$ such that*

$$(2.39) \qquad |u_\pm(z,n)| \leq C_\pm \exp(-\gamma^\pm n), \quad \pm n \in \mathbb{N}.$$

For $\gamma^\pm$ we can choose

$$(2.40) \qquad \gamma^\pm = \ln\left(1 + (1-\varepsilon)\frac{\sup_{\beta\in\mathbb{R}\cup\{\infty\}} \text{dist}(\sigma(H_\pm^\beta), z)}{2\sup_{n\in\mathbb{N}} |a(\pm n)|}\right), \quad \varepsilon > 0$$

(the corresponding constant $C_\pm$ will depend on ε).

Proof. By the definition of $H_\pm^\beta$ it suffices to consider the case $\beta = \infty$, $\delta = \mathrm{dist}(\sigma(H_+), z) > 0$ (otherwise alter $b(1)$ or shift the interval). Recall $G_+(z, 1, n) = u_+(z, n)/(-a(0)u_+(z, 0))$ and consider ($\gamma > 0$)

$$\begin{aligned}
\mathrm{e}^{\gamma(n-1)} G_+(z, 1, n) &= \langle \delta_1, P_{-\gamma}(H_+ - z)^{-1} P_\gamma \delta_n \rangle = \langle \delta_1, (P_{-\gamma} H_+ P_\gamma - z)^{-1} \delta_n \rangle \\
&= \langle \delta_1, (H_+ - z + Q_\gamma)^{-1} \delta_n \rangle,
\end{aligned} \tag{2.41}$$

where

$$(P_\gamma u)(n) = \mathrm{e}^{\gamma n} u(n) \tag{2.42}$$

and

$$(Q_\gamma u)(n) = a(n)(\mathrm{e}^\gamma - 1)u(n+1) + a(n-1)(\mathrm{e}^{-\gamma} - 1)u(n-1). \tag{2.43}$$

Moreover, choosing $\gamma = \ln(1 + (1-\varepsilon)\delta/(2 \sup_{n \in \mathbb{N}} |a(\pm n)|))$ we have

$$\|Q_\gamma\| \leq 2 \sup_{n \in \mathbb{N}} |a(\pm n)|(\mathrm{e}^\gamma - 1) = (1-\varepsilon)\delta. \tag{2.44}$$

Using the first resolvent equation

$$\begin{aligned}
\|(H_+ - z + Q_\gamma)^{-1}\| &\leq \|(H_+ - z)^{-1}\| + \|(H_+ - z)^{-1}\| \\
&\quad \times \|Q_\gamma\| \|(H_+ - z + Q_\gamma)^{-1}\|
\end{aligned} \tag{2.45}$$

and $\|(H_+ - z)^{-1}\| = \delta^{-1}$ implies the desired result

$$|\mathrm{e}^{\gamma(n-1)} G_+(z, 1, n)| \leq \|(H_+ - z + Q_\gamma)^{-1}\| \leq \frac{1}{\delta\varepsilon}. \tag{2.46}$$

$\square$

This result also shows that for any solution $u(z, n)$ which is not a multiple of $u_\pm(z, .)$, we have

$$\sqrt{|u(z,n)|^2 + |u(z, n+1)|^2} \geq \frac{\mathrm{const}}{|a(n)|} \mathrm{e}^{\gamma^\pm n} \tag{2.47}$$

since otherwise the Wronskian of u and $u_\pm(z)$ would vanish as $n \to \pm\infty$. Hence, (cf. (1.34))

$$\{\lambda \in \mathbb{R} | \underline{\gamma}^\pm(\lambda) = 0\} \subseteq \sigma_{ess}(H_\pm) \tag{2.48}$$

since $\underline{\gamma}^\pm(\lambda) > \gamma^\pm$, $\lambda \notin \sigma_{ess}(H_\pm)$.

2.3. Positive solutions

In this section we want to investigate solutions of $\tau u = \lambda u$ with $\lambda \leq \sigma(H)$. We will assume

$$a(n) < 0 \tag{2.49}$$

throughout this section. As a warm-up we consider the case $\lambda < \sigma(H)$.

Lemma 2.6. *Suppose $a(n) < 0$ and let $\lambda < \sigma(H)$. Then we can assume*

$$u_\pm(\lambda, n) > 0, \quad n \in \mathbb{Z}, \tag{2.50}$$

and we have

$$(n - n_0) s(\lambda, n, n_0) > 0, \quad n \in \mathbb{Z} \setminus \{n_0\}. \tag{2.51}$$

Proof. From $(H - \lambda) > 0$ one infers $(H_{+,n} - \lambda) > 0$ and hence

$$\text{(2.52)} \qquad 0 < \langle \delta_{n+1}, (H_{+,n} - \lambda)^{-1} \delta_{n+1} \rangle = \frac{u_+(\lambda, n+1)}{-a(n) u_+(\lambda, n)},$$

showing that $u_+(\lambda)$ can be chosen to be positive. Furthermore, for $n > n_0$ we obtain

$$\text{(2.53)} \qquad 0 < \langle \delta_n, (H_{+,n_0} - \lambda)^{-1} \delta_n \rangle = \frac{u_+(\lambda, n) s(\lambda, n, n_0)}{-a(n_0) u_+(\lambda, n_0)},$$

implying $s(\lambda, n, n_0) > 0$ for $n > n_0$. The remaining case is similar. □

The general case $\lambda \leq \sigma(H)$ requires an additional investigation. We first prove that (2.51) also holds for $\lambda \leq \sigma(H)$ with the aid of the following lemma.

Lemma 2.7. *Suppose $a(n) < 0$ and let $u \not\equiv 0$ solve $\tau u = u$. Then $u(n) \geq 0$, $n \geq n_0$ (resp. $n \leq n_0$), implies $u(n) > 0$, $n > n_0$ (resp. $n < n_0$). Similarly, $u(n) \geq 0$, $n \in \mathbb{Z}$, implies $u > 0$, $n \in \mathbb{Z}$.*

Proof. Fix $n > n_0$ (resp. $n < n_0$). Then, $u(n) = 0$ implies $u(n \pm 1) > 0$ (since u cannot vanish at two consecutive points) contradicting $0 = (b(n) - \lambda) u(n) = -a(n) u(n+1) - a(n-1) u(n-1) > 0$. □

The desired result now follows from

$$\text{(2.54)} \qquad s(\lambda, n, n_0) = \lim_{\varepsilon \downarrow 0} s(\lambda - \varepsilon, n, n_0) \geq 0, \quad n > n_0,$$

and the lemma. In addition, we note

$$\text{(2.55)} \qquad b(n) - \lambda = -a(n) s(\lambda, n+1, n-1) > 0, \quad \lambda \leq \sigma(H).$$

The following corollary is simple but powerful.

Corollary 2.8. *Suppose $u_j(\lambda, n)$, $j = 1, 2$, are two solutions of $\tau u = \lambda u$, $\lambda \leq \sigma(H)$, with $u_1(\lambda, n_0) = u_2(\lambda, n_0)$ for some $n_0 \in \mathbb{Z}$. Then if*

$$\text{(2.56)} \qquad (n - n_0)\bigl(u_1(\lambda, n) - u_2(\lambda, n)\bigr) > 0 \quad (resp. < 0)$$

holds for one $n \in \mathbb{Z} \backslash \{n_0\}$, then it holds for all. If $u_1(\lambda, n) = u_2(\lambda, n)$ for one $n \in \mathbb{Z} \backslash \{n_0\}$, then u_1 and u_2 are equal.

Proof. Use $u_1(\lambda, n) - u_2(\lambda, n) = c\, s(\lambda, n, n_0)$ for some $c \in \mathbb{R}$. □

In particular, this says that for $\lambda \leq \sigma(H)$ solutions $u(\lambda, n)$ can change sign (resp. vanish) at most once (since $s(z, n, n_0)$ does).

For the sequence

$$\text{(2.57)} \qquad u_m(\lambda, n) = \frac{s(\lambda, n, m)}{s(\lambda, 0, m)}, \quad m \in \mathbb{N},$$

this corollary implies that

$$\text{(2.58)} \qquad \phi_m(\lambda) = u_m(\lambda, 1) = \frac{s(\lambda, 1, m)}{s(\lambda, 0, m)}$$

is increasing with m since we have $u_{m+1}(\lambda, m) > 0 = u_m(\lambda, m)$. Next, since $a(0) s(\lambda, 1, m) + a(-1) s(\lambda, -1, m) = (\lambda - b(0)) s(\lambda, 0, m)$ implies

$$\text{(2.59)} \qquad \phi_m(\lambda) < \frac{\lambda - b(0)}{a(0)},$$

2.3. Positive solutions

we can define

(2.60)
$$\phi_+(\lambda) = \lim_{m\to\infty} \phi_m(\lambda), \quad u_+(\lambda, n) = \lim_{m\to\infty} u_m(\lambda, n) = c(\lambda, n) + \phi_+(\lambda) s(\lambda, n).$$

By construction we have $u_+(\lambda, n) > u_m(\lambda, n)$, $n \in \mathbb{N}$, implying $u_+(\lambda, n) > 0$, $n \in \mathbb{N}$. For $n < 0$ we have at least $u_+(\lambda, n) \geq 0$ since $u_m(\lambda, n) > 0$ and thus $u_+(\lambda, n) > 0$, $n \in \mathbb{Z}$, by Lemma 2.7.

Let $u(\lambda, n)$ be a solution with $u(\lambda, 0) = 1$, $u(\lambda, 1) = \phi(\lambda)$. Then, by the above analysis, we infer that $u(\lambda, n) > 0$, $n \in \mathbb{N}$, is equivalent to $\phi(\lambda) \geq \phi_+(\lambda)$ and hence $u(\lambda, n) \geq u_+(\lambda, n)$, $n \in \mathbb{N}$ (with equality holding if and only if $u(\lambda) = u_+(\lambda)$). In this sense $u_+(\lambda)$ is the **minimal positive solution** (also **principal** or **recessive solution**) near $+\infty$. In particular, since every solution can change sign at most once, we infer that if there is a square summable solution near $+\infty$, then it must be equal to $u_+(\lambda, n)$ (up to a constant multiple), justifying our notation.

Moreover, if $u(\lambda)$ is different from $u_+(\lambda)$, then constancy of the Wronskian

(2.61)
$$\frac{u_+(\lambda, n)}{u(\lambda, n)} - \frac{u_+(\lambda, n+1)}{u(\lambda, n+1)} = \frac{W(u(\lambda), u_+(\lambda))}{-a(n) u(\lambda, n) u(\lambda, n+1)}$$

together with $W(u(\lambda), u_+(\lambda)) = a(0)(\phi_+(\lambda) - \phi(\lambda)) > 0$ shows that the sequence $u_+(\lambda, n)/u(\lambda, n)$ is decreasing for all $u(\lambda) \neq u_+(\lambda)$ if and only if $u_+(\lambda)$ is minimal. Moreover, we claim

(2.62)
$$\lim_{n\to\infty} \frac{u_+(\lambda, n)}{u(\lambda, n)} = 0.$$

In fact, suppose $\lim_{n\to\infty} u_+(\lambda, n)/u(\lambda, n) = \varepsilon > 0$. Then $u_+(\lambda, n) > \varepsilon u(\lambda, n)$, $n \in \mathbb{N}$, and hence $u_\varepsilon = (u_+(\lambda, n) - \varepsilon u(\lambda, n))/(1-\varepsilon) > 0$, $n \in \mathbb{N}$. But $u_+(\lambda, n) < u_\varepsilon(\lambda, n)$ implies $u(\lambda, n) < u_+(\lambda, n)$, a contradiction.

Conversely, (2.62) for one $u(\lambda)$ uniquely characterizes $u_+(\lambda, n)$ since (2.62) remains valid if we replace $u(\lambda)$ by any positive linear combination of $u(\lambda)$ and $u_+(\lambda)$.

Summing up (2.61) shows that

(2.63)
$$\sum_{j=0}^{\infty} \frac{1}{-a(j) u(\lambda, j) u(\lambda, j+1)} = \frac{1}{a(0)(\phi_+(\lambda) - \phi(\lambda))} < \infty.$$

Moreover, the sequence

(2.64)
$$v(\lambda, n) = u(\lambda, n) \sum_{j=n}^{\infty} \frac{1}{-a(j) u(\lambda, j) u(\lambda, j+1)} > 0$$

solves $\tau u = \lambda u$ (cf. (1.51)) and equals $u_+(\lambda, n)$ up to a constant multiple since $\lim_{n\to\infty} v(\lambda, n)/u(\lambda, n) = 0$.

By reflection we obtain a corresponding minimal positive solution $u_-(\lambda, n)$ near $-\infty$. Let us summarize some of the results obtained thus far.

Lemma 2.9. *Suppose $a(n) < 0$, $\lambda \leq \sigma(H)$ and let $u(\lambda, n)$ be a solution with $u(\lambda, n) > 0$, $\pm n \geq 0$. Then the following conditions are equivalent.*

(i). *$u(\lambda, n)$ is minimal near $\pm\infty$.*

(ii). *We have*

$$\text{(2.65)} \qquad \frac{u(\lambda,n)}{v(\lambda,n)} \leq \frac{v(\lambda,0)}{u(\lambda,0)}, \quad \pm n \geq 0,$$

for any solution $v(\lambda,n)$ with $v(\lambda,n) > 0$, $\pm n \geq 0$.

(iii). *We have*

$$\text{(2.66)} \qquad \lim_{n \to \pm\infty} \frac{u(\lambda,n)}{v(\lambda,n)} = 0.$$

for one solution $v(\lambda,n)$ with $v(\lambda,n) > 0$, $\pm n \geq 0$.

(iv). *We have*

$$\text{(2.67)} \qquad \sum_{j \in \pm \mathbb{N}} \frac{1}{-a(j)u(j)u(j+1)} = \infty.$$

Recall that minimality says that for a solution $u(\lambda, n)$ with $u(\lambda, 0) = 1$, $u(\lambda, 1) = \phi(\lambda)$ to be positive on $\mathbb{N}$, we need $\phi(\lambda) \geq \phi_+(\lambda)$. Similarly, for $u(\lambda, n)$ to be positive on $-\mathbb{N}$ we need $\phi(\lambda) \leq \phi_-(\lambda)$. In summary, $u(n) > 0$, $n \in \mathbb{Z}$, if and only if $\phi_+(\lambda) \leq \phi(\lambda) \leq \phi_-(\lambda)$ and thus any positive solution can (up to constant multiples) be written as

$$\text{(2.68)} \qquad u(\lambda, n) = \frac{1-\sigma}{2} u_-(\lambda, n) + \frac{1+\sigma}{2} u_+(\lambda, n), \quad \sigma \in [-1, 1].$$

Two cases may occur

(i). $u_-(\lambda, n), u_+(\lambda, n)$ are linearly dependent (i.e., $\phi_+(\lambda) = \phi_-(\lambda)$) and there is only one (up to constant multiples) positive solution. In this case $H - \lambda$ is called **critical**.

(ii). $u_-(\lambda, n), u_+(\lambda, n)$ are linearly independent and

$$\text{(2.69)} \qquad u_\sigma(\lambda, n) = \frac{1+\sigma}{2} u_+(\lambda, n) + \frac{1-\sigma}{2} u_-(\lambda, n),$$

is positive if and only if $\sigma \in [-1, 1]$. In this case $H - \lambda$ is called **subcritical**

If $H - \lambda > \sigma(H)$, then $H - \lambda$ is always subcritical by Lemma 2.6. To emphasize this situation one sometimes calls $H - \lambda$ **supercritical** if $H - \lambda > \sigma(H)$.

In case (ii) it is not hard to show using (2.62) that for two positive solutions $u_j(\lambda, n)$, $j = 1, 2$, we have

$$\text{(2.70)} \qquad u_\sigma(\lambda, n) = \frac{1+\sigma}{2} u_1(\lambda, n) + \frac{1-\sigma}{2} u_2(\lambda, n) > 0 \quad \Leftrightarrow \quad \sigma \in [-1, 1],$$

if and only if the solutions $u_{1,2}$ equal $u_\pm$ up to constant multiples.

Remark 2.10. Assuming $a(n) < 0$, the requirement $H - \lambda \geq 0$ is also necessary for a positive solution to exist. In fact, any positive solution can be used to factor $H - \lambda = A^*A \geq 0$ (cf. Corollary 11.2).

Similarly, if $a(n) > 0$, the requirement $H - \lambda \leq 0$ is sufficient and necessary for a positive solution to exist (consider $-(H - \lambda) \geq 0$). If $a(n)$ is not of a fixed sign, no general statement is possible.

2.4. Weyl circles

In this section we will advocate a different approach for Weyl m-functions. There are two main reasons for doing this. First of all this approach is similar to Weyl's original one for differential equations and second it will provide an alternative characterization of Weyl m-functions as limits needed later.

Let $s_\beta(z,.), c_\beta(z,.)$ be the fundamental system of (1.19) corresponding to the initial conditions

$$
\begin{aligned}
s_\beta(z,0) &= -\sin(\alpha), & s_\beta(z,1) &= \cos(\alpha), \\
c_\beta(z,0) &= \cos(\alpha), & c_\beta(z,1) &= \sin(\alpha),
\end{aligned}
\tag{2.71}
$$

where $\beta = \cot(\alpha)$. Clearly,

$$
W(c_\beta(z), s_\beta(z)) = a(0).
\tag{2.72}
$$

The general idea is to approximate $H_\pm^\beta$ by finite matrices. We will choose the boundary conditions associated with β, β_N at $n = 0, N$, respectively. The corresponding matrix $H_{0,N}^{\beta,\beta_N}$ will have eigenvalues $\lambda_j(N)$, $1 \leq j \leq \tilde{N}$, and corresponding eigenvectors $s_\beta(\lambda_j(N), n)$ (since $s_\beta(z,n)$ fulfills the boundary condition at 0). We note that $\lambda_j(N)$, $1 \leq j \leq \tilde{N}$, depend also on β.

If we set

$$
\tilde{m}_N^\beta(z) = \operatorname{sgn}(N) \frac{W_N(s_\beta(\lambda_j(N)), c_\beta(z))}{a(0) W_N(s_\beta(\lambda_j(N)), s_\beta(z))}, \qquad N \in \mathbb{Z}\setminus\{0\}
\tag{2.73}
$$

(independent of the eigenvalue chosen), then the solution

$$
u_N(z,n) = a(0)^{-1} c_\beta(z,n) - \operatorname{sgn}(N) \tilde{m}_N^\beta(z) s_\beta(z,n)
\tag{2.74}
$$

satisfies the boundary condition at N, that is, $W_N(s_\beta(\lambda_1), u_N(z)) = 0$. The function $\tilde{m}_N^\beta(z)$ is rational (w.r.t. z) and has poles at the eigenvalues $\lambda_j(N)$. In particular, they are the analogs of the Weyl $\tilde{m}$-functions for finite Jacobi operators. Hence it suggests itself to consider the limit $N \to \pm\infty$, where our hope is to obtain the Weyl $\tilde{m}$-functions for the Jacobi operators $H_\pm^\beta$.

We fix $\lambda_0 \in \mathbb{R}$ and set

$$
\beta_N = -\frac{s_\beta(\lambda_0, N+1)}{s_\beta(\lambda_0, N)}
\tag{2.75}
$$

implying $\lambda_0 = \lambda_j(N)$ for one j. It will turn out that the above choice for β_N is not really necessary since $\tilde{m}_N^\beta(z)$ will, as a consequence of boundedness of H, converge for arbitrary sequences β_N. However, the above choice also works in the case of unbounded operators.

Before we turn to the limit, let us derive some additional properties for finite N. The limits

$$
\begin{aligned}
\lim_{z \to \lambda_j(N)} W_N(s_\beta(\lambda_j(N)), c_\beta(z)) &= -a(0), \\
\lim_{z \to \lambda_j(N)} \frac{W_N(s_\beta(\lambda_j(N)), s_\beta(z))}{z - \lambda_j(N)} &= W_N(s_\beta(\lambda_j(N)), s_\beta'(\lambda_j(N)))
\end{aligned}
\tag{2.76}
$$

imply that all poles of $m_N(z,\beta)$ are simple. Using (2.29) to evaluate (2.76) one infers that the negative residue at $\lambda_j(N)$ is given by

$$(2.77) \qquad \gamma_N^\beta(\lambda_j(N)) = \Big(\sum_{n=N+1}^{\tilde{N}_0} s_\beta(\lambda_j(N), n)^2 \Big)^{-1}, \quad N \gtrless 0.$$

The numbers $\gamma_N^\beta(\lambda_j(N))$ are called norming constants. Hence one gets

$$(2.78) \qquad \tilde{m}_N^\beta(z) = -\sum_{j=1}^{\tilde{N}} \frac{\gamma_N^\beta(\lambda_j(N))}{z - \lambda_j(N)} + \begin{cases} \frac{\beta^{\mp 1}}{a(0)}, & \beta^{\mp 1} \neq \infty \\ \frac{z - b(\substack{1\\0})}{a(0)^2}, & \beta^{\mp 1} = \infty \end{cases}, \quad N \gtrless 0,$$

and with the help of (2.24) we obtain

$$(2.79) \qquad \sum_{n=N+1}^{\tilde{N}_0} |u_N(z,n)|^2 = \frac{\mathrm{Im}(\tilde{m}_N^\beta(z))}{\mathrm{Im}(z)}, \quad N \gtrless 0,$$

that is, $m_N^\beta(z)$ are Herglotz functions.

Now we turn to the limits $N \to \pm\infty$. Fix $z \in \mathbb{C}\backslash\mathbb{R}$. Observe that if λ_1 varies in $\mathbb{R}$, then β_N takes all values in $\mathbb{R} \cup \{\infty\}$ $\tilde{N}$ times. Rewriting (2.73) shows that

$$(2.80) \qquad \tilde{m}_N^\beta(z) = \frac{\mathrm{sgn}(N)}{a(0)} \frac{c_\beta(z,N)\beta_N + c_\beta(z,N+1)}{s_\beta(z,N)\beta_N + s_\beta(z,N+1)}$$

is a Möbius transformation and hence the values of $\tilde{m}_N^\beta(z)$ for different $\beta_N \in \mathbb{R} \cup \{\infty\}$ lie on a circle (also called Weyl circle) in the complex plane (note that $z \in \mathbb{R}$ would correspond to the degenerate circle $\mathbb{R} \cup \{\infty\}$). The center of this circle is

$$(2.81) \qquad c_N = \mathrm{sgn}(N) \frac{W_N(c_\beta(z), \overline{s_\beta(z)})}{2a(0)\mathrm{Im}(z)[s_\beta(z)]_N}$$

and the radius is

$$(2.82) \qquad r_N = \left| \frac{W(c_\beta(z), s_\beta(z))}{2a(0)\mathrm{Im}(z)[s_\beta(z)]_N} \right| = \frac{1}{2|\mathrm{Im}(z)[s_\beta(z)]_N|}.$$

Using

$$(2.83) \qquad [a(0)^{-1}c_\beta(z) - \mathrm{sgn}(N)\tilde{m}s_\beta(z)]_N = [s_\beta(z)]_N \big(|\tilde{m} - c_N|^2 - r_N^2\big)$$

we see that this circle is equivalently characterized by the equation

$$(2.84) \qquad \{\tilde{m} \in \mathbb{C} | [a(0)^{-1}c_\beta(z) + \mathrm{sgn}(N)\tilde{m}s_\beta(z)]_N = 0\}.$$

Since $[.]_N$, $N > 0$, is decreasing with respect to N, the circle corresponding to $N+1$ lies inside the circle corresponding to N. Hence these circles tend to a limit point for any sequence $(\beta_N \in \mathbb{R} \cup \{\infty\})_{N \in \mathbb{N}}$ since

$$(2.85) \qquad \lim_{N \to \infty} -[s_\beta(z)]_N = \sum_{n=0}^{\infty} |s_\beta(z,n)|^2 = \infty$$

2.4. Weyl circles

(otherwise H_+^β would have a non-real eigenvalue). Similarly for $N < 0$. Thus the pointwise convergence of $\tilde{m}_N^\beta(z)$ is clear and we may define

$$\tilde{m}_\pm^\beta(z) = \lim_{N \to \pm\infty} \tilde{m}_N^\beta(z). \tag{2.86}$$

Moreover, the above sequences are locally bounded in z (fix N and take all circles corresponding to a (sufficiently small) neighborhood of any point z and note that all following circles lie inside the ones corresponding to N) and by Vitali's theorem ([**229**], p. 168) they converge uniformly on every compact set in $\mathbb{C}_\pm = \{z \in \mathbb{C} | \pm \text{Im}(z) > 0\}$, implying that $\tilde{m}_\pm^\beta(z)$ are again holomorphic on $\mathbb{C}_\pm$.

Remark 2.11. (i). Since $\tilde{m}_N^\beta(z)$ converges for arbitrary choice of the sequence β_N we even have

$$\tilde{m}_\pm^\beta(z) = \lim_{N \to \pm\infty} \frac{c_\beta(z,N)}{a(0)s_\beta(z,N)}. \tag{2.87}$$

Moreover, this approach is related to (2.60). Using (1.62), (1.63) shows $\phi_m(\lambda) = -c(\lambda,m)/s(\lambda,m)$ and establishes the equivalence.

(ii). That the Weyl circles converge to a point is a consequence of the boundedness of $a(n), b(n)$. In the general case the limit could be a circle or a point (independent of $z \in \mathbb{C}\backslash\mathbb{R}$). Accordingly one says that τ is limit circle or limit point at $\pm\infty$. (See Section 2.6 for further information on unbounded operators.)

As anticipated by our notation, $\tilde{m}_\pm^\beta(z)$ are closely related to $m_\pm^\beta(z)$. This will be shown next. We claim that

$$u_\pm(z,n) = a(0)^{-1}c_\beta(z,n) \mp \tilde{m}_\pm^\beta(z)s_\beta(z,n) \tag{2.88}$$

is square summable near $\pm\infty$. Since $\tilde{m}_+^\beta$ lies in the interior of all Weyl circles the limit $\lim_{N\to\infty}[u_+(z)]_N \geq 0$ must exist and hence $u_+ \in \ell_+^2(\mathbb{Z})$ by (2.24). Moreover, $u_+ \in \ell_+^2(\mathbb{Z})$ implies $[u_+]_\infty = 0$. Similarly for $u_-(z)$. In addition, (cf. (2.4))

$$\sum_{n=-\infty}^{\substack{\infty \\ -1}} |u_\pm(z,n)|^2 = \frac{\text{Im}(\tilde{m}_\pm^\beta(z))}{\text{Im}(z)}, \tag{2.89}$$

implies that $\tilde{m}_\pm^\beta(z)$ are Herglotz functions (note that $u_\pm(z,n)$ depends on β because of the normalization $u_\pm(z,0) = a(0)^{-1}\cos(\alpha) \pm \tilde{m}_\pm^\beta(z)\sin(\alpha)$). In particular, their Herglotz representation reads

$$\tilde{m}_\pm^\beta(z) = \frac{\beta^{\mp 1}}{a(0)} + \int_\mathbb{R} \frac{d\tilde{\rho}_\pm^\beta(\lambda)}{\lambda - z}, \qquad \beta \neq \begin{matrix} 0 \\ \infty \end{matrix}. \tag{2.90}$$

This finally establishes the connection

$$\tilde{m}_\pm^\infty(z) = \tilde{m}_\pm(z) = \frac{\mp u_\pm(z,1)}{a(0)u_\pm(z,0)} \tag{2.91}$$

as expected. Furthermore, $\tilde{m}_\pm^{\beta_1}(z)$ can be expressed in terms of $\tilde{m}_\pm^{\beta_2}(z)$ (use that $u_\pm$ is unique up to a constant) by

$$\tilde{m}_\pm^{\beta_1}(z) = \pm\frac{1}{a(0)}\frac{a(0)\cos(\alpha_2 - \alpha_1)\tilde{m}_\pm^{\beta_2}(z) \mp \sin(\alpha_2 - \alpha_1)}{a(0)\sin(\alpha_2 - \alpha_1)\tilde{m}_\pm^{\beta_2}(z) \pm \cos(\alpha_2 - \alpha_1)}, \tag{2.92}$$

where $\beta_{1,2} = \cot(\alpha_{1,2})$. Specializing to the case $\beta_1 = \beta$, $\beta_2 = \infty$ we infer

$$(2.93) \quad \tilde{m}_+^\beta(z,n) = \frac{\beta m_+(z) + a(0)^{-1}}{\beta - a(0)m_+(z)}, \quad \tilde{m}_-^\beta(z,n) = \frac{m_-(z,1) + a(0)^{-1}\beta}{1 - \beta a(0)m_-(z,1)}$$

which should be compared with (2.16), (2.17), respectively.

2.5. Canonical forms of Jacobi operators and the moment problem

The aim of this section is to derive canonical forms for H, $H_\pm$ and to relate the spectra of these operators to the corresponding measures encountered in the previous sections.

Since $s(z,n)$ is a polynomial in z we infer by induction (cf. Lemma 1.2)

$$(2.94) \quad s(H_+, n)\delta_1 = \sum_{j=0}^{n} s_{j,n}(0) H_+^j \delta_1 = \delta_n,$$

implying that δ_1 is a cyclic vector for H_+. We recall the measure

$$(2.95) \quad d\rho_+(\lambda) = d\langle \delta_1, P_{(-\infty,\lambda]}(H_+)\delta_1\rangle$$

and consider the Hilbert space $L^2(\mathbb{R}, d\rho_+)$. Since $d\rho_+$ is supported on $\sigma(H_+)$ this space is the same as the space $L^2(\sigma(H_+), d\rho_+)$. The scalar product is given by

$$(2.96) \quad \langle f, g\rangle_{L^2} = \int_\mathbb{R} \overline{f(\lambda)} g(\lambda)\, d\rho_+(\lambda).$$

If f, g are polynomials we can evaluate their scalar product without even knowing $d\rho_+(\lambda)$ since

$$(2.97) \quad \langle f, g\rangle_{L^2} = \langle f(H_+)\delta_1, g(H_+)\delta_1\rangle.$$

Applying this relation in the special case $f(\lambda) = s(\lambda, m)$, $g(\lambda) = s(\lambda, n)$, we obtain from equation (2.94) that the polynomials $s(z,n)$, $n \in \mathbb{N}$, are orthogonal with respect to this scalar product, that is,

$$(2.98) \quad \langle s(\lambda, m), s(\lambda, n)\rangle_{L^2} = \int_\mathbb{R} s(\lambda, m) s(\lambda, n)\, d\rho_+(\lambda) = \delta_{m,n}.$$

We will see in Theorem 4.5 that $s(\lambda, n)$ has $n - 1$ distinct real roots which interlace the roots of $s(\lambda, n+1)$.

Now consider the following transformation U from the set $\ell_0(\mathbb{N})$ onto the set of all polynomials (**eigenfunction expansion**)

$$(Uf)(\lambda) = \sum_{n=1}^{\infty} f(n) s(\lambda, n),$$

$$(2.99) \quad (U^{-1}F)(n) = \int_\mathbb{R} s(\lambda, n) F(\lambda) d\rho_+(\lambda).$$

A simple calculation for $F(\lambda) = (Uf)(\lambda)$ shows that U is unitary,

$$(2.100) \quad \sum_{n=1}^{\infty} |f(n)|^2 = \int_\mathbb{R} |F(\lambda)|^2 d\rho_+(\lambda).$$

This leads us to the following result.

2.5. Canonical forms and the moment problem

Theorem 2.12. *The unitary transformation*

$$
(2.101) \quad \tilde{U}: \begin{array}{ccc} \ell^2(\mathbb{N}) & \to & L^2(\mathbb{R}, d\rho_+) \\ f(n) & \mapsto & \sum_{n=1}^{\infty} f(n) s(\lambda, n) \end{array}
$$

(where the sum is to be understood as norm limit) maps the operator H_+ to the multiplication operator by λ. More explicitly,

$$
(2.102) \quad H_+ = \tilde{U}^{-1} \tilde{H} \tilde{U},
$$

where

$$
(2.103) \quad \tilde{H}: \begin{array}{ccc} L^2(\mathbb{R}, d\rho_+) & \to & L^2(\mathbb{R}, d\rho_+) \\ F(\lambda) & \mapsto & \lambda F(\lambda) \end{array}.
$$

Proof. Since $d\rho_+$ is compactly supported the set of all polynomials is dense in $L^2(\mathbb{R}, d\rho_+)$ (Lemma B.1) and U extends to the unitary transformation $\tilde{U}$. The rest follows from

$$
\tilde{H} F(\lambda) = \tilde{U} H_+ \tilde{U}^{-1} F(\lambda) = \tilde{U} H_+ \int_{\mathbb{R}} s(\lambda, n) F(\lambda) d\rho_+(\lambda)
$$
$$
(2.104) \quad = \tilde{U} \int_{\mathbb{R}} \lambda s(\lambda, n) F(\lambda) d\rho_+(\lambda) = \lambda F(\lambda).
$$
$\square$

This implies that the spectrum of H_+ can be characterized as follows (see Lemma B.5). Let the Lebesgue decomposition of $d\rho_+$ be given by

$$
(2.105) \quad d\rho_+ = d\rho_{+,pp} + d\rho_{+,ac} + d\rho_{+,sc},
$$

where pp, ac, and sc refer to the pure point, absolutely continuous, and singularly continuous part of the measure ρ_+ (with respect to Lebesgue measure), respectively.

Then the **pure point, absolutely continuous**, and **singular continuous spectra** of H_+ are given by (see Lemma B.5)

$$
\sigma(H_+) = \{\lambda \in \mathbb{R} | \lambda \text{ is a growth point of } \rho_+\},
$$
$$
\sigma_{pp}(H_+) = \{\lambda \in \mathbb{R} | \lambda \text{ is a growth point of } \rho_{+,pp}\},
$$
$$
\sigma_{ac}(H_+) = \{\lambda \in \mathbb{R} | \lambda \text{ is a growth point of } \rho_{+,ac}\},
$$
$$
(2.106) \quad \sigma_{sc}(H_+) = \{\lambda \in \mathbb{R} | \lambda \text{ is a growth point of } \rho_{+,sc}\}.
$$

Recall that $\sigma_{pp}(H_+)$ is in general not equal to the **point spectrum** $\sigma_p(H_+)$ (i.e., the set of eigenvalues of H_+). However, we have at least

$$
(2.107) \quad \sigma_{pp}(H_+) = \overline{\sigma_p(H_+)},
$$

where the bar denotes closure.

An additional decomposition in continuous and singular part with respect to the Hausdorff measure dh^α (see Appendix B) will be of importance as well,

$$
(2.108) \quad d\rho_+ = d\rho_{+,\alpha c} + d\rho_{+,\alpha s}.
$$

The corresponding spectra are defined analogously

$$
\sigma_{\alpha c}(H_+) = \{\lambda \in \mathbb{R} | \lambda \text{ is a growth point of } \rho_{+,\alpha c}\},
$$
$$
\sigma_{\alpha s}(H_+) = \{\lambda \in \mathbb{R} | \lambda \text{ is a growth point of } \rho_{+,\alpha s}\}.
$$

They will be used in Section 3.3.

Finally, we show how a^2, b can be reconstructed from the measure ρ_+. In fact, even the moments $m_{+,j}$, $j \in \mathbb{N}$, are sufficient for this task. This is generally known as **(Hamburger) moment problem**.

Suppose we have a given sequence m_j, $j \in \mathbb{N}_0$, such that

$$(2.109) \qquad C(k) = \det \begin{pmatrix} m_0 & m_1 & \cdots & m_{k-1} \\ m_1 & m_2 & \cdots & m_k \\ \vdots & \vdots & \ddots & \vdots \\ m_{k-1} & m_k & \cdots & m_{2k-2} \end{pmatrix} > 0, \quad k \in \mathbb{N}.$$

Without restriction we will assume $m_0 = 1$. Using this we can define a sesquilinear form on the set of polynomials as follows

$$(2.110) \qquad \langle P(\lambda), Q(\lambda) \rangle_{L^2} = \sum_{j,k=0}^{\infty} m_{j+k} \overline{p_j} q_k,$$

where $P(z) = \sum_{j=0}^{\infty} p_j z^j$, $Q(z) = \sum_{j=0}^{\infty} q_j z^j$ (note that all sums are finite). The polynomials

$$(2.111) \quad s(z,k) = \frac{1}{\sqrt{C(k-1)C(k)}} \det \begin{pmatrix} m_0 & m_1 & \cdots & m_{k-1} \\ m_1 & m_2 & \cdots & m_k \\ \vdots & \vdots & \ddots & \vdots \\ m_{k-2} & m_{k-1} & \cdots & m_{2k-3} \\ 1 & z & \cdots & z^{k-1} \end{pmatrix}, k \in \mathbb{N}$$

(set $C(0) = 1$), form a basis for the set of polynomials which is immediate from

$$(2.112) \qquad s(z,k) = \sqrt{\frac{C(k-1)}{C(k)}} \left(z^{k-1} + \frac{D(k-1)}{C(k-1)} z^{k-2} + O(z^{k-3}) \right),$$

where $D(0) = 0$, $D(1) = m_1$, and

$$(2.113) \qquad D(k) = \det \begin{pmatrix} m_0 & m_1 & \cdots & m_{k-2} & m_k \\ m_1 & m_2 & \cdots & m_{k-1} & m_{k+1} \\ \vdots & \vdots & \ddots & \vdots & \vdots \\ m_{k-1} & m_k & \cdots & m_{2k-3} & m_{2k-1} \end{pmatrix}, k \in \mathbb{N}.$$

Moreover, this basis is orthonormal, that is,

$$(2.114) \qquad \langle s(\lambda, j), s(\lambda, k) \rangle_{L^2} = \delta_{j,k},$$

since

$$\langle s(\lambda, k), \lambda^j \rangle = \frac{1}{\sqrt{C(k-1)C(k)}} \det \begin{pmatrix} m_0 & m_1 & \cdots & m_{k-1} \\ m_1 & m_2 & \cdots & m_k \\ \vdots & \vdots & \ddots & \vdots \\ m_{k-2} & m_{k-1} & \cdots & m_{2k-3} \\ m_j & m_{j+1} & \cdots & m_{j+k-1} \end{pmatrix}$$

$$(2.115) \qquad = \begin{cases} 0, & 0 \leq j \leq k-2 \\ \sqrt{\frac{C(k)}{C(k-1)}}, & j = k-1 \end{cases}.$$

In particular, the sesquilinear form (2.110) is positive definite and hence an inner product (note that $C(k) > 0$ is also necessary for this).

2.5. Canonical forms and the moment problem

Expanding the polynomial $zs(z,k)$ in terms of $s(z,j)$, $j \in \mathbb{N}$, we infer

$$zs(z,k) = \sum_{j=0}^{k+1} \langle s(\lambda,j), \lambda s(\lambda,k) \rangle_{L^2} s(z,j)$$

(2.116)
$$= a(k)s(z,k+1) + b(k)s(z,k) + a(k-1)s(z,k-1)$$

(set $s(z,0) = 0$) with

(2.117) $\quad a(k) = \langle s(\lambda, k+1), \lambda s(\lambda, k) \rangle_{L^2}, \quad b(k) = \langle s(\lambda, k), \lambda s(\lambda, k) \rangle_{L^2}, \quad k \in \mathbb{N}.$

In addition, comparing powers of z in (2.116) shows

(2.118) $\quad a(k) = \dfrac{\sqrt{C(k-1)C(k+1)}}{C(k)}, \quad b(k) = \dfrac{D(k)}{C(k)} - \dfrac{D(k-1)}{C(k-1)}.$

In terms of our original setting this says that given the measure $d\rho_+$ (or its moments, $m_{+,j}$, $j \in \mathbb{N}$) we can compute $s(\lambda, n)$, $n \in \mathbb{N}$, via orthonormalization of the set λ^n, $n \in \mathbb{N}_0$. This fixes $s(\lambda, n)$ up to a sign if we require $s(\lambda, n)$ real-valued. Then we can compute $a(n), b(n)$ as above (up to the sign of $a(n)$ which changes if we change the sign of $s(\lambda, n)$). Summarizing, $d\rho_+$ uniquely determines $a(n)^2$ and $b(n)$. Since knowing $d\rho_+(\lambda)$ is equivalent to knowing $m_+(z)$, the same is true for $m_+(z)$ (compare also the proof of Theorem 2.29). In fact, we have an even stronger result.

Theorem 2.13. *Suppose that the bounded measure $d\rho_+$ is not supported on a finite set. Then there exists a unique bounded Jacobi operator H_+ having $d\rho_+$ as spectral measure.*

Proof. We have already seen that the necessary and sufficient condition for our reconstruction procedure to work is that the sesquilinear form generated by the moments m_j of $d\rho_+$ is positive definite. Pick any nonzero polynomial $P(\lambda)$. Due to our assumption we can find $\varepsilon > 0$ and an interval I such that $\rho_+(I) \neq 0$ and $P(\lambda)^2 \geq \varepsilon$, $\lambda \in I$. Hence $\langle P(\lambda), P(\lambda) \rangle \geq \varepsilon \rho_+(I)$.

As a consequence we can define $a(n), b(n), s(\lambda, n)$, and the unitary transform $\tilde{U}$ as before. By construction $H_+ = \tilde{U}^{-1}\tilde{H}\tilde{U}$ is a bounded Jacobi operator associated with $a(n), b(n)$. That ρ_+ is the spectral measure of H_+ follows from (using $(\tilde{U}\delta_1)(\lambda) = 1$)

(2.119) $\quad \langle \delta_1, P_\Lambda(H_+)\delta_1 \rangle = \langle \tilde{U}\delta_1, P_\Lambda(\tilde{H})\tilde{U}\delta_1 \rangle = \int_\mathbb{R} \chi_\Lambda(\lambda) d\rho_+(\lambda)$

for any Borel set $\Lambda \subseteq \mathbb{R}$. $\square$

If $d\rho_+$ is supported on N points, the reconstruction procedure will break down after N steps (i.e., $C(N+1) = 0$) and we get a finite Jacobi matrix with $d\rho_+$ as spectral measure.

We also remark

(2.120) $\quad c(z,n) = -a(0) \displaystyle\int_\mathbb{R} \dfrac{s(z,n) - s(\lambda,n)}{z - \lambda} d\rho_+(\lambda), \quad n \in \mathbb{N},$

since one easily verifies $\tau_+ c(z,n) = zc(z,n) - a(0)\delta_0(n)$, $n \in \mathbb{N}$ (use (2.98) with $m=1$). Moreover, this implies

$$(2.121) \qquad u_+(z,n) = \frac{c(z,n)}{a(0)} - m_+(z)s(z,n) = \int_{\mathbb{R}} \frac{s(\lambda,n)}{z-\lambda} d\rho_+(\lambda)$$

and it is not hard to verify

$$(2.122) \qquad G_+(z,n,m) = \int_{\mathbb{R}} \frac{s(\lambda,n)s(\lambda,m)}{\lambda - z} d\rho_+(\lambda).$$

The Jacobi operator H can be treated along the same lines. Since we essentially repeat the analysis of H_+ we will be more sketchy.

Consider the vector valued polynomials

$$(2.123) \qquad \underline{S}(z,n) = \Big(c(z,n), s(z,n)\Big).$$

The analog of (2.94) reads

$$(2.124) \qquad s(H,n)\delta_1 + c(H,n)\delta_0 = \delta_n.$$

This is obvious for $n=0,1$ and the rest follows from induction upon applying H to (2.124). We introduce the spectral measures

$$(2.125) \qquad d\rho_{i,j}(.) = d\langle \delta_i, P_{(-\infty,\lambda]}(H)\delta_j \rangle,$$

and the (hermitian) matrix valued measure

$$(2.126) \qquad d\rho = \begin{pmatrix} d\rho_{0,0} & d\rho_{0,1} \\ d\rho_{1,0} & d\rho_{1,1} \end{pmatrix}.$$

The diagonal part consists of positive measures and the off-diagonal part can be written as the difference of two positive measures

$$(2.127) \qquad d\rho_{0,1}(\lambda) = d\rho_{1,0}(\lambda) = d\rho_{0,1,+}(\lambda) - d\rho_{0,1,-}(\lambda),$$

where

$$(2.128) \qquad \begin{aligned} d\rho_{0,1,+}(\lambda) &= \frac{1}{2}d\langle (\delta_0+\delta_1), P_{(-\infty,\lambda]}(H)(\delta_0+\delta_1)\rangle, \\ d\rho_{0,1,-}(\lambda) &= \frac{1}{2}(d\langle \delta_0, P_{(-\infty,\lambda]}(H)\delta_0\rangle + d\langle \delta_1, P_{(-\infty,\lambda]}(H)\delta_1\rangle). \end{aligned}$$

Moreover, $d\rho$ is a positive matrix measure and we have a corresponding Hilbert space $L^2(\mathbb{R}, \mathbb{C}^2, d\rho)$ with scalar product given by

$$(2.129) \qquad \langle \underline{F}, \underline{G}\rangle_{L^2} = \sum_{i,j=0}^{1} \int_{\mathbb{R}} \overline{F_i(\lambda)} G_j(\lambda) d\rho_{i,j}(\lambda) \equiv \int_{\mathbb{R}} \overline{\underline{F}(\lambda)} \underline{G}(\lambda) d\rho(\lambda)$$

and if $\underline{F}, \underline{G}$ are vector valued polynomials, then

$$(2.130) \qquad \langle \underline{F}, \underline{G}\rangle_{L^2} = \langle F_0(H)\delta_0 + F_1(H)\delta_1, G_0(H)\delta_0 + G_1(H)\delta_1\rangle.$$

By (2.124) the vector valued polynomials $\underline{S}(\lambda,n)$ are orthogonal with respect to $d\rho$,

$$(2.131) \qquad \langle \underline{S}(.,m), \underline{S}(.,n)\rangle_{L^2} = \delta_{m,n}.$$

The formulas analogous to (2.117) then read

$$(2.132) \quad a(n) = \langle \underline{S}(\lambda,n+1), \lambda \underline{S}(\lambda,n)\rangle_{L^2}, \quad b(n) = \langle \underline{S}(\lambda,n), \lambda \underline{S}(\lambda,n)\rangle_{L^2}, \quad n \in \mathbb{Z}.$$

2.5. Canonical forms and the moment problem

Next, we consider the following transformation U from the set $\ell_0(\mathbb{Z})$ onto the set of vector-valued polynomials (**eigenfunction expansion**)

$$(Uf)(\lambda) = \sum_{n\in\mathbb{Z}} f(n)\underline{S}(\lambda, n),$$

(2.133)
$$(U^{-1}\underline{F})(n) = \int_{\mathbb{R}} \underline{S}(\lambda, n)\underline{F}(\lambda)d\rho(\lambda).$$

Again a simple calculation for $\underline{F}(\lambda) = (Uf)(\lambda)$ shows that U is unitary,

(2.134)
$$\sum_{n\in\mathbb{Z}} |f(n)|^2 = \int_{\mathbb{R}} \overline{\underline{F}(\lambda)}\underline{F}(\lambda)d\rho(\lambda).$$

Extending U to a unitary transformation $\tilde{U}$ we obtain as in the case of H_+ the following

Theorem 2.14. *The unitary transformation*

(2.135)
$$\tilde{U} : \begin{array}{ccc} \ell^2(\mathbb{Z}) & \to & L^2(\mathbb{R}, \mathbb{C}^2, d\rho) \\ f(n) & \mapsto & \sum_{n=-\infty}^{\infty} f(n)\underline{S}(\lambda, n) \end{array}$$

(where the sum is to be understood as norm limit) maps the operator H to the multiplication operator by λ, that is,

(2.136)
$$\tilde{H} = \tilde{U}H\tilde{U}^{-1},$$

where

(2.137)
$$\tilde{H} : \begin{array}{ccc} L^2(\mathbb{R}, \mathbb{C}^2, d\rho) & \to & L^2(\mathbb{R}, \mathbb{C}^2, d\rho) \\ \underline{F}(\lambda) & \mapsto & \lambda\underline{F}(\lambda) \end{array}.$$

For the Green function of H we obtain

(2.138)
$$G(z, n, m) = \int_{\mathbb{R}} \frac{\underline{S}(\lambda, n)\underline{S}(\lambda, m)}{\lambda - z} d\rho(\lambda).$$

By Lemma B.13, in order to characterize the spectrum of H one only needs to consider the trace $d\rho^{tr}$ of $d\rho$ given by

(2.139)
$$d\rho^{tr} = d\rho_{0,0} + d\rho_{1,1}.$$

Let the Lebesgue decomposition (cf. (2.105)) of $d\rho^{tr}$ be given by

(2.140)
$$d\rho^{tr} = d\rho^{tr}_{pp} + d\rho^{tr}_{ac} + d\rho^{tr}_{sc},$$

then the **pure point**, **absolutely continuous**, and **singular continuous spectra** of H are given by

$$\sigma(H) = \{\lambda \in \mathbb{R} | \lambda \text{ is a growth point of } \rho^{tr}\},$$
$$\sigma_{pp}(H) = \{\lambda \in \mathbb{R} | \lambda \text{ is a growth point of } \rho^{tr}_{pp}\},$$
$$\sigma_{ac}(H) = \{\lambda \in \mathbb{R} | \lambda \text{ is a growth point of } \rho^{tr}_{ac}\},$$
(2.141)
$$\sigma_{sc}(H) = \{\lambda \in \mathbb{R} | \lambda \text{ is a growth point of } \rho^{tr}_{sc}\}.$$

The **Weyl matrix** $M(z)$ is defined as

$$M(z) = \int_{-\infty}^{\infty} \frac{d\rho(\lambda)}{\lambda - z} - \frac{1}{2a(0)} \begin{pmatrix} 0 & 1 \\ 1 & 0 \end{pmatrix}$$

(2.142)
$$= \begin{pmatrix} g(z,0) & \frac{h(z,0)}{2a(0)} \\ \frac{h(z,0)}{2a(0)} & g(z,1) \end{pmatrix}, \quad z \in \mathbb{C}\backslash\sigma(H).$$

Explicit evaluation yields

(2.143) $$M(z) = \frac{1}{\tilde{m}_+(z) + \tilde{m}_-(z)} \begin{pmatrix} -\frac{1}{a(0)^2} & \frac{\tilde{m}_+(z)-\tilde{m}_-(z)}{2a(0)} \\ \frac{\tilde{m}_+(z)-\tilde{m}_-(z)}{2a(0)} & \tilde{m}_+(z)\tilde{m}_-(z) \end{pmatrix},$$

and the determinant reads

(2.144) $$\det M(z) = -\frac{1}{4a(0)^2}.$$

In terms of the original Weyl m-functions we obtain

$$g(z,0) = \frac{-1}{z - b(0) - a(0)^2 m_+(z) - a(-1)^2 m_-(z)},$$

(2.145) $$h(z,0) = \big(z - b(0) + a(0)^2 m_+(z) - a(-1)^2 m_-(z)\big) g(z,0).$$

Finally, notice that we can replace $c(z,n)$, $s(z,n)$ by any other pair of linearly independent solutions. For example, we could use $c_\beta(z,n)$, $s_\beta(z,n)$. As in (2.123) we define

(2.146) $$\underline{S}_\beta(z,n) = \big(c_\beta(z,n), s_\beta(z,n)\big) = U_\alpha \underline{S}(z,n),$$

where U_α is rotation by the angle α ($\beta = \cot\alpha$), that is,

(2.147) $$U_\alpha = \begin{pmatrix} \cos\alpha & \sin\alpha \\ -\sin\alpha & \cos\alpha \end{pmatrix}.$$

Hence all objects need to be rotated by the angle α. For instance, introducing $M^\beta(z) = U_\alpha M(z) U_\alpha^{-1}$ we infer

(2.148) $$M^\beta(z) = \frac{1}{\tilde{m}_+^\beta(z) + \tilde{m}_-^\beta(z)} \begin{pmatrix} -\frac{1}{a(0)^2} & \frac{\tilde{m}_+^\beta(z)-\tilde{m}_-^\beta(z)}{2a(0)} \\ \frac{\tilde{m}_+^\beta(z)-\tilde{m}_-^\beta(z)}{2a(0)} & \tilde{m}_+^\beta(z)\tilde{m}_-^\beta(z) \end{pmatrix}.$$

Note also

(2.149) $$\frac{-1}{a(0)^2(\tilde{m}_+^\beta(z) + \tilde{m}_-^\beta(z))} = \sin^2(\alpha)\,\gamma^\beta(z,0).$$

If we restrict ourselves to the absolutely continuous part of the spectrum, we can do even more. We abbreviate $\tilde{m}_\pm(\lambda) = \lim_{\varepsilon\downarrow 0} \tilde{m}_\pm(\lambda + i\varepsilon)$, $\lambda \in \mathbb{R}$, implying (cf. Appendix B)

$$d\rho_{0,0,ac}(\lambda) = \frac{\text{Im}(\tilde{m}_+(\lambda) + \tilde{m}_-(\lambda))}{\pi a(0)^2 |\tilde{m}_+(\lambda) + \tilde{m}_-(\lambda)|^2} d\lambda,$$

$$d\rho_{0,1,ac}(\lambda) = \frac{\text{Im}(\tilde{m}_+(\lambda))\text{Re}(\tilde{m}_-(\lambda)) - \text{Re}(\tilde{m}_+(\lambda))\text{Im}(\tilde{m}_-(\lambda))}{\pi a(0)|\tilde{m}_+(\lambda) + \tilde{m}_-(\lambda)|^2} d\lambda,$$

(2.150) $$d\rho_{1,1,ac}(\lambda) = \frac{\text{Im}(|\tilde{m}_-(\lambda)|^2 \tilde{m}_+(\lambda) + |\tilde{m}_+(\lambda)|^2 \tilde{m}_-(\lambda))}{\pi |\tilde{m}_+(\lambda) + \tilde{m}_-(\lambda)|^2} d\lambda.$$

2.6. Some remarks on unbounded operators

Choosing the new basis

$$
\begin{pmatrix} u_+(\lambda,n) \\ u_-(\lambda,n) \end{pmatrix} = V(\lambda) \begin{pmatrix} c(\lambda,n) \\ s(\lambda,n) \end{pmatrix} \tag{2.151}
$$

with

$$
V(\lambda) = \frac{1}{\tilde{m}_+(\lambda) + \tilde{m}_-(\lambda)} \begin{pmatrix} \frac{1}{a(0)} & -\tilde{m}_+(\lambda) \\ \frac{1}{a(0)} & \tilde{m}_-(\lambda) \end{pmatrix} \tag{2.152}
$$

we get

$$
(V^{-1}(\lambda))^* \, d\rho_{ac}(\lambda) \, V^{-1}(\lambda) = \begin{pmatrix} \operatorname{Im}(\tilde{m}_-(\lambda)) & 0 \\ 0 & \operatorname{Im}(\tilde{m}_+(\lambda)) \end{pmatrix} d\lambda. \tag{2.153}
$$

Note that $V(\lambda)$ is not unitary. We will show how to diagonalize all of $d\rho$ in Section 3.1.

2.6. Some remarks on unbounded operators

In this section we temporarily drop the boundedness assumption on the coefficients a, b. This renders H unbounded and implies that we are no longer able to define H on all of $\ell^2(\mathbb{Z})$. Nevertheless we can define the **minimal and maximal operator** allied with τ as follows

$$
H_{min} : \begin{array}{c} \mathfrak{D}(H_{min}) \to \ell^2(\mathbb{Z}) \\ f \mapsto \tau f \end{array}, \quad H_{max} : \begin{array}{c} \mathfrak{D}(H_{max}) \to \ell^2(\mathbb{Z}) \\ f \mapsto \tau f \end{array}, \tag{2.154}
$$

with

$$
\mathfrak{D}(H_{min}) = \ell_0(\mathbb{Z}), \quad \mathfrak{D}(H_{max}) = \{f \in \ell^2(\mathbb{Z}) | \tau f \in \ell^2(\mathbb{Z})\}. \tag{2.155}
$$

By Green's formula (1.20) we have $H_{min}^* = H_{max}$ and

$$
H_{max}^* = \overline{H_{min}} : \begin{array}{c} \mathfrak{D}(H_{max}^*) \to \ell^2(\mathbb{Z}) \\ f \mapsto \tau f \end{array}, \tag{2.156}
$$

with

$$
\mathfrak{D}(H_{max}^*) = \{f \in \mathfrak{D}(H_{max}) | \lim_{n \to \pm\infty} W_n(\overline{f}, g) = 0, \, g \in \mathfrak{D}(H_{max})\}. \tag{2.157}
$$

Here $H^*, \overline{H}$ denote the adjoint, closure of an operator H, respectively. We also remark that $\lim_{n \to \pm\infty} W_n(\overline{f}, g)$ exists for $f, g \in \mathfrak{D}(H_{max})$ as can be easily shown using (1.20). Similar definitions apply to $H_\pm$.

Since we might have $H_{max}^* \neq H_{max}$, H_{max} might not be self-adjoint in general. The key to this problem will be the **limit point** (l.p.), **limit circle** (l.c.) classification alluded to in Remark 2.11 (ii). To make things precise, we call τ l.p. at $\pm\infty$ if $s(z_0,.) \notin \ell^2(\pm\mathbb{N})$ for some $z_0 \in \mathbb{C}\backslash\mathbb{R}$. Otherwise τ is called l.c. at $\pm\infty$.

In order to draw some first consequences from this definition we note that all considerations of Section 2.4 do not use boundedness of a, b except for (2.85). However, if $\sum_{n \in \mathbb{N}} s(z_0, n) < \infty$ (considering only $\beta = \infty$ for simplicity), then the circle corresponding to $\tilde{m}_N(z_0)$ converges to a circle instead of a point as $N \to \infty$. If $\tilde{m}_\pm(z_0)$ is defined to be any point on this limiting circle, everything else remains unchanged. In particular, $u_+(z_0,.) \in \ell^2(\mathbb{N})$ and $s(z_0,.) \in \ell^2(\mathbb{N})$ shows that every solution of $\tau u = z_0 u$ is in $\ell^2(\mathbb{N})$ if τ is l.c. at $+\infty$.

This enables us to reveal the connections between the l.p. / l.c. characterization and the self-adjointness of unbounded Jacobi operators. We first consider $H_\pm$.

We recall that $H_{min,+}$ is symmetric and that $(H_{min,+})^* = H_{max,+}$. We need to investigate the deficiency indices $d_\pm = \dim \mathrm{Ker}(H_{max,+} - z_\pm)$, $z_\pm \in \mathbb{C}_\pm$, of $H_{min,+}$. They are independent of $z_\pm \in \mathbb{C}_\pm$ and equal (i.e., $d_- = d_+$) since $H_{min,+}$ is real (cf. [192], [241], Chapter 8). Thus, to compute $d_\pm$ it suffices to consider $\mathrm{Ker}(H_{max,+} - z_0)$. Since any element of $\mathrm{Ker}(H_{max,+} - z_0)$ is a multiple of $s(z_0)$ we infer $d_- = d_+ = 0$ if $s(z_0) \notin \ell^2(\mathbb{N})$ and $d_- = d_+ = 1$ if $s(z_0) \in \ell^2(\mathbb{N})$. This shows that $H_{max,+}$ is self-adjoint if and only if τ is l.p. at $+\infty$. Moreover, $s(z_0) \in \ell^2(\mathbb{N})$ implies $s(z) \in \ell^2(\mathbb{N})$ for all $z \in \mathbb{C}\backslash\mathbb{R}$ since $d_\pm = 1$ independent of $z \in \mathbb{C}\backslash\mathbb{R}$. Or, put differently, the l.p. / l.c. definition is independent of $z \in \mathbb{C}\backslash\mathbb{R}$.

If τ is l.c. at $\pm\infty$, then our considerations imply that all solutions of (1.19) for all $z \in \mathbb{C}\backslash\mathbb{R}$ are square summable near $\pm\infty$, respectively. This is even true for all $z \in \mathbb{C}$.

Lemma 2.15. *Suppose that all solutions of $\tau u = zu$ are square summable near $\pm\infty$ for one value $z = z_0 \in \mathbb{C}$. Then this is true for all $z \in \mathbb{C}$.*

Proof. If u fulfills (1.19), we may apply (1.48) to $(\tau - z_0)u = (z - z_0)u$,

$$u(n) = u(n_0)c(z_0, n, n_0) + u(n_0 + 1)s(z_0, n, n_0)$$

(2.158)
$$- \frac{z - z_0}{a(0)} \sum_{j=n_0+1}^{n}{}^* \big(c(z_0, n)s(z_0, j) - c(z_0, j)s(z_0, n)\big)u(j).$$

By assumption, there exists a constant $M \geq 0$ such that

(2.159)
$$\sum_{j=n_0+1}^{\infty} |c(z_0, j)|^2 \leq M, \quad \sum_{j=n_0+1}^{\infty} |s(z_0, j)|^2 \leq M.$$

Invoking the Cauchy-Schwarz inequality we obtain the estimate ($n > n_0$)

$$\Big| \sum_{j=n_0+1}^{n} \big(c(z_0, n)s(z_0, j) - c(z_0, j)s(z_0, n)\big)u(j) \Big|^2$$

$$\leq \sum_{j=n_0+1}^{n} |c(z_0, n)s(z_0, j) - c(z_0, j)s(z_0, n)|^2 \sum_{j=n_0+1}^{n} |u(j)|^2$$

(2.160)
$$\leq M\big(|c(z_0, n)|^2 + |s(z_0, n)|^2\big) \sum_{j=n_0+1}^{n} |u(j)|^2.$$

Since n_0 is arbitrary we may choose n_0 in (2.159) so large, that we have $4a(0)^{-1}|z - z_0|M^2 \leq 1$. Again using Cauchy-Schwarz

$$\sum_{j=n_0+1}^{n} |u(j)|^2 \leq (|u(n_0)|^2 + |u(n_0 + 1)|^2)\tilde{M} + \frac{2|z - z_0|}{a(0)} M^2 \sum_{j=n_0+1}^{n} |u(j)|^2$$

(2.161)
$$\leq (|u(n_0)|^2 + |u(n_0 + 1)|^2)\tilde{M} + \frac{1}{2} \sum_{j=n_0+1}^{n} |u(j)|^2,$$

2.6. Some remarks on unbounded operators

where $\tilde{M}$ is chosen such that $\sum_{j=n_0+1}^{\infty} |c(z_0,j,n_0)|^2 \leq \tilde{M}$, $\sum_{j=n_0+1}^{\infty} |s(z_0,j,n_0)|^2 \leq \tilde{M}$ holds. Solving for the left hand side finishes the proof,

$$(2.162) \qquad \sum_{j=n_0+1}^{n} |u(j)|^2 \leq 2(|u(n_0)|^2 + |u(n_0+1)|^2)\tilde{M}.$$

□

In summary, we have the following lemma.

Lemma 2.16. *The operator $H_{max,\pm}$ is self-adjoint if and only if one of the following statements holds.*

(i) *τ is l.p. at $\pm\infty$.*
(ii) *There is a solution of (1.19) for some $z \in \mathbb{C}$ (and hence for all) which is not square summable near $\pm\infty$.*
(iii) *$W_{\pm\infty}(f,g) = 0$, for all $f,g \in \mathfrak{D}(H_{max,\pm})$.*

To simplify notation, we will only consider the endpoint $+\infty$ in the following. The necessary modifications for $-\infty$ are straightforward.

Next, let us show a simple but useful criterion for τ being *l.p.* at $+\infty$. If τ is *l.c.* at $+\infty$, we can use the Wronskian

$$(2.163) \qquad \frac{a(0)}{a(n)} = c(z,n)s(z,n+1) - c(z,n+1)s(z,n),$$

to get (using the Cauchy-Schwarz inequality)

$$(2.164) \qquad \sum_{n \in \mathbb{N}} \frac{1}{|a(n)|} \leq \frac{2}{|a(0)|}\sqrt{\sum_{n \in \mathbb{N}} |c(z,n)|^2 \sum_{n \in \mathbb{N}} |s(z,n)|^2}.$$

This shows that a sufficient condition for τ to be *l.p.* at $+\infty$ is

$$(2.165) \qquad \sum_{n \in \mathbb{N}} \frac{1}{|a(n)|} = \infty.$$

The remaining question is: What happens in the *l.c.* case? Obviously we need some suitable boundary conditions. The boundary condition

$$(2.166) \qquad BC_{n_0,\alpha}(f) = \cos(\alpha)f(n_0) + \sin(\alpha)f(n_0+1) = 0$$

of Remark 1.9 makes no sense if $n_0 = \pm\infty$. However, (2.166) can be written as $W_{n_0}(v,f) = 0$, where v is any sequence satisfying $BC_{n_0,\alpha}(v) = 0$ and $|v(n_0)| + |v(n_0+1)| \neq 0$. Moreover, all different boundary conditions can be obtained by picking v such that $W_{n_0}(\bar{v},v) = 0$ and $W_{n_0}(v,f) \neq 0$ for some f. This latter characterization of the boundary condition may be generalized.

We define the set of boundary conditions for τ at $+\infty$ by

$$(2.167) \quad BC_+(\tau) = \{v \in \mathfrak{D}(H_{max,+}) |\ W_{+\infty}(\bar{v},v) = 0,\ W_{+\infty}(\bar{v},f) \neq 0 \text{ for some}$$
$$f \in \mathfrak{D}(H_{max,+}) \text{ if } \tau \text{ is } l.c. \text{ at } \pm\infty\}.$$

Observe that the first requirement holds if v is real. The second is void if τ is *l.p.* (at $+\infty$). Otherwise, if τ is *l.c.*, there is at least one *real* v for which it holds (if not, (iii) of Lemma 2.16 implies that τ is *l.p.*). Two sequences $v_{1,2} \in BC_+(\tau)$ are called equivalent if $W_{+\infty}(v_1,v_2) = 0$.

Lemma 2.17. *Let $v \in BC_+(\tau)$ and set*

(2.168) $$\mathfrak{D}_+(v) = \{f \in \mathfrak{D}(H_{max,+}) | W_{+\infty}(v, f) = 0\}.$$

Then

(i) $W_{+\infty}(v, f) = 0 \Leftrightarrow W_{+\infty}(v, \overline{f}) = 0$,
(ii) $W_{+\infty}(g, f) = 0$ *for* $f, g \in \mathfrak{D}_+(v)$.

Moreover, $W_{+\infty}(v_1, v_2) = 0$ is equivalent to $\mathfrak{D}_+(v_1) = \mathfrak{D}_+(v_2)$.

Proof. For all $f_1, \ldots, f_4 \in \mathfrak{D}(H_{max})$ we can take the limits $n \to +\infty$ in the Plücker identity

(2.169) $$W_n(f_1, f_2)W_n(f_3, f_4) + W_n(f_1, f_3)W_n(f_4, f_2) + W_n(f_1, f_4)W_n(f_2, f_3) = 0.$$

Now choose $f_1 = v$, $f_2 = \hat{f}$, $f_3 = \overline{v}$, $f_4 = f$ to conclude $W_{+\infty}(v, f) = 0$ implies $W_{+\infty}(\overline{v}, f) = 0$. Then choose $f_1 = v$, $f_2 = \hat{f}$, $f_3 = f$, $f_4 = \overline{g}$ to show (ii). The last assertion follows from (ii) upon choosing $v = v_1$, $g = v_2$. $\square$

Combining this lemma with Green's formula (1.20) shows

Theorem 2.18. *Choose $v \in BC_+(\tau)$, then*

(2.170) $$\begin{aligned} H_+ : \mathfrak{D}_+(v) &\to \ell^2(\mathbb{N}) \\ f &\mapsto \tau f \end{aligned}$$

is a self-adjoint extension of $H_{min,+}$.

In the *l.p.* case the boundary condition $W_{+\infty}(v, f) = 0$ is of course always fulfilled and thus $\mathfrak{D}_+(v) = \mathfrak{D}(H_{max,+})$ for any $v \in BC_+(\tau)$.

Clearly, we can also define self-adjoint operators $H_{n_0,+}$ and $H^\beta_{n_0,+}$ corresponding to H_+ as we did in Section 1.2 for the bounded case.

Now, that we have found self-adjoint extensions, let us come back to the Weyl m-functions. We fix $v(n) = s_\beta(\lambda, n)$, $\lambda \in \mathbb{R}$, for the boundary condition at $+\infty$. Observe that

(2.171) $$\tilde{m}^\beta_+(z) = \frac{1}{a(0)} \lim_{n \to +\infty} \frac{W_n(s_\beta(\lambda), c_\beta(z))}{W_n(s_\beta(\lambda), s_\beta(z))}$$

lies on the limiting circle. This is clear if τ is *l.p.* at $+\infty$. Otherwise, τ *l.c.* at $+\infty$ implies $c_\beta(z), s_\beta(z) \in \mathfrak{D}(H_{max,+})$ and both Wronskians converge to a limit as pointed out earlier in this section. Moreover, if $W_n(s_\beta(\lambda), s_\beta(z)) = 0$, then $s_\beta(z) \in \mathfrak{D}(H^\beta_+)$ and hence $z \in \sigma(H^\beta_+)$. In particular, $W_n(s_\beta(\lambda), s_\beta(z)) \neq 0$ for $z \in \mathbb{C}\backslash\mathbb{R}$ and we can call $\tilde{m}^\beta_+(z)$ the Weyl $\tilde{m}$-function of H^β_+.

In addition, the function

(2.172) $$u_+(z, n) = \frac{c_\beta(z, n)}{a(0)} \mp \tilde{m}^\beta_+(z) s_\beta(z, n)$$

is in $\ell^2_+(\mathbb{Z})$ and satisfies the boundary condition

(2.173) $$W_{+\infty}(s_\beta(\lambda), u_\pm(z)) = 0.$$

The boundary condition uniquely characterizes $u_+(z, n)$ up to a constant in the *l.c.* case.

We have seen that the question of self-adjointness is simple if τ is *l.p.*. One the other hand, the spectrum gets simple if τ is *l.c.*.

2.6. Some remarks on unbounded operators

Lemma 2.19. *If τ is l.c. at $+\infty$, then the resolvent of H_+ is a Hilbert-Schmidt operator. In particular, this implies that H_+ has purely discrete spectrum, $\sigma(H_+) = \sigma_d(H_+)$, and*

(2.174)
$$\sum_{\lambda \in \sigma_d(H_+)} \frac{1}{1+\lambda^2} < \infty.$$

Proof. The result is a consequence of the following estimate

$$\sum_{(n,m)\in\mathbb{Z}^2} |G_+(z,m,n)|^2 = \frac{1}{|W|^2} \sum_{n\in\mathbb{Z}} \left(|u_+(z,n)|^2 \sum_{m<n} |s(z,m)|^2 \right.$$
$$\left. + |s(z,n)|^2 \sum_{m\geq n} |u_+(z,m)|^2 \right) \leq \frac{2}{|W|^2} \|u_+(z)\|^2 \|s(z)\|^2,$$

where $W = W(s(z), u_+(z))$. $\square$

Our next goal is to find a good parameterization of all self-adjoint extensions if τ is l.c. at $+\infty$.

First of all note that any real solution of $\tau u = \lambda u$, $\lambda \in \mathbb{R}$, is in $BC_+(\tau)$ (since $W(u, \tilde{u}) \neq 0$ for any linearly independent solution of $\tau u = \lambda u$). Now fix

(2.175) $\qquad v_\alpha(n) = \cos(\alpha) c(0,n) + \sin(\alpha) s(0,n), \qquad \alpha \in [0, \pi),$

and note that different α's imply different extensions since $W(v_{\alpha_1}, v_{\alpha_2}) = \sin(\alpha_2 - \alpha_1)/a(0)$.

Lemma 2.20. *All self-adjoint extensions of $H_{min,+}$ correspond to some v_α with unique $\alpha \in [0, \pi)$.*

Proof. Let H_+ be a self-adjoint extension of $H_{min,+}$ and $\lambda_0 \in \sigma(H_+)$ be an eigenvalue with corresponding eigenfunction $s(\lambda_0, n)$. Using Green's formula (1.20) with $f = s(\lambda_0)$, $g \in \mathfrak{D}(H_+)$ we see $\mathfrak{D}(H_+) \subseteq \mathfrak{D}_+(s(\lambda_0))$ and hence $\mathfrak{D}(H_+) = \mathfrak{D}_+(s(\lambda_0))$ by maximality of self-adjoint operators. Let $\alpha \in [0, \pi)$ be the unique value for which

(2.176)
$$W_{+\infty}(v_\alpha, s(\lambda_0)) = \cos(\alpha) W_{+\infty}(c(0), s(\lambda_0)) + \sin(\alpha) W_{+\infty}(s(0), s(\lambda_0)) = 0.$$

Then $\mathfrak{D}_+(s(\lambda_0)) = \mathfrak{D}_+(v_\alpha)$. $\square$

Now we turn back again to operators on $\ell^2(\mathbb{Z})$. We use that corresponding definitions and results hold for the other endpoint $-\infty$ as well. With this in mind we have

Theorem 2.21. *Choose $v_\pm \in BC_\pm(\tau)$ as above, then the operator H with domain*

(2.177) $\qquad \mathfrak{D}(H) = \{f \in \mathfrak{D}(H_{max}) | W_{+\infty}(v_+, f) = W_{-\infty}(v_-, f) = 0\}$

is self-adjoint.

Again, if τ is l.p. at $\pm\infty$, the corresponding boundary condition is void and can be omitted. We also note that if τ is l.c. at both $\pm\infty$, then we have not found all self-adjoint extensions since we only consider separated boundary conditions (i.e., one for each endpoint) and not coupled ones which connect the behavior of functions at $-\infty$ and $+\infty$.

As before we have

Lemma 2.22. *If τ is l.c. at both $\pm\infty$, then the resolvent of H is a Hilbert-Schmidt operator. In particular, the spectrum of H is purely discrete.*

Most results found for bounded Jacobi operators still hold with minor modifications. One result that requires some changes is Theorem 2.13.

Theorem 2.23. *A measure $d\rho_+$ which is not supported on a finite set is the spectral measure of a unique Jacobi operator H_+ if and only if the set of polynomials is dense in $L^2(\mathbb{R}, d\rho_+)$.*

Proof. If the set of polynomials is dense in $L^2(\mathbb{R}, d\rho_+)$ we can use the same proof as in Theorem 2.13 to show existence of a unique Jacobi operator with $d\rho_+$ as spectral measure.

Conversely, let H_+ be given, and let U be a unitary transform mapping H_+ to multiplication by λ in $L^2(\mathbb{R}, d\rho_+)$ (which exists by the spectral theorem). Then $|(U\delta_1)(\lambda)|^2 = 1$ since

$$(2.178) \qquad \int_\Lambda d\rho_+ = \langle \delta_1, P_\Lambda(H_+)\delta_1 \rangle = \langle U\delta_1, P_\Lambda(\lambda)U\delta_1 \rangle = \int_\Lambda |(U\delta_1)(\lambda)|^2 d\rho_+(\lambda)$$

for any Borel set $\Lambda \subseteq \mathbb{R}$. So, by another unitary transformation, we can assume $(U\delta_1)(\lambda) = 1$. And since the span of $(H_+)^j \delta_1$, $j \in \mathbb{N}_0$, is dense in $\ell^2(\mathbb{Z})$, so is the span of $(U(H_+)^j \delta_1)(\lambda) = \lambda^j$ in $L^2(\mathbb{R}, d\rho_+)$. $\square$

A measure $d\rho_+$ which is not supported on a finite set and for which the set of polynomials is dense in $L^2(\mathbb{R}, d\rho_+)$ will be called **Jacobi measure**.

There are some interesting consequences for the moment problem.

Lemma 2.24. *A set $\{m_{+,j}\}_{j \in \mathbb{N}}$ forms the moments of a Jacobi measure if and only if (2.109) holds.*

Moreover,

$$(2.179) \qquad \mathrm{supp}(\rho_+) \subseteq [-R, R] \quad \Leftrightarrow \quad |m_{+,j}| \leq R^j, \quad j \in \mathbb{N}.$$

Proof. If (2.109) holds we get sequences $a(n)$, $b(n)$ by (2.118). The spectral measure of any self-adjoint extension has $m_{+,j}$ as moments.

If $\mathrm{supp}(\rho_+) \subseteq [-R, R]$, then $|m_{+,j}| \leq \int |\lambda|^j d\rho_+(\lambda) \leq R^j \int d\rho_+(\lambda) = R^j$. Conversely, if $\mathrm{supp}(\rho_+) \not\subseteq [-R, R]$, then there is an $\varepsilon > 0$ such that $C_\varepsilon = \rho_+(\{\lambda | |\lambda| > R + \varepsilon\}) > 0$ and hence $|m_{+,2j}| \geq \int_{|\lambda| > R + \varepsilon} \lambda^{2j} d\rho_+(\lambda) \geq C_\varepsilon (R + \varepsilon)^{2j}$. $\square$

Finally, let us look at uniqueness.

Theorem 2.25. *A measure is uniquely determined by its moments if and only if the associated Jacobi difference expression τ (defined via (2.118)) is l.p. at $+\infty$.*

Proof. Our assumption implies that $H_{min,+}$ is essentially self-adjoint and hence $(H_{min,+} - z)\mathfrak{D}(H_{min,+})$ is dense in $\ell^2(\mathbb{N})$ for any $z \in \mathbb{C}_\pm$. Denote by $\mathfrak{H}_0$ the set of polynomials on $\mathbb{R}$ and by $\mathfrak{H}$ the closure of $\mathfrak{H}_0$ with respect to the scalar product (2.110). Then $(\lambda - z)\mathfrak{H}_0$ is dense in $\mathfrak{H}$ and hence there is a sequence of polynomials $P_{z,n}(\lambda)$, $z \in \mathbb{C}_\pm$, such that $(\lambda - z)P_{z,n}(\lambda)$ converges to 1 in $\mathfrak{H}$. Let ρ be a measure with correct moments. Then

$$(2.180) \qquad \int_\mathbb{R} |P_{z,n}(\lambda) - \frac{1}{\lambda - z}|^2 d\rho(\lambda) \leq \int_\mathbb{R} \frac{|\lambda - z|^2}{\mathrm{Im}(z)^2} |P_{z,n}(\lambda) - \frac{1}{\lambda - z}|^2 d\rho(\lambda)$$

shows that $P_{z,n}(\lambda)$ converges to $(\lambda - z)^{-1}$ in $L^2(\mathbb{R}, d\rho)$ and consequently the Borel transform

$$\int \frac{d\rho(\lambda)}{\lambda - z} = \langle 1, \frac{1}{\lambda - z} \rangle_{L^2} = \lim_{n \to \infty} \langle 1, P_{z,n} \rangle \tag{2.181}$$

is uniquely determined by the moments. Since ρ is uniquely determined by its Borel transform we are done. $\square$

We know that τ is *l.p.* at $+\infty$ if the moments are polynomially bounded by (2.179). However, a weaker bound on the growth of the moments also suffices to ensure the *l.p.* case.

Lemma 2.26. *Suppose*

$$|m_{+,j}| \leq CR^j j!, \quad j \in \mathbb{N}, \tag{2.182}$$

then τ associated with $\{m_{+,j}\}_{j \in \mathbb{N}}$ is l.p. at $+\infty$.

Proof. Our estimate implies that $e^{iz\lambda} \in L^1(\mathbb{R}, d\rho)$ for $|\text{Im}(z)| < 1/R$. Hence the Fourier transform

$$\int_{\mathbb{R}} e^{iz\lambda} d\rho(\lambda) = \sum_{j=0}^{\infty} m_{+,j} \frac{(iz)^j}{j!} \tag{2.183}$$

is holomorphic in the strip $|\text{Im}(z)| < 1/R$. This shows that the Fourier transform is uniquely determined by the moments and so is the Borel transform and hence the measure (see (B.9)). $\square$

2.7. Inverse spectral theory

In this section we present a simple recursive method of reconstructing the sequences a^2, b when the Weyl matrix (cf. (2.142))

$$M(z, n) = \begin{pmatrix} g(z, n) & \frac{h(z,n)}{2a(n)} \\ \frac{h(z,n)}{2a(n)} & g(z, n+1) \end{pmatrix}, \quad z \in \mathbb{C} \backslash \sigma(H), \tag{2.184}$$

is known for one fixed $n \in \mathbb{Z}$. As a consequence, we are led to several uniqueness results.

By virtue of the Neumann series for the resolvent of H we infer (cf. (6.2) below and Section 6.1 for more details)

$$\begin{aligned} g(z, n) &= -\frac{1}{z} - \frac{b(n)}{z^2} + O(\frac{1}{z^3}), \\ h(z, n) &= -1 - \frac{2a(n)^2}{z^2} + O(\frac{1}{z^3}). \end{aligned} \tag{2.185}$$

Hence $a(n)^2, b(n)$ can be easily recovered as follows

$$\begin{aligned} b(n) &= -\lim_{z \to \infty} z(1 + zg(z, n)), \\ a(n)^2 &= -\frac{1}{2} \lim_{z \to \infty} z^2 (1 + h(z, n)). \end{aligned} \tag{2.186}$$

Furthermore, we have the useful identities (use (1.100))

$$4a(n)^2 g(z, n) g(z, n+1) = h(z, n)^2 - 1 \tag{2.187}$$

and

$$h(z, n+1) + h(z, n) = 2(z - b(n+1))g(z, n+1), \quad (2.188)$$

which show that $g(z, n)$ and $h(z, n)$ together with $a(n)^2$ and $b(n)$ can be determined recursively if, say, $g(z, n_0)$ and $h(z, n_0)$ are given.

In addition, we infer that $a(n)^2$, $g(z, n)$, $g(z, n+1)$ determine $h(z, n)$ up to one sign,

$$h(z, n) = \left(1 + 4a(n)^2 g(z, n) g(z, n+1)\right)^{1/2}, \quad (2.189)$$

since $h(z, n)$ is holomorphic with respect to $z \in \mathbb{C} \backslash \sigma(H)$. The remaining sign can be determined from the asymptotic behavior $h(z, n) = -1 + O(z^{-2})$.

Hence we have proved the important result that $M(z, n_0)$ determines the sequences a^2, b. In fact, we have proved the slightly stronger result:

Theorem 2.27. *One of the following set of data*
(i) $g(., n_0)$ and $h(., n_0)$
(ii) $g(., n_0+1)$ and $h(., n_0)$
(iii) $g(., n_0)$, $g(., n_0+1)$, and $a(n_0)^2$
for one fixed $n_0 \in \mathbb{Z}$ uniquely determines the sequences a^2 and b.

Remark 2.28. (i) Let me emphasize that the two diagonal elements $g(z, n_0)$ and $g(z, n_0+1)$ alone plus $a(n_0)^2$ are sufficient to reconstruct $a(n)^2$, $b(n)$. This is in contradistinction to the case of one-dimensional Schrödinger operators, where the diagonal elements of the Weyl matrix determine the potential only up to reflection.

You might wonder how the Weyl matrix of the operator H_R associated with the (at n_0) reflected coefficients a_R, b_R (cf. Lemma 1.7) look like. Since reflection at n_0 exchanges $m_\pm(z, n_0)$ (i.e., $m_{R,\pm}(z, n_0) = m_\mp(z, n_0)$) we infer

$$\begin{aligned} g_R(z, n_0) &= g(z, n_0), \\ h_R(z, n_0) &= -h(z, n_0) + 2(z - b(n_0))g(z, n_0), \\ g_R(z, n_0+1) &= \frac{a(n_0)^2}{a(n_0-1)^2} g(z, n_0+1) + \frac{z - b(n_0)}{a(n_0-1)^2} \Big(h(z, n_0) \\ &\quad + (z - b(n_0))g(z, n_0) \Big), \end{aligned} \quad (2.190)$$

in obvious notation.
(ii) Remark 6.3(ii) will show that the sign of $a(n)$ cannot be determined from either $g(z, n_0)$, $h(z, n_0)$, or $g(z, n_0+1)$.

The off-diagonal Green function can be recovered as follows

$$G(z, n+k, n) = g(z, n) \prod_{j=n}^{n+k-1} \frac{1 + h(z, j)}{2a(j) g(z, j)}, \quad k > 0, \quad (2.191)$$

and we remark

$$\begin{aligned} a(n)^2 g(z, n+1) - a(n-1)^2 g(z, n-1) + (z - b(n))^2 g(z, n) \\ = (z - b(n)) h(z, n). \end{aligned} \quad (2.192)$$

2.7. Inverse spectral theory

A similar procedure works for H_+. The asymptotic expansion

$$(2.193) \quad m_+(z,n) = -\frac{1}{z} - \frac{b(n+1)}{z^2} - \frac{a(n+1)^2 + b(n+1)^2}{z^3} + O(z^{-4})$$

shows that $a(n+1)^2, b(n+1)$ can be recovered from $m_+(z,n)$. In addition, (2.11) shows that $m_+(z,n_0)$ determines $a(n)^2, b(n), m_+(z,n), n > n_0$. Similarly, (by reflection) $m_-(z,n_0)$ determines $a(n-1)^2, b(n), m_-(z,n-1), n < n_0$. Hence both $m_\pm(z,n_0)$ determine $a(n)^2, b(n)$ except for $a(n_0-1)^2, a(n_0)^2, b(n_0)$. However, since $a(n_0-1)^2, a(n_0)^2, b(n_0)$, and $m_-(z,n_0)$ can be computed from $\tilde{m}_-(z,n_0)$ we conclude:

Theorem 2.29. *The quantities $\tilde{m}_+(z,n_0)$ and $\tilde{m}_-(z,n_0)$ uniquely determine $a(n)^2$ and $b(n)$ for all $n \in \mathbb{Z}$.*

Next, we recall the function $\gamma^\beta(z,n)$ introduced in (1.107) with asymptotic expansion

$$(2.194) \quad \gamma^\beta(z,n) = -\frac{\beta}{a(n)} - \frac{1+\beta^2}{z} - \frac{b(n+1) + 2\beta a(n) + \beta^2 b(n)}{z^2} + O(\frac{1}{z^3}).$$

Our goal is to prove

Theorem 2.30. *Let $\beta_1, \beta_2 \in \mathbb{R} \cup \{\infty\}$ with $\beta_1 \neq \beta_2$ be given. Then $\gamma^{\beta_j}(.,n_0)$, $j=1,2$, for one fixed $n_0 \in \mathbb{Z}$ uniquely determine $a(n)^2, b(n)$ for all $n \in \mathbb{Z}$ (set $\gamma^\infty(z,n) = g(z,n)$) unless $(\beta_1, \beta_2) = (0,\infty), (\infty, 0)$. In the latter case $a(n_0)^2$ is needed in addition. More explicitly, we have*

$$g(z,n) = \frac{\gamma^{\beta_1}(z,n) + \gamma^{\beta_2}(z,n) + 2R(z)}{(\beta_2 - \beta_1)^2},$$

$$g(z,n+1) = \frac{\beta_2^2 \gamma^{\beta_1}(z,n) + \beta_1^2 \gamma^{\beta_2}(z,n) + 2\beta_1\beta_2 R(z)}{(\beta_2 - \beta_1)^2},$$

$$(2.195) \quad h(z,n) = \frac{\beta_2 \gamma^{\beta_1}(z,n) + \beta_1 \gamma^{\beta_2}(z,n) + (\beta_1 + \beta_2) R(z)}{(-2a(n))^{-1}(\beta_2 - \beta_1)^2},$$

where $R(z)$ is the branch of

$$(2.196) \quad R(z) = \left(\frac{(\beta_2 - \beta_1)^2}{4a(n)^2} + \gamma^{\beta_1}(z,n)\gamma^{\beta_2}(z,n)\right)^{1/2} = \frac{\beta_1 + \beta_2}{2a(n)} + O(\frac{1}{z}),$$

which is holomorphic for $z \in \mathbb{C}\backslash\sigma(H)$ and has asymptotic behavior as indicated. If one of the numbers β_1, β_2 equals ∞, one has to replace all formulas by their limit using $g(z,n) = \lim_{\beta \to \infty} \beta^{-2} \gamma^\beta(z,n)$.

Proof. Clearly, if $(\beta_1, \beta_2) \neq (0,\infty), (\infty, 0)$, we can determine $a(n)$ from equation (2.194). Hence by Theorem 2.27 it suffices to show (2.195). Since the first equation follows from (2.187) and the other two, it remains to establish the last two equations in (2.195). For this we prove that the system

$$(2.197) \quad (g^+)^2 + 2\frac{\beta_j}{2a(n)} hg^+ + \frac{\beta_j^2}{4a(n)^2}(h^2 - 1) = g^+ \gamma^{\beta_j}(z,n), \quad j=1,2,$$

has a unique solution $(g^+, h) = (g(z,n+1), h(z,n))$ for $|z|$ large enough which is holomorphic with respect to z and satisfies the asymptotic requirements (2.185).

We first consider the case $\beta_j \neq 0, \infty$. Changing to new variables (x_1, x_2), $x_j = (2a(n)/\beta_j)g^+ + h$, our system reads

$$(2.198) \qquad x_j^2 - 1 = \frac{\beta_1 \beta_2}{\beta_j^2} \frac{2a(n)\gamma^{\beta_j}(z,n)}{\beta_2 - \beta_1}(x_1 - x_2), \quad j = 1, 2.$$

Picking $|z|$ large enough we can assume $\gamma^{\beta_j}(z,n) \neq 0$ and the solution set of the new system is given by the intersection of two parabolas. In particular, (2.197) has at most four solutions. Two of them are clearly $g^+ = 0$, $h = \pm 1$. But they do not have the correct asymptotic behavior and hence are of no interest to us. The remaining two solutions are given by the last two equations of (2.195) with the branch of $R(z)$ arbitrarily. However, we only get correct asymptotics ($g^+ = -z^{-1} + O(z^{-2})$ resp. $h = -1 + O(z^{-2})$) if we fix the branch as in (2.196). This shows that $g(z, n+1)$, $h(z,n)$ can be reconstructed from γ^{β_1}, γ^{β_2} and we are done. The remaining cases can be treated similarly. □

Finally, we want to give an alternative characterization of the sequences $g(z)$, $h(z)$ respectively $\tilde{m}_\pm(z)$. This characterization will come handy in Section 12.2.

Theorem 2.31. *Consider the two conditions*
(i). Suppose the sequences $g(z)$, $h(z)$ are holomorphic near ∞ and satisfy

$$(2.199) \quad \begin{aligned} 4a^2 g(z) g^+(z) &= h^2(z) - 1, & g(z) &= -z^{-1} + O(z^{-2}), \\ h^+(z) + h(z) &= 2(z - b^+) g^+(z), & h(z) &= -1 + O(z^{-1}). \end{aligned}$$

(ii). Suppose the sequences $\tilde{m}_\pm(z)$ are meromorphic near ∞ and satisfy

$$(2.200) \qquad a^2 \tilde{m}_\pm(z) + \frac{1}{\tilde{m}_\pm^-(z)} = \mp(z - b), \quad \begin{aligned} \tilde{m}_+(z) &= -z^{-1} + O(z^{-2}) \\ \tilde{m}_-(z) &= a^2 z + O(z^0) \end{aligned}.$$

Then (i) (resp. (ii)) is necessary and sufficient for $g(z)$, $h(z)$ (resp. $\tilde{m}_\pm(z)$) to be the corresponding coefficients of the Weyl M-matrix (resp. the Weyl $\tilde{m}$-functions) of H.

Moreover, if $g(z)$, $h(z)$ and $\tilde{m}_\pm(z)$ are related by

$$(2.201) \quad g(z,n) = \frac{-a(n)^{-2}}{\tilde{m}_+(z,n) + \tilde{m}_-(z,n)}, \quad g(z, n+1) = \frac{\tilde{m}_+(z,n) \tilde{m}_-(z,n)}{\tilde{m}_+(z,n) + \tilde{m}_-(z,n)},$$
$$h(z,n) = \frac{\tilde{m}_+(z,n) - \tilde{m}_-(z,n)}{\tilde{m}_+(z,n) + \tilde{m}_-(z,n)},$$

respectively

$$(2.202) \qquad \tilde{m}_\pm(z,n) = -\frac{1 \pm h(z,n)}{2a(n)^2 g(z,n)} = \frac{2g(z, n+1)}{1 \mp h(z,n)},$$

then one condition implies the other.

Proof. Necessity has been established earlier in this section. The relation between the two sets of data is straightforward and hence it suffices to consider (e.g.) $\tilde{m}_\pm(z)$. Since $\tilde{m}_\pm(z)$ are both meromorphic near ∞ they are uniquely determined by their Laurent expansion around ∞. But the coefficients of the Laurent expansion can be determined uniquely using the recursion relations for $\tilde{m}_\pm(z)$ (see Lemma 6.7 for more details). □

Chapter 3

Qualitative theory of spectra

In the previous chapter we have derived several tools for investigating the spectra of Jacobi operators. However, given $a(n)$, $b(n)$ it will, generally speaking, not be possible to compute $m_\pm(z)$ or $G(z, m, n)$ explicitly and to read off some spectral properties of H from $a(n)$, $b(n)$ directly. In this section we will try to find more explicit criteria. The main idea will be to relate spectral properties of H with properties of solutions of $\tau u = \lambda u$. Once this is done, we can relate properties of the solutions with properties of the coefficients a, b. In this way we can read off spectral features of H from a, b without solving the equation $\tau u = \lambda u$.

The endeavor of characterizing the spectrum of H is usually split up into three parts:

1. Locating the essential spectrum.
2. Determining which parts of the essential spectrum are absolutely continuous.
3. Trying to give a bound on the number of eigenvalues.

We will freely use the results and notation from Appendix B. The reader might want to review some of this material first.

3.1. Locating the spectrum and spectral multiplicities

We begin by trying to locate the spectrum of H_+. By Lemma B.5, $\lambda \in \sigma(H_+)$ if and only if $m_+(z)$ is not holomorphic in a neighborhood of λ. Moreover, by (B.27) we even have

(3.1) $$\sigma(H_+) = \overline{\{\lambda \in \mathbb{R} | \lim_{\varepsilon \downarrow 0} \operatorname{Im}(m_+(\lambda + i\varepsilon)) > 0\}}.$$

Note that by equation (B.41), the point spectrum of H_+ is given by

(3.2) $$\sigma_p(H_+) = \{\lambda_0 \in \mathbb{R} | \lim_{\varepsilon \downarrow 0} \frac{\varepsilon}{i} m_+(\lambda_0 + i\varepsilon) > 0\}.$$

Now let us relate the spectrum to the asymptotic behavior of $s(\lambda, n)$. Let us first show that $s(\lambda, n)$ does not grow too fast for λ in the spectrum. More precisely, we have

Lemma 3.1. *Let $f \in \ell^2(\mathbb{N})$, then the set*

(3.3) $$S_{f,+}(\tau) = \{\lambda \in \mathbb{R} | f(n) s(\lambda, n) \in \ell^2(\mathbb{N})\},$$

is of full H_+ spectral measure, that is, $\rho_+(S_{f,+}(\tau)) = 1$. The same is true for the set

(3.4) $$S_+(\tau) = \{\lambda \in \mathbb{R} | \liminf_{n \to \infty} \frac{1}{n} \sum_{m=1}^{n} |s(\lambda, m)|^2 < \infty\}.$$

Proof. Since the polynomials $s(\lambda, n)$ are orthonormal, one verifies

(3.5) $$\|f\|^2 = \int_{\mathbb{R}} \sum_{n \in \mathbb{N}} |f(n) s(\lambda, n)|^2 d\rho_+(\lambda).$$

Hence $\sum_{n \in \mathbb{N}} |f(n) s(\lambda, n)|^2 < \infty$ a.e. with respect to $d\rho_+$. Similarly, by Fatou's lemma we have

(3.6) $$\int_{\mathbb{R}} \liminf_{n \to \infty} \frac{1}{n} \sum_{m=1}^{n} |s(\lambda, m)|^2 d\rho_+(\lambda) < \liminf_{n \to \infty} \int_{\mathbb{R}} \frac{1}{n} \sum_{m=1}^{n} |s(\lambda, m)|^2 d\rho_+(\lambda) = 1,$$

which implies the second claim. $\square$

Choosing $f(n) = n^{-1/2-\delta}$, $\delta > 0$, we see for example

(3.7) $$\sup_{n \in \mathbb{N}} \frac{|s(\lambda, n)|}{n^{1/2+\delta}} < \infty, \quad \lambda \text{ a.e. } d\rho_+.$$

As a consequence we obtain

Theorem 3.2. *Let $\delta > 0$, then*

$$\sigma(H_+) = \overline{S_+(\tau)} = \overline{\left\{\lambda \in \mathbb{R} | \sup_{n \in \mathbb{N}} \frac{|s(\lambda, n)|}{n^{1/2+\delta}} < \infty\right\}}$$

(3.8) $$= \overline{\left\{\lambda \in \mathbb{R} | \limsup_{n \to \infty} \frac{1}{n} \ln(1 + |s(\lambda, n)|) = 0\right\}}.$$

The spectrum of H is simple.

Proof. Denote the last tow sets (without closure) in (3.8) by S_1, S_2. Then we need to show $\sigma(H_+) = \overline{S_+} = \overline{S_1} = \overline{S_2}$. By equation (2.47) we have $\overline{S_+}, \overline{S_1} \subseteq \overline{S_2} \subseteq \sigma(H_+)$ and since $S_{+,1,2}$ are all of full spectral measure, the claim follows as in the proof of (B.27).

That the spectrum of H_+ is simple is immediate from Theorem 2.12. $\square$

Now let us turn to H. Here we have $\lambda \in \sigma(H)$ if and only if $g(z, 0) + g(z, 1)$ is not holomorphic in z near λ and

(3.9) $$\sigma(H) = \overline{\{\lambda \in \mathbb{R} | \lim_{\varepsilon \downarrow 0} \operatorname{Im}(g(\lambda + i\varepsilon, 0) + g(\lambda + i\varepsilon, 1)) > 0\}}.$$

To obtain again a characterization in terms of solutions, we need the analog of Lemma 3.1.

3.1. Locating the spectrum and spectral multiplicities

Lemma 3.3. *Let $f \in \ell^2(\mathbb{Z})$, then the set*

(3.10) $\quad S_f(\tau) = \{\lambda \in \mathbb{R} | f(n) s_\beta(\lambda, n) \in \ell^2(\mathbb{Z}) \text{ for some } \beta = \beta(\lambda)\}$

is of full H spectral measure, that is, $\rho^{tr}(S_f(\tau)) = 1$. The same is true for the set

(3.11) $\quad S(\tau) = \{\lambda \in \mathbb{R} | \liminf_{n\to\infty} \frac{1}{n} \sum_{m=1}^{n} |s_\beta(\lambda, m)|^2 < \infty \text{ for some } \beta = \beta(\lambda)\}.$

Proof. We use $U(\lambda)$ from (B.62) which diagonalizes $d\rho$ and change to the new pair of solutions

(3.12) $\quad \underline{U}(\lambda, n) = U(\lambda) \underline{S}(\lambda, n).$

Now, proceeding as in the proof of Lemma 3.1, we obtain

(3.13) $\quad \|f\| = \int_{\mathbb{R}} \left(r_1(\lambda) \sum_{n \in \mathbb{Z}} |f(n) U_1(z,n)|^2 + r_2(\lambda) \sum_{n \in \mathbb{Z}} |f(n) U_2(z,n)|^2 \right) d\rho^{tr}(\lambda).$

Hence for a.e. λ with respect to $d\rho^{tr}$ we have $\sum_{n\in\mathbb{Z}} r_i(\lambda) |f(n) U_i(\lambda, n)|^2 < \infty$ for $i = 1, 2$. Since $r_1(\lambda) + r_2(\lambda) = 1$, $r_i(\lambda) U_i(\lambda, n)$ is nonzero for one i. The real part is the required solution. The second part follows from Fatou's lemma as in the proof of the previous lemma. $\square$

As before, choosing $f(n) = (1 + |n|)^{-1/2-\delta}$, $\delta > 0$, we obtain the following result.

Theorem 3.4. *Let $\delta > 0$, then*

$$\sigma(H) = \overline{S(\tau)} = \overline{\left\{ \lambda \in \mathbb{R} | \sup_{n \in \mathbb{Z}} \frac{|s_\beta(\lambda, n)|}{n^{1/2+\delta}} < \infty \text{ for some } \beta = \beta(\lambda) \right\}}$$

(3.14) $\quad = \overline{\left\{ \lambda \in \mathbb{R} | \limsup_{n\to\infty} \frac{1}{n} \ln(1 + |s_\beta(\lambda, n)|) = 0 \text{ for some } \beta = \beta(\lambda) \right\}}.$

The spectral multiplicity of H is at most two.

Before scrutinizing the spectral multiplicity of H, let me remark a few things.

Remark 3.5. (i). The point spectrum of H can be characterized by

(3.15) $\quad \sigma_p(H) = \{ \lambda \in \mathbb{R} | \lim_{\varepsilon \downarrow 0} \frac{\varepsilon}{i} \big(g(\lambda + i\varepsilon, 0) + g(\lambda + i\varepsilon, 1) \big) > 0 \}.$

At each discrete eigenvalue $\lambda \in \sigma_d(H)$ either $\tilde{m}_+(z,n)$ and $\tilde{m}_-(z,n)$ both have a pole or $\tilde{m}_-(z,n) + \tilde{m}_+(z,n) = 0$.

(ii). It is interesting to know which parts of the spectrum of H can be read off from $d\rho_{0,0}$ alone, that is, from $g(z,0)$. Since $g(z,0) = -a(0)^{-2}(\tilde{m}_-(z) + \tilde{m}_+(z))^{-1}$, we know $\sigma_{ess}(H_-) \cup \sigma_{ess}(H_+)$ and it will follow from Lemma 3.7 that $g(z,0)$ determines the essential spectrum of H. Moreover, if $\lambda \in \sigma_d(H)$, we either have $u_-(\lambda, 0), u_+(\lambda, 0) \neq 0$ and hence $\tilde{m}_-(z,n) + \tilde{m}_+(z,n) = 0$ (implying a pole of $g(z,0)$ at $z = \lambda$) or $u_-(\lambda, 0), u_+(\lambda, 0) = 0$ and hence poles for both $\tilde{m}_\pm(z,n)$ (implying a zero of $g(z,0)$ at $z = \lambda$). Summarizing, $g(z,n)$ determines $\sigma(H)$ except for the discrete eigenvalues which are also (discrete) simultaneous eigenvalues of H_- and H_+.

To say more about the spectral multiplicity of H, we need to investigate the determinant of $R = d\rho/d\rho^{tr}$. By (B.61) we infer

(3.16) $$R_{i,j}(\lambda) = \lim_{\varepsilon \downarrow 0} \frac{\text{Im}(M_{i,j}(\lambda + i\varepsilon))}{\text{Im}(g(\lambda + i\varepsilon, 0) + g(\lambda + i\varepsilon, 1))}.$$

Hence, using (2.143) we see

(3.17) $$\det R(\lambda) = \lim_{\varepsilon \downarrow 0} \frac{\text{Im}(\tilde{m}_+(\lambda + i\varepsilon))\text{Im}(\tilde{m}_-(\lambda + i\varepsilon))}{a(0)^2|\tilde{m}_+(\lambda + i\varepsilon) + \tilde{m}_-(\lambda + i\varepsilon)|^2} \frac{1}{\text{Im}(g(\lambda + i\varepsilon, 0) + g(\lambda + i\varepsilon, 1))^2}$$

Observe that the first factor is bounded by $(2a(0))^{-2}$. Now Lemma B.14 immediately gives

Lemma 3.6. *The singular spectrum of H has spectral multiplicity one. The absolutely continuous spectrum of H has multiplicity two on $\sigma_{ac}(H_+) \cap \sigma_{ac}(H_-)$ and multiplicity one on $\sigma_{ac}(H) \backslash (\sigma_{ac}(H_+) \cap \sigma_{ac}(H_-))$.*

Proof. Recall from Appendix B that the set

(3.18) $$M_s = \{\lambda \in \mathbb{R} | \lim_{\varepsilon \downarrow 0} \text{Im}(g(\lambda + i\varepsilon, 0) + g(\lambda + i\varepsilon, 1)) = \infty\}$$

is a minimal support for the singular spectrum (singular continuous plus pure point spectrum) of H. Hence the singular part of H is given by $H_s = H_{pp} \oplus H_{sc} = P_{M_s}(H)H$ and the absolutely continuous part is given by $H_{ac} = (1 - P_{M_s}(H))H$. So we see that the singular part has multiplicity one by Lemma B.14.

For the absolutely continuous part use that

(3.19) $$M_{ac,\pm} = \{\lambda \in \mathbb{R} | 0 < \lim_{\varepsilon \downarrow 0} \text{Im}(\tilde{m}_\pm(\lambda + i\varepsilon)) < \infty\}$$

are minimal supports for the absolutely continuous spectrum of $H_\pm$. Again the remaining result follows from Lemma B.14. $\square$

3.2. Locating the essential spectrum

The objective of this section is to locate the **essential spectrum** (which by the way is always nonempty since H is bounded) for certain classes of Jacobi operators. Since our operators have spectral multiplicity at most two, the essential spectrum is the set of all accumulation points of the spectrum.

We first show that it suffices to consider $H_\pm$. In fact, even H_+ is enough since the corresponding results for H_- follow from reflection.

If we embed $H_{n_0}^\infty$ into $\ell^2(\mathbb{Z})$ by defining (e.g.) $H_{n_0}^\infty u(n_0) = 0$, we see from (1.106) that the kernel of the resolvent corresponding to this new operator $\tilde{H}_{n_0}^\infty$ is given by

(3.20) $$\tilde{G}_{n_0}^\infty(z, n, m) = G(z, n, m) - \frac{G(z, n, n_0)G(z, n_0, m)}{G(z, n_0, n_0)} - \frac{1}{z}\delta_{n_0}(n)\delta_{n_0}(m).$$

Hence $\tilde{H}_{n_0}^\infty$ is a rank two resolvent perturbation of H implying $\sigma_{ess}(H) = \sigma_{ess}(\tilde{H}_{n_0}^\infty)$. However, the last term in (3.20) is clearly artificial and it is more natural to consider $H_{n_0}^\infty$ as rank one perturbation.

3.2. Locating the essential spectrum

Lemma 3.7. *For any Jacobi operator H we have*

$$\sigma_{ess}(H) = \sigma_{ess}(H_0^\infty) = \sigma_{ess}(H_-) \cup \sigma_{ess}(H_+) \tag{3.21}$$

and, in addition, $\sigma_{ess}(H_\pm) = \sigma_{ess}(H_{\pm,n_0}^\beta)$ for any $n_0 \in \mathbb{Z}$, $\beta \in \mathbb{R} \cup \{\infty\}$.

Moreover, for $\lambda_0 < \lambda_1$ we have

$$\dim \operatorname{Ran} P_{(\lambda_0,\lambda_1)}(H_{n_0}^\infty) - 1 \leq \dim \operatorname{Ran} P_{(\lambda_0,\lambda_1)}(H) \leq \dim \operatorname{Ran} P_{(\lambda_0,\lambda_1)}(H_{n_0}^\infty) + 1. \tag{3.22}$$

Proof. The first part follows since finite rank perturbations have equivalent essential spectra.

For the second claim, there is nothing to prove if $(\lambda_0, \lambda_1) \cap \sigma_{ess}(H) \neq \emptyset$. If this is not the case, let us view $g(., n_0)$ as a function from $(\lambda_0, \lambda_1) \to \mathbb{R} \cup \{\infty\} \cong S^1$. Then $g(\lambda, n_0) = \infty$ implies that λ is a simple eigenvalue of H but not of $H_{n_0}^\infty$. On the other hand, if $g(\lambda, n_0) = 0$, then either λ is a single eigenvalue of $H_{n_0}^\infty$ (but not of H) or a double eigenvalue of $H_{n_0}^\infty$ and a single eigenvalue of H. Combining this observation with the fact that $g(., n_0)$ is monotone (cf. equation (2.37)) finishes the proof. $\square$

Note that $\sigma(H_+) \cap \sigma(H_+^\beta) = \sigma_{ess}(H_+)$ for $\beta \neq \infty$.

The following lemma shows that λ is in the essential spectrum of H if there exists a non square summable solution which does not grow exponentially. This is a strengthening of (2.48).

Lemma 3.8. *Suppose there is a solution of $\tau u = \lambda u$ such that*

$$\lim_{k \to \infty} \frac{1}{n_k} \ln(u(2n_k)^2 + u(2n_k - 1)^2) = 0 \tag{3.23}$$

for some increasing sequence $0 < n_k < n_{k+1}$. Then $\lambda \in \sigma_{ess}(H_+)$. Similarly for H_-.

Proof. Normalizing $u(0)^2 + u(1)^2 = 1$ we have $u(n) = s_\beta(\lambda, n)$ (up to sign). Altering $b(1)$ we can assume $\beta = \infty$ and shifting $b(n)$ it is no restriction to set $\lambda = 0$ (i.e., $u(n) = s(0, n)$). We will try to construct a Weyl sequence (cf. [**192**], Thm. VII.12) as follows. Set

$$u_N(n) = \begin{cases} \frac{s(0,n)}{\sqrt{\sum_{j=1}^N s(0,j)^2}}, & n \leq N \\ 0, & n > N \end{cases} \tag{3.24}$$

Then $\|u_N\| = 1$ and

$$H_+ u_N(n) = \frac{1}{\sqrt{\sum_{j=1}^N s(0,j)^2}} \Big(a(N) s(0, N) \delta_N(n)$$
$$+ (a(N-1) s(0, N-1) + b(N) s(0, N)) \delta_{N+1}(n) \Big). \tag{3.25}$$

Hence

$$\|H_+ u_{2N}\|^2 \leq \operatorname{const} \frac{f(N)}{\sum_{j=1}^N f(j)}, \tag{3.26}$$

where $f(n) = s(0, 2n)^2 + s(0, 2n-1)^2$. If the right hand side does not tend to zero, we must have

$$(3.27) \qquad f(n) \geq \varepsilon \sum_{j=1}^{n} f(j), \qquad \varepsilon > 0,$$

and hence $f(n) \geq (1-\varepsilon)^{1-n}$. But this contradicts our assumption (3.23)

$$(3.28) \qquad 0 = \lim_{k \to \infty} \frac{1}{n_k} \ln f(n_k) \geq \lim_{k \to \infty} \frac{1-n_k}{n_k} \ln(1-\varepsilon) = -\ln(1-\varepsilon) > 0.$$

Hence $\|H_+ u_{2N}\| \to 0$ implying $0 \in \sigma(H_+)$ by Weyl's theorem. If 0 were a discrete eigenvalue, $s(0, n)$ would decrease exponentially by Lemma 2.5 and the limit in (3.23) would be negative. So $0 \in \sigma_{ess}(H_+)$ as claimed. □

Next, we note that the essential spectrum depends only on the asymptotic behavior of the sequences $a(n)$ and $b(n)$.

Lemma 3.9. *Let $H_{1,2}$ be two Jacobi operators and suppose*

$$(3.29) \qquad \lim_{n \to \infty} \big(|a_2(n)| - |a_1(n)|\big) = \lim_{n \to \infty} \big(b_2(n) - b_1(n)\big) = 0.$$

Then

$$(3.30) \qquad \sigma_{ess}(H_{1,+}) = \sigma_{ess}(H_{2,+}).$$

Proof. Clearly we can assume $a_{1,2}(n) > 0$. The claim now follows easily from the fact that, by assumption, $H_{2,+} - H_{1,+}$ is compact (approximate $H_{2,+} - H_{1,+}$ by finite matrices and use that norm limits of compact operators are compact). □

Finally, let us have a closer look at the location of the essential spectrum. We abbreviate

$$(3.31) \qquad c_\pm(n) = b(n) \pm \big(|a(n)| + |a(n-1)|\big)$$

and

$$(3.32) \qquad \underline{c}_- = \liminf_{n \to \infty} |c_-(n)|, \qquad \overline{c}_+ = \limsup_{n \to \infty} |c_+(n)|.$$

Lemma 3.10. *We have*

$$(3.33) \qquad \sigma_{ess}(H_+) \subseteq [\underline{c}_-, \overline{c}_+].$$

If, in addition, $a(n), b(n)$ are slowly oscillating, that is,

$$(3.34) \qquad \lim_{n \to \infty} (|a(n+1)| - |a(n)|) = 0, \qquad \lim_{n \to \infty} (b(n+1) - b(n)) = 0,$$

then we have

$$(3.35) \qquad \sigma_{ess}(H_+) = [\underline{c}_-, \overline{c}_+].$$

Proof. Clearly we can assume $a(n) > 0$. But we may even assume $c_+(n) \leq \overline{c}_+$ and $c_-(n) \geq \underline{c}_-$. In fact, consider $\hat{H}_+$ associated with the sequences $\hat{a}(n) = \min\{a(n), \overline{c}_+ - a(n-1) - b(n), -\underline{c}_- - a(n-1) + b(n)\}$, $\hat{b}(n) = b(n)$. Since $a(n) - \hat{a}(n) = \max\{0, c_+(n) - \overline{c}_+, \underline{c}_- - c_-(n)\}$ we infer $\sigma_{ess}(H_+) = \sigma_{ess}(\hat{H}_+)$ from Lemma 3.9.

For the first claim we use

$$\langle f, H_+ f\rangle = -a(0)|f(1)|^2 + \sum_{n=1}^{\infty}\Big(-a(n)|f(n+1) - f(n)|^2$$
(3.36)
$$+ (a(n-1) + a(n) + b(n))|f(n)|^2\Big)$$

and proceed as in Lemma 1.8, proving (3.33).

Suppose (3.34) holds and fix $\lambda \in (\underline{c}_-, \overline{c}_+)$. To prove (3.35) we need to show $\lambda \in \sigma_{ess}(H_+)$. By Weyl's criterion (cf. [**192**], Thm. VII.12) it suffices to exhibit an orthonormal sequence $u_N \in \ell(\mathbb{N})$ with $\|(H_+ - \lambda)u_N\| \to 0$ as $N \to \infty$. For each $N \in \mathbb{N}$ we can find $n_N > N(N-1)/2$ such that
(i). there is a $\lambda_N \in [c_-(n_N), c_+(n_N)]$ such that $|\lambda - \lambda_N| \leq 1/\sqrt{N}$ and
(ii). $|a(n) - a_N| + |a(n-1) - a_N| + |b(n) - b_N| \leq 1/\sqrt{N}$ for $n_N \leq n \leq n_N + N$, where $a_N = (a(n_N) + a(n_N - 1))/2)$, $b_N = b(n_N)$.

If we set

(3.37)
$$k_N = \frac{\lambda_N - b_N + \sqrt{(\lambda_N - b_N)^2 - 4a_N^2}}{2a_N},$$

we have $|k_N| = 1$ since $|\lambda_N - b_N| \leq 2a_N$. Now define $u_N(n) = (k_N)^n/\sqrt{N}$ for $n_N \leq n < n_N + N$ and $u_N(n) = 0$ otherwise. Then $\|u_N\| = 1$ and $u_N(n)u_M(n) = 0$, $N \neq M$. Moreover, we have

$$\|(H_+ - \lambda)u_N\| \leq \|(H_{N,+} - \lambda_N)u_N\| + \|(\lambda - \lambda_N)u_N\| + \|(H_+ - H_{N,+})u_N\|$$
(3.38)
$$\leq \frac{4\|a\|_\infty}{\sqrt{N}} + \frac{1}{\sqrt{N}} + \frac{1}{\sqrt{N}} \to 0$$

as $N \to \infty$, where in the last estimate we have used $\tau_N(k_N)^n = \lambda_N(k_N)^n$, (i), and (ii), respectively. This proves $\lambda \in \sigma_{ess}(H_+)$ and hence (3.35). $\square$

3.3. Locating the absolutely continuous spectrum

In this section we will try to locate the absolutely continuous spectrum of H in some situations. We will first investigate the half line operator H_+ whose spectral properties are related to the boundary behavior of $m_+(\lambda + i\varepsilon)$ as $\varepsilon \downarrow 0$ (see Appendix B). Our strategy in this section will be to relate this boundary behavior to the behavior of solutions of $\tau u = \lambda u$.

But first we start with a preliminary lemma which shows that the absolutely continuous spectrum of H is completely determined by those of H_- and H_+.

Lemma 3.11. *For any $n_0 \in \mathbb{Z}$, $\beta \in \mathbb{R} \cup \{\infty\}$, the absolutely continuous part of the spectral measures corresponding to $H_\pm$ and $H_{\pm,n_0}^\beta$, respectively, are equivalent. For fixed n_0 and different β, the singular parts are mutually disjoint. In particular,*

(3.39)
$$\sigma_{ac}(H_\pm) = \sigma_{ac}(H_{\pm,n_0}^\beta).$$

The same is true for the trace measure $d\rho^{tr}$ corresponding to H and $d\tilde{\rho}_+ + d\tilde{\rho}_-$. In particular,

(3.40)
$$\sigma_{ac}(H) = \sigma_{ac}(H_-) \cup \sigma_{ac}(H_+).$$

In addition, let us abbreviate $g(\lambda, n) = \limsup_{\varepsilon \downarrow 0} \text{Im}(g(\lambda + i\varepsilon, n))$, $m_\pm(\lambda, n) = \limsup_{\varepsilon \downarrow 0} \text{Im}(m_\pm(\lambda + i\varepsilon, n))$, and $\xi(\lambda, n) = \pi^{-1} \limsup_{\varepsilon \downarrow 0} \arg g(\lambda + i\varepsilon, n)$. Then for any fixed $n_0 \in \mathbb{Z}$ the following sets

$$
\begin{aligned}
M_{ac,1} &= \{\lambda \in \mathbb{R} \,|\, 0 < \text{Im}(g(\lambda, n_0))\}, \\
M_{ac,2} &= \{\lambda \in \mathbb{R} \,|\, 0 < \xi(\lambda, n_0) < 1\}, \\
M_{ac,3} &= \bigcup_{\sigma \in \{\pm\}} \{\lambda \in \mathbb{R} \,|\, 0 < \text{Im}(m_\sigma(\lambda, n_0))\}
\end{aligned}
$$
(3.41)

are minimal supports for the absolutely continuous spectrum of H.

Proof. Using (2.16) and (i) of Theorem B.8 we infer

$$d\rho^\beta_{+,n_0,ac}(\lambda) = \frac{\beta^2}{|\beta - a(n_0) m_+(\lambda, n_0)|^2} d\rho_{+,n_0,ac}(\lambda) \qquad (3.42)$$

and since $H_{+,n_0} = H^0_{+,n_0+1}$ the equivalence of the absolutely continuous parts of the spectral measures follows for the $+$ sign. That the singular parts are disjoint follows from (2.92) since $m_+^{\beta_1}(\lambda) = \infty$ implies $m_+^{\beta_2}(\lambda) \neq \infty$ for all $\beta_2 \neq \beta_1$. The $-$ part is similar.

It remains to establish the claims for H. It is no restriction to set $n_0 = 0$. By Theorem B.8 (ii) the set L of all $\lambda \in \mathbb{R}$ for which both limits $\lim_{\varepsilon \downarrow 0} m_\pm(\lambda + i\varepsilon)$ exist and are finite is of full Lebesgue measure. For $\lambda \in L$ we can compute $\lim_{\varepsilon \downarrow 0} M(\lambda + i\varepsilon)$ from (2.143). Hence we obtain $M_{ac,1} \cap L = M_{ac,2} \cap L = M_{ac,3} \cap L$. Moreover, $M_{ac,1} \cap L = \{\lambda | 0 < \text{Im}(g(\lambda, 0) + g(\lambda, 1))\}$ and the claim is immediate from Lemma B.7 and Lemma B.12. $\square$

In particular, this says that if $\limsup_{\varepsilon \downarrow 0} \text{Im}(m_+(\lambda + i\varepsilon)) < \infty$, $\lambda \in (\lambda_0, \lambda_1)$, then the spectrum of H_+ is purely absolutely continuous in (λ_0, λ_1). Note that we could replace $m_\pm(z, n)$ by $\tilde{m}_\pm(z, n)$ in the above theorem.

Now recall the split-up (2.108), $d\rho_+ = d\rho_{+,\alpha c} + d\rho_{+,\alpha s}$, with respect to the Hausdorff measure h^α, $\alpha \in [0, 1]$. In view of Theorem B.9 a support $C^\alpha(\rho_+)$ of $d\rho_{+,\alpha c}$ can be found by investigating the boundary values of $m_+(\lambda + i\varepsilon)$ as $\varepsilon \downarrow 0$. We set

$$\|f\|_{(1,n)} = \sqrt{\sum_{j=1}^n |f(j)|^2} \qquad (3.43)$$

and extend this definition for real $n \geq 1$ by linear interpolation of $\|f\|^2_{(1,n)}$. That is, we have $\|f\|^2_{(1,n)} = \int_0^n |f(\llbracket x \rrbracket + 1)|^2 dx$, where $\llbracket . \rrbracket$ denotes the integer part (see (4.14)).

Lemma 3.12. *Let* $\varepsilon = a(0)(2\|s(\lambda)\|_{(1,n)} \|c(\lambda)\|_{(1,n)})^{-1}$, *then*

$$5 - \sqrt{24} \leq |a(0) m_+(\lambda + i\varepsilon)| \frac{\|s(\lambda)\|_{(1,n)}}{\|c(\lambda)\|_{(1,n)}} \leq 5 + \sqrt{24}. \qquad (3.44)$$

3.3. Locating the absolutely continuous spectrum

Proof. Let $n \in \mathbb{N}$, then by (2.158)

$$u_+(\lambda + i\varepsilon, n) = \frac{c(\lambda, n)}{a(0)} - m_+(\lambda + i\varepsilon)s(\lambda, n)$$

(3.45)
$$- \frac{i\varepsilon}{a(0)} \sum_{j=1}^{n} \big(c(\lambda, n)s(\lambda, j) - c(\lambda, j)s(\lambda, n)\big)u_+(\lambda + i\varepsilon, j).$$

Hence one obtains after a little calculation (as in the proof of Lemma 2.15)

$$\|u_+(\lambda + i\varepsilon)\|_{(1,n)} \geq \|\frac{c(\lambda)}{a(0)} - m_+(\lambda + i\varepsilon)s(\lambda)\|_{(1,n)}$$

(3.46)
$$- 2\varepsilon \|s(\lambda)\|_{(1,n)} \|\frac{c(\lambda)}{a(0)}\|_{(1,n)} \|u_+(\lambda + i\varepsilon)\|_{(1,n)}.$$

Using the definition of ε and (2.89) we obtain

$$\|\frac{c(\lambda)}{a(0)} - m_+(\lambda + i\varepsilon)s(\lambda)\|^2_{(1,n)} \leq 4\|u_+(\lambda + i\varepsilon)\|^2_{(1,n)}$$

$$\leq 4\|u_+(\lambda + i\varepsilon)\|^2_{(1,\infty)}$$

$$\leq \frac{4}{\varepsilon} \operatorname{Im}\big(m_+(\lambda + i\varepsilon)\big)$$

(3.47)
$$\leq 8\|s(\lambda)\|_{(1,n)} \|\frac{c(\lambda)}{a(0)}\|_{(1,n)} \|u_+(\lambda + i\varepsilon, n)\|_{(1,n)}.$$

Combining this estimate with

(3.48) $\quad \|\frac{c(\lambda)}{a(0)} - m_+(\lambda + i\varepsilon)s(\lambda)\|^2_{(1,n)} \geq \big(\|\frac{c(\lambda)}{a(0)}\|_{(1,n)} - |m_+(\lambda + i\varepsilon)|\|s(\lambda)\|_{(1,n)}\big)^2$

shows $(1-x)^2 \leq 8x$, where $x = |a(0)m_+(\lambda + i\varepsilon)| \|s(\lambda)\|_{(1,n)} \|c(\lambda)\|^{-1}_{(1,n)}$. □

We now introduce the concept of α-**subordinacy**, $\alpha \in [0,1]$. A nonzero solution u of $\tau u = zu$ is called α-**subordinate** at $+\infty$ with respect to another solution v if

(3.49)
$$\liminf_{n \to +\infty} \frac{\|u\|^{2-\alpha}_{(1,n)}}{\|v\|^{\alpha}_{(1,n)}} = 0.$$

Moreover, u is called α-subordinate at $+\infty$ if (3.49) holds for for any linearly independent solution v.

An α_1-subordinate solution is also an α_2-subordinate solution for any $\alpha_1 \leq \alpha_2$ and if there is no α_2-subordinate solution, then there is also no α_1-subordinate solution.

For example, if $u \in \ell^2_\pm(\mathbb{N})$, then u is α-subordinate for $\alpha \in (0,1]$ (there can be no 0-subordinate solution). In fact, let $v(\lambda)$ be a second linearly independent solution. Since $W_n(u(\lambda), v(\lambda))$ is a nonzero constant, we infer that $v(n)$ cannot be bounded since otherwise $W_n(u(\lambda), v(\lambda)) \to 0$ as $n \to \pm\infty$. Moreover, by (2.47) we see that the set of all $\lambda \in \mathbb{R}$ for which no α-subordinate ($\alpha \in (0,1]$) solution exists must be a subset of the essential spectrum.

A similar definition applies to $-\infty$.

The following theorem is an immediate consequence of our previous lemma and Theorem B.9.

Theorem 3.13. *Let $\alpha \in [0,1]$, $\lambda \in \mathbb{R}$. Then $\lambda \in C^\alpha(\rho_+)$ if $s(\lambda)$ is not α-subordinate with respect to $c(\lambda)$. If $\alpha \in [0,1)$, the converse is also true.*

In particular, if no α-subordinate solution at $+\infty$ exists for any $\lambda \in (\lambda_1, \lambda_2)$, then ρ_+ is purely α-continuous on (λ_1, λ_2), that is, $[\lambda_1, \lambda_2] \subseteq \sigma_{\alpha c}(H_+)$ and $(\lambda_1, \lambda_2) \cap \sigma_{\alpha s}(H_+) = \emptyset$, $\alpha \in (0,1]$.

The case $\alpha = 1$ deserves special attention, since 1-continuous is the same as absolutely continuous. We will call u subordinate with respect to v if

$$(3.50) \qquad \lim_{n \to +\infty} \frac{\|u\|_{(1,n)}}{\|v\|_{(1,n)}} = 0.$$

We use the ordinary limit (instead of lim inf) in the case $\alpha = 1$ for convenience later on. We will infer from Lemma 3.14 below that, for our purpose, this makes no difference.

It is easy to see that if u is subordinate with respect to v, then it is subordinate with respect to any linearly independent solution. In particular, a subordinate solution is unique up to a constant. Moreover, if a solution u of $\tau u = \lambda u$, $\lambda \in \mathbb{R}$, is subordinate, then it is real up to a constant, since both the real and the imaginary part are subordinate. For $z \in \mathbb{C}\backslash\mathbb{R}$ we know that there is always a subordinate solution near $\pm\infty$, namely $u_\pm(z,n)$. The following result considers the case $z \in \mathbb{R}$.

Lemma 3.14. *Let $\lambda \in \mathbb{R}$. There is a near $+\infty$ subordinate solution $u(\lambda)$ if and only if $m_+(\lambda + i\varepsilon)$ converges to a limit in $\mathbb{R} \cup \{\infty\}$ as $\varepsilon \downarrow 0$. Moreover,*

$$(3.51) \qquad \lim_{\varepsilon \downarrow 0} m_+(\lambda + i\varepsilon) = -\frac{u(\lambda, 1)}{a(0) u(\lambda, 0)}$$

in this case. A similar result holds near $-\infty$.

Proof. If $u(\lambda) = s(\lambda)$ is subordinate or $m_+(\lambda + i0) = \infty$, the claim follows from Lemma 3.12. Similarly, if $u(\lambda) = c(\lambda)$ is subordinate or $m_+(\lambda + i0) = 0$. If $u(\lambda) = s_\beta(\lambda)$ is subordinate or $m_+(\lambda + i0) = \beta/a(0)$, we can reduce it to the first case using the transform $b(1) \to b(1) - a(0)/\beta$ (see (2.16)). $\square$

We are interested in $N_\pm(\tau)$, the set of all $\lambda \in \mathbb{R}$ for which no subordinate solution exists, that is,

$$(3.52) \qquad N_\pm(\tau) = \{\lambda \in \mathbb{R} | \text{No solution of } \tau u = \lambda u \text{ is subordinate near } \pm\infty\}.$$

Since the set, for which the limit $\lim_{\varepsilon \downarrow 0} m_+(\lambda + i\varepsilon)$ does not exist, is of zero spectral and Lebesgue measure (Theorem B.8 (ii)), changing the lim in (3.50) to a lim inf will affect $N_\pm(\tau)$ only on such a set (which is irrelevant for our purpose).

Then, as consequence of the previous lemma, we have

Theorem 3.15. *The set $N_\pm(\tau)$ is a minimal support for the absolutely continuous spectrum of $H_\pm$. In particular,*

$$(3.53) \qquad \sigma_{ac}(H_\pm) = \overline{N_\pm(\tau)}^{ess}.$$

Proof. By Theorem B.8 (ii) we may assume $m_+(\lambda + i0)$ exists and is finite almost everywhere. But for those values λ we have $0 < \text{Im}(m_\pm(\lambda + i0)) < \infty$ if and only if $\lambda \in N_\pm(\tau)$ and the result follows from Lemma B.7. $\square$

Using Lemma 3.1 we can now show

3.3. Locating the absolutely continuous spectrum

Theorem 3.16. *Let $f \in \ell^2(\mathbb{Z})$ and*

(3.54) $\quad B_{f,\pm}(\tau) = \{\lambda \in \mathbb{R} | f(n)u(n) \in \ell^2(\pm\mathbb{N}) \text{ for all solutions of } \tau u = \lambda u\}.$

Then we have

(3.55) $\quad\quad\quad\quad\quad\quad \sigma_{ac}(H_\pm) \subseteq \overline{B_{f,\pm}(\tau)}^{ess}.$

Choosing $f(n) = (1 + |n|)^{-1/2-\delta}$, $\delta > 0$, we even see

(3.56) $\quad\quad\quad\quad\quad\quad \sigma_{ac}(H_\pm) \subseteq \overline{\{\lambda \in \mathbb{R} | \overline{\gamma}^\pm(\lambda) = 0\}}^{ess}.$

In addition, let

(3.57) $\quad B_\pm(\tau) = \{\lambda \in \mathbb{R} | \liminf_{n \to \infty} \frac{1}{n} \sum_{m=1}^{n} u(m) < \infty \text{ for all solutions of } \tau u = \lambda u\}.$

then $B_\pm(\tau) \subseteq N_\pm(\tau)$ and

(3.58) $\quad\quad\quad\quad\quad\quad \sigma_{ac}(H_\pm) = \overline{B_\pm(\tau)}^{ess}.$

Proof. We only consider the $+$ sign. By Lemma 3.1 the sets

(3.59) $\quad\quad\quad\quad S_{f,\pm}^\beta(\tau) = \{\lambda \in \mathbb{R} | f(n) s_\beta(\lambda, n) \in \ell^2(\pm\mathbb{N})\}$

are of full $H_\pm^\beta$ spectral measure. Next, note that $B_{f,+}(\tau) = S_{f,+}^\infty(\tau) \cap S_{f,+}^\beta(\tau)$ if $\beta \neq \infty$. Moreover, since $d\rho_{+,ac}$ and $d\rho_{+,ac}^\beta$ are equivalent, we infer $\rho_{+,ac}(\mathbb{R}\backslash S_{f,+}^\beta(\tau)) = \rho_{+,ac}^\beta(\mathbb{R}\backslash S_{f,+}^\beta(\tau)) = 0$. Thus $S_{f,+}^\infty(\tau)$ and $S_{f,+}^\beta(\tau)$ are both supports for $d\rho_{+,ac}$ and so is their intersection.

The second assertion follows since $\lambda \in S_{f,+}(\tau)$ implies $\gamma^+(\lambda) = 0$ for our particular choice of f.

Finally, as before $B_\pm(\tau)$ is a support for $\sigma_{ac}(H_\pm)$ and hence, by the previous theorem, it suffices to verify $B_\pm(\tau) \subseteq N_\pm(\tau)$. Suppose $\lambda \in B_+(\tau)$ is such that a subordinate solution u exists, that is, $\lim_{n \to +\infty} \|u\|_{(1,n)} / \|v\|_{(1,n)} = 0$. By assumption, there is a subsequence n_j such that $n_j^{-1} \|v\|_{(1,n_j)}$ is bounded. Hence $n_j^{-1} \|u\|_{(1,n_j)}$ must tend to zero. But this contradicts constancy of the Wronskian

(3.60) $\quad 0 < |W(u,v)| = \frac{1}{n_j} \sum_{m=1}^{n_j} W_n(u,v) \leq const \sqrt{\frac{\|u\|_{(1,n_j+1)}}{n_j} \frac{\|v\|_{(1,n_j+1)}}{n_j}} \to 0$

(use Cauchy-Schwarz). The case $\lambda \in B_-(\tau)$ is similar. $\square$

Now, we want to extend our results to operators H on $\ell^2(\mathbb{Z})$. The spectrum of H is related to the boundary behavior of $g(\lambda + i\varepsilon, n)$, $n = 0, 1$, which again is related to the boundary behavior of $\tilde{m}_\pm(\lambda + i\varepsilon)$ by equation (2.201).

Lemma 3.17. *Suppose $\liminf_{\varepsilon \downarrow 0} \operatorname{Im}(m_\sigma(\lambda + i\varepsilon)) > 0$ and $\limsup_{\varepsilon \downarrow 0} |m_\sigma(\lambda + i\varepsilon)| < \infty$ for $\lambda \in (\lambda_1, \lambda_2)$, $\sigma = -$ or $\sigma = +$. Then $[\lambda_1, \lambda_2] \subseteq \sigma_{ac}(H)$ and $(\lambda_1, \lambda_2) \cap \sigma_s(H) = \emptyset$. Here $\sigma_s(H) = \sigma_{sc}(H) \cup \sigma_{pp}(H)$.*

Proof. Consider (e.g.) $\sigma = +$. Our assumption implies that $[\lambda_1, \lambda_2] \subseteq \sigma_{ess}(H_+) \subseteq \sigma_{ess}(H)$. Moreover, using (2.201) one easily shows

(3.61) $\quad\quad\quad\quad \limsup_{\varepsilon \downarrow 0} \operatorname{Im}(g(\lambda + i\varepsilon, n)) < \infty, \quad n = 0, 1.$

Together with Lemma B.7 this implies that the singular spectrum of H is not supported on (λ_1, λ_2). □

This last result is the key to our next theorem.

Theorem 3.18. *On every subinterval* $(\lambda_1, \lambda_2) \subseteq N_\pm(\tau)$, *the spectrum of* $H_\pm^\beta$ *is purely absolutely continuous in the sense that*

(3.62) $\qquad [\lambda_1, \lambda_2] \subseteq \sigma_{ac}(H_\pm^\beta), \qquad (\lambda_1, \lambda_2) \cap \sigma_s(H_\pm^\beta) = \emptyset.$

In addition, the spectrum of H *is purely absolutely continuous on every subinterval* $(\lambda_1, \lambda_2) \subseteq N_-(\tau) \cup N_+(\tau)$, *that is,*

(3.63) $\qquad [\lambda_1, \lambda_2] \subseteq \sigma_{ac}(H), \qquad (\lambda_1, \lambda_2) \cap \sigma_s(H) = \emptyset.$

Proof. Clearly Lemma 3.12 (and all other considerations) holds for any other boundary condition β. Hence only the claim concerning H needs to be proven. If $\lambda \in N_+(\tau)$, we know $\limsup_{\varepsilon \downarrow 0} |\tilde{m}_+^\beta(\lambda + i\varepsilon)| = C(\beta) < \infty$ and we need to show $\liminf_{\varepsilon \downarrow 0} \operatorname{Im}(\tilde{m}_+(\lambda + i\varepsilon)) > 0$. Suppose that we can find a sequence $\varepsilon_n \to 0$ such that $\operatorname{Im}(\tilde{m}_+(\lambda + i\varepsilon_n)) \to 0$ as $n \to \infty$. Now set $\tilde{m}_+(\lambda + i\varepsilon_n) = x_n + iy_n$ and $\beta_n = a(0)x_n$. Since $\limsup |x_n| < \infty$ we can pass to a subsequence such that x_n converges. This implies $(x_n, y_n, \beta_n) \to (x_0, 0, \beta_0 = a(0)x_0)$ and by (2.92)

$$|\tilde{m}_+^{\beta_0}(\lambda + i\varepsilon_n)| = \frac{1}{|a(0)|} \left| \frac{1 + a(0)\beta_0(x_n + iy_n)}{\beta_0 - a(0)(x_n + iy_n)} \right|$$

(3.64)
$$\geq \frac{1}{|a(0)|} \frac{|1 + \beta_0\beta_n|}{\sqrt{(\beta_0 - \beta_n)^2 + y_n^2}}$$

contradicting our assumption. The case $\lambda \in N_-(\tau)$ is similar. □

The relation between the absolutely continuous spectrum and (non-)existence of subordinate solutions is often referred to as **principle of subordinacy**.

Finally, let us show how these results can be applied. Note that $\lambda \in B_\pm(\tau)$ if all solutions of $\tau u = \lambda u$ are bounded near $\pm\infty$.

Theorem 3.19. *Suppose the sequences* $a(n) > 0, b(n)$ *satisfy*

(3.65) $\qquad \sum_{n \in \mathbb{Z}} |a(n+1) - a(n)| + |b(n+1) - b(n)| < \infty$

and set

(3.66) $\qquad \lim_{n \to \pm\infty} a(n) = a_\pm, \qquad \lim_{n \to \pm\infty} b(n) = b_\pm.$

Then for any $\lambda \in (-2a_\pm + b_\pm, 2a_\pm + b_\pm)$ *all solutions of* $\tau u = \lambda u$ *are bounded near* $\pm\infty$. *Hence we have*

$\sigma_{ess}(H_\pm) = \sigma_{ac}(H_\pm) = [-2a_\pm + b_\pm, 2a_\pm + b_\pm], \quad \sigma_{sc}(H_\pm) = \emptyset,$

(3.67) $\qquad \sigma_{pp}(H_\pm) \subset (-\infty, -2a_\pm + b_\pm] \cup [2a_\pm + b_\pm, \infty).$

Moreover,

$\sigma_{ess}(H) = \sigma_{ac}(H) = \sigma_{ac}(H_+) \cup \sigma_{ac}(H_-), \quad \sigma_{sc}(H) = \emptyset,$

(3.68) $\qquad \sigma_{pp}(H) \subset \overline{\mathbb{R} \backslash \sigma_{ess}(H)}.$

Proof. After the transformation $z \to 2a_+ z + b_+$ we can assume $a_+ = 1/2$, $b_+ = 0$. Consider the quantity

$$(3.69) \qquad K(n) = a(n)\big(u(n+1)^2 + u(n)^2\big) + (b(n) - \lambda)u(n)u(n+1),$$

where u is a solution of $\tau u = \lambda u$. For $\lambda \in (-1, 1)$ we can find $N \in \mathbb{N}$, $\varepsilon > 0$ such that

$$(3.70) \qquad \frac{\lambda - b(n)}{2a(n)} \leq 1 - \varepsilon, \quad n \geq N.$$

Whence

$$\frac{K(n)}{a(n)} \geq u(n+1)^2 + u(n)^2 - (1-\varepsilon)2u(n)u(n+1)$$
$$= \varepsilon(u(n+1)^2 + u(n)^2) + (1-\varepsilon)(u(n+1) - u(n))^2$$
$$(3.71) \qquad \geq \varepsilon(u(n+1)^2 + u(n)^2).$$

Thus to show u bounded near $+\infty$ it suffices to show this result for K. One computes

$$K(n+1) - K(n) = \frac{a(n+1) - a(n)}{a(n+1)}\Big(K(n) + (a(n+1) + a(n))u(n+1)^2\Big)$$
$$(3.72) \qquad + \frac{a(n)(b(n+1) - b(n))}{a(n+1)}u(n)u(n+1)$$

which implies using (3.71)

$$(3.73) \qquad K(n+1) \leq (1 + C(n))K(n), \quad C(n) \in \ell^1(\mathbb{Z}),$$

where

$$(3.74) \qquad C(n) = \frac{|a(n+1) - a(n)|}{a(n+1)}\Big(1 + \frac{a(n+1) + a(n)}{\varepsilon a(n)}\Big) + \frac{|b(n+1) - b(n)|}{2\varepsilon a(n+1)}.$$

Hence

$$(3.75) \qquad K(n) \leq K(n_0) \prod_{m=n_0}^{\infty} (1 + C(m)), \quad n \geq n_0,$$

is bounded. A similar argument shows u bounded near $-\infty$. The rest follows from Lemma 3.10 and Theorem 3.18. □

Note that (3.65) does not exclude eigenvalues at the boundary of the essential spectrum. For example, taking

$$(3.76) \qquad a(n) = \frac{1}{2}, \qquad b(n) = \frac{2 - 3n^2}{4 + n^4},$$

there is an eigenvalue at 1 with corresponding eigenfunction $u_\pm(1, n) = (1+n^2)^{-1}$. The reason is that the first moment $\sum_{n \in \mathbb{Z}} |nb(n)|$ is not finite. We will learn more about this case in Chapter 10.

3.4. A bound on the number of eigenvalues

In this section we will derive a Birman-Schwinger type bound on the number of eigenvalues below the essential spectrum of certain Jacobi operators.

Theorem 3.20. *Let H_0, H be operators in a Hilbert space $\mathfrak{H}$. Suppose $H_0 \geq \lambda_0$ and $\lambda_0 \in \sigma_{ess}(H_0)$. Let H be such that $H - H_0$ is Hilbert-Schmidt, then, for $\lambda < \lambda_0$ we have*

$$\dim \operatorname{Ran} P_{(-\infty, \lambda)}(H) \leq \operatorname{tr}\big((H - H_0)(H_0 - \lambda)^{-1}\big)^2. \tag{3.77}$$

Similarly, if $H_0 \leq \lambda_0$ and $\lambda_0 \in \sigma_{ess}(H_0)$. Then

$$\dim \operatorname{Ran} P_{(\lambda, \infty)}(H) \leq \operatorname{tr}\big((H - H_0)(\lambda - H_0)^{-1}\big)^2 \tag{3.78}$$

for $\lambda > \lambda_0$.

Proof. Since $H - H_0$ is compact we have $\sigma_{ess}(H) = \sigma_{ess}(H_0)$. Using the same notation as in [**195**], Theorem XIII.1 we infer as in the proof of the proposition after Theorem XIII.2 that $\mu_n(H_0 + \beta(H - H_0))$ is monotone nonincreasing with respect to $\beta \in [0, \infty)$. Now we use that

$$(H_0 + \beta(H - H_0))u = \lambda u, \qquad u \in \mathfrak{H}, \quad \lambda < \lambda_0, \tag{3.79}$$

is equivalent to

$$v = \beta K_\lambda v, \qquad v \in \mathfrak{H}, \tag{3.80}$$

where $K_\lambda = \sqrt{(H_0 - \lambda)^{-1}}(H_0 - H)\sqrt{(H_0 - \lambda)^{-1}}$ and $u = \sqrt{(H_0 - \lambda)^{-1}}v$. Notice that K_λ is a self-adjoint operator. By [**192**], Theorem VI.25 we have $\operatorname{tr} K_\lambda^2 = \operatorname{tr}((H_0 - H)(H - \lambda)^{-1})^2 < \infty$. Denote the nonzero eigenvalues of K_λ by $1/\beta_m$. Then each eigenvalue $\mu_n(H_0 + \beta(H - H_0))$ with $\mu_n(H) < \lambda$ passes λ when $\beta = \beta_m \in (0, 1)$. Hence

$$\dim \operatorname{Ran} P_{(-\infty, \lambda)}(H) \leq \sum_{n \in \mathbb{N}} \frac{1}{\beta_n^2} = \operatorname{tr} K_\lambda^2 \tag{3.81}$$

concluding the proof of the first assertion. Using $P_{(-\infty, \lambda)}(-H) = P_{(-\lambda, \infty)}(H)$ one obtains the second. $\square$

Finally, let us consider a case where this estimate can be made somewhat more explicit. Namely, we will assume that we can find a function $\tilde{G}_0(\lambda, n, m)$ such that

$$|G_0(\lambda, n, m)| \leq \tilde{G}_0(\lambda, n, m) \quad \text{and} \quad \tilde{G}_0(\lambda, m + n, m) \leq \tilde{G}_0(\lambda, m, m). \tag{3.82}$$

This assumption seems reasonable since $G_0(\lambda, n, .) \in \ell^2(\mathbb{Z})$. Then we have

$$|G_0(\lambda, n, m)| \leq \min\big(\tilde{g}_0(\lambda, n), \tilde{g}_0(\lambda, m)\big) \leq \sqrt{\tilde{g}_0(\lambda, n)}\sqrt{\tilde{g}_0(\lambda, m)}, \tag{3.83}$$

where $\tilde{g}_0(\lambda, n) = \tilde{G}_0(\lambda, n, n)$.

Corollary 3.21. *Let H_0 and H be two given Jacobi operators for which the assumptions of Theorem 3.20 hold. Moreover, suppose there is a $\tilde{g}_0(\lambda, n)$ such that*

$$|G_0(\lambda, n, m)| \leq \sqrt{\tilde{g}_0(\lambda, n)}\sqrt{\tilde{g}_0(\lambda, m)}. \tag{3.84}$$

3.4. A bound on the number of eigenvalues

Then we have

$$\dim \operatorname{Ran} P_{(\lambda, \pm\infty)}(H) \leq \sum_{n\in\mathbb{Z}} 2|a(n) - a_0(n)|\sqrt{\tilde{g}_0(\lambda, n+1)}\sqrt{\tilde{g}_0(\lambda, n)}$$

(3.85)
$$+ |b(n) - b_0(n)|\tilde{g}_0(\lambda, n)$$

for $\lambda < \lambda_0 \leq H_0$, $\lambda > \lambda_0 \geq H_0$, *respectively.*

Proof. The kernel of $(H - H_0)(H_0 - \lambda)^{-1}$ is given by

$$K(\lambda, n, m) = (a(n) - a_0(n))G(\lambda, n+1, m) + (b(n) - b_0(n))G(\lambda, n, m)$$

(3.86)
$$+ (a(n-1) - a_0(n-1))G(\lambda, n-1, m)$$

which can be estimated by

$$\tilde{K}(\lambda, n, m) = |a(n) - a_0(n)|\sqrt{\tilde{g}_0(\lambda, n+1)}\sqrt{\tilde{g}_0(\lambda, m)}$$
$$+ |b(n) - b_0(n)|\sqrt{\tilde{g}_0(\lambda, n)}\sqrt{\tilde{g}_0(\lambda, m)}$$

(3.87)
$$+ |a(n-1) - a_0(n-1)|\sqrt{\tilde{g}_0(\lambda, n-1)}\sqrt{\tilde{g}_0(\lambda, m)}.$$

Now

$$\operatorname{tr}\big((H - H_0)(\lambda - H_0)^{-1}\big)^2 = \sum_{n,m\in\mathbb{Z}} K(\lambda, n, m)K(\lambda, m, n)$$

$$\leq \sum_{n,m\in\mathbb{Z}} \tilde{K}(\lambda, n, m)\tilde{K}(\lambda, m, n)$$

(3.88)
$$= \sum_{n,m\in\mathbb{Z}} \tilde{K}(\lambda, n, n)\tilde{K}(\lambda, m, m) = \Big(\sum_{n\in\mathbb{Z}} \tilde{K}(\lambda, n, n)\Big)^2$$

concludes the proof. □

Applications of these results will be given in Section 10.2.

Chapter 4

Oscillation theory

There is a close connection between the number of sign flips of solutions of Jacobi equations and the spectra of the corresponding Jacobi operators. The investigation of this interconnection was started by Sturm (in the case of differential equations) and is now known as oscillation theory.

4.1. Prüfer variables and Sturm's separation theorem

In this section we will study oscillation properties of solutions of (1.19). To be more precise, we are interested in the number of sign flips of solutions. It will be convenient to assume

(4.1) $$a(n) < 0, \quad b(n) \in \mathbb{R}.$$

In the sequel, a solution of (1.19) will always mean a real-valued, nonzero solution of (1.19). Given a solution $u(\lambda, .)$ of $\tau u = \lambda u$, $\lambda \in \mathbb{R}$, we introduce **Prüfer variables** $\rho_u(\lambda, .), \theta_u(\lambda, .)$ via

(4.2) $$\begin{aligned} u(\lambda, n) &= \rho_u(\lambda, n) \sin \theta_u(\lambda, n), \\ u(\lambda, n+1) &= \rho_u(\lambda, n) \cos \theta_u(\lambda, n). \end{aligned}$$

Notice that the Prüfer angle $\theta_u(\lambda, n)$ is only defined up to an additive integer multiple of 2π (which depends on n).

Inserting (4.2) into $(\tau - \lambda)u = 0$ yields

(4.3) $$a(n) \cot \theta_u(\lambda, n) + a(n-1) \tan \theta_u(\lambda, n-1) = \lambda - b(n)$$

and

(4.4) $$\rho_u(\lambda, n) \sin \theta_u(\lambda, n) = \rho_u(\lambda, n-1) \cos \theta_u(\lambda, n-1).$$

Equation (4.3) is a discrete Riccati equation (cf. (1.52)) for $\cot \theta_u(n)$ and (4.4) can be solved if $\theta_u(n)$ is known provided it is replaced by

(4.5) $$a(n)\rho_u(\lambda, n) = a(n-1)\rho_u(\lambda, n-1) = 0$$

if $\sin\theta_u(\lambda, n) = \cos\theta_u(\lambda, n-1) = 0$ (use $\tau u = \lambda u$ and (4.8) below). Explicitly,

$$(4.6) \qquad \rho_u(\lambda, n) = \rho_u(\lambda, 0) \prod_{m=0}^{n-1}{}^* \frac{\cos\theta_u(\lambda, m)}{\sin\theta_u(\lambda, m+1)},$$

where $\cos\theta_u(\lambda, m)/\sin\theta_u(\lambda, m+1)$ has to be replaced by $a(m+1)/a(m)$ whenever $\sin\theta_u(\lambda, m+1) = 0$.

The Wronskian of two solutions $u_i(\lambda_i, n)$, $i = 1, 2$, reads

$$(4.7) \quad W_n(u_1(\lambda_1), u_2(\lambda_2)) = a(n)\rho_{u_1}(\lambda_1, n)\rho_{u_2}(\lambda_2, n)\sin(\theta_{u_1}(\lambda_1, n) - \theta_{u_2}(\lambda_2, n)).$$

The next lemma considers zeros of solutions and their Wronskians more closely. In particular, we will show that a solutions (resp. their Wronskians) must switch sign at a zero.

Lemma 4.1. *Let $u_{1,2}$ be solutions of $\tau u_{1,2} = \lambda_{1,2} u_{1,2}$ corresponding to $\lambda_1 \neq \lambda_2$, respectively. Then*

$$(4.8) \qquad u_1(n) = 0 \quad\Rightarrow\quad u_1(n-1)u_1(n+1) < 0.$$

Moreover, suppose $W_n(u_1, u_2) = 0$ but $W_{n-1}(u_1, u_2)W_{n+1}(u_1, u_2) \neq 0$, then

$$(4.9) \qquad W_{n-1}(u_1, u_2)W_{n+1}(u_1, u_2) < 0.$$

Otherwise, if $W_n(u_1, u_2) = W_{n+1}(u_1, u_2) = 0$, then necessarily

$$(4.10) \qquad u_1(n+1) = u_2(n+1) = 0 \quad\text{and}\quad W_{n-1}(u_1, u_2)W_{n+2}(u_1, u_2) < 0.$$

Proof. The fact $u_1(n) = 0$ implies $u_1(n-1)u_1(n+1) \neq 0$ (otherwise u_1 vanishes identically) and $a(n)u_1(n+1) = -a(n-1)u_1(n-1)$ (from $\tau u_1 = \lambda u_1$) shows $u_1(n-1)u_1(n+1) < 0$.

Next, from (1.20) we infer

$$(4.11) \qquad W_{n+1}(u_1, u_2) - W_n(u_1, u_2) = (\lambda_2 - \lambda_1)u_1(n+1)u_2(n+1).$$

And $W_n(u_1, u_2) = 0$ is equivalent to $u_1(n) = c\, u_2(n)$, $u_1(n+1) = c\, u_2(n+1)$ for some $c \neq 0$. Hence applying the above formula gives

$$(4.12) \qquad W_{n-1}(u_1, u_2)W_{n+1}(u_1, u_2) = -c^2(\lambda_2 - \lambda_1)^2 u_1(n)^2 u_1(n+1)^2,$$

proving the first claim. If $W_n(u_1, u_2)$, $W_{n+1}(u_1, u_2)$ are both zero, we must have $u_1(n+1) = u_2(n+1) = 0$ and as before $W_{n-1}(u_1, u_2)W_{n+2}(u_1, u_2) = -(\lambda_2 - \lambda_1)^2 u_1(n) u_1(n+2) u_2(n) u_2(n+2)$. Hence the claim follows from the first part (4.8). □

We can make the Prüfer angle $\theta_u(\lambda, .)$ unique by fixing, for instance, $\theta_u(\lambda, 0)$ and requiring

$$(4.13) \qquad [\![\theta_u(\lambda, n)/\pi]\!] \leq [\![\theta_u(\lambda, n+1)/\pi]\!] \leq [\![\theta_u(\lambda, n)/\pi]\!] + 1,$$

where

$$(4.14) \qquad [\![x]\!] = \sup\{n \in \mathbb{Z}\,|\,n < x\}.$$

Since our solutions $u(\lambda, n)$ will usually depend continuously on λ, the same should be true for $\theta_u(\lambda, n)$. In particular, continuity in λ should be compatible with the requirement (4.13).

4.1. Prüfer variables and Sturm's theorem

Lemma 4.2. *Let $\Lambda \subseteq \mathbb{R}$ be an interval. Suppose $u(\lambda, n)$ is continuous with respect to $\lambda \in \Lambda$ and (4.13) holds for one $\lambda_0 \in \Lambda$. Then it holds for all $\lambda \in \Lambda$ if we require $\theta_u(., n) \in C(\Lambda, \mathbb{R})$.*

Proof. Fix n and set

(4.15) $\quad \theta_u(\lambda, n) = k\pi + \delta(\lambda), \quad \theta_u(\lambda, n+1) = k\pi + \Delta(\lambda), \quad k \in \mathbb{Z},$

where $\delta(\lambda) \in (0, \pi]$, $\Delta(\lambda) \in (0, 2\pi]$. If (4.13) should break down, then by continuity we must have one of the following cases for some $\lambda_1 \in \Lambda$. (i) $\delta(\lambda_1) = 0$ and $\Delta(\lambda_1) \in (\pi, 2\pi)$, (ii) $\delta(\lambda_1) = \pi$ and $\Delta(\lambda_1) \in (0, \pi)$, (iii) $\Delta(\lambda_1) = 0$ and $\delta(\lambda_1) \in (0, \pi)$, (iv) $\Delta(\lambda_1) = 2\pi$ and $\delta(\lambda_1) \in (0, \pi)$. Abbreviate $R = \rho(\lambda_1, n)\rho(\lambda_1, n+1)$. Case (i) implies $0 > \sin(\Delta(\lambda_1)) = \cos(k\pi)\sin(k\pi + \Delta(\lambda_1)) = R^{-1}u(\lambda_1, n+1)^2 > 0$, contradicting (i). Case (ii) is similar. Case (iii) implies $\delta(\lambda_1) = \pi/2$ and hence $1 = \sin(k\pi + \pi/2)\cos(k\pi) = R^{-1}u(\lambda_1, n)u(\lambda_1, n+2)$ contradicting (4.8). Again, case (iv) is similar. $\square$

Let us call a point $n \in \mathbb{Z}$ a **node** of a solution u if either $u(n) = 0$ or $a(n)u(n)u(n+1) > 0$. Then, $[\![\theta_u(n)/\pi]\!] = [\![\theta_u(n+1)/\pi]\!]$ implies no node at n. Conversely, if $[\![\theta_u(n+1)/\pi]\!] = [\![\theta_u(n)/\pi]\!] + 1$, then n is a node by (4.8). Denote by $\#(u)$ the total number of nodes of u and by $\#_{(m,n)}(u)$ the number of nodes of u between m and n. More precisely, we will say that a node n_0 of u lies between m and n if either $m < n_0 < n$ or if $n_0 = m$ but $u(m) \neq 0$. Hence we conclude that the Prüfer angle of a solution counts the number of nodes.

Lemma 4.3. *Let $m < n$. Then we have for any solution u*

(4.16) $\quad \#_{(m,n)}(u) = [\![\theta_u(n)/\pi]\!] - \lim_{\varepsilon \downarrow 0} [\![\theta_u(m)/\pi + \varepsilon]\!]$

and

(4.17) $\quad \#(u) = \lim_{n \to \infty} \left([\![\theta_u(n)/\pi]\!] - [\![\theta_u(-n)/\pi]\!] \right).$

Next, we prove the analog of **Sturm's separation theorem** for differential equations.

Lemma 4.4. *Let $u_{1,2}$ be solutions of $\tau u = \lambda u$ corresponding to $\lambda_1 \leq \lambda_2$. Suppose $m < n$ are two consecutive points which are either nodes of u_1 or zeros of $W.(u_1, u_2)$ (the cases $m = -\infty$ or $n = +\infty$ are allowed if u_1 and u_2 are both in $\ell^2_{\pm}(\mathbb{Z})$ and $W_{\pm\infty}(u_1, u_2) = 0$, respectively) such that u_1 has no further node between m and n. Then u_2 has at least one node between m and $n+1$. Moreover, suppose $m_1 < \cdots < m_k$ are consecutive nodes of u_1. Then u_2 has at least $k - 1$ nodes between m_1 and m_k. Hence we even have*

(4.18) $\quad \#_{(m,n)}(u_2) \geq \#_{(m,n)}(u_1) - 1.$

Proof. Suppose u_2 has no node between m and $n+1$. Hence we may assume (perhaps after flipping signs) that $u_1(j) > 0$ for $m < j < n$, $u_1(n) \geq 0$, and $u_2(j) > 0$ for $m \leq j \leq n$. Moreover, $u_1(m) \leq 0$, $u_1(n+1) < 0$, and $u_2(n+1) \geq 0$ provided m, n are finite. By Green's formula (1.20)

(4.19) $\quad 0 \leq (\lambda_2 - \lambda_1) \sum_{j=m+1}^{n} u_1(j)u_2(j) = W_n(u_1, u_2) - W_m(u_1, u_2).$

Evaluating the Wronskians shows $W_n(u_1, u_2) < 0$, $W_m(u_1, u_2) > 0$, which is a contradiction.

It remains to prove the last part. We will use induction on k. The case $k = 1$ is trivial and $k = 2$ has already been proven. Denote the nodes of u_2 which are lower or equal than m_{k+1} by $n_k > n_{k-1} > \cdots$. If $n_k > m_k$ we are done since there are $k - 1$ nodes n such that $m_1 \leq n \leq m_k$ by induction hypothesis. Otherwise we can find k_0, $0 \leq k_0 \leq k$, such that $m_j = n_j$ for $1 + k_0 \leq j \leq k$. If $k_0 = 0$, we are done and hence we can suppose $k_0 \geq 1$. By induction hypothesis it suffices to show that there are $k - k_0$ nodes n of u_2 with $m_{k_0} \leq n \leq m_{k+1}$. By assumption $m_j = n_j$, $1 + k_0 \leq j \leq k$, are the only nodes n of u_2 such that $m_{k_0} \leq n \leq m_{k+1}$. Abbreviate $m = m_{k_0}$, $n = m_{k+1}$ and assume without restriction $u_1(m+1) > 0$, $u_2(m) > 0$. Since the nodes of u_1 and u_2 coincide we infer $0 < \sum_{j=m+1}^{n} u_1(j)u_2(j)$ and we can proceed as in the first part to obtain a contradiction. □

We call τ **oscillatory** if one solution of $\tau u = 0$ has an infinite number of nodes. In addition, we call τ oscillatory at $\pm\infty$ if one solution of $\tau u = 0$ has an infinite number of nodes near $\pm\infty$. We remark that if one solution of $(\tau - \lambda)u = 0$ has infinitely many nodes, so has any other (corresponding to the same λ) by (4.18). Furthermore, $\tau - \lambda_1$ oscillatory implies $\tau - \lambda_2$ oscillatory for all $\lambda_2 \geq \lambda_1$ (again by (4.18)).

Now we turn to the special solution $s(\lambda, n)$ characterized by the initial conditions $s(\lambda, 0) = 0$, $s(\lambda, 1) = 1$. As in Lemma 2.4 we infer

$$(4.20) \qquad W_n(s(\lambda), s'(\lambda)) = \sum_{j=1}^{n}{}^{*} s(\lambda, j)^2.$$

Here the prime denotes the derivative with respect to λ. Evaluating the above equation using Prüfer variables shows

$$(4.21) \qquad \theta'_s(\lambda, n) = \frac{1}{-a(n)\rho_s(\lambda, n)^2} \sum_{j=1}^{n}{}^{*} s(\lambda, j)^2.$$

In particular, $\theta'_s(\lambda, n) < 0$ for $n < -1$, $\theta'_s(\lambda, -1) = \theta'_s(\lambda, 0) = 0$, and $\theta'_s(\lambda, n) > 0$ for $n > 0$. Equation (4.21) implies that nodes of $s(\lambda, n)$ for $n \in \mathbb{N}$ move monotonically to the left without colliding. In addition, since $s(\lambda, n)$ cannot pick up nodes locally by (4.8), all nodes must enter at ∞ and since $\theta'_s(\lambda, 0) = 0$ they are trapped inside $(0, \infty)$.

We will normalize $\theta_s(\lambda, 0) = 0$ implying $\theta_s(\lambda, -1) = -\pi/2$. Since $s(\lambda, n)$ is a polynomial in λ, we easily infer $s(\lambda, n_0) \gtrless 0$ for fixed $n_0 \gtrless 0$ and λ sufficiently small (see (1.68) and (1.69)). This implies

$$(4.22) \qquad -\pi < \theta_s(\lambda, n_0) < -\pi/2,\ n_0 < -1,\quad 0 < \theta_s(\lambda, n_0) < \pi,\ n_0 \geq 1,$$

for fixed n and λ sufficiently small. Moreover, dividing (4.3) by λ and letting $\lambda \to -\infty$ using (4.22) shows

$$(4.23) \qquad \lim_{\lambda \to -\infty} \frac{\cot(\theta_s(\lambda, n))^{\pm 1}}{\lambda} = \frac{1}{a(n)}, \quad n \begin{array}{c} \geq +1 \\ < -1 \end{array},$$

and hence

(4.24) $$\theta_s(\lambda, n) = \begin{cases} -\frac{\pi}{2} - \frac{a(n)}{\lambda} + o(\frac{1}{\lambda}), & n < -1 \\ \frac{a(n)}{\lambda} + o(\frac{1}{\lambda}), & n \geq 1 \end{cases},$$

as $\lambda \to -\infty$.

Now what happens with $\theta_s(\lambda, n)$ as λ increases? Suppose $n \geq 1$ for simplicity. We already know that for λ small, $\theta_s(\lambda, n)$ starts near 0. Then it increases as λ increases by (4.21). It crosses $\pi/2$ at the first zero of $s(\lambda, n+1)$. Next it crosses π at the first zero of $s(\lambda, n)$. This process continues until it finally crosses $n\pi$ at the last zero of $s(\lambda, n+1)$. Here we have assumed that $s(\lambda, n)$, $n \in \mathbb{N}$, has precisely $n-1$ distinct real zeros. However, this is easily seen to hold, since $s(\lambda_0, n) = 0$ implies that λ_0 is an eigenvalue for $H_{0,n}$ (see (1.65)). Hence λ_0 must be real and simple since $H_{0,n}$ is self-adjoint and its spectrum is simple (cf. Remark 1.10), respectively. Let us summarize these findings. The interlacing of zeros is a classical result from the theory orthogonal polynomials.

Theorem 4.5. *The polynomial $s(\lambda, n)$, $n \in \mathbb{N}$, has $n - 1$ real and distinct roots denoted by*

(4.25) $$\lambda_1^n < \lambda_2^n < \cdots \lambda_{n-1}^n.$$

The zeros of $s(\lambda, n)$ and $s(\lambda, n+1)$ are interlacing, that is,

(4.26) $$\lambda_1^{n+1} < \lambda_1^n < \lambda_2^{n+1} < \cdots < \lambda_{n-1}^n < \lambda_n^{n+1}.$$

Moreover,

(4.27) $$\sigma(H_{0,n}) = \{\lambda_j^n\}_{j=1}^{n-1}.$$

Proceeding as for (4.24) we also note

(4.28) $$\theta_s(\lambda, n) = \begin{cases} \frac{(2n+1)\pi}{2} + \frac{a(n)}{\lambda} + o(\frac{1}{\lambda}), & n < -1 \\ n\pi - \frac{a(n)}{\lambda} + o(\frac{1}{\lambda}), & n \geq 1 \end{cases},$$

as $\lambda \to +\infty$.

Analogously, let $u_\pm(\lambda, n)$ be solutions of $\tau u = \lambda u$ as in Lemma 2.2. Then Lemma 2.4 implies

$$\theta'_+(\lambda, n) = \frac{\sum_{j=n+1}^{\infty} u_+(\lambda, j)^2}{a(n)\rho_+(\lambda, n)^2} < 0,$$

(4.29) $$\theta'_-(\lambda, n) = \frac{\sum_{j=-\infty}^{n} u_-(\lambda, j)^2}{-a(n)\rho_-(\lambda, n)^2} > 0,$$

where we have abbreviated $\rho_{u_\pm} = \rho_\pm$, $\theta_{u_\pm} = \theta_\pm$.

Since H is bounded from below we can normalize

(4.30) $$0 < \theta_\mp(\lambda, n) < \pi/2, \quad n \in \mathbb{Z}, \quad \lambda < \inf \sigma(H),$$

and we get as before

(4.31) $$\theta_-(\lambda, n) = \frac{a(n)}{\lambda} + o(\frac{1}{\lambda}), \quad \theta_+(\lambda, n) = \frac{\pi}{2} - \frac{a(n)}{\lambda} + o(\frac{1}{\lambda}), \quad n \in \mathbb{Z},$$

as $\lambda \to -\infty$.

4.2. Classical oscillation theory

Before we come to the first applications we recall a lemma from functional analysis. It will be one of our main ingredients in the following theorems.

Lemma 4.6. *Let H be a bounded self-adjoint operator and η_j, $1 \leq j \leq k$, be linearly independent elements of a (separable) Hilbert space $\mathfrak{H}$.*
(i). Let $\lambda \in \mathbb{R}$. If

$$\langle \eta, H\eta \rangle < \lambda \|\eta\|^2 \tag{4.32}$$

for any nonzero linear combination $\eta = \sum_{j=1}^{k} c_j \eta_j$, then

$$\dim \operatorname{Ran} P_{(-\infty,\lambda)}(H) \geq k. \tag{4.33}$$

Similarly, $\langle \eta, H\eta \rangle > \lambda \|\eta\|^2$ implies $\dim \operatorname{Ran} P_{(\lambda,\infty)}(H) \geq k$.
(ii). Let $\lambda_1 < \lambda_2$. If

$$\left\| \left(H - \frac{\lambda_2 + \lambda_1}{2}\right)\eta \right\| < \frac{\lambda_2 - \lambda_1}{2} \|\eta\| \tag{4.34}$$

for any nonzero linear combination $\eta = \sum_{j=1}^{k} c_j \eta_j$, then

$$\dim \operatorname{Ran} P_{(\lambda_1,\lambda_2)}(H) \geq k. \tag{4.35}$$

Proof. (i). Let $M = \operatorname{span}\{\eta_j\} \subseteq \mathfrak{H}$. We claim $\dim P_{(-\infty,\lambda)}(H)M = \dim M = k$. For this it suffices to show $\operatorname{Ker} P_{(-\infty,\lambda)}(H)|_M = \{0\}$. Suppose $P_{(-\infty,\lambda)}(H)\eta = 0$, $\eta \neq 0$. Then, abbreviating $d\rho_\eta(x) = d\langle \eta, P_{(-\infty,x)}(H)\eta \rangle$, we see that for any nonzero linear combination η

$$\begin{aligned}
\langle \eta, H\eta \rangle &= \int_{\mathbb{R}} x \, d\rho_\eta(x) = \int_{[\lambda,\infty)} x \, d\rho_\eta(x) \\
&\geq \lambda \int_{[\lambda,\infty)} d\rho_\eta(x) = \lambda \|\eta\|^2.
\end{aligned} \tag{4.36}$$

This contradicts our assumption (4.32). (ii). Using the same notation as before we need to show $\operatorname{Ker} P_{(\lambda_1,\lambda_2)}(H)|_M = \{0\}$. If $P_{(\lambda_1,\lambda_2)}(H)\eta = 0$, $\eta \neq 0$, then,

$$\begin{aligned}
\left\|\left(H - \frac{\lambda_2+\lambda_1}{2}\right)\eta\right\|^2 &= \int_{\mathbb{R}} \left(x - \frac{\lambda_2+\lambda_1}{2}\right)^2 d\rho_\eta(x) = \int_\Lambda x^2 d\rho_\eta\left(x + \frac{\lambda_2+\lambda_1}{2}\right) \\
&\geq \frac{(\lambda_2-\lambda_1)^2}{4} \int_\Lambda d\rho_\eta\left(x + \frac{\lambda_2+\lambda_1}{2}\right) = \frac{(\lambda_2-\lambda_1)^2}{4} \|\eta\|^2,
\end{aligned} \tag{4.37}$$

where $\Lambda = (-\infty, -(\lambda_2-\lambda_1)/2] \cup [(\lambda_2-\lambda_1)/2, \infty)$. But this is a contradiction as before. $\square$

Using this result we can now show that the number of nodes of $s(\lambda, n)$ equals the number of eigenvalues below λ. We begin with the case of finite and half-line operators.

Theorem 4.7. *Let $\lambda \in \mathbb{R}$. Then we have*

$$\dim \operatorname{Ran} P_{(-\infty,\lambda)}(H_{0,n}) = \#_{(0,n)}(s(\lambda)), \quad n > 1, \tag{4.38}$$

and

$$\dim \operatorname{Ran} P_{(-\infty,\lambda)}(H_+) = \#_{(0,+\infty)}(s(\lambda)). \tag{4.39}$$

The same theorem holds if $+$ is replaced by $-$.

4.2. Classical oscillation theory

Proof. We only carry out the proof for the plus sign (the other part following from reflection). By virtue of (4.21), (4.24), and Lemma 4.3 we infer

$$\dim \operatorname{Ran} P_{(-\infty,\lambda)}(H_{0,n}) = [\![\theta_s(\lambda,n)/\pi]\!] = \#_{(0,n)}(s(\lambda)), \quad n > 1, \tag{4.40}$$

since $\lambda \in \sigma(H_{0,n})$ if and only if $\theta_s(\lambda,n) = 0 \mod \pi$. Let $k = \#(s(\lambda))$ if $\#(s(\lambda)) < \infty$, otherwise the following argument works for arbitrary $k \in \mathbb{N}$. If we pick n so large that k nodes of $s(\lambda)$ are to the left of n, we have k eigenvalues $\hat{\lambda}_1 < \cdots < \hat{\lambda}_k < \lambda$ of $H_{0,n}$. Taking an arbitrary linear combination $\eta(m) = \sum_{j=1}^{k} c_j s(\hat{\lambda}_j, n)$, $c_j \in \mathbb{C}$, for $m < n$ and $\eta(m) = 0$ for $m \geq n$ a straightforward calculation (using orthogonality of $s(\hat{\lambda}_j)$) verifies

$$\langle \eta, H_+ \eta \rangle < \lambda \|\eta\|^2. \tag{4.41}$$

Invoking Lemma 4.6 shows

$$\dim \operatorname{Ran} P_{(-\infty,\lambda)}(H_+) \geq k. \tag{4.42}$$

For the reversed inequality we can assume $k = \#(s(\lambda)) < \infty$. Consider $\tilde{H}_{0,n} = H_{0,n} \oplus \lambda \mathbb{1}$ on $\ell^2(0,n) \oplus \ell^2(n-1,\infty)$. Since $\tilde{H}_{0,n} \to H$ strongly as $n \to \infty$ this implies ([**116**], Lemma 5.2)

$$\dim \operatorname{Ran} P_{(-\infty,\lambda)}(H_+) \leq \lim_{n\to\infty} \dim \operatorname{Ran} P_{(-\infty,\lambda)}(H_{0,n}) = k \tag{4.43}$$

completing the proof. □

Remark 4.8. (i). Consider the following example

$$a(n) = -\frac{1}{2}, \quad n \in \mathbb{N},$$

$$b(1) = -1, b(2) = -b_2, b(3) = -\frac{1}{2}, b(n) = 0, \quad n \geq 4. \tag{4.44}$$

The essential spectrum of H_+ is given by $\sigma_{ess}(H_+) = [-1,1]$ and one might expect that H_+ has no eigenvalues below the essential spectrum if $b_2 \to -\infty$. However, since we have

$$s(-1,0) = 0, s(-1,1) = 1, s(-1,2) = 0, s(-1,n) = -1, n \geq 3, \tag{4.45}$$

Theorem 4.7 shows that, independent of $b_2 \in \mathbb{R}$, there is always precisely one eigenvalue below the essential spectrum.

(ii). By a simple transformation we obtain the corresponding result for H_{+,n_0}^{β}, $\beta \neq 0$,

$$\dim \operatorname{Ran} P_{(-\infty,\lambda)}(H_{+,n_0}^{\beta}) = \#_{(0,+\infty)}(s_\beta(\lambda,.,n_0)), \tag{4.46}$$

where $s_\beta(\lambda,.,n_0)$ is the solution satisfying the boundary condition in (1.94) (see (2.71)). Similar modifications apply to Theorems 4.13, 4.16, and 4.17 below.

As a consequence of Theorem 4.7 we infer the following connection between being oscillatory or not and the infimum of the essential spectrum.

Corollary 4.9. *We have*

$$\dim \operatorname{Ran} P_{(-\infty,\lambda)}(H_\pm) < \infty \tag{4.47}$$

if and only if $\tau - \lambda$ is non-oscillatory near $\pm\infty$ and hence

$$\inf \sigma_{ess}(H_\pm) = \inf\{\lambda \in \mathbb{R} \,|\, (\tau - \lambda) \text{ is oscillatory at } \pm\infty\}. \tag{4.48}$$

Moreover, let $\lambda_0 < \cdots < \lambda_k < \ldots$ be the eigenvalues of $H_\pm$ below the essential spectrum of $H_\pm$. Then the eigenfunction corresponding to λ_k has precisely k nodes inside $(0, \pm\infty)$.

In a similar way we obtain

Theorem 4.10. *Let $\lambda < \inf \sigma_{ess}(H)$. Then*
$$\dim \operatorname{Ran} P_{(-\infty,\lambda)}(H) = \#(u_+(\lambda)) = \#(u_-(\lambda)). \tag{4.49}$$

Proof. Again it suffices to prove the first equality. By virtue of (4.29) and (4.31) we infer
$$\dim \operatorname{Ran} P_{(-\infty,\lambda)}(H_{-,n}) = [\![\theta_-(\lambda,n)/\pi]\!], \quad n \in \mathbb{Z}. \tag{4.50}$$
We first want to show that $[\![\theta_-(\lambda,n)/\pi]\!] = \#_{(-\infty,n)}(u_-(\lambda))$ or equivalently that $\lim_{n \to \infty} [\![\theta_-(\lambda,n)/\pi]\!] = 0$. Suppose $\lim_{n \to \infty} [\![\theta_-(\lambda_1,n)/\pi]\!] = k \geq 1$ for some $\lambda_1 \in \mathbb{R}$ (saying that $u_-(.,n)$ looses at least one node at $-\infty$). In this case we can find n such that $\theta_-(\lambda_1,n) > k\pi$ for $m \geq n$. Now pick λ_0 such that $\theta_-(\lambda_0,n) = k\pi$. Then $u_-(\lambda_0,.)$ has a node at n but no node between $-\infty$ and n (by Lemma 4.3). Now apply Lemma 4.4 to $u_-(\lambda_0,.)$, $u_-(\lambda_1,.)$ to obtain a contradiction. The rest follows as in the proof of Theorem 4.7. □

As before we obtain

Corollary 4.11. *We have*
$$\dim \operatorname{Ran} P_{(-\infty,\lambda)}(H) < \infty \tag{4.51}$$
if and only if $\tau - \lambda$ is non-oscillatory and hence
$$\inf \sigma_{ess}(H) = \inf\{\lambda \in \mathbb{R} \mid (\tau - \lambda) \text{ is oscillatory}\}. \tag{4.52}$$
Furthermore, let $\lambda_0 < \cdots < \lambda_k < \ldots$ be the eigenvalues of H below the essential spectrum of H. Then the eigenfunction corresponding to λ_k has precisely $k - 1$ nodes.

Remark 4.12. (i). Corresponding results for the projection $P_{(\lambda,\infty)}(H)$ can be obtained from $P_{(\lambda,\infty)}(H) = P_{(-\infty,-\lambda)}(-H)$. In fact, it suffices to change the definition of a node according to $u(n) = 0$ or $a(n)u(n)u(n+1) < 0$ and $P_{(-\infty,\lambda)}(H)$ to $P_{(\lambda,\infty)}(H)$ in all results of this section.
(ii). Defining $u_\pm(\lambda,n)$ as in Remark 2.3 for $\lambda = \inf \sigma_{ess}(H)$, one sees that Theorem 4.10 holds for $\lambda \leq \inf \sigma_{ess}(H)$.

Theorem 4.13. *Let $\lambda_1 < \lambda_2$. Suppose $\tau - \lambda_2$ is oscillatory near $+\infty$. Then*
$$\dim \operatorname{Ran} P_{(\lambda_1,\lambda_2)}(H_+) = \liminf_{n \to +\infty} \Big(\#_{(0,n)}(s(\lambda_2)) - \#_{(0,n)}(s(\lambda_1)) \Big). \tag{4.53}$$

The same theorem holds if $+$ is replaced by $-$.

Proof. As before we only carry out the proof for the plus sign. Abbreviate $\Delta(n) = [\![\theta_s(\lambda_2,n)/\pi]\!] - [\![\theta_s(\lambda_1,n)/\pi]\!] = \#_{(0,n)}(s(\lambda_2)) - \#_{(0,n)}(s(\lambda_1))$. By (4.40) we infer
$$\dim \operatorname{Ran} P_{[\lambda_1,\lambda_2)}(H_{0,n}) = \Delta(n), \quad n > 2. \tag{4.54}$$
Let $k = \liminf \Delta(n)$ if $\limsup \Delta(n) < \infty$ and $k \in \mathbb{N}$ otherwise. We contend that there exists $n \in \mathbb{N}$ such that
$$\dim \operatorname{Ran} P_{(\lambda_1,\lambda_2)}(H_{0,n}) \geq k. \tag{4.55}$$

In fact, if $k = \limsup \Delta(n) < \infty$, it follows that $\Delta(n)$ is eventually equal to k and since $\lambda_1 \notin \sigma(H_{0,m}) \cap \sigma(H_{0,m+1})$, $m \in \mathbb{N}$, we are done in this case. Otherwise we can pick n such that $\dim \operatorname{Ran} P_{[\lambda_1,\lambda_2)}(H_{0,n}) \geq k+1$. Hence $H_{0,n}$ has at least k eigenvalues $\hat{\lambda}_j$ with $\lambda_1 < \hat{\lambda}_1 < \cdots < \hat{\lambda}_k < \lambda_2$. Again let $\eta(m) = \sum_{j=1}^{k} c_j s(\hat{\lambda}_j, n)$, $c_j \in \mathbb{C}$ for $m < n$ and $\eta(m) = 0$ for $n \geq m$ be an arbitrary linear combination. Then

$$(4.56) \qquad \|(H_+ - \frac{\lambda_2 + \lambda_1}{2})\eta\| < \frac{\lambda_2 - \lambda_1}{2}\|\eta\|$$

together with the Lemma 4.6 implies

$$(4.57) \qquad \dim \operatorname{Ran} P_{(\lambda_1,\lambda_2)}(H_+) \geq k.$$

To prove the second inequality we use $\tilde{H}_{0,n} = H_{0,n} \oplus \lambda_2 \mathbb{1} \to H_+$ strongly as $n \to \infty$ and proceed as before

$$(4.58) \qquad \dim \operatorname{Ran} P_{(\lambda_1,\lambda_2)}(H_+) \leq \liminf_{n\to\infty} P_{[\lambda_1,\lambda_2)}(\tilde{H}_{0,n}) = k$$

since $P_{[\lambda_1,\lambda_2)}(\tilde{H}_{0,n}) = P_{[\lambda_1,\lambda_2)}(H_{0,n})$. $\square$

4.3. Renormalized oscillation theory

The objective of this section is to look at the nodes of the Wronskian of two solutions $u_{1,2}$ corresponding to $\lambda_{1,2}$, respectively. We call $n \in \mathbb{Z}$ a node of the Wronskian if $W_n(u_1, u_2) = 0$ and $W_{n+1}(u_1, u_2) \neq 0$ or if $W_n(u_1, u_2) W_{n+1}(u_1, u_2) < 0$. Again we will say that a node n_0 of $W(u_1, u_2)$ lies between m and n if either $m < n_0 < n$ or if $n_0 = m$ but $W_{n_0}(u_1, u_2) \neq 0$. We abbreviate

$$(4.59) \qquad \Delta_{u_1,u_2}(n) = (\theta_{u_2}(n) - \theta_{u_1}(n)) \mod 2\pi.$$

and require

$$(4.60) \qquad [\![\Delta_{u_1,u_2}(n)/\pi]\!] \leq [\![\Delta_{u_1,u_2}(n+1)/\pi]\!] \leq [\![\Delta_{u_1,u_2}(n)/\pi]\!] + 1.$$

We will fix $\lambda_1 \in \mathbb{R}$ and a corresponding solution u_1 and choose a second solution $u(\lambda, n)$ with $\lambda \in [\lambda_1, \lambda_2]$. Now let us consider

$$(4.61) \qquad W_n(u_1, u(\lambda)) = -a(n)\rho_{u_1}(n)\rho_u(\lambda, n) \sin(\Delta_{u_1,u}(\lambda, n))$$

as a function of $\lambda \in [\lambda_1, \lambda_2]$. As in the case of solutions, we first show that the normalization (4.60) is compatible with continuity in λ.

Lemma 4.14. *Suppose* $\Delta_{u_1,u}(\lambda_1,.)$ *satisfies (4.60), then we have*

$$(4.62) \qquad \Delta_{u_1,u}(\lambda, n) = \theta_u(\lambda, n) - \theta_{u_1}(n),$$

where $\theta_u(\lambda,.)$, $\theta_{u_1}(.)$ *both satisfy (4.13). That is,* $\Delta_{u_1,u}(.,n) \in C([\lambda_1, \lambda_2], \mathbb{R})$ *and (4.60) holds for all* $\Delta_{u_1,u}(\lambda,.)$ *with* $\lambda \in [\lambda_1, \lambda_2]$. *In particular, the second inequality in (4.13) is attained if and only if n is a node of $W_{.}(u_1, u(\lambda))$. Moreover, denote by $\#_{(m,n)}W(u_1, u_2)$ the total number of nodes of $W_{.}(u_1, u_2)$ between m and n. Then*

$$(4.63) \qquad \#_{(m,n)}W(u_1, u_2) = [\![\Delta_{u_1,u_2}(n)/\pi]\!] - \lim_{\varepsilon \downarrow 0}[\![\Delta_{u_1,u_2}(m)/\pi + \varepsilon]\!]$$

and

$$(4.64) \qquad \begin{aligned} \#W(u_1, u_2) &= \#_{(-\infty,\infty)}W(u_1, u_2) \\ &= \lim_{n\to\infty} \left([\![\Delta_{u_1,u_2}(n)/\pi]\!] - [\![\Delta_{u_1,u_2}(-n)/\pi]\!]\right). \end{aligned}$$

Proof. We fix n and set

(4.65) $$\Delta_{u_1,u}(\lambda, n) = k\pi + \delta(\lambda), \quad \Delta_{u_1,u}(\lambda, n+1) = k\pi + \Delta(\lambda),$$

where $k \in \mathbb{Z}, \delta(\lambda_1) \in (0, \pi]$ and $\Delta(\lambda_1) \in (0, 2\pi]$. Clearly (4.62) holds for $\lambda = \lambda_1$ since $W(u_1, u(\lambda_1))$ is constant. If (4.60) should break down, we must have one of the following cases for some $\lambda_0 \geq \lambda_1$. (i) $\delta(\lambda_0) = 0$, $\Delta(\lambda_0) \in (\pi, 2\pi]$, or (ii) $\delta(\lambda_0) = \pi$, $\Delta(\lambda_0) \in (0, \pi]$, or (iii) $\Delta(\lambda_0) = 2\pi$, $\delta(\lambda_0) \in (\pi, \pi]$, or (iv) $\Delta(\lambda_0) = 0$, $\delta(\lambda_0) \in (\pi, \pi]$. For notational convenience let us set $\delta = \delta(\lambda_0), \Delta = \Delta(\lambda_0)$ and $\theta_{u_1}(n) = \theta_1(n), \theta_u(\lambda_0, n) = \theta_2(n)$. Furthermore, we can assume $\theta_{1,2}(n) = k_{1,2}\pi + \delta_{1,2}$, $\theta_{1,2}(n+1) = k_{1,2}\pi + \Delta_{1,2}$ with $k_{1,2} \in \mathbb{Z}, \delta_{1,2} \in (0, \pi]$ and $\Delta_{1,2} \in (0, 2\pi]$.

Suppose (i). Then

(4.66) $$W_{n+1}(u_1, u(\lambda_0)) = (\lambda_0 - \lambda_1)u_1(n+1)u(\lambda_0, n+1).$$

Inserting Prüfer variables shows

(4.67) $$\sin(\Delta_2 - \Delta_1) = \rho \cos^2(\delta_1) \geq 0$$

for some $\rho > 0$ since $\delta = 0$ implies $\delta_1 = \delta_2$. Moreover, $k = (k_2 - k_1) \mod 2$ and $k\pi + \Delta = (k_2 - k_1)\pi + \Delta_2 - \Delta_1$ implies $\Delta = (\Delta_2 - \Delta_1) \mod 2\pi$. Hence we have $\sin \Delta \geq 0$ and $\Delta \in (\pi, 2\pi]$ implies $\Delta = 2\pi$. But this says $\delta_1 = \delta_2 = \pi/2$ and $\Delta_1 = \Delta_2 = \pi$. Since we have at least $\delta(\lambda_2 - \varepsilon) > 0$ and hence $\delta_2(\lambda_2 - \varepsilon) > \pi/2$, $\Delta_2(\lambda_2 - \varepsilon) > \pi$ for $\varepsilon > 0$ sufficiently small. Thus from $\Delta(\lambda_2 - \varepsilon) \in (\pi, 2\pi)$ we get

(4.68) $$0 > \sin\Delta(\lambda_2 - \varepsilon) = \sin(\Delta_2(\lambda_2 - \varepsilon) - \pi) > 0,$$

contradicting (i).

Suppose (ii). Again by (4.66) we have $\sin(\Delta_2 - \Delta_1) \geq 0$ since $\delta_1 = \delta_2$. But now $(k+1) = (k_1 - k_2) \mod 2$. Furthermore, $\sin(\Delta_2 - \Delta_1) = -\sin(\Delta) \geq 0$ says $\Delta = \pi$ since $\Delta \in (0, \pi]$. Again this implies $\delta_1 = \delta_2 = \pi/2$ and $\Delta_1 = \Delta_2 = \pi$. But since $\delta(\lambda)$ increases/decreases precisely if $\Delta(\lambda)$ increases/decreases for λ near λ_0, (4.60) stays valid.

Suppose (iii) or (iv). Then

(4.69) $$W_n(u_1, u(\lambda_0)) = -(\lambda_0 - \lambda_1)u_1(n+1)u(\lambda_0, n+1).$$

Inserting Prüfer variables gives

(4.70) $$\sin(\delta_2 - \delta_1) = -\rho \sin(\Delta_1)\sin(\Delta_2)$$

for some $\rho > 0$. We first assume $\delta_2 > \delta_1$. In this case we infer $k = (k_2 - k_1) \mod 2$ implying $\Delta_2 - \Delta_1 = 0 \mod 2\pi$ contradicting (4.70). Next assume $\delta_2 \leq \delta_1$. Then we obtain $(k+1) = (k_2 - k_1) \mod 2$ implying $\Delta_2 - \Delta_1 = \pi \mod 2\pi$ and hence $\sin(\delta_2 - \delta_1) \geq 0$ from (4.70). Thus we get $\delta_1 = \delta_2 = \pi/2$ $\Delta_1 = \Delta_2 = \pi$, and hence $\Delta_2 - \Delta_1 = 0 \mod 2\pi$ contradicting (iii), (iv). This settles (4.62).

Furthermore, if $\Delta(\lambda) \in (0, \pi]$, we have no node at n since $\delta(\lambda) = \pi$ implies $\Delta(\lambda) = \pi$ by (ii). Conversely, if $\Delta(\lambda) \in (\pi, 2\pi]$ we have a node at n since $\Delta(\lambda) = 2\pi$ is impossible by (iii). The rest being straightforward. $\square$

Equations (4.16), (4.62), and (4.63) imply

Corollary 4.15. *Let $\lambda_1 \leq \lambda_2$ and suppose $u_{1,2}$ satisfy $\tau u_{1,2} = \lambda_{1,2} u_{1,2}$, respectively. Then we have*

(4.71) $$|\#_{(n,m)}W(u_1, u_2) - (\#_{(n,m)}(u_2) - \#_{(n,m)}(u_1))| \leq 2$$

4.3. Renormalized oscillation theory

Now we come to a renormalized version of Theorem 4.13. We first need the result for a finite interval.

Theorem 4.16. *Fix $n_1 < n_2$ and $\lambda_1 < \lambda_2$. Then*

$$(4.72) \qquad \dim \operatorname{Ran} P_{(\lambda_1, \lambda_2)}(H_{n_1, n_2}) = \#_{(n_1, n_2)} W(s(\lambda_1, ., n_1), s(\lambda_2, ., n_2)).$$

Proof. We abbreviate

$$(4.73) \qquad \Delta(\lambda, n) = \Delta_{s(\lambda_1 \ldots n_1), s(\lambda \ldots n_2)}(n)$$

and normalize (perhaps after flipping the sign of $s(\lambda_1, ., n_1)$) $\Delta(\lambda_1, n) \in (0, \pi]$. From (4.21) we infer

$$(4.74) \qquad \dim \operatorname{Ran} P_{(\lambda_1, \lambda_2)}(H_{n_1, n_2}) = -\lim_{\varepsilon \downarrow 0} [\![\Delta(\lambda_2, n_1)/\pi + \varepsilon]\!]$$

since $\lambda \in \sigma(H_{n_1, n_2})$ is equivalent to $\Delta(\lambda, n_1) = 0 \mod \pi$. Using (4.63) completes the proof. $\square$

Theorem 4.17. *Fix $\lambda_1 < \lambda_2$. Then*

$$(4.75) \qquad \dim \operatorname{Ran} P_{(\lambda_1, \lambda_2)}(H_+) = \#_{(0, +\infty)} W(s(\lambda_1), s(\lambda_2)).$$

The same theorem holds if $+$ is replaced by $-$.

Proof. Again, we only prove the result for H_+. Set $k = \#_{(0,\infty)} W(s(\lambda_1), s(\lambda_2))$ provided this number is finite and $k \in \mathbb{N}$ otherwise. We abbreviate

$$(4.76) \qquad \Delta(\lambda, n) = \Delta_{s(\lambda_1), s(\lambda)}(n)$$

and normalize $\Delta(\lambda_1, n) = 0$ implying $\Delta(\lambda, n) > 0$ for $\lambda > \lambda_1$. Hence, if we choose n so large that all k nodes are to the left of n, we have

$$(4.77) \qquad \Delta(\lambda, n) > k\pi.$$

Thus we can find $\lambda_1 < \hat\lambda_1 < \cdots < \hat\lambda_k < \lambda_2$ with $\Delta(\hat\lambda_j, n) = j\pi$. Now define

$$(4.78) \qquad \eta_j(m) = \begin{cases} s(\hat\lambda_j, m) - \rho_j s(\lambda_1, m), & m \leq n \\ 0, & m \geq n \end{cases},$$

where $\rho_j \neq 0$ is chosen such that $s(\hat\lambda_j, m) = \rho_j s(\lambda_1, m)$ for $m = n, n+1$. Furthermore observe that

$$(4.79) \qquad \tau\eta_j(m) = \begin{cases} \hat\lambda_j s(\hat\lambda_j, m) - \lambda_1 \rho_1 s(\lambda_1, m), & m \leq n \\ 0, & m \geq n \end{cases}$$

and that $s(\lambda_1, m)$, $s(\hat\lambda_j, .)$, $1 \leq j \leq k$, are orthogonal on $1, \ldots, n$. Next, let $\eta = \sum_{j=1}^{k} c_j \eta_j$, $c_j \in \mathbb{C}$, be an arbitrary linear combination, then a short calculation verifies

$$(4.80) \qquad \|(H_+ - \frac{\lambda_2 + \lambda_1}{2})\eta\| < \frac{\lambda_2 - \lambda_1}{2}\|\eta\|.$$

Invoking Lemma 4.6 gives

$$(4.81) \qquad \dim \operatorname{Ran} P_{(\lambda_1, \lambda_2)}(H_+) \geq k.$$

To prove the reversed inequality is only necessary if $\#_{(0,\infty)}W(s(\lambda_1),s(\lambda_2)) < \infty$. In this case we look at $H_{0,n}^{\infty,\beta}$ with $\beta = s(\lambda_2, n+1)/s(\lambda_2, n)$. By Theorem 4.16 and Remark 4.8 (ii) we have

$$\dim \operatorname{Ran} P_{(\lambda_1,\lambda_2)}(\tilde{H}_{0,n}^{\infty,\beta}) = \#_{(0,n)}W(s(\lambda_1), s(\lambda_2)). \tag{4.82}$$

Now use strong convergence of $\tilde{H}_{0,n}^{\infty,\beta} = H_{0,n}^{\infty,\beta} \oplus \lambda_1 \mathbb{1}$ to H_+ as $n \to \infty$ to obtain

$$\dim \operatorname{Ran} P_{(\lambda_1,\lambda_2)}(H_+) \leq \liminf_{n\to\infty} \dim \operatorname{Ran} P_{(\lambda_1,\lambda_2)}(\tilde{H}_{0,n}^{\infty,\beta}) = k \tag{4.83}$$

completing the proof. □

As a consequence we obtain the analog of Corollary 4.9.

Corollary 4.18. *Let $u_{1,2}$ satisfy $\tau u_{1,2} = \lambda_{1,2} u_{1,2}$. Then*

$$\#_{(0,\pm\infty)}W(u_1, u_2) < \infty \quad \Leftrightarrow \quad \dim \operatorname{Ran} P_{(\lambda_1,\lambda_2)}(H_\pm) < \infty. \tag{4.84}$$

Proof. By (4.18) and Corollary 4.15 we learn that $\#_{(0,\pm\infty)}W(u_1, u_2)$ is finite if and only if $\#_{(0,\pm\infty)}W(s(\lambda_1), s(\lambda_2))$ is finite. □

Finally, we turn to our main result for Jacobi operators H on $\mathbb{Z}$.

Theorem 4.19. *Fix $\lambda_1 < \lambda_2$ and suppose $[\lambda_1, \lambda_2] \cap \sigma_{ess}(H) = \emptyset$. Then*

$$\begin{aligned} \dim \operatorname{Ran} P_{(\lambda_1,\lambda_2)}(H) &= \#W(u_\mp(\lambda_1), u_\pm(\lambda_2)) \\ &= \#W(u_\pm(\lambda_1), u_\pm(\lambda_2)). \end{aligned} \tag{4.85}$$

Proof. Since the proof is similar to the proof of Theorem 4.17 we will only outline the first part. Let $k = \#W(u_+(\lambda_1), u_-(\lambda_2))$ if this number is finite and $k \in \mathbb{N}$ otherwise. Pick $n > 0$ so large that all zeros of the Wronskian are between $-n$ and n. We abbreviate

$$\Delta(\lambda, n) = \Delta_{u_+(\lambda_1), u_-(\lambda)}(n) \tag{4.86}$$

and normalize $\Delta(\lambda_1, n) \in [0, \pi)$ implying $\Delta(\lambda, n) > 0$ for $\lambda > \lambda_1$. Hence, if we choose $n \in \mathbb{N}$ so large that all k nodes are between $-n$ and n, we can assume

$$\Delta(\lambda, n) > k\pi. \tag{4.87}$$

Thus we can find $\lambda_1 < \hat{\lambda}_1 < \cdots < \hat{\lambda}_k < \lambda_2$ with $\Delta(\hat{\lambda}_j, n) = 0 \mod \pi$. Now define

$$\eta_j(m) = \begin{cases} u_-(\hat{\lambda}_j, m) & m \leq n \\ \rho_j u_+(\lambda_1, m) & m \geq n \end{cases}, \tag{4.88}$$

where $\rho_j \neq 0$ is chosen such that $u_-(\hat{\lambda}_j, m) = \rho_j u_+(\lambda_1, m)$ for $m = n, n+1$. Now proceed as in the previous theorems. □

Again, we infer as a consequence.

Corollary 4.20. *Let $u_{1,2}$ satisfy $\tau u_{1,2} = \lambda_{1,2} u_{1,2}$. Then*

$$\#W(u_1, u_2) < \infty \quad \Leftrightarrow \quad \dim \operatorname{Ran} P_{(\lambda_1,\lambda_2)}(H) < \infty. \tag{4.89}$$

Proof. Follows from Corollary 4.18 and $\dim \operatorname{Ran} P_{(\lambda_1,\lambda_2)}(H)$ finite if and only if both $\dim \operatorname{Ran} P_{(\lambda_1,\lambda_2)}(H_-)$ and $\dim \operatorname{Ran} P_{(\lambda_1,\lambda_2)}(H_+)$ finite (see (3.22)). □

Remark 4.21. (i). Lemma 1.6 shows that all theorems remain valid if our hypothesis $a(n) < 0$ is replaced by $a(n) \neq 0$.
(ii). Defining $u_\pm(\lambda, n)$ as in Remark 2.3 for λ at the boundary of $\sigma_{ess}(H)$, one sees that Theorem 4.19 holds for $[\lambda_1, \lambda_2] \cap \sigma_{ess}(H) \subseteq \{\lambda_1, \lambda_2\}$.

Chapter 5

Random Jacobi operators

Up to this point we have only considered the case of Jacobi operators where the sequences a and b are assumed to be known. However, this assumption is often not fulfilled in physical applications. For example, look at the one-dimensional crystal model in Section 1.5. In a more realistic setting such a crystal will contain impurities, the precise locations of which are in general unknown. All one might know is that these impurities occur with a certain probability. This leads us to the study of Jacobi operators, where the coefficients are random variables. These **random Jacobi operators** serve also as a model in solid state physics of disordered systems, such as alloys, glasses, and amorphous materials in the so called tight binding approximation.

5.1. Random Jacobi operators

To begin with, let us introduce the proper setting for this new model. We consider the probability space $(\Omega, \mathcal{F}, \mu)$, where $\mathcal{F}$ is a σ-algebra on Ω and μ a probability measure (i.e., $\mu(\Omega) = 1$) on $(\Omega, \mathcal{F})$. A random variable f is a measurable functions on Ω and its expectation is $\mathbb{E}(f) = \int_\Omega f(\omega) d\mu(\omega)$.

If you are not familiar with the basics of probability (respectively measure) theory, you might want to refer to any textbook (e.g., [**39**]) first.

We will choose

(5.1) $$\Omega = \Omega_0^{\mathbb{Z}} = \{\omega = (\omega(j))_{j \in \mathbb{Z}} | \omega(j) \in \Omega_0\},$$

where Ω_0 is a bounded Borel subset of $(\mathbb{R}\backslash\{0\}) \times \mathbb{R}$, and $\mathcal{F}$ is the σ-algebra generated by the cylinder sets (i.e., by sets of the form $\{\omega \in \Omega | \omega(j) \in B_i, 1 \leq j \leq n\}$ with B_i Borel subsets of Ω_0).

On Ω we have a **discrete invertible dynamical system**, namely the shift operators

(5.2) $$T^i : \begin{array}{rcl} \Omega & \to & \Omega \\ \omega(j) & \mapsto & \omega(j-i) \end{array}.$$

Since the shift of a cylinder set is again one, $T(= T^1)$ is measurable.

The measure μ is called **invariant** (or **stationary**) if $\mu(TF) = \mu(F)$ for any $F \in \mathcal{F}$. Alternatively, T is called measure preserving. In addition, our dynamical system T is called **ergodic** (with respect to μ) if any shift invariant set $F = TF$ has probability $\mu(F)$ zero or one.

Given $(\Omega, \mathcal{F}, \mu)$ as above we can define random variables

(5.3)
$$\begin{aligned} a : \Omega &\to \ell^\infty(\mathbb{Z}, \mathbb{R}\backslash\{0\}) \\ \omega(j) &\mapsto a_\omega(n) = \omega_1(n) \\ b : \Omega &\to \ell^\infty(\mathbb{Z}, \mathbb{R}\backslash\{0\}) \\ \omega(j) &\mapsto b_\omega(n) = \omega_2(n) \end{aligned}$$

and a corresponding random Jacobi operator

(5.4) $$H_\omega = a_\omega S^+ + a_\omega^- S^- + b_\omega.$$

Since we assumed Ω_0 bounded, there is a constant C (independent of ω) such that $\|H_\omega\| \leq C$. The simplest example is to take a probability measure μ_0 on Ω_0 and consider the product measure $\mu = \times_{i \in \mathbb{Z}} \mu_0$. It is easy to see that μ is ergodic. This is known as **Anderson model**.

Now what are the questions we should ask about H_ω? Clearly, if we look for features shared by all operators H_ω, $\omega \in \Omega$, we could as well think of ω as being fixed. So we will look for properties which only hold with a certain probability. For example, properties which hold almost surely (a.s.), that is with probability one.

From now on we will assume the following hypothesis throughout this entire chapter.

Hypothesis H. 5.1. Let $(\Omega, \mathcal{F}, \mu)$ be as described in (5.1) and let T, as introduced in (5.2), be ergodic with respect to μ.

As a warm up we recall a simple lemma concerning invariant random variables. A random variable f is called invariant (with respect to T^i) if $f(T^i\omega) = f(\omega)$.

Lemma 5.2. *An invariant random variable $f : \Omega \to \mathbb{R} \cup \{\infty\}$ is constant almost surely (i.e., there is a set with probability one on which f is constant).*

Proof. Let $I_i^n = [\frac{i}{2^n}, \frac{i+1}{2^n})$, $I_\infty^n = \{\infty\}$ and $\Omega_i^n = f^{-1}(I_i^n) = \{\omega | \frac{i}{2^n} \leq f < \frac{i+1}{2^n}\}$, $\Omega_\infty^n = f^{-1}(\infty)$. The sets Ω_i^n, $i \in \mathbb{Z} \cup \{\infty\}$, are disjoint and $\mu(\Omega_i^n) \in \{0, 1\}$ due to ergodicity. Moreover, since

(5.5) $$\sum_{i \in \mathbb{Z} \cup \{\infty\}} \mu(\Omega_i^n) = \mu(\bigcup_{i \in \mathbb{Z} \cup \{\infty\}} \Omega_i^n) = \mu(\Omega) = 1$$

there exists $i_n \in \mathbb{Z} \cup \{\infty\}$ such that $\mu(\Omega_{i_n}^n) = 1$ and $\mu(\Omega_i^n) = 0$, $i \neq i_n$. And since the nesting intervals $I_{i_{n+1}}^{n+1} \subset I_{i_n}^n$ must converge to a point $\{x_0\} = \bigcap_{n \in \mathbb{N}} I_{i_n}^n$ we have

(5.6) $$\mu(f^{-1}(x_0)) = \mu(\bigcap_{n \in \mathbb{N}} \Omega_{i_n}^n) = \lim_{n \to \infty} \mu(\Omega_{i_n}^n) = 1.$$

□

This result clearly suggests to look at properties of H_ω which are invariant under the shift T^i. Since we have

(5.7) $$H_{T^i\omega} = S^i H_\omega S^{-i},$$

5.1. Random Jacobi operators

we see that $H_{T^i\omega}$ and H_ω are unitarily equivalent. Hence, one might expect that the spectrum is *constant* a.s.. Since the spectrum itself is no random variable we need to look for random variables which contain information on the location of the spectrum. Promising candidates are traces of the spectral projections

(5.8) $$\operatorname{tr}(P_\Lambda(H_\omega)) = \dim \operatorname{Ran} P_\Lambda(H_\omega).$$

Based on this observation we obtain the following theorem.

Theorem 5.3. *There exists a set $\Sigma \subset \mathbb{R}$ such that*

(5.9) $$\sigma(H_\omega) = \Sigma \quad a.s..$$

In addition

(5.10) $$\sigma_d(H_\omega) = \emptyset \quad a.s..$$

Proof. We start by claiming that the function $\omega \mapsto P_\Lambda(H_\omega)$ is weakly measurable (i.e., $\langle f, P_\Lambda(H_\omega) g \rangle$ is measurable for any $f, g \in \ell^2(\mathbb{Z})$; by the polarization identity it suffices to consider the case $f = g$). In fact, $\omega \mapsto H_\omega$ is weakly measurable and so is any polynomial of H_ω, since products of weakly measurable functions are again weakly measurable. Approximating $P_\Lambda(H_\omega)$ by (ω-independent) polynomials in the strong topology, the claim follows. As a consequence, the trace

(5.11) $$\operatorname{tr}(P_\Lambda(H_\omega)) = \sum_{n \in \mathbb{Z}} \langle \delta_n, P_\Lambda(H_\omega) \delta_n \rangle$$

is an invariant (by (5.7)) random variable and hence a.s. constant by the lemma.

Moreover,

(5.12) $$\begin{aligned}\operatorname{tr}(P_\Lambda(H_\omega)) &= \mathbb{E}(\operatorname{tr}(P_\Lambda(H_\omega))) \\ &= \mathbb{E}\Big(\sum_{n \in \mathbb{Z}} \langle \delta_n, P_\Lambda(H_\omega)\delta_n \rangle\Big) = \sum_{n \in \mathbb{Z}} \mathbb{E}(\langle \delta_n, P_\Lambda(H_\omega)\delta_n \rangle) \\ &= \sum_{n \in \mathbb{Z}} \mathbb{E}(\langle \delta_0, P_\Lambda(H_{T^n\omega})\delta_0 \rangle) = \sum_{n \in \mathbb{Z}} \mathbb{E}(\langle \delta_0, P_\Lambda(H_\omega)\delta_0 \rangle)\end{aligned}$$

is zero or infinity depending on $\mathbb{E}(\langle \delta_0, P_\Lambda(H_\omega)\delta_0 \rangle) = 0$ or not. Here we have used invariance of μ with respect to T^n in the last equality.

For any pair of rational numbers $(p, q) \in \mathbb{Q}^2$ we set $d_{(p,q)} = 0, \infty$ if the almost sure value of $\operatorname{tr} P_{(p,q)}(H_\omega)$ is $0, \infty$, respectively. Set

(5.13) $$\tilde{\Omega} = \bigcap_{(p,q) \in \mathbb{Q}^2} \Omega_{(p,q)}, \quad \Omega_{(p,q)} = \{\omega | \operatorname{tr} P_{(p,q)}(H_\omega) = d_{(p,q)}\}.$$

Then $\mu(\tilde{\Omega}) = 1$ for $\mu(\Omega_{(p,q)}) = 1$ and the intersection is countable.

Now for $\omega_1, \omega_2 \in \tilde{\Omega}$ we claim $\sigma(H_{\omega_1}) = \sigma(H_{\omega_2})$. Indeed, if $\lambda \notin \sigma(H_{\omega_1})$, there exist two rational numbers $\lambda_1 < \lambda < \lambda_2$ such that

(5.14) $$0 = \dim \operatorname{Ran} P_{(\lambda_1,\lambda_2)}(H_{\omega_1}) = d_{(\lambda_1,\lambda_2)} = \dim \operatorname{Ran} P_{(\lambda_1,\lambda_2)}(H_{\omega_1})$$

implying $\lambda \notin \sigma(H_{\omega_2})$. Interchanging ω_1 and ω_2 finishes the first part.

For the remaining one suppose $\lambda \in \sigma_d(H_\omega)$. Then

(5.15) $$0 < \dim \operatorname{Ran} P_{(\lambda_1,\lambda_2)}(H_\omega) < \infty$$

with $\lambda_1 < \lambda < \lambda_2$ sufficiently close to λ. This is impossible for $\omega \in \tilde{\Omega}$. So $\sigma_d(H_\omega) = \emptyset$, $\omega \in \tilde{\Omega}$. $\square$

Considering the corresponding projections onto the absolutely continuous, singular continuous, and pure point subspaces we expect a similar result for the corresponding spectra. The tricky part is to establish measurability of these projections.

Theorem 5.4. *There exist sets Σ_{ac}, Σ_{sc}, and Σ_{pp} such that*
$$\sigma_{ac}(H_\omega) = \Sigma_{ac} \quad a.s.,$$
$$\sigma_{sc}(H_\omega) = \Sigma_{sc} \quad a.s.,$$
(5.16)
$$\sigma_{pp}(H_\omega) = \Sigma_{pp} \quad a.s..$$

Proof. Given the proof of the previous theorem, we only need to show that two of the projections $P^{ac}(H)$, $P^{sc}(H)$, and $P^{pp}(H)$ are weakly measurable (for fixed ω, which is omitted here for notational simplicity). Indeed, we have $P_\Lambda^{ac}(H) = P^{ac}(H)P_\Lambda(H)$, etc.. We will show that $P^c(H) = P^{ac}(H) + P^{sc}(H)$ and $P^s(H) = P^{sc}(H) + P^{pp}(H)$ are weakly measurable.

By the RAGE Theorem ([50], Theorem 5.8) we have

(5.17)
$$\lim_{T \to \infty} \frac{1}{T} \left\| \int_0^T e^{itH} \chi_N P^c(H) e^{-itH} \right\| = 0,$$

where $\chi_N(n) = 1$ if $|n| \leq N$ and $\chi_N(n) = 0$ if $|n| > N$. Hence

$$\langle f, P^c(H)g \rangle = \frac{1}{T} \int_0^T \langle f, e^{itH}(\mathbb{1} - \chi_N + \chi_N) P^c(H) e^{-itH} g \rangle dt$$

$$= \lim_{T \to \infty} \frac{1}{T} \int_0^T \langle f, e^{itH}(\mathbb{1} - \chi_N) P^c(H) e^{-itH} g \rangle dt$$

$$= \lim_{T \to \infty} \frac{1}{T} \int_0^T \langle f, e^{itH}(\mathbb{1} - \chi_N) e^{-itH} g \rangle dt$$

(5.18)
$$- \lim_{T \to \infty} \frac{1}{T} \int_0^T \langle f, e^{itH}(\mathbb{1} - \chi_N) P^{pp}(H) e^{-itH} g \rangle dt$$

Now we estimate the last term using $P^{pp}(H)g = \sum_j g_j$, where g_j are eigenfunctions of H, that is, $H g_j = \lambda_j g_j$.

$$\langle f, e^{itH}(\mathbb{1} - \chi_N) P^{pp}(H) e^{-itH} g \rangle \leq \|e^{-itH} f\| \|(\mathbb{1} - \chi_N) e^{-itH} P^{pp}(H) g\|$$
(5.19)
$$\leq \|f\| \sum_j \|(\mathbb{1} - \chi_N) e^{-it\lambda_j} g_j\| \leq \|f\| \sum_j \|(\mathbb{1} - \chi_N) g_j\|.$$

Hence we have

(5.20) $$\langle f, P^c(H)g \rangle = \lim_{N \to \infty} \lim_{T \to \infty} \frac{1}{T} \int_0^T \langle f, e^{itH}(\mathbb{1} - \chi_N) e^{-itH} g \rangle dt$$

which is measurable since $e^{itH} = \sum_{j=0}^\infty (itH)^j/j!$.

Finally, $P^s(H)$ is measurable because of

(5.21) $$\langle f, P^s(H)g \rangle = \lim_{n \to \infty} \sup_{I \in \mathcal{I}, |I| < 1/n} \langle f, P_I(H)g \rangle$$

(see Lemma B.6). □

Since eigenvalues of H_ω are simple we get another interesting result.

Theorem 5.5. *For any $\lambda \in \mathbb{R}$ we have $\mu(\{\omega | \lambda \in \sigma_p(H_\omega)\}) = 0$.*

5.2. The Lyapunov exponent and the density of states

Proof. Since $\text{tr} P_{\{\lambda\}}(H_\omega)$ equals zero or infinity a.s. its probability of being one is zero. $\square$

This does not say that Σ_{pp} is empty. But if it is non-empty, it must be locally uncountable.

5.2. The Lyapunov exponent and the density of states

In this section we will need that a_ω is bounded away from zero.

Hypothesis H. 5.6. Suppose $\Omega_0 \subseteq [M^{-1}, M] \times [-M, M]$ for some $M \geq 1$, that is,

$$\frac{1}{M} \leq a_\omega \leq M, \qquad -M \leq b_\omega \leq M. \tag{5.22}$$

In particular, we have $\|H_\omega\| \leq 3M$.

We recall the definition of the Lyapunov exponent of H_ω from Section 1.1

$$\gamma_\omega^\pm(z) = \lim_{n \to \pm\infty} \frac{1}{|n|} \ln \|\Phi_\omega(z, n)\| \tag{5.23}$$

provided this limit exists. For its investigation we need some standard results from ergodic theory.

A sequence of random variables $f_n : \Omega \to \mathbb{R}$ is called a **subadditive process** if

$$f_{m+n}(\omega) \leq f_m(\omega) + f_n(T^m \omega), \tag{5.24}$$

where T is measure preserving. If equality holds, the process is called **additive**. Then we have the **subadditive ergodic theorem** by Kingman [148] (see [214] for a simple proof).

Theorem 5.7. If $\{f_n\}_{n \in \mathbb{N}}$ is a subadditive process satisfying $\mathbb{E}(|f_n|) < \infty$, $n \in \mathbb{N}$, and $\inf_n \mathbb{E}(f_n)/n > -\infty$, then $f_n(\omega)/n$ converges almost surely and, if T is ergodic,

$$\lim_{n \to \infty} \frac{1}{n} f_n(\omega) = \lim_{n \to \infty} \frac{1}{n} \mathbb{E}(f_n) = \inf_{n \in \mathbb{N}} \frac{1}{n} \mathbb{E}(f_n) \quad a.s.. \tag{5.25}$$

In the additive case we have $f_n(\omega) = \sum_{m=0}^{n-1} f_1(T^m \omega)$ and the theorem says that $\lim_{n \to \infty} f_n(\omega)/n = \mathbb{E}(f_1)$ almost surely if $\mathbb{E}(|f_1|) < \infty$. This is known as **Birkhoff's ergodic theorem**.

Now we can prove the following result.

Theorem 5.8. For each $z \in \mathbb{C}$ there exists a number

$$\gamma(z) = \inf_{n \in \mathbb{N}} \frac{1}{n} \mathbb{E}(\ln \|\Phi_\omega(z, n)\|) \tag{5.26}$$

(independent of ω) such that

$$\lim_{n \to \pm\infty} \frac{1}{|n|} \ln \|\Phi_\omega(z, n)\| = \gamma(z) \quad a.s.. \tag{5.27}$$

Proof. Abbreviate $f_n(\omega) = \ln \|\Phi_\omega(z, n)\|$, then f_n is a subadditive process since by (1.32)

$$\begin{aligned} f_{m+n}(\omega) &= \ln \|\Phi_\omega(z, n+m, m)\Phi_\omega(z, m)\| \\ &\leq f_n(T^{-m}\omega) + f_m(\omega). \end{aligned} \tag{5.28}$$

Moreover, for $n > 0$ (recall (1.33))

$$\frac{1}{M} \leq \|\Phi_\omega(z, n)\| \leq \prod_{j=1}^{n}{}^* \|U_\omega(z, j)\| \leq (M(M + |z|))^n \tag{5.29}$$

implies $\mathbb{E}(|f_n|) \leq nM(M + |z|)$ and $\inf_{n \in \mathbb{N}} \mathbb{E}(f_n)/n \geq 0$. Since the same considerations apply to $n < 0$ we infer from Theorem 5.7

$$\lim_{n \to \pm\infty} \frac{f_n(\omega)}{|n|} = \inf_{n \in \pm\mathbb{N}} \frac{1}{|n|} \mathbb{E}(f_n) \tag{5.30}$$

almost surely. It remains to show that both limits are equal, provided both exist. This follows using (1.32), (1.45), and invariance of our measure

$$\begin{aligned}
\mathbb{E}(f_{-n}) &= \mathbb{E}(\ln \|\Phi_\omega(z, -n, 0)\|) = \mathbb{E}(\ln \|\Phi_\omega(z, 0, -n)^{-1}\|) \\
&= \mathbb{E}(\ln \|\Phi_{T^n\omega}(z, n, 0)^{-1}\|) = \mathbb{E}(\ln \|\Phi_\omega(z, n, 0)^{-1}\|) \\
&= \mathbb{E}(\ln a_\omega(n) - \ln a_\omega(0) + \ln \|\Phi_\omega(z, n, 0)\|) = \mathbb{E}(f_n).
\end{aligned} \tag{5.31}$$

$\square$

In particular this result says that for fixed $z \in \mathbb{C}$ we have $\gamma_\omega^+(z) = \gamma_\omega^-(z) = \gamma(z)$ almost surely.

We call $\gamma(z)$ the Lyapunov exponent. It has several interesting properties. First of all, by the considerations after (1.36) we have $\gamma(z) \geq 0$. Before we can proceed, we need to review the concept of a subharmonic function.

A measurable function $f : \mathbb{C} \to \mathbb{R} \cup \{-\infty\}$ is called **submean** if $\max(f, 0)$ is locally integrable and

$$f(z_0) \leq \frac{1}{2\pi i r^2} \int_{|z - z_0| \leq r} f(z)\, d\bar{z} \wedge dz, \quad r > 0. \tag{5.32}$$

It is called **uppersemicontinuous** (u.s.c.) if for any sequence $z_n \to z_0$ we have

$$\limsup_{z_n \to z_0} f(z) \leq f(z_0). \tag{5.33}$$

If f is both, submean and u.s.c., it is called **subharmonic**. As an immediate consequence we obtain that

$$f(z_0) = \lim_{r \downarrow 0} \frac{1}{2\pi i r^2} \int_{|z - z_0| \leq r} f(z)\, d\bar{z} \wedge dz \tag{5.34}$$

if f is subharmonic. The properties of being submean or subharmonic are preserved by certain limiting operations.

Lemma 5.9. (i). *Suppose f_n are submean and $f_{n+1} \leq f_n$. Then $f(z) = \inf_n f_n(z)$ is submean. If, in addition, f_n are subharmonic, then so is f.*
(ii). *Suppose f_n are submean and $\sup_n \max(f_n, 0)$ is locally integrable. Then $\sup_n f_n$ and $\limsup_n f_n$ are submean.*

Proof. (i). Switching to $f_n - f_1$ it is no restriction to assume $f_n \leq 0$ and hence the first assertion follows from the monotone convergence theorem. The second one

5.2. The Lyapunov exponent and the density of states

follows since the infimum of u.s.c. functions is again u.s.c..
(ii). Since

$$
\sup_n f_n(z_0) \leq \frac{1}{2\pi i r^2} \sup_n \int_{|z-z_0|\leq r} f_n(z)\, d\bar{z} \wedge dz
$$

(5.35)
$$
\leq \frac{1}{2\pi i r^2} \int_{|z-z_0|\leq r} \sup_n f_n(z)\, d\bar{z} \wedge dz,
$$

we see that $\sup_n f_n$ is submean. Combining this result with the previous one and $\limsup_n f_n = \inf_N \sup_{n\geq N} f_n$ finishes the proof. $\square$

As a first observation note that the norm $\|\Phi_\omega(z,n)\|$ is subharmonic. In fact, $\|\Phi_\omega(z,n)\| \geq 1/M$ is continuous and submean by

(5.36)
$$
\|\Phi_\omega(z_0,n)\| = \frac{1}{2\pi i r^2}\left\|\int_{|z-z_0|\leq r}\Phi_\omega(z,n) d\bar{z}\wedge dz\right\| \leq \frac{1}{2\pi i r^2}\int_{|z-z_0|\leq r}\|\Phi_\omega(z,n)\| d\bar{z}\wedge dz.
$$

Moreover, since the logarithm is convex, $\ln \|\Phi_\omega(z,n)\|$ is submean by Jensen's inequality.

This is already half of the proof that $\gamma(z)$ is subharmonic.

Lemma 5.10. *The Lyapunov exponent $\gamma(z)$ is subharmonic. Furthermore, the upper Lyapunov exponents $\overline{\gamma}_\omega^\pm(z)$ of H_ω are submean.*

Proof. By Fubini's theorem the function

(5.37)
$$
\mathbb{E}(\ln\|\Phi_\omega(z,n)\|)
$$

is submean. By the estimate (5.29) and the dominated convergence theorem it is also continuous and hence subharmonic. Thus, $\gamma(z)$ is the limit of a monotone not increasing sequence of subharmonic functions

(5.38)
$$
\frac{1}{2^n}\mathbb{E}(\ln\|\Phi_\omega(z,2^n)\|) \searrow \gamma(z).
$$

The rest follows from our previous lemma. $\square$

This result looks pretty technical at the first sight, however, in connection with (5.34) it says that in order to show $\gamma(z) = \tilde{\gamma}(z)$ for some subharmonic function $\tilde{\gamma}(z)$ it suffices to show equality for $z \in \mathbb{C}\backslash\mathbb{R}$. This will be used in what follows since the case $z \in \mathbb{C}\backslash\mathbb{R}$ is usually easier to handle as we will see below.

For $z \in \mathbb{C}\backslash\mathbb{R}$ we know the asymptotic behavior of the solutions of $\tau_\omega u = zu$. There are solutions $u_{\omega,\pm}(z,n)$ which decay exponentially as $n \to \pm\infty$ and grow exponentially in the other direction. Hence it is not hard to see that

(5.39)
$$
\gamma(z) = \lim_{n\to\pm\infty} \frac{1}{2n}\ln\left(|u_\omega(z,n)|^2 + |u_\omega(z,n+1)|^2\right)
$$

for any solution $u_\omega(z,n)$ which is not a multiple of $u_{\omega,\pm}(z,n)$ and the limits exist whenever the limits in (5.27) exist. Based on this observation we obtain.

Theorem 5.11. Let $E_0 = \min \Sigma$ and $E_\infty = \max \Sigma$ with Σ from (5.9). For $z \in \mathbb{C}\backslash\mathbb{R}$ we have

$$\gamma(z) = \lim_{n\to\mp\infty} \frac{1}{|n|} \ln |u_{\omega,\pm}(z,n)| = -\lim_{n\to\pm\infty} \frac{1}{|n|} \ln |u_{\omega,\pm}(z,n)|$$
(5.40)
$$= -\mathbb{E}(\ln a_\omega) - \mathbb{E}(\ln |m_{\omega,\pm}(z)|)$$

almost surely. In addition,

(5.41) $$\gamma(z) = -\mathbb{E}(\ln a_\omega) + \ln|E_\infty - z| - \mathrm{Re}\int_{E_0}^{E_\infty} \frac{\xi(\lambda)}{\lambda - z} d\lambda,$$

where $\xi(\lambda) = \mathbb{E}(\xi_{\omega,\pm}(\lambda,n))$ and $\xi_{\omega,\pm}(\lambda,n)$ are the ξ-functions associated with the Weyl m-functions $m_{\omega,\pm}(z,n)$ (see Theorem B.11).

Proof. We first compute

$$\lim_{n\to\infty} \frac{1}{2n} \ln\left(|u_{\omega,+}(z,n)|^2 + |u_{\omega,+}(z,n+1)|^2\right) =$$

$$= \lim_{n\to\infty} \frac{1}{2n} \left(\ln(1 + |m_{\omega,+}(z,n)|^2) + 2\ln|u_{\omega,+}(z,n+1)|\right)$$

(5.42)
$$= \lim_{n\to\infty} \frac{1}{n} \sum_{j=0}^{n-1} \ln|a(j)m_{\omega,+}(z,j)|,$$

where we have used (2.3). Since $m_{\omega,+}(z,n)$ is measurable and satisfies (2.3) we can invoke Birkhoff's ergodic theorem to see that the limit is $\mathbb{E}(\ln a_\omega) + \mathbb{E}(\ln|m_{\omega,+}(z)|)$. Moreover, using (compare (6.56) below)

(5.43) $$\ln m_{\omega,+}(z,n) = -\ln(E_{\omega,\infty} - z) + \int_{E_{\omega,0}}^{E_{\omega,\infty}} \frac{\xi_{\omega,+}(\lambda,n)}{\lambda - z} d\lambda,$$

and Fubini's theorem we see

(5.44) $$\mathbb{E}(\ln|m_{\omega,+}(z)|) = -\ln|E_\infty - z| + \mathrm{Re}\int_{E_0}^{E_\infty} \frac{\mathbb{E}(\xi_{\omega,+}(\lambda,n))}{\lambda - z} d\lambda.$$

Computing the remaining three limits in a similar fashion the result follows from (5.39). □

Note that (5.41) implies

(5.45) $$\gamma(\lambda + i\varepsilon) = \gamma(\lambda - i\varepsilon), \qquad \varepsilon \in \mathbb{R}.$$

Next we want to shed some additional light on the function $\xi(\lambda)$ which appeared in the last theorem. To do this we first introduce the so-called density of states. We consider the finite operators $H_{\omega,0,n}$ with Dirichlet boundary conditions at 0 and n. The integrated density of states for this operator is defined by

(5.46) $$N_n^\omega(\lambda) = \frac{1}{n} \dim \mathrm{Ran} P_{(-\infty,\lambda)}(H_{\omega,0,n+1}) = \frac{1}{n} \mathrm{tr} P_{(-\infty,\lambda)}(H_{\omega,0,n+1}).$$

That is we take the number of eigenstates below λ and divide it by the total number of eigenstates. In particular, $N_n^\omega(\lambda)$ is zero below the spectrum of $H_{\omega,0,n+1}$, one above the spectrum of $H_{\omega,0,n+1}$, and increasing in between. By Theorem 4.7 we also have

(5.47) $$N_n^\omega(\lambda) = \frac{1}{n} \#_{(0,n+1)}(s_\omega(\lambda))$$

5.2. The Lyapunov exponent and the density of states

which is why $N_n^\omega(\lambda)$ is also called rotation number.

Our hope is that $N_n^\omega(\lambda)$ will tend to some limit as $n \to \infty$.

If the process

$$f_n(\omega) = \operatorname{tr} P_{(-\infty,\lambda)}(H_{\omega,0,n+1}) \tag{5.48}$$

were subadditive, the desired result would follow directly from Kingman's Theorem. Unfortunately this is not true! To see this observe

$$H_{\omega,0,n+m+1} = H_{\omega,0,m+1} \oplus H_{T^{-m}\omega,0,n+1} + M_{\omega,m}, \tag{5.49}$$

where $M_{\omega,m} = a_\omega(m)(\langle \delta_m, .\rangle \delta_{m+1} + \langle \delta_{m+1}, .\rangle \delta_m)$. In the simplest case $m = n = 1$ the above equation reads

$$\begin{pmatrix} b_1 & a_1 \\ a_1 & b_2 \end{pmatrix} = \begin{pmatrix} b_1 & 0 \\ 0 & 0 \end{pmatrix} \oplus \begin{pmatrix} 0 & 0 \\ 0 & b_2 \end{pmatrix} + a_1 \begin{pmatrix} 0 & 1 \\ 1 & 0 \end{pmatrix}. \tag{5.50}$$

Let $b_+ = \max(b_1, b_2)$ and $b_- = \min(b_1, b_2)$. Then the eigenvalues of this matrix are $b_- - c$ and $b_+ + c$ for suitable $c > 0$. Hence

$$f_2(\omega) = f_1(\omega) + f_1(T^{-1}\omega) + \begin{cases} 0, & \lambda \leq b_- - c \\ 1, & b_- - c < \lambda \leq b_- \\ 0, & b_- < \lambda \leq b_+ \\ -1, & b_+ < \lambda \leq b_+ + c \\ 0, & b_+ + c < \lambda \end{cases} \tag{5.51}$$

shows that neither $f_n(\omega)$ nor $-f_n(\omega)$ is subadditive in general.

However, since $M_{\omega,m}$ is a rank two perturbation, we have at least

$$f_{n+m}(\omega) \leq f_m(\omega) + f_n(T^{-m}\omega) + 2 \tag{5.52}$$

and hence $\tilde{f}_n(\omega) = f_n(\omega) + 2$ is subadditive. Clearly $f_n(\omega)/n$ converges if and only if $\tilde{f}_n(\omega)/n$ converges in which case the limits are equal. Summarizing,

$$\lim_{n \to \infty} N_n^\omega(\lambda) = N(\lambda) \tag{5.53}$$

almost surely. The quantity $N(\lambda)$ is called **integrated density of states**. Because of (5.47) it is also known as **rotation number**. Some of its properties are collected in the following lemma.

Lemma 5.12. *There is a set $\tilde{\Omega} \subseteq \Omega$ of probability one such that for continuous f,*

$$\lim_{n \to \infty} \int f(\lambda) dN_n^\omega(\lambda) = \int f(\lambda) dN(\lambda) = \mathbb{E}(\langle \delta_0, f(H_\omega)\delta_0\rangle), \quad \omega \in \tilde{\Omega}. \tag{5.54}$$

In other words, dN_n^ω converges weakly to dN almost surely. Moreover, $N(\lambda)$ is continuous and (5.54) also holds for $f = \chi_{(-\infty,\lambda_0)}$.

Proof. Note that $f \mapsto \mathbb{E}(\langle \delta_0, f(H_\omega)\delta_0\rangle)$ is a positive linear functional on the space $C([-3M, 3M])$ and hence by the Riez-Markov Theorem there is a (positive probability) measure $d\tilde{N}$ such that

$$\int f(\lambda) d\tilde{N}(\lambda) = \mathbb{E}(\langle \delta_0, f(H_\omega)\delta_0\rangle). \tag{5.55}$$

We begin by showing that (5.54) holds with dN replaced by $d\tilde{N}$. Let us choose $f(\lambda) = \lambda^k$, $k \in \mathbb{N}_0$, first and consider the process

(5.56)
$$f_n(\omega) = \operatorname{tr}(H_{\omega,0,n+1})^k = \sum_{m=1}^{n} \langle \delta_m, (H_{\omega,0,n+1})^k \delta_m \rangle.$$

Since each summand depends on n (unless $k = 0, 1$) it is not clear that this process is (sub)additive. Hence we will try to replace it by

(5.57)
$$\tilde{f}_n(\omega) = \operatorname{tr}(H_\omega)^k = \sum_{m=1}^{n} \langle \delta_m, (H_\omega)^k \delta_m \rangle,$$

which is additive. By Birkhoff's theorem we have $\lim_{n \to \infty} \tilde{f}_n(\omega)/n = \int \lambda^k d\tilde{N}(\lambda)$ for ω in a set $\tilde{\Omega}_k$ of probability one. Now what about the error we make when replacing f_n by $\tilde{f}_n$? The matrix elements $\langle \delta_m, (H_\omega)^k \delta_m \rangle$ and $\langle \delta_m, (H_\omega)^k \delta_m \rangle$ only differ for $m = n, n-1, \ldots, n - n(k)$ and the difference can be estimated using $a_\omega \leq M$ and $|b_\omega| \leq M$. Hence we obtain

(5.58)
$$|\tilde{f}_n(\omega) - f_n(\omega)| \leq C_k(M)$$

and so $f_n(\omega)/n$ converges if and only if $\tilde{f}_n(\omega)/n$ converges in which case the limits are equal. Choosing $\tilde{\Omega} = \cap_{k=0}^{\infty} \tilde{\Omega}_k$ we see that (5.54) holds if f is a polynomial.

Now let f be continuous and choose a polynomial $f_\varepsilon(\lambda)$ such that the difference $|f(\lambda) - f_\varepsilon(\lambda)|$ is at most $\varepsilon/2$ for all $\lambda \in [-3M, 3M]$. Then we have

(5.59)
$$\left| \int f(\lambda) d\tilde{N}(\lambda) - \int f(\lambda) dN_n^\omega(\lambda) \right| \leq \left| \int f_\varepsilon(\lambda) dN_n^\omega(\lambda) - \int f_\varepsilon(\lambda) d\tilde{N}(\lambda) \right| +$$
$$+ \int |f(\lambda) - f_\varepsilon(\lambda)| dN_n^\omega(\lambda) + \int |f(\lambda) - f_\varepsilon(\lambda)| d\tilde{N}(\lambda).$$

Performing the limit $n \to \infty$ we see

(5.60)
$$\lim_{n \to \infty} \left| \int f(\lambda) d\tilde{N}(\lambda) - \int f(\lambda) dN_n^\omega(\lambda) \right| \leq \varepsilon$$

and letting $\varepsilon \downarrow 0$ establishes (5.54) for continuous f.

Let $f_\varepsilon(\lambda) = \varepsilon^2/(\lambda^2 + \varepsilon^2)$. Using dominated convergence we obtain

(5.61)
$$\tilde{N}(\{\lambda_0\}) = \lim_{\varepsilon \downarrow 0} \int f_\varepsilon(\lambda) d\tilde{N}(\lambda) = \lim_{\varepsilon \downarrow 0} \mathbb{E}(\langle \delta_0, f_\varepsilon(H_\omega) \delta_0 \rangle)$$
$$= \mathbb{E}(\lim_{\varepsilon \downarrow 0} \langle \delta_0, f_\varepsilon(H_\omega) \delta_0 \rangle) = \mathbb{E}(\langle \delta_0, P_{\{\lambda_0\}}(H_\omega) \delta_0 \rangle) = 0,$$

where the last equality follows from Theorem 5.5. This shows that $\tilde{N}(\lambda)$ is continuous.

To show (5.54) for $f(\lambda) = \chi_{(-\infty, \lambda_0)}(\lambda)$ we approximate f by continuous functions f_ε given by $f_\varepsilon(\lambda) = 1 - (\lambda - \lambda_0)/\delta$ for $\lambda_0 < \lambda < \lambda_0 + \delta$ and $f_\varepsilon(\lambda) = f(\lambda)$ else. The parameter δ is chosen such that $\tilde{N}((\lambda_0 - \delta, \lambda_0 - \delta)) \leq \varepsilon/2$ (this is possible since $\tilde{N}$ is continuous). Now observe

(5.62)
$$\lim_{n \to \infty} \int |f(\lambda) - f_\varepsilon(\lambda)| dN_n^\omega(\lambda) \leq \lim_{n \to \infty} \int g_\varepsilon(\lambda) dN_n^\omega(\lambda) = \int g_\varepsilon(\lambda) d\tilde{N}(\lambda) \leq \frac{\varepsilon}{2},$$

where $g_\varepsilon(\lambda) = (1 - |\lambda - \lambda_0|/\delta) \chi_{(\lambda_0 - \delta, \lambda_0 + \delta)}(\lambda)$, and proceed as before.

5.2. The Lyapunov exponent and the density of states

Equation (5.54) for $f = \chi_{(-\infty,\lambda)}$ finally tells us $N(\lambda) = \tilde{N}(\lambda)$. □

Remark 5.13. (i). The proof of the previous lemma shows that

$$(5.63) \quad N(\lambda) = \lim_{n \to \infty} \frac{1}{n} \text{tr}(\chi_{(0,n+1)} P_{(-\infty,\lambda)}(H_\omega)) = \frac{1}{n} \sum_{m=1}^{\infty} \langle \delta_m, P_{(-\infty,\lambda)}(H_\omega) \delta_m \rangle,$$

where the right hand side is an additive process.
(ii). We can replace $H_{\omega,0,n+1}$ by $H_{\omega,-n-1,n+1}$, that is,

$$(5.64) \quad N(\lambda) = \lim_{n \to \infty} \frac{1}{2n+1} \text{tr} P_{(-\infty,\lambda)}(H_{\omega,-n-1,n+1})),$$

or we could choose different boundary conditions.

In particular,

$$(5.65) \quad N(\lambda) = \lim_{n \to \infty} \frac{1}{n} \#_{(0,n+1)}(u_\omega(\lambda))$$

for any real (nontrivial) solution $\tau_\omega u_\omega(\lambda) = \lambda u_\omega(\lambda)$ (compare (4.18)).

Note that $N(\lambda)$ can also be viewed as the average of the spectral measures $d\rho_{\omega,n,n}(\lambda)$ since by (5.54) for $f = \chi_{(-\infty,\lambda)}$,

$$(5.66) \quad N(\lambda) = \mathbb{E}(\langle \delta_0, P_{(-\infty,\lambda)}(H_\omega) \delta_0 \rangle) = \mathbb{E}(\rho_{\omega,n,n}((-\infty,\lambda))).$$

In particular, we obtain the following corollary.

Corollary 5.14. *We have $\sigma(dN) = \Sigma$. Moreover, $N(\lambda) = 0$ for $\lambda \leq E_0$ and $N(\lambda) = 1$ for $\lambda \geq E_\infty$.*

Another consequence is the **Thouless formula**.

Theorem 5.15. *For any $z \in \mathbb{C}$ we have*

$$(5.67) \quad \gamma(z) = -\mathbb{E}(\ln a_\omega) + \int_{E_0}^{E_\infty} \ln|\lambda - z| dN(\lambda)$$

and $N(\lambda) = \xi(\lambda)$.

Proof. Invoking (1.65) we get

$$(5.68) \quad \frac{1}{n} \ln|s_\omega(\lambda,n)| = \frac{1}{n} \sum_{m=1}^{n} \ln a_\omega(m) + \int \ln|\lambda - z| dN_n^\omega(\lambda)$$

for $z \in \mathbb{C}\backslash\mathbb{R}$. Using (5.39) with $u_\omega = s_\omega$ to compute $\gamma(z)$ proves (5.67) for $z \in \mathbb{C}\backslash\mathbb{R}$. Since both quantities are subharmonic (replace $\ln|\lambda - z|$ by $\ln(\varepsilon + |\lambda - z|)$ and let $\varepsilon \downarrow 0$), equality holds for all $z \in \mathbb{C}$. Comparing (5.41) and (5.67) we obtain $N(\lambda) = \xi(\lambda)$ after an integration by parts. □

There are a few more useful formulas. Set

$$(5.69) \quad \overline{g}(z) = \mathbb{E}(g_\omega(z,n)) = \mathbb{E}(g_\omega(z,0)) = \int \frac{dN(\lambda)}{\lambda - z},$$

then

$$(5.70) \quad \frac{\partial}{\partial \varepsilon} \gamma(\lambda + i\varepsilon) = \int \frac{\varepsilon\, dN(x)}{(x-\lambda)^2 - \varepsilon^2} = \text{Im}(\overline{g}(\lambda + i\varepsilon)), \quad |\varepsilon| > 0.$$

In particular, the moments of dN are the expectations of the coefficients of the Laurent expansion of $g_\omega(z,0)$ around infinity (compare Section 6.1 below)

(5.71) $$\mathbb{E}(\langle \delta_0, (H_\omega)^j \delta_0 \rangle) = \int \lambda^j dN(\lambda).$$

Explicitly we have ($\int dN(\lambda) = 1$)

(5.72) $$\int \lambda\, dN(\lambda) = \mathbb{E}(b_\omega(0)), \quad \int \lambda^2 dN(\lambda) = 2\mathbb{E}(a_\omega(0)^2) + \mathbb{E}(b_\omega(0)^2), \text{ etc..}$$

In addition, solving (2.11) for the $\frac{1}{m_{\omega,\pm}(z,n)}$ term, taking imaginary parts, logs, and expectations we find

(5.73) $$2\gamma(\lambda + i\varepsilon) = \mathbb{E}\left(\ln(1 + \frac{\varepsilon}{a_\omega(\begin{smallmatrix}0\\-1\end{smallmatrix})\mathrm{Im}(m_{\omega,\pm}(\lambda + i\varepsilon))}) \right), \quad |\varepsilon| > 0.$$

Finally, let us try to apply the results found in this section. To begin with, note that we know $\underline{\gamma}_\omega^\pm(z) = \overline{\gamma}_\omega^\pm(z) = \gamma(z)$ a.s. only for fixed $z \in \mathbb{C}$. On the other hand, to say something about H_ω we need information on the (upper, lower) Lyapunov exponents of H_ω for all (or at least *most* z).

Lemma 5.16. *The Lebesgue measure of the complement of the set $\{\lambda \in \mathbb{R} | \underline{\gamma}_\omega^\pm(\lambda) = \overline{\gamma}_\omega^\pm(\lambda) = \gamma(\lambda)\}$ is zero almost surely. The same is true for $\lambda \in \mathbb{C}$. Moreover,*

(5.74) $$\overline{\gamma}_\omega^\pm(z) \leq \gamma(z) \text{ for all } z \text{ a.s..}$$

Proof. The set $L = \{(\lambda, \omega) | \underline{\gamma}_\omega^\pm(z) = \overline{\gamma}_\omega^\pm(z) = \gamma(z)\}$ is measurable in the product space $\mathbb{R} \times \Omega$. The bad set is $B = (\mathbb{R} \times \Omega) \backslash L$ and the corresponding sections are $B_\omega = \{\lambda | (\lambda, \omega) \in B\}$, $B_\lambda = \{\omega | (\lambda, \omega) \in B\}$. Since we know $\mu(B_\lambda) = 0$ we obtain from Fubini's theorem

(5.75) $$(\lambda \times \mu)(B) = \int_\Omega |B_\omega| d\mu(\omega) = \int_\mathbb{R} \mu(B_\lambda) d\lambda = 0$$

that $|B_\omega| = 0$ a.s. as claimed. Similar for $\lambda \in \mathbb{C}$. To prove the remaining assertion, we use that $\overline{\gamma}_\omega^\pm(z)$ is submean

(5.76) $$\begin{aligned}\overline{\gamma}_\omega^\pm(z_0) &\leq \frac{1}{2\pi \mathrm{i} r^2} \int_{|z-z_0| \leq r} \overline{\gamma}_\omega^\pm(z)\, d\overline{z} \wedge dz \\ &= \frac{1}{2\pi \mathrm{i} r^2} \int_{|z-z_0| \leq r} \gamma(z)\, d\overline{z} \wedge dz = \gamma(z_0).\end{aligned}$$

$\square$

Now we are ready to prove the following theorem which should be compared with Theorem 3.16.

Theorem 5.17. *The absolutely continuous spectrum of H_ω is given by*

(5.77) $$\Sigma_{ac} = \overline{\{\lambda \in \mathbb{R} | \gamma(\lambda) = 0\}}^{ess}$$

almost surely.

Proof. Combining Theorem 3.16 and the previous lemma we have immediately $\Sigma_{ac} \subseteq \overline{\{\lambda \in \mathbb{R} | \gamma(\lambda) = 0\}}^{ess}$. The converse is a little harder.

5.2. The Lyapunov exponent and the density of states

Abbreviate $N = \{\lambda \in \mathbb{R} | \gamma(\lambda) = 0\}$. Using $x(1 + x/2)^{-1} \leq \ln(1 + x)$, $x \geq 0$, we obtain from (5.73)

$$\mathbb{E}\left(\frac{\varepsilon}{n_{\omega,\pm}(\lambda + i\varepsilon)}\right) \leq \mathbb{E}\left(\ln\left(1 + \frac{\varepsilon}{n_{\omega,\pm}(\lambda + i\varepsilon) - \frac{\varepsilon}{2}}\right)\right) = 2\gamma(\lambda + i\varepsilon), \quad (5.78)$$

where $n_{\omega,\pm}(z) = a\binom{0}{-1}\mathrm{Im}(m_{\omega,\pm}(z)) + \frac{\mathrm{Im}(z)}{2}$. Moreover, we have by (5.70) and Fatou's lemma

$$\mathbb{E}\left(\liminf_{\varepsilon\downarrow 0} \frac{1}{n_{\omega,\pm}(\lambda + i\varepsilon)}\right) \leq \liminf_{\varepsilon\downarrow 0} \mathbb{E}\left(\frac{1}{n_{\omega,\pm}(\lambda + i\varepsilon)}\right)$$
$$= 2\liminf_{\varepsilon\downarrow 0} \frac{\gamma(\lambda + i\varepsilon)}{\varepsilon} = \mathrm{Im}(\overline{g}(\lambda + i0)) < \infty \quad (5.79)$$

for $\lambda \in N$ for which the limit $\lim_{\varepsilon\downarrow 0} \mathrm{Im}(\overline{g}(\lambda + i\varepsilon))$ exists and is finite. Denote this set by N' and observe $|N \setminus N'| = 0$ (see Theorem B.8 (ii)). Thus we see $\mu(\{\omega | \limsup_{\varepsilon\downarrow 0} \mathrm{Im}(m_{\omega,\pm}(\lambda + i\varepsilon)) = 0\}) = 0$ for $\lambda \in N'$. And as in the proof of the previous lemma this implies $|\{\lambda | \limsup_{\varepsilon\downarrow 0} \mathrm{Im}(m_{\omega,\pm}(\lambda + i\varepsilon)) = 0\} \cap N| = 0$ almost surely. A look at Lemma B.7 completes the proof. $\square$

Next, let us extend this result.

Lemma 5.18. *If $\gamma(\lambda) = 0$ for λ in a set N of positive Lebesgue measure, then we have*

$$a(0)\mathrm{Im}(m_{\omega,+}(\lambda + i0)) = a(-1)\mathrm{Im}(m_{\omega,-}(\lambda + i0)),$$
$$\mathrm{Re}\left(\frac{-1}{g_\omega(\lambda + i0, 0)}\right) = 0. \quad (5.80)$$

almost surely for a.e. $\lambda \in N$.

Proof. Let us abbreviate $n_\pm = n_{\omega,\pm}(\lambda + i\varepsilon)$ (see the notation from the previous proof) and $n = n_\omega(\lambda + i\varepsilon) = -g_\omega(\lambda + i\varepsilon, 0)^{-1}$. An elementary calculation using $\mathrm{Im}(n) = n_+ + n_-$ verifies

$$\left(\frac{1}{n_+} + \frac{1}{n_-}\right)\frac{(n_+ - n_-)^2 + (\mathrm{Re}(n))^2}{|n|^2} = \left(\frac{1}{n_+} + \frac{1}{n_-}\right)\left(1 - 4\frac{n_+ n_-}{|n|^2}\right)$$
$$= \frac{1}{n_+} + \frac{1}{n_-} + 4\mathrm{Im}\left(\frac{1}{n}\right). \quad (5.81)$$

Invoking (5.70) and (5.78) we see

$$\mathbb{E}\left(\left(\frac{1}{n_+} + \frac{1}{n_-}\right)\frac{(n_+ - n_-)^2 + (\mathrm{Re}(n))^2}{|n|^2}\right) \leq 4\left(\frac{\gamma(\lambda + i\varepsilon)}{\varepsilon} - \frac{\partial \gamma(\lambda + i\varepsilon)}{\partial \varepsilon}\right). \quad (5.82)$$

Moreover, since for a.e. λ the limits $n_{\omega,\pm}(\lambda + i0)$ and $n_\omega(\lambda + i0)$ exist by Theorem B.8 (ii), we are done. $\square$

The condition (5.80) says precisely that $\tilde{m}_{\omega,+}(\lambda + i0) = -\overline{\tilde{m}_{\omega,+}(\lambda + i0)}$. This should be compared with the reflectionless condition in Lemma 8.1 (iii) below.

Theorem 5.19. *If $\gamma(\lambda)$ vanishes on an open interval $I = (\lambda_1, \lambda_2)$, the spectrum of H_ω is purely absolutely continuous on I almost surely, that is, $(\Sigma_{sc} \cup \Sigma_{pp}) \cap I = \emptyset$.*

Proof. By (5.80) and Lemma B.4 we see that $-g_\omega(\lambda + i0, 0)^{-1}$ is holomorphic on I and nonzero. So the same is true for $g_\omega(\lambda + i0, 0)$ and the result follows from (3.41). $\square$

As a consequence we have that the class of random Jacobi operators where $\gamma(\lambda)$ vanishes on a set of positive Lebesgue measure is rather special.

Theorem 5.20. *If γ vanishes on a set of positive Lebesgue measure, then for almost surely $\omega \in \Omega$, the sequences $(a_\omega(n), b_\omega(n))_{n \in \mathbb{Z}}$ are determined by either $(a_\omega(n), b_\omega(n))_{n \geq n_0}$ or $(a_\omega(n), b_\omega(n))_{n \leq n_0}$ for any $n_0 \in \mathbb{Z}$.*

Proof. It is no restriction to assume $n_0 = 0$. By (5.80) and Lemma B.4 we learn that $m_{\omega,+}(z)$ uniquely determines $m_{\omega,-}(z)$ and vice versa. Hence the claim follows from Theorem 2.29. □

Sequences $(a_\omega(n), b_\omega(n))_{n \in \mathbb{Z}}$ of the above type are referred to as **deterministic sequences**. Hence the theorem says that for nondeterministic sequences we have $\Sigma_{ac} = \emptyset$. In particular this holds for the Anderson model. An example for deterministic sequences are almost periodic ones which will be discussed in the next section.

5.3. Almost periodic Jacobi operators

Now we will consider the case where the coefficients a, b are almost periodic sequences. We review some basic facts first.

Let $c^0 = (a^0, b^0) \in \ell^\infty(\mathbb{Z}, \mathbb{R}_+) \times \ell^\infty(\mathbb{Z})$ and let us abbreviate $c^m = S^m c^0 = (S^m a^0, S^m b^0)$ (S^m being the shift operator from (1.8)). We set $\Omega(c^0) = \{c^m\}_{m \in \mathbb{Z}}$ and call the closure the hull of c^0, $\mathrm{hull}(c^0) = \overline{\Omega(c^0)}$. The sequence c^0 is called **almost periodic** if its hull is compact. Note that compactness implies that (H.5.1) holds for almost periodic sequences.

For an almost periodic sequence c^0 we can define a product

$$(5.83) \qquad \begin{aligned} \circ : \Omega(c^0) \times \Omega(c^0) &\to \Omega(c^0) \\ (c^m, c^n) &\mapsto c^m \circ c^n = c^{m+n} \end{aligned},$$

which makes $\Omega(c^0)$ an abelian group. Moreover, this product is continuous since

$$(5.84) \qquad \begin{aligned} \|c^{n_1} \circ c^{m_1} - c^{n_0} \circ c^{m_0}\| &\leq \|c^{n_1+m_1} - c^{n_1+m_0}\| + \|c^{n_1+m_0} - c^{n_0+m_0}\| \\ &= \|c^{m_1} - c^{m_0}\| + \|c^{n_1} - c^{n_0}\|, \end{aligned}$$

where we have used the triangle inequality and translation invariance of the norm in $\ell^\infty(\mathbb{Z}) \times \ell^\infty(\mathbb{Z})$. Hence our product can be uniquely extended to the hull of c^0 such that $(\mathrm{hull}(c^0), \circ)$ becomes a compact topological group. On such a compact topological group we have a unique invariant

$$(5.85) \qquad \int f(cc') d\mu(c') = \int f(c') d\mu(c'), \qquad c \in \mathrm{hull}(c^0),$$

Baire measure μ, the **Haar measure**, if we normalize $\mu(\mathrm{hull}(c^0)) = 1$. The invariance property (5.85) implies

$$(5.86) \qquad \mu(S^n F) = \mu(F), \qquad F \subseteq \mathrm{hull}(c^0),$$

since $S^n c = c^n \circ c$. If we knew that μ is ergodic, we could apply the theory of the first section with $\Omega = \mathrm{hull}(c^0)$. Now let $S^n F = F$, $n \in \mathbb{Z}$, and consider

$$(5.87) \qquad \tilde{\mu}(G) = \int_G \chi_F d\mu.$$

5.3. Almost periodic Jacobi operators

Then $\tilde{\mu}$ also satisfies (5.85) and hence must be equal to μ up to normalization by uniqueness of the Haar measure. Hence $\tilde{\mu}(G) = \tilde{\mu}(\text{hull}(c^0))\mu(G)$ and using $\tilde{\mu}(F) = \tilde{\mu}(\text{hull}(c^0)) = \mu(F)$ we obtain for $G = F$ that $\mu(F) = \mu(F)^2$ implying $\mu(F) \in \{0,1\}$. Thus μ is ergodic.

As a consequence we note

Theorem 5.21. *Let (a,b) be almost periodic. For all Jacobi operators corresponding to sequences in the hull of (a,b), the spectrum is the same and the discrete spectrum is empty.*

Proof. By Theorem 5.3 the claim holds for a set of full Haar measure. Since such a set is dense we can approximate any point (a,b) in the hull by a sequence (a_n, b_n) within this set in the sup norm. Hence the corresponding operators H_n converge to H in norm and the result follows. $\square$

Since the absolutely continuous part is not stable under small perturbations, Theorem 5.4 cannot be directly extended to the the entire hull of an almost periodic sequence.

Theorem 5.22. *Let (a,b) be almost periodic. For all Jacobi operators corresponding to sequences in a subset of full Haar measure of $\text{hull}((a,b))$, the pure point, singular continuous, absolutely continuous spectrum is the same, respectively.*

One way of constructing examples of almost periodic operators is to choose periodic functions $p_1(x)$, $p_2(x)$ (with period one) and consider

(5.88) $\quad (a_\theta(n), b_\theta(n)) = (p_1(2\pi\alpha_1 n + \theta_1), p_1(2\pi\alpha_2 n + \theta_2)), \quad \alpha \in \mathbb{R}^2, \theta \in [0, 2\pi)^2.$

Note that such sequences are even quasi periodic by construction.

We will restrict our attention to the most prominent model where $p_1(x) = 1$ and $p_2(x) = \beta \cos(x)$. The corresponding operator H_θ is known as **almost Mathieu operator** and given by

(5.89) $\quad (a_\theta(n), b_\theta(n)) = (1, \beta \cos(2\pi\alpha n + \theta)), \quad \alpha \in \mathbb{R}, \theta \in [0, 2\pi).$

Since $b_{\theta+\pi} = -b_\theta$ we can assume $\beta > 0$ without loss of generality. We need to distinguish two cases.

If $\alpha \in \mathbb{Q}$, say $\alpha = \frac{M}{N}$ (with N minimal), then $b_\theta(n+N) = b_\theta(n)$ and

(5.90) $\quad \text{hull}(1, b_\theta) = \{(1, b_{\theta+2\pi\alpha n})\}_{n=1}^N.$

The underlying group is $\mathbb{Z}_N$ and the Haar measure is the normalized counting measure. This is a special case of periodic sequences and will be investigated in Chapter 7.

The case we are interested in here is $\alpha \notin \mathbb{Q}$. Then

(5.91) $\quad \text{hull}(1, b_\theta) = \{(1, b_\theta) | \theta \in [0, 2\pi)\},$

where the right hand side is compact as the continuous image of the circle. The underlying group is $S^1 = \{e^{i\theta} | \theta \in [0, 2\pi)\}$ and the Haar measure is $\frac{d\theta}{2\pi}$. Note that the shift satisfies $S^m b_\theta = b_{\theta+2\pi\alpha m}$.

We start our investigation with a bound for the Lyapunov exponent from below.

Lemma 5.23. *Suppose $\alpha \notin \mathbb{Q}$, then the Lyapunov exponent of H_θ satisfies*

(5.92) $\quad\quad\quad\quad\quad\quad\quad \gamma(z) \geq \ln \frac{\beta}{2}.$

Proof. Observe

(5.93) $$b_\theta(n) = \frac{\beta}{2}(e^{2\pi i\alpha n}w + e^{2\pi i\alpha n}w^{-1}), \quad w = e^{2\pi i\theta},$$

and (cf. (1.29))

(5.94) $$U_\theta(z,n) = \frac{1}{w}\begin{pmatrix} 0 & w \\ -w & zw - \frac{\beta}{2}(e^{2\pi i\alpha n}w^2 + e^{2\pi i\alpha n}) \end{pmatrix}.$$

Hence $f_n(w) = w^n \Phi_\theta(z,n)$ is holomorphic with respect to w and we have by Jensen's inequality that

(5.95) $$\frac{1}{2\pi}\int_0^{2\pi} \ln\|f_n(e^{2\pi i\theta})\|d\theta \geq \ln\|f_n(0)\| = n\ln\frac{\beta}{2}.$$

Using (5.26) we compute

(5.96) $$\gamma(z) = \inf_{n\in\mathbb{N}}\frac{1}{n}\mathbb{E}(\ln\|f_n\|) = \inf_{n\in\mathbb{N}}\frac{1}{n}\int_0^{2\pi}\ln\|f_n(e^{2\pi i\theta})\|\frac{d\theta}{2\pi} \geq \ln\frac{\beta}{2}$$

since $\|\Phi_\theta(z,n)\| = \|f_n(w)\|$. $\square$

In particular, applying Theorem 5.17 we obtain

Theorem 5.24. *If $\beta > 2$ and $\alpha \notin \mathbb{Q}$, then $\sigma_{ac}(H_\theta) = \emptyset$ for a.e. θ.*

Moreover, we can even exclude point spectrum if α is well approximated by rational numbers and hence b_θ is well approximated by periodic sequences. The key observation is the following lemma.

Lemma 5.25. *Let $a(n) \geq \delta^{-1} > 0$. Suppose there are periodic sequences a_k and b_k, $k \in \mathbb{N}$, with (not necessarily minimal) period N_k tending to infinity as $k \to \infty$ such that*

(5.97) $$\sup_{|n|\leq 2N_k+1}|a(n) - a_k(n)| \leq (C_k)^{N_k}, \quad \sup_{|n|\leq 2N_k}|b(n) - b_k(n)| \leq (C_k)^{N_k}$$

for some positive sequence C_k tending to zero as $k \to \infty$. Then any (nontrivial) solution of $\tau u = zu$ satisfies (see the notation in (1.28))

(5.98) $$\limsup_{n\to\infty}\frac{\|\underline{u}(n)\| + \|\underline{u}(-n)\|}{\|\underline{u}(0)\|} \geq 1.$$

Proof. It is no restriction to assume $\delta^{-1} \leq |a_k(n)| \leq \sup|a(n)|$ and $\inf b(n) \leq b_k(n) \leq \sup b(n)$. Setting $M(z) = \delta\max(\|a\|_\infty, \|b-z\|_\infty)$ we obtain for the transfer matrices

(5.99) $$\|\Phi(z,n,m)\| \leq M(z)^{|n-m|}, \quad \|\Phi_k(z,n,m)\| \leq M(z)^{|n-m|}.$$

Now using

(5.100) $$\Phi(z,n) - \Phi_k(z,n) = \sum_{m=1}^{n}{}^* \Phi_k(z,n,m)(U(z,m) - U_k(z,m))\Phi(z,m-1)$$

we see

(5.101) $$\sup_{|n|\leq 2N_k}\|\Phi(z,n) - \Phi_k(z,n)\| \leq 2\delta N_k M(z)(M(z)^2 C_k)^{N_k}$$

which tends to zero as $k \to \infty$.

5.3. Almost periodic Jacobi operators

Abbreviate $x_j = \|\underline{u}_k(jN_k)\| = \|\Phi_k(z, jN_k)\underline{u}(0)\|$. Periodicity of a_k, b_k implies $\det \Phi_k(z, N_k) = 1$ and $\Phi_k(z, jN_k) = \Phi_k(z, N_k)^j$ (compare also Section 7.1). Hence using the characteristic equation $(A^2 + \text{tr}(A)A + \det(A) = 0)$ and the triangle inequality it is not hard to show (set $A = \Phi_k(z, N_k)^{\pm 1}$) that $x_1 + x_{-1} \geq \tau$ and $x_{\pm 2} + \tau x_{\pm 1} \geq 1$, where $\tau = \text{tr}\Phi_k(z, N_k)$. Hence $\max(x_2 + x_{-2}, x_1 + x_{-1}) \geq 1$ proving the claim. $\square$

In order for b_θ to satisfy the assumptions of the previous lemma we need an appropriate condition for α. A suitable one is that α is a **Liouville number**, that is, $\alpha \in \mathbb{R}\backslash\mathbb{Q}$ and for any $k \in \mathbb{N}$ there exists $M_k, N_k \in \mathbb{N}$ such that

$$(5.102) \qquad \left|\alpha - \frac{M_k}{N_k}\right| \leq \frac{1}{k^{N_k}}.$$

Theorem 5.26. Suppose α is a Liouville number, then $\sigma_{pp}(H_\theta) = \emptyset$.

Proof. For $b_{\theta,k}(n) = \beta\cos(2\pi\frac{M_k}{N_k}n + \theta)$ we estimate

$$\sup_{|n|\leq 2N_k} |b_\theta(n) - b_{\theta,k}(n)| = \beta \sup_{|n|\leq 2N_k} |\cos(2\pi\alpha n + \theta) - \cos(2\pi\frac{M_k}{N_k}n + \theta)|$$

$$(5.103) \qquad \leq 2\pi\beta \sup_{|n|\leq 2N_k} |n|\left|\alpha - \frac{M_k}{N_k}\right| \leq \frac{4\pi\beta N_k}{k^{N_k}}$$

and the result follows from the previous lemma. $\square$

Combining Theorem 5.24 and Theorem 5.26 we obtain the

Corollary 5.27. *The spectrum of H_θ is purely singular continuous for a.e. θ if α is a Liouville number and $\beta > 2$.*

Chapter 6

Trace formulas

Trace formulas are an important tool for both spectral and inverse spectral theory. As a first ingredient we consider asymptotic expansions which have already turned out useful in Section 2.7 and are of independent interest. In the remaining section we will develop the theory of xi functions and trace formulas based on Krein's spectral shift theory.

6.1. Asymptotic expansions

Our aim is to derive asymptotic expansions for $g(z, n) = G(z, n, n)$ and $\gamma^\beta(z, n)$. Since both quantities are defined as expectations of resolvents

(6.1) $$\begin{aligned} g(z, n) &= \langle \delta_n, (H-z)^{-1} \delta_n \rangle, \\ \gamma^\beta(z, n) &= (1+\beta^2)\langle \delta_n^\beta, (H-z)^{-1} \delta_n^\beta \rangle - \frac{\beta}{a(n)}, \end{aligned}$$

all we have to do is invoking **Neumann's expansion** for the resolvent

(6.2) $$\begin{aligned} (H-z)^{-1} &= -\sum_{j=0}^{N-1} \frac{H^j}{z^{j+1}} + \frac{1}{z^N} H^N (H-z)^{-1} \\ &= -\sum_{j=0}^{\infty} \frac{H^j}{z^{j+1}}, \quad |z| > \|H\|. \end{aligned}$$

In summary, we obtain the following result as simple consequence (see also Section 2.1) of the spectral theorem.

Lemma 6.1. *Suppose $\delta \in \ell^2(\mathbb{Z})$ with $\|\delta\| = 1$. Then*

(6.3) $$g(z) = \langle \delta, (H-z)^{-1} \delta \rangle$$

is Herglotz, that is,

(6.4) $$g(z) = \int_\mathbb{R} \frac{1}{\lambda - z} d\rho_\delta(\lambda),$$

where $d\rho_\delta(\lambda) = d\langle\delta, P_{(-\infty,\lambda]}(H)\delta\rangle$ is the spectral measure of H associated to the sequence δ. Moreover,

(6.5) $$\mathrm{Im}(g(z)) = \mathrm{Im}(z)\|(H-z)^{-1}\delta\|^2,$$
$$g(\bar{z}) = \overline{g(z)}, \qquad |g(z)| \le \|(H-z)^{-1}\| \le \frac{1}{|\mathrm{Im}(z)|},$$

and

(6.6) $$g(z) = -\sum_{j=0}^{\infty} \frac{\langle\delta, H^j\delta\rangle}{z^{j+1}}.$$

We note that solving (1.107) for $h(z,n)$ shows that $h(z,n)$ can be written as the difference of two Herglotz functions.

Lemma 6.1 implies the following **asymptotic expansions** for $g(z,n)$, $h(z,n)$, and $\gamma^\beta(z,n)$.

Theorem 6.2. *The quantities $g(z,n)$, $h(z,n)$, and $\gamma^\beta(z,n)$ have the following Laurent expansions*

$$g(z,n) = -\sum_{j=0}^{\infty} \frac{g_j(n)}{z^{j+1}}, \quad g_0 = 1,$$

$$h(z,n) = -1 - \sum_{j=0}^{\infty} \frac{h_j(n)}{z^{j+1}}, \quad h_0 = 0,$$

(6.7) $$\gamma^\beta(z,n) = -\frac{\beta}{a(n)} - \sum_{j=0}^{\infty} \frac{\gamma_j^\beta(n)}{z^{j+1}}, \quad \gamma_0^\beta = 1 + \beta^2.$$

Moreover, the coefficients are given by

$$g_j(n) = \langle\delta_n, H^j\delta_n\rangle,$$
$$h_j(n) = 2a(n)\langle\delta_{n+1}, H^j\delta_n\rangle,$$
$$\gamma_j^\beta(n) = \langle(\delta_{n+1} + \beta\delta_n), H^j(\delta_{n+1} + \beta\delta_n)\rangle$$
(6.8) $$= g_j(n+1) + \frac{\beta}{a(n)}h_j(n) + \beta^2 g_j(n), \quad j \in \mathbb{N}_0.$$

Remark 6.3. *Using the unitary transform U_ε of Lemma 1.6 shows that $g_j(n)$, $h_j(n)$ do not depend on the sign of $a(n)$, that is, they only depend on $a(n)^2$.*

The next lemma shows how to compute g_j, h_j recursively.

Lemma 6.4. *The coefficients $g_j(n)$ and $h_j(n)$ for $j \in \mathbb{N}_0$ satisfy the following recursion relations*

$$g_{j+1} = \frac{h_j + h_j^-}{2} + bg_j,$$
(6.9) $$h_{j+1} - h_{j+1}^- = 2\Big(a^2 g_j^+ - (a^-)^2 g_j^-\Big) + b\Big(h_j - h_j^-\Big).$$

Proof. The first equation follows from

(6.10) $$g_{j+1}(n) = \langle H\delta_n, H^j\delta_n\rangle = \frac{h_j(n) + h_j(n-1)}{2} + b(n)g_j(n)$$

6.1. Asymptotic expansions

using $H\delta_n = a(n)\delta_{n+1} + a(n-1)\delta_{n-1} + b(n)\delta_n$. Similarly,

$$h_{j+1}(n) = b(n)h_j(n) + 2a(n)^2 g_j(n+1) + 2a(n-1)a(n)\langle\delta_{n+1}, H^j\delta_{n-1}\rangle$$
(6.11)
$$= b(n+1)h_j(n) + 2a^2 g_j(n) + 2a(n)a(n+1)\langle\delta_{n+2}, H^j\delta_n\rangle$$

and eliminating $\langle\delta_{n+1}, H^j\delta_{n-1}\rangle$ completes the proof. $\square$

The system (6.9) does not determine $g_j(n)$, $h_j(n)$ uniquely since it requires solving a first order recurrence relation at each step, producing an unknown summation constant each time

$$h_{j+1}(n) - h_{j+1}(n_0) = \sum_{m=n_0+1}^{n}{}^{*} \Big(2a(m)^2 g_j(m+1) - 2a(m-1)^2 g_j(m-1)$$
(6.12)
$$+ b(m)h_j(m) - b(m)h_j(m-1)\Big).$$

To determine the constant observe that the right hand side must consists of two summands, the first one involving $a(n+k)$, $b(n+k)$ with $|k|<j$ and the second one involving $a(n_0+k)$, $b(n_0+k)$ with $|k|<j$. This follows since the left hand side is of this form (use $H\delta_n = a(n)\delta_{n+1} + a(n-1)\delta_{n-1} + b(n)\delta_n$ and induction). In particular, the first summand must be equal to $h_{j+1}(n)$ and the second to $h_{j+1}(n_0)$. This determines the unknown constant.

Since this procedure is not very straightforward, it is desirable to avoid these constants right from the outset. This can be done at the expense of giving up linearity. If we compare powers of z in (2.187) we obtain

(6.13)
$$h_{j+1} = 2a^2 \sum_{\ell=0}^{j} g_{j-\ell} g_\ell^+ - \frac{1}{2}\sum_{\ell=0}^{j} h_{j-\ell} h_\ell, \quad j \in \mathbb{N},$$

which determines g_j, h_j recursively together with the first equation of (6.9). Explicitly we obtain

(6.14)
$$g_0 = 1, \quad g_1 = b, \quad g_2 = a^2 + (a^-)^2 + b^2,$$
$$h_0 = 0, \quad h_1 = 2a^2, \quad h_2 = 2a^2(b^+ + b)$$

and hence from (6.8)

$$\gamma_0^\beta = 1 + \beta^2, \quad \gamma_1^\beta = b^+ + 2a\beta + b\beta^2,$$
(6.15)
$$\gamma_2^\beta = (a^+)^2 + a^2 + (b^+)^2 + 2a(b^+ + b)\beta + (a^2 + (a^-)^2 + b^2)\beta^2.$$

Remark 6.5. (i). Let us advocate another approach which produces a recursion for g_j only. Inserting the expansion (6.7) for $g(z,n)$ into (1.109) and comparing coefficients of z^j one infers

$$g_0 = 1, \quad g_1 = b, \quad g_2 = a^2 + (a^-)^2 + b^2,$$
$$g_3 = a^2(b^+ + 2b) + (a^-)^2(2b + b^-) + b^3,$$
$$g_{j+1} = 2b g_j - a^2 g_{j-1}^+ + (a^-)^2 g_{j-1}^- - b^2 g_{j-1} - \frac{1}{2}\sum_{\ell=0}^{j-1} k_{j-\ell-1} k_\ell$$
(6.16)
$$+ 2a^2\Big(\sum_{\ell=0}^{j-1} g_{j-\ell-1} g_\ell^+ - 2b\sum_{\ell=0}^{j-2} g_{j-\ell-2} g_\ell^+ + b^2 \sum_{\ell=0}^{j-3} g_{j-\ell-3} g_\ell^+\Big),$$

for $j \geq 3$, where $k_0(n) = -b(n)$ and

(6.17) $\quad k_j = a^2 g_{j-1}^+ - (a^-)^2 g_{j-1}^- + b^2 g_{j-1} - 2b g_j + g_{j+1}, \quad j \in \mathbb{N}.$

If $g_j(n)$ is known, $h_j(n)$ can be determined using

(6.18) $\quad h_{j+1} = b h_j + g_{j+2} - 2b g_{j+1} + a^2 g_j^+ - (a^-)^2 g_j^- + b^2 g_j, \quad j \in \mathbb{N}_0.$

Equation (6.18) follows after inserting the first equation of (6.9) into the second. (ii). Analogously, one can get a recurrence relation for γ_j^β using (1.112). Since this approach gets too cumbersome we omit further details at this point.

Next, we turn to Weyl m-functions. As before we obtain

Lemma 6.6. *The quantities $m_\pm(z, n)$ have the Laurent expansions*

(6.19) $\quad m_\pm(z, n) = -\sum_{j=0}^\infty \frac{m_{\pm,j}(n)}{z^{j+1}}, \quad m_{\pm,0}(n) = 1.$

The coefficients $m_{\pm,j}(n)$ are given by

(6.20) $\quad m_{\pm,j}(n) = \langle \delta_{n\pm 1}, (H_{\pm,n})^j \delta_{n\pm 1} \rangle, \quad j \in \mathbb{N},$

and satisfy

$$m_{\pm,0} = 1, \quad m_{\pm,1} = b^\pm,$$

(6.21) $\quad m_{\pm,j+1} = b^\pm m_{\pm,j} + (a^{\pm\pm})^2 \sum_{\ell=0}^{j-1} m_{\pm,j-\ell-1} m_{\pm,\ell}^+, \quad j \in \mathbb{N}.$

Finally, we consider the solutions $\phi_\pm(z, n) = \mp a(n) \tilde{m}_\pm(z, n)$ of the Riccati equation (1.52). We will show uniqueness of solutions with particular asymptotic behavior.

Lemma 6.7. *The Riccati equation (1.52) has unique solutions $\phi_\pm(z, n)$ having the Laurent expansions*

(6.22) $\quad \phi_\pm(z, n) = \left(\frac{a(n)}{z}\right)^{\pm 1} \sum_{j=0}^\infty \frac{\phi_{\pm,j}(n)}{z^j}.$

The coefficients $\phi_{\pm,j}(n)$ satisfy

$$\phi_{+,0} = 1, \quad \phi_{+,1} = b^+,$$

(6.23) $\quad \phi_{+,j+1} = b^+ \phi_{+,j} + (a^+)^2 \sum_{\ell=0}^{j-1} \phi_{+,j-\ell-1} \phi_{+,\ell}^+, \; j \geq 1,$

and

$$\phi_{-,0} = 1, \quad \phi_{-,1} = -b,$$

(6.24) $\quad \phi_{-,j+1} = -b \phi_{-,j}^- - \sum_{\ell=1}^j \phi_{-,j-\ell+1} \phi_{-,\ell}^- - (a^-)^2 \delta_{1,j}, \; j \geq 1.$

Proof. Existence is clear from

(6.25) $\quad \phi_+(z, n) = -a(n) m_+(z, n), \quad \phi_-(z, n) = \frac{z + b(n) + a(n-1)^2 m_-(z, n)}{a(n)},$

and the previous lemma. Moreover, inserting (6.22) into the Riccati equation (1.52) produces the relations (6.23), (6.24) which determine the coefficients $\phi_{\pm,j}$ uniquely. Since $\phi_\pm(z)$ is uniquely determined by its Laurent expansion we are done. $\square$

Explicitly one computes

$$\phi_-(z,.) = \frac{z}{a}\Big(1 - \frac{b}{z} - \frac{(a^-)^2}{z^2} - \frac{(a^-)^2 b^-}{z^3} - \frac{(a^-)^2((a^{--})^2 + (b^-)^2)}{z^4}$$
$$+ O(\frac{1}{z^5})\Big),$$

$$\phi_+(z,.) = \frac{a}{z}\Big(1 + \frac{b^+}{z} + \frac{(a^+)^2 + (b^+)^2}{z^2} + \frac{(a^+)^2(2b^+ + b^{++}) + (b^+)^3}{z^3}$$
$$+ \frac{((a^+)^2 + (b^+)^2)^2 + (a^+)^2((b^+ + b^{++})^2 + (a^{++})^2)}{z^4}$$
(6.26) $$+ O(\frac{1}{z^5})\Big).$$

This lemma has some interesting consequences for $u_\pm(z, n)$. If we have a solution $\phi_\pm(z, n)$ of the Riccati equation with asymptotic behavior (6.22), we know that $u_\pm(z, n, n_0) = c(z, n, n_0) + \phi_\pm(z, n_0) s(z, n, n_0)$ are square summable near $\pm\infty$. Moreover, we obtain

$$u_-(z, n, n_0) = \Big(\prod_{j=n_0}^{n-1}{}^* \frac{z}{a(j)}\Big)\Big(1 - \frac{1}{z}\sum_{j=n_0}^{n-1}{}^* b(j) + O(\frac{1}{z^2})\Big),$$

(6.27) $$u_+(z, n, n_0) = \Big(\prod_{j=n_0}^{n-1}{}^* \frac{a(j)}{z}\Big)\Big(1 + \frac{1}{z}\sum_{j=n_0+1}^{n}{}^* b(j) + O(\frac{1}{z^2})\Big).$$

Remark 6.8. If we reflect at a point n_0 (cf. (1.82)) we obtain from (6.8)

(6.28) $$g_{R,j}(n_0 + k) = g_j(n_0 - k), \qquad h_{R,j}(n_0 + k) = h_j(n_0 - k - 1)$$

and thus

(6.29) $$\gamma_{R,j}^\beta(n_0 + k) = \beta^2 \gamma_j^{1/\beta}(n_0 - k).$$

In addition, since we have $m_{R,+}(z, n_0 + k) = m_-(z, n_0 - k)$ we infer

(6.30) $$\phi_{+,\ell+2}(n_0 + k) = -a_R(n_0 - k)^2 \phi_{R,-,\ell}(n_0 - k).$$

6.2. General trace formulas and xi functions

The objective of the present section is to develop trace formulas for arbitrary Jacobi operators. Applications will be given at the end of Section 7.5 and in Section 8.1. Our main ingredient will be **Krein's spectral shift theory** (for the special case of rank one perturbations). Hence we will recall some basic results first.

Consider the following rank one perturbation

(6.31) $$H_{n,\theta} = H + \theta \langle \delta_n, . \rangle \delta_n, \qquad \theta \geq 0,$$

and abbreviate

(6.32) $$g_\theta(z, n) = \langle \delta_n, (H_{n,\theta} - z)^{-1} \delta_n \rangle.$$

By virtue of the second resolvent formula one infers

(6.33) $\quad (H_{n,\theta} - z)^{-1} = (H - z)^{-1} - \theta \langle (H_{n,\theta} - \bar{z})^{-1}\delta_n, .\rangle (H - z)^{-1}\delta_n$

and hence

(6.34) $$g_\theta(z, n) = \frac{g(z, n)}{1 + \theta g(z, n)}.$$

Furthermore, we compute $(H_{n,\theta} - z)^{-1}\delta_n = (1 + \theta g_\theta(z, n))(H - z)^{-1}\delta_n$ and thus

(6.35) $\quad (H_{n,\theta} - z)^{-1} = (H - z)^{-1} - \dfrac{\theta}{1 + \theta g(z, n)} \langle (H - \bar{z})^{-1}\delta_n, .\rangle (H - z)^{-1}\delta_n$

(a special case of Krein's resolvent formula). We have $\sigma_{ess}(H_{n,\theta}) = \sigma_{ess}(H)$ and an investigation of the discrete poles of (6.35) shows

(6.36) $$\sigma_d(H_{n,\theta}) = \{\lambda \in \mathbb{R} \backslash \sigma_{ess}(H) | g(\lambda, n) = -\frac{1}{\theta}\}.$$

Moreover, since $g(\lambda, n)$ is increasing (w.r.t. λ) in each spectral gap, there can be at most one discrete eigenvalue of $H_{n,\theta}$ in each spectral gap.

Computing traces we obtain

(6.37) $$\begin{aligned}\operatorname{tr}\Big((H_{n,\theta} - z)^{-1} - (H - z)^{-1}\Big) &= -\frac{\theta}{1 + \theta g(z, n)} \langle \delta_n, (H - z)^{-2}\delta_n\rangle \\ &= -\frac{d}{dz} \ln(1 + \theta g(z, n)).\end{aligned}$$

Next, note that $g(z, n)$ and hence $1 + \theta g(z, n)$ is Herglotz. Hence we have the following exponential representation (cf. Lemma B.12 (iii))

(6.38) $$1 + \theta g(z, n) = \exp\left(\int_\mathbb{R} \frac{\xi_\theta(\lambda, n)}{\lambda - z} d\lambda\right),$$

where

(6.39) $$\xi_\theta(\lambda, n) = \frac{1}{\pi} \lim_{\varepsilon \downarrow 0} \arg\Big(1 + \theta g(\lambda + i\varepsilon, n)\Big).$$

Now we can rewrite (6.37) as

(6.40) $$\operatorname{tr}\Big((H_{n,\theta} - z)^{-1} - (H - z)^{-1}\Big) = -\int_\mathbb{R} \frac{\xi_\theta(\lambda, n)}{(\lambda - z)^2} d\lambda.$$

Moreover, $\xi_\theta(\lambda, n)$ (which is only defined a.e.) is compactly supported (see also Lemma B.12 (iii)) and $\int_\mathbb{R} \xi_\theta(\lambda, n) d\lambda = \theta$. Comparing the Laurent expansions around ∞ of both sides in (6.40) gives (formally)

(6.41) $$\operatorname{tr}\Big((H_{n,\theta})^\ell - H^\ell\Big) = \ell \int_\mathbb{R} \lambda^{\ell-1} \xi_\theta(\lambda, n) d\lambda.$$

However, (6.41) can be easily obtained rigorously by comparing coefficients in (6.35) and then taking traces of the resulting equations.

More general, Krein's result implies

(6.42) $$\operatorname{tr}\Big(f(H_{n,\theta}) - f(H)\Big) = \int_\mathbb{R} f'(\lambda) \xi_\theta(\lambda, n) d\lambda$$

for continuously differentiable f (f can be assumed compactly supported since only values in the support of $\xi_\theta(., n)$ contribute). The functions $1 + \theta g(z, n)$ and $\xi_\theta(\lambda, n)$

are called **perturbation determinant** and **spectral shift function** of the pair $(H, H_{n,\theta})$, respectively.

Let us collect some parts of these results in our first theorem.

Theorem 6.9. *Let $\xi_\theta(\lambda, n)$ be defined as above. Then we have*

$$(6.43) \qquad b_\theta^{(\ell)}(n) = \operatorname{tr}\left((H_{n,\theta})^{\ell+1} - H^{\ell+1}\right) = (\ell+1) \int_\mathbb{R} \lambda^\ell \xi_\theta(\lambda, n) d\lambda,$$

with (cf. (6.8))

$$(6.44) \qquad b_\theta^{(0)}(n) = \theta, \quad b_\theta^{(\ell)}(n) = \theta(\ell+1)g_\ell(n) + \theta \sum_{j=1}^{\ell} g_{\ell-j}(n) b_\theta^{(j-1)}(n), \quad \ell \in \mathbb{N}.$$

Proof. The claim follows after expanding both sides of

$$(6.45) \qquad \ln(1 + \theta g(z, n)) = \int_\mathbb{R} \frac{\xi_\theta(\lambda, n)}{\lambda - z} d\lambda$$

and comparing coefficients using the following connections between the series of $g(z)$ and $\ln(1 + g(z))$ (cf., e.g., [185]). Let $g(z)$ have the Laurent expansion

$$(6.46) \qquad g(z) = \sum_{\ell=1}^{\infty} \frac{g_\ell}{z^\ell}$$

as $z \to \infty$. Then we have

$$(6.47) \qquad \ln(1 + g(z)) = \sum_{\ell=1}^{\infty} \frac{c_\ell}{z^\ell},$$

where

$$(6.48) \qquad c_1 = g_1, \quad c_\ell = g_\ell - \sum_{j=1}^{\ell-1} \frac{j}{\ell} g_{\ell-j} c_j, \quad \ell \geq 2.$$

$\square$

In the special case $\ell = 1$ we obtain

$$(6.49) \qquad b(n) = \frac{1}{\theta} \int_\mathbb{R} \lambda \xi_\theta(\lambda, n) d\lambda - \frac{\theta}{2}.$$

Next, we want to investigate the case where the coupling constant θ tends to ∞. It is clear that $H_{\theta,n} f$ will only converge if $f(n) = 0$. On the other hand,

$$(6.50) \qquad (H_{n,\theta} - z)^{-1} \to (H - z)^{-1} - \frac{1}{g(z,n)} \langle (H - \bar{z})^{-1} \delta_n, . \rangle (H - z)^{-1} \delta_n$$

converges (in norm) as $\theta \to \infty$. Comparison with (1.106) shows $(H_{n,\theta} - z)^{-1} \to (H_n^\infty - z)^{-1}$ if we embed $(H_n^\infty - z)^{-1}$ into $\ell^2(\mathbb{Z})$ using $(H_n^\infty - z)^{-1} \delta_n = 0$, that is, $(H_n^\infty - z)^{-1} = (\mathbb{1} - P_n^\infty)(H - z)^{-1}(\mathbb{1} - P_n^\infty)$ (cf. Remark 1.11). In particular, (6.40) is still valid

$$(6.51) \qquad \operatorname{tr}\left((H_n^\infty - z)^{-1} - (H - z)^{-1}\right) = -\int_\mathbb{R} \frac{\xi(\lambda, n)}{(\lambda - z)^2} d\lambda.$$

Here the **xi function** $\xi(\lambda, n)$ is defined by

$$(6.52) \quad g(z,n) = |g(\mathrm{i},n)| \exp\left(\int_{\mathbb{R}} \left(\frac{1}{\lambda - z} - \frac{\lambda}{1+\lambda^2}\right)\xi(\lambda,n)d\lambda\right), \quad z \in \mathbb{C}\backslash\sigma(H),$$

respectively by

$$(6.53) \quad \xi(\lambda, n) = \frac{1}{\pi} \lim_{\varepsilon \downarrow 0} \arg g(\lambda + \mathrm{i}\varepsilon, n), \qquad \arg(.) \in (-\pi, \pi].$$

Since $\xi(., n)$ is *not* compactly supported, the asymptotic expansions require a somewhat more careful analysis. We abbreviate

$$(6.54) \qquad E_0 = \inf \sigma(H), \qquad E_\infty = \sup \sigma(H),$$

and note that $g(\lambda, n) > 0$ for $\lambda < E_0$, which follows from $(H - \lambda) > 0$ (implying $(H-\lambda)^{-1} > 0$). Similarly, $g(\lambda, n) < 0$ for $\lambda > E_\infty$ follows from $(H - \lambda) < 0$. Thus $\xi(\lambda, n)$ satisfies $0 \le \xi(\lambda, n) \le 1$,

$$(6.55) \quad \int_{\mathbb{R}} \frac{\xi(\lambda,n)}{1+\lambda^2} d\lambda = \arg g(\mathrm{i},n), \text{ and } \xi(\lambda,n) = \begin{cases} 0 \text{ for } z < E_0 \\ 1 \text{ for } z > E_\infty \end{cases}.$$

Using (6.55) together with the asymptotic behavior of $g(.,n)$ we infer

$$(6.56) \quad g(z,n) = \frac{1}{E_\infty - z} \exp\left(\int_{E_0}^{E_\infty} \frac{\xi(\lambda,n)d\lambda}{\lambda - z}\right).$$

As before we set $H_n^\infty \delta_n = 0$ implying

$$(6.57) \quad \begin{aligned} \mathrm{tr}\Big(f(H_n^\infty) - f(H)\Big) &= \int_{\mathbb{R}} f'(\lambda)\xi(\lambda,n)d\lambda \\ &= \int_{E_0}^{E_\infty} f'(\lambda)\xi(\lambda,n)d\lambda - f(E_\infty) \end{aligned}$$

for continuously differentiable f with compact support (again compact support is no restriction since only values in $[E_0, E_\infty]$ contribute).

Theorem 6.10. *Let $\xi(\lambda, n)$ be defined as above. Then we have the following trace formula*

$$(6.58) \quad b^{(\ell)}(n) = \mathrm{tr}\Big(H^\ell - (H_n^\infty)^\ell\Big) = E_\infty^\ell - \ell \int_{E_0}^{E_\infty} \lambda^{\ell-1}\xi(\lambda,n)d\lambda,$$

where

$$(6.59) \quad \begin{aligned} b^{(1)}(n) &= b(n), \\ b^{(\ell)}(n) &= \ell\, g_\ell(n) - \sum_{j=1}^{\ell-1} g_{\ell-j}(n) b^{(j)}(n), \quad \ell > 1. \end{aligned}$$

Proof. The claim follows after expanding both sides of

$$(6.60) \quad \ln\Big((E_\infty - z)g(z,n)\Big) = \int_{E_0}^{E_\infty} \frac{\xi(\lambda,n)d\lambda}{\lambda - z}$$

and comparing coefficients as before. $\square$

6.2. General trace formulas and xi functions

The special case $\ell = 1$ of equation (6.58) reads

$$(6.61) \quad b(n) = E_\infty - \int_{E_0}^{E_\infty} \xi(\lambda, n) d\lambda = \frac{E_0 + E_\infty}{2} + \frac{1}{2}\int_{E_0}^{E_\infty}(1 - 2\xi(\lambda,n))d\lambda.$$

Similarly, $H_{n_0}^\beta$ can be obtained as the limit of the operator $H + \theta\langle \delta_n^\beta, . \rangle \delta_n^\beta$ as $\theta \to \infty$. We have

$$(6.62) \quad \gamma^\beta(z, n) = -\frac{\beta}{a(n)} \exp\Big(\int_{\mathbb{R}} \frac{\xi^\beta(\lambda, n)d\lambda}{\lambda - z}\Big), \quad z \in \mathbb{C}\setminus\sigma(H_n^\beta), \beta \in \mathbb{R}\setminus\{0\},$$

where

$$(6.63) \quad \xi^\beta(\lambda, n) = \frac{1}{\pi} \lim_{\varepsilon \downarrow 0} \arg\big(\gamma^\beta(\lambda + i\varepsilon, n)\big) - \delta^\beta, \quad \delta^\beta = \begin{cases} 0, & \beta a(n) < 0 \\ 1, & \beta a(n) > 0 \end{cases}$$

and $0 \leq \operatorname{sgn}(-a(n)\beta)\,\xi^\beta(\lambda, n) \leq 1$. The function $\xi^\beta(\lambda, n)$ is compactly supported and we get as before

Theorem 6.11. *Let $\xi^\beta(\lambda, n)$ be defined as above. Then we have*

$$(6.64) \quad b^{\beta,(\ell)}(n) = (\ell + 1)\frac{\beta}{a(n)}\int_{\mathbb{R}} \lambda^\ell \xi^\beta(\lambda, n)d\lambda, \quad \ell \in \mathbb{N},$$

where

$$b^{\beta,(0)}(n) = 1 + \beta^2,$$

$$(6.65) \quad b^{\beta,(\ell)}(n) = (\ell+1)\gamma_\ell^\beta(n) - \frac{\beta}{a(n)}\sum_{j=1}^\ell \gamma_{\ell-j}^\beta(n) b^{\beta,(j-1)}(n), \quad \ell \in \mathbb{N}.$$

Specializing to $\ell = 0$ in (6.64) we obtain

$$(6.66) \quad a(n) = \frac{1}{\beta + \beta^{-1}} \int_{\mathbb{R}} \xi^\beta(\lambda, n) d\lambda.$$

Finally, observe that $\xi^\beta(\lambda, n_0)$ for two values of β and one fixed n_0 determines the sequences $a(n), b(n)$ for all $n \in \mathbb{Z}$.

Lemma 6.12. *Let $\beta_1 \neq \beta_2 \in \mathbb{R} \cup \{\infty\}$ be given. Then $(\beta_j, \xi^{\beta_j}(., n_0))$, $j = 1, 2$, for one fixed $n_0 \in \mathbb{Z}$ uniquely determines $a(n)^2, b(n)$ for all $n \in \mathbb{Z}$.*

Proof. If we know β and $\xi^\beta(z, n_0)$, we know $a(n_0)$ by (6.66) and $\gamma^\beta(z, n_0)$ by (6.62). Hence the result follows from Theorem 2.30. $\square$

Chapter 7

Jacobi operators with periodic coefficients

Some of the most interesting Jacobi operators are those with periodic coefficients a, b. In this chapter we will develop Floquet theory to investigate this case. This will allow us to give a complete characterization of the qualitative behavior of solutions and hence also a complete spectral characterization. Especially, we will see that all important quantities associated with periodic sequences have a simple polynomial dependence with respect to the spectral parameter z and can be computed explicitly. In fact, this feature is shared by a much larger class of Jacobi operators to be considered in the next chapter. In addition, this chapter also provides the motivation for definitions being made there.

7.1. Floquet theory

For this entire chapter we will assume a, b to be periodic with period N (for $N = 1$ see Sections 1.3 and 7.4).

Hypothesis H. 7.1. Suppose there is an $N \in \mathbb{N}$ such that

$$(7.1) \qquad a(n + N) = a(n), \qquad b(n + N) = b(n).$$

In addition, we agree to abbreviate

$$(7.2) \qquad A = \prod_{j=1}^{N} a(n_0 + j) = \prod_{j=1}^{N} a(j), \qquad B = \sum_{j=1}^{N} b(n_0 + j) = \sum_{j=1}^{N} b(j).$$

We start by recalling the transfer matrix $\Phi(z, n, n_0)$ (cf. (1.30)) which transfers initial conditions $u(n_0)$, $u(n_0+1)$ at n_0 of a solution u into the corresponding values $u(n)$, $u(n+1)$ at n. It suggests itself to investigate what happens if we move on N steps, that is, to look at the **monodromy matrix**

$$(7.3) \qquad M(z, n_0) = \Phi(z, n_0 + N, n_0).$$

A first naive guess would be that all initial conditions return to their starting values after N steps (i.e., $M(z,n_0) = \mathbb{1}$) and hence all solutions are periodic. However, this is too much to hope for since it already fails for $N = 1$ (see Section 1.3).

On the other hand, since it does not matter whether we start our N steps at n_0, at $n_0 + N$, or even $n_0 + \ell N$, $\ell \in \mathbb{Z}$, we infer that $M(z, n_0)$ is periodic, that is, $M(z, n_0 + N) = M(z, n_0)$, $n_0 \in \mathbb{Z}$. Moreover, we even have $\Phi(z, n_0 + \ell N, n_0) = M(z, n_0)^\ell$. Thus $\Phi(z, n, n_0)$ exhibits an exponential behavior if we move on N points in each step. If we factor out this exponential term, the remainder should be periodic.

For this purpose we rewrite $M(z, n_0)$ a little bit. Using periodicity (7.1) we obtain from (1.33)

(7.4) $$\det M(z, n_0) = 1$$

and hence we can find a periodic matrix $Q(z, n_0)$ such that

(7.5) $$M(z, n_0) = \exp\bigl(\mathrm{i} N Q(z, n_0)\bigr), \quad \operatorname{tr} Q(z, n_0) = 0.$$

Now we can write

(7.6) $$\Phi(z, n, n_0) = P(z, n, n_0) \exp\bigl(\mathrm{i}(n - n_0) Q(z, n_0)\bigr), \quad P(z, n_0, n_0) = \mathbb{1}.$$

A simple calculation

(7.7) $$\begin{aligned} P(z, n+N, n_0) &= \Phi(z, n+N, n_0) M(z, n_0) \exp\bigl(-\mathrm{i}(n - n_0) Q(z, n_0)\bigr) \\ &= \Phi(z, n+N, n_0 + N) \exp\bigl(-\mathrm{i}(n - n_0) Q(z, n_0)\bigr) \\ &= \Phi(z, n, n_0) \exp\bigl(-\mathrm{i}(n - n_0) Q(z, n_0)\bigr) = P(z, n, n_0) \end{aligned}$$

shows that $P(z, n, n_0)$ is indeed periodic as anticipated.

Next we want to turn this result into a corresponding result for the solutions of $\tau u = zu$. The key is clearly to investigate the Jordan canonical form of $M(z, n_0)$ (resp. $Q(z, n_0)$).

From (1.32) we see

(7.8) $$\begin{aligned} M(z, n_1) &= \Phi(z, n_1 + N, n_0 + N) M(z, n_0) \Phi(z, n_0, n_1) \\ &= \Phi(z, n_1, n_0) M(z, n_0) \Phi(z, n_1, n_0)^{-1} \end{aligned}$$

and hence the **discriminant**

(7.9) $$\Delta(z) = \frac{1}{2} \operatorname{tr} M(z, n_0) = \frac{1}{2}\bigl(c(z, n_0 + N, n_0) + s(z, n_0 + N + 1, n_0)\bigr),$$

plus the eigenvalues $m^\pm(z)$ (resp. $\pm q(z)$) of $M(z, n_0)$ (resp. $Q(z, n_0)$)

(7.10) $$m^\pm(z) = \exp(\pm \mathrm{i} N q(z)) = \Delta(z) \pm (\Delta(z)^2 - 1)^{1/2}, \quad m^+(z) m^-(z) = 1,$$

are independent of n_0. The branch in the above root is fixed as follows

(7.11) $$(\Delta(z)^2 - 1)^{1/2} = \frac{-1}{2A} \prod_{j=0}^{2N-1} \sqrt{z - E_j},$$

where $(E_j)_{j=0}^{2N-1}$ are the zeros of $\Delta(z)^2 - 1$ (cf. Appendix A.7). This definition implies the following expansion for large z

(7.12) $$m^\pm(z) = (2\Delta(z))^{\mp 1}\Bigl(1 + O\bigl(\frac{1}{z^{2N}}\bigr)\Bigr) = \Bigl(\frac{A}{z^N}\Bigr)^{\pm 1}\Bigl(1 \pm \frac{B}{z} + O\bigl(\frac{1}{z^2}\bigr)\Bigr)$$

and will ensure $|m^+(z)| \leq 1$, $|m^-(z)| \geq 1$ later on.

7.1. Floquet theory

The eigenvalues $m^\pm(z)$ are called **Floquet multipliers** and $q(z)$ is called **Floquet momentum**.

For later use we note

$$m^+(z) - m^-(z) = 2\mathrm{i}\sin(Nq(z)) = 2(\Delta(z)^2 - 1)^{1/2},$$
(7.13)
$$m^+(z) + m^-(z) = 2\cos(Nq(z)) = 2\Delta(z).$$

Remark 7.2. It is natural to look at $(\Delta(z)^2-1)^{1/2}$ as meromorphic function on the corresponding Riemann surface. The two functions associated with $\pm(\Delta(z)^2-1)^{1/2}$ can then be viewed as the two branches of this single function on the Riemann surface. This will be exploited in Chapter 8 and Chapter 9.

Now we are able compute the Jordan canonical form of $M(z, n_0)$. We need to distinguish three cases.

Case 1). $\Delta(z)^2 \neq 1$ (and hence $m^+(z) \neq m^-(z)$). If $s(z, n_0 + N, n_0) \neq 0$ we set

(7.14)
$$e^\pm(z, n_0) = \begin{pmatrix} 1 \\ \phi_\pm(z, n_0) \end{pmatrix},$$

where

(7.15)
$$\phi_\pm(z, n) = \frac{m^\pm(z) - c(z, n_0 + N, n_0)}{s(z, n_0 + N, n_0)} = \frac{c(z, n_0 + N + 1, n_0)}{m^\pm(z) - s(z, n_0 + N + 1, n_0)}.$$

If $s(\mu, n_0 + N, n_0) = 0$, then $s(\mu, n, n_0)$ is periodic with respect to n and $s(\mu, n_0 + N, n_0) = m^\sigma(\mu)$ for some $\sigma \in \{\pm\}$. Moreover, $\det M(\mu, n_0) = 1$ says $c(\mu, n_0 + N, n_0) = s(\mu, n_0 + N + 1, n_0)^{-1} = m^{-\sigma}(\mu)$ and hence $\phi_{-\sigma}(z, n_0)$ tends to a finite limit $\phi_{-\sigma}(\mu, n_0)$ as $z \to \mu$. Thus we can set $e^\sigma(\mu, n_0) = (0, 1)$ and $e^{-\sigma}(\mu, n_0) = (1, \phi_{-\sigma}(\mu, n_0))$.

Now the matrix

(7.16)
$$U(z, n_0) = \big(e^+(z, n_0), e^-(z, n_0)\big)$$

will transform $M(z, n_0)$ into

(7.17)
$$U(z, n_0)^{-1} M(z, n_0) U(z, n_0) = \begin{pmatrix} m^+(z) & 0 \\ 0 & m^-(z) \end{pmatrix}.$$

Moreover, we have two corresponding solutions (**Floquet solutions**)

(7.18)
$$u_\pm(n, z) = e_1^\pm(z, n_0) c(z, n, n_0) + e_2^\pm(z, n_0) s(z, n, n_0),$$

satisfying

(7.19)
$$u_\pm(z, n + N) = m^\pm(z) u_\pm(z, n).$$

They are linearly independent and unique up to multiples.

Case 2). $\Delta(z) = \pm 1$ (and hence $m^+(z) = m^-(z) = \pm 1$) and $M(z, n_0)$ has two linearly independent eigenvectors. We necessarily have

(7.20)
$$M(z, n_0) = \pm \mathbb{1}$$

and there is nothing to be done. All solutions satisfy

(7.21)
$$u(z, n + N) = \pm u(z, n).$$

Case 3). $\Delta(z) = \pm 1$ (and hence $m^+(z) = m^-(z) = \pm 1$) and $M(z, n_0)$ has only one eigenvector. If $s(z, n_0 + N, n_0) \neq 0$ we set

$$(7.22) \qquad e(z, n_0) = \begin{pmatrix} s(z, n_0 + N, n_0) \\ \frac{s(z, n_0 + N + 1, n_0) - c(z, n_0 + N, n_0)}{2} \end{pmatrix}, \quad \hat{e}(z, n_0) = \begin{pmatrix} 0 \\ 1 \end{pmatrix}$$

or, if $c(z, n_0 + N + 1, n_0) \neq 0$ we set

$$(7.23) \qquad e(z, n_0) = \begin{pmatrix} \frac{c(z, n_0 + N, n_0) - s(z, n_0 + N + 1, n_0)}{2} \\ -c(z, n_0 + N + 1, n_0) \end{pmatrix}, \quad \hat{e}(z, n_0) = \begin{pmatrix} 1 \\ 0 \end{pmatrix}.$$

If both are zero we have Case 2.

Then $U(z, n_0) = (e(z, n_0), \hat{e}(z, n_0))$ will transform $M(z, n_0)$ into

$$(7.24) \qquad U(z, n_0)^{-1} M(z, n_0) U(z, n_0) = \begin{pmatrix} \pm 1 & 1 \\ 0 & \pm 1 \end{pmatrix}.$$

Moreover, there are solutions $u(z)$, $\hat{u}(z)$ such that

$$(7.25) \qquad u(z, n + N) = \pm u(z, n), \quad \hat{u}(z, n + N) = \pm \hat{u}(z, n) + u(z, n).$$

Summarizing these results we obtain Floquet's theorem.

Theorem 7.3. *The solutions of $\tau u = zu$ can be characterized as follows.*
(i). If $\Delta(z)^2 \neq 1$ there exist two solutions satisfying

$$(7.26) \qquad u_\pm(z, n) = p_\pm(z, n) e^{\pm i q(z) n}, \quad p_\pm(z, n + N) = p_\pm(z, n).$$

(ii). If $\Delta(z)^2 = \pm 1$, then either all solutions satisfy

$$(7.27) \qquad u(z, n + N) = \pm u(z, n)$$

or there are two solutions satisfying

$$(7.28) \qquad u(z, n) = p(z, n), \quad \hat{u}(z, n) = \hat{p}(z, n) + n\, p(z, n)$$

with $p(z, n + N) = \pm p(z, n)$, $\hat{p}(z, n + N) = \pm \hat{p}(z, n)$.

If we normalize the Floquet solutions by $u_\pm(z, n_0) = 1$ they are called **Floquet functions** $\psi_\pm(z, n, n_0)$ and a straightforward calculation yields

$$(7.29) \qquad \begin{aligned} \psi_\pm(z, n, n_0) &= c(z, n, n_0) + \phi_\pm(z, n_0) s(z, n, n_0), \\ \phi_\pm(z, n_0) &= \frac{\psi_\pm(z, n + 1, n_0)}{\psi_\pm(z, n, n_0)}. \end{aligned}$$

Observe that $\phi_\pm(z, n_0)$ is periodic, $\phi_\pm(z, n_0 + N) = \phi_\pm(z, n_0)$, and satisfies the Riccati equation (1.52).

Moreover, using our expansion (7.12) we get for large z

$$(7.30) \qquad \begin{aligned} \phi_-(z, n) &= \frac{s(z, n + N + 1, n)}{s(z, n + N, n)} + O(z^{1-2N}), \\ \phi_+(z, n) &= -\frac{c(z, n + N, n)}{s(z, n + N, n)} + O(z^{1-2N}). \end{aligned}$$

7.2. Connections with the spectra of finite Jacobi operators

In this section we want to gain further insight by considering finite Jacobi matrices. The results will also shed some light on our trace formulas.

First, we look at the operators $H_{n_1,n_2}^{\beta_1,\beta_2}$. We are mainly interested in the special case $\hat{H}_{n_0}^{\beta} = H_{n_0,n_0+N}^{\beta,\beta}$ associated with the following Jacobi matrices (cf. (1.64))

$$\hat{H}_{n_0}^{\infty} = J_{n_0,n_0+N}, \qquad \hat{H}_{n_0}^{0} = J_{n_0+1,n_0+N+1},$$

(7.31) $$\hat{H}_{n_0}^{\beta} = \begin{pmatrix} \hat{b}_{n_0+1} & a_{n_0+1} & & & \\ a_{n_0+1} & b_{n_0+2} & \ddots & & \\ & \ddots & \ddots & \ddots & \\ & & \ddots & b_{n_0+N-1} & a_{n_0+N-1} \\ & & & a_{n_0+N-1} & \hat{b}_{n_0+N} \end{pmatrix},$$

(abbreviating $a(n)$, $b(n)$ as a_n, b_n) with $\hat{b}_{n_0+1} = b_{n_0+1} - a_{n_0}\beta^{-1}$, $\hat{b}_{n_0+N} = b_{n_0} - a_{n_0}\beta$, and $\beta \in \mathbb{R}\setminus\{0\}$.

Second, we look at the operators

(7.32) $$\tilde{H}_{n_0}^{\theta} = \begin{pmatrix} b_{n_0+1} & a_{n_0+1} & & & e^{-i\theta}a_{n_0} \\ a_{n_0+1} & b_{n_0+2} & \ddots & & \\ & \ddots & \ddots & \ddots & \\ & & \ddots & b_{n_0+N-1} & a_{n_0+N-1} \\ e^{i\theta}a_{n_0} & & & a_{n_0+N-1} & b_{n_0+N} \end{pmatrix},$$

with $\theta \in [0, 2\pi)$. These latter operators are associated with the following coupled boundary conditions ($\theta \in [0, 2\pi)$)

(7.33) $$\begin{aligned} u(n_0)\exp(i\theta) - u(n_0 + N) &= 0, \\ u(n_0 + 1)\exp(i\theta) - u(n_0 + 1 + N) &= 0. \end{aligned}$$

The spectrum of $\hat{H}_{n_0}^{\beta}$ is given by the zeros $\{\lambda_j^{\beta}(n_0)\}_{j=1}^{\tilde{N}}$ (where $\tilde{N} = N - 1$ for $\beta = 0, \infty$ and $\tilde{N} = N$ otherwise) of the characteristic equation

(7.34) $$\det(z - \hat{H}_{n_0}^{\beta}) = \prod_{j=1}^{\tilde{N}}(z - \lambda_j^{\beta}(n_0)).$$

In the special case $\beta = \infty$ we set $\lambda_j^{\infty}(n_0) = \mu_j(n_0)$ and consequently $\lambda_j^{0}(n_0) = \mu_j(n_0 + 1)$ by (7.31).

As in (1.65) we can give an alternate characterization of the eigenvalues $\lambda_j^{\beta}(n_0)$. Recall from Remark 1.9 that $\hat{H}_{n_0}^{\beta}$ can be obtained from H by imposing boundary conditions at n_0, n_0+N. Hence any eigenfunction must satisfy these boundary conditions. The solution satisfying the boundary condition at n_0 is (up to a constant) given by

(7.35) $$s_{\beta}(z, n, n_0) = \sin(\alpha)c(z, n, n_0) - \cos(\alpha)s(z, n, n_0), \quad \beta = \cot(\alpha).$$

It is an eigenfunction of $\hat{H}_{n_0}^\beta$ if and only if it also satisfies the boundary condition at $n_0 + N$. Thus we infer for $\beta \in \mathbb{R}\backslash\{0\}$

$$\cos(\alpha)s_\beta(z, n_0 + N, n_0) + \sin(\alpha)s_\beta(z, n_0 + N + 1, n_0)$$
(7.36)
$$= \frac{-(\beta + \beta^{-1})^{-1}}{A} \prod_{j=1}^N (z - \lambda_j^\beta(n_0)),$$

where the leading coefficient follows from (1.68). For $\beta = 0, \infty$ we have

$$c(z, n_0 + N + 1, n_0) = \frac{-a(n_0)}{A} \prod_{j=1}^{N-1} (z - \mu_j(n_0 + 1)),$$
(7.37)
$$s(z, n_0 + N, n_0) = \frac{a(n_0)}{A} \prod_{j=1}^{N-1} (z - \mu_j(n_0)).$$

Since the function satisfying the boundary condition at one side is unique (up to a constant), the spectrum $\sigma(\hat{H}_{n_0}^\beta)$ is simple (cf. Remark 1.10).

For later use note that standard linear algebra immediately yields the following **trace relations** (recall (7.2))

(7.38)
$$\sum_{j=1}^{N-1} \mu_j(n_0) = \text{tr}\hat{H}_{n_0}^\infty = B - b(n_0),$$
$$\sum_{j=1}^N \lambda_j^\beta(n_0) = \text{tr}\hat{H}_{n_0}^\beta = B - (\beta + \beta^{-1})a(n_0).$$

Higher powers can be computed from (1.67).

Next, we turn to $\tilde{H}_{n_0}^\theta$. Again the spectrum is given by the zeros $\{E_j^\theta\}_{j=1}^N$ of the characteristic equation

(7.39)
$$\det(z - \tilde{H}_{n_0}^\theta) = \prod_{j=1}^N (z - E_j^\theta).$$

Our notation suggests that $\sigma(\tilde{H}_{n_0}^\theta)$ is independent of n_0. In fact, extending the eigenfunctions of $\tilde{H}_{n_0}^\theta$ to a solution of (1.19) on $\mathbb{Z}$, we get a solution with the property $u(n + N) = e^{i\theta}u(n)$. Conversely, any solution of this kind gives rise to an eigenfunction of $\tilde{H}_{n_0}^\theta$. In other words, the spectrum of $\tilde{H}_{n_0}^\theta$ can be characterized by

(7.40) $\lambda \in \sigma(\tilde{H}_{n_0}^\theta) \Leftrightarrow (m^+(\lambda) - e^{i\theta})(m^-(\lambda) - e^{i\theta}) = 0 \Leftrightarrow \Delta(\lambda) = \cos(\theta),$

where the independence of n_0 is evident. Moreover, we have $\sigma(\tilde{H}_{n_0}^\theta) = \sigma(\tilde{H}_{n_0}^{-\theta})$ ($-\theta$ is to be understood mod2π) since the corresponding operators are anti-unitarily equivalent.

In the special case $\theta = 0, \pi$ we set $\tilde{H}_{n_0}^0 = \tilde{H}_{n_0}^+$, $\tilde{H}_{n_0}^\pi = \tilde{H}_{n_0}^-$ and similarly for the eigenvalues $E_j^0 = E_j^+$, $E_j^\pi = E_j^-$.

By the above analysis, $E_j^\pm$ are the zeros of $\Delta(z) \mp 1$ and we may write

(7.41)
$$\Delta(z) \mp 1 = \frac{1}{2A} \prod_{j=1}^N (z - E_j^\pm).$$

7.2. Connections with the spectra of finite Jacobi operators

Since the $N-1$ zeros of $d\Delta(z)/dz$ must interlace the N zeros of both $\Delta(z) \mp 1$ we infer

(7.42) $\quad E_1^{\pm} < E_1^{\mp} \leq E_2^{\mp} < E_2^{\pm} \leq E_3^{\pm} < \cdots \leq E_N^{\pm(-1)^{N-1}} < E_N^{\pm(-1)^N}$

for $\mathrm{sgn}(A) = \pm(-1)^N$. The numbers in the above sequence are denoted by E_j, $0 \leq j \leq 2N-1$. A typical discriminant is depicted below.

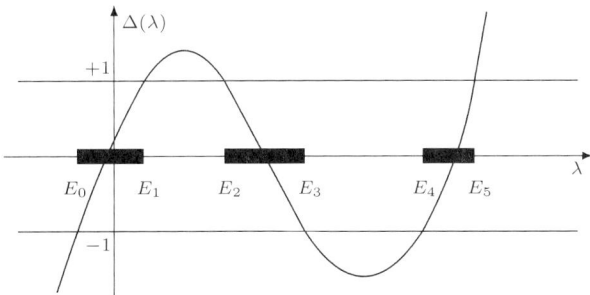

This also shows (using (7.42)) that for $\theta \in (0,\pi) \cup (\pi, 2\pi)$ the spectrum is simple. The case when the spectrum of $\tilde{H}_n^{\pm}$ is degenerated can be easily characterized.

Lemma 7.4. *The following conditions for $E \in \mathbb{R}$ are equivalent.*

(i). E is a double eigenvalue of $\tilde{H}_n^{\pm}$.
(ii). All solutions of $\tau u = E u$ satisfy $u(E, n+N) = \pm u(E, n)$.
(iii). $M(E, n) = \pm \mathbb{1}$.
(iv). $E = E_{2j-1} = E_{2j} \in \sigma(\hat{H}_n^{\beta})$ for all $\beta \in \mathbb{R} \cup \{\infty\}$.

In particular, if one of these conditions holds for one $n \in \mathbb{Z}$, it automatically holds for all.

The trace relation for $\tilde{H}_n^{\theta}$ reads

(7.43) $\quad \displaystyle\sum_{j=1}^{N} E_j^{\theta} = \mathrm{tr}\,\tilde{H}_n^{\theta} = B, \quad \text{or} \quad \sum_{j=0}^{2N-1} E_j = 2B.$

It can be used to eliminate B in the previous two trace relations (7.38) which produces the same trace relations already found in Section 6.2. This will become clear in Chapter 8 (see (8.11)).

Now, let us use this information to investigate H. Our choice for the branch in (7.11) implies $|m^+(z)| < 1$ for $|z|$ large (see also (7.12) and Appendix A.7). Thus ψ_+ is square summable near $+\infty$ for $|z|$ large. But since ψ_+ is holomorphic for $z \in \mathbb{C}\backslash\mathbb{R}$ it is square summable for all $z \in \mathbb{C}\backslash\mathbb{R}$ (see Lemma 2.2). So we must have $|m^+(z)| < 1$ for $z \in \mathbb{C}\backslash\mathbb{R}$ and hence by continuity $|m^+(z)| \leq 1$ for $z \in \mathbb{C}$.

In particular, we can use $\psi_{\pm}$ to compute $g(z,n)$. For the Wronskian of the Floquet functions we obtain

(7.44) $\quad W(\psi_-, \psi_+) = \dfrac{2a(n_0)(\Delta(z)^2 - 1)^{1/2}}{s(z, n_0 + N, n_0)} = a(n_0)\Big(\phi_+(z,n_0) - \phi_-(z,n_0)\Big)$

and if we evaluate (1.99) at $m = n = n_0$ ($u_{\pm} = \psi_{\pm}$) we get from (7.37)

(7.45) $\quad g(z, n_0) = \dfrac{a(n_0)^{-1}}{\phi_+(z,n_0) - \phi_-(z,n_0)} = \dfrac{\prod_{j=1}^{N-1}(z - \mu_j(n_0))}{-\prod_{j=0}^{2N-1}\sqrt{z - E_j}}.$

Since n_0 is arbitrary we must have

$$(7.46) \quad \psi_+(z,n,n_0)\psi_-(z,n,n_0) = \frac{a(n_0)}{a(n)} \frac{s(z,n+N,n)}{s(z,n_0+N,n_0)} = \prod_{j=1}^{N-1} \frac{z-\mu_j(n)}{z-\mu_j(n_0)}.$$

Similarly, we compute

$$\gamma^\beta(z,n) = \frac{(\phi_+(z,n)+\beta)(\phi_-(z,n)+\beta)}{a(n)(\phi_+(z,n)-\phi_-(z,n))}$$

$$= \frac{(m^+(z) - \sin^{-1}(\alpha)s_\beta(z,n+N,n))(m^-(z) - \sin^{-1}(\alpha)s_\beta(z,n+N,n))}{2a(n)s(z,n+N,n)(\Delta(z)^2-1)^{1/2}}$$

$$= \frac{\cos(\alpha)s_\beta(z,n+N,n) + \sin(\alpha)s_\beta(z,n+N+1,n)}{-2\sin^2(\alpha)a(n)(\Delta(z)^2-1)^{1/2}}$$

$$(7.47) \quad = -\frac{\beta}{a(n)} \frac{\prod_{j=1}^{N}(z-\lambda_j^\beta(n))}{\prod_{j=0}^{2N-1}\sqrt{z-E_j}},$$

where we have used (7.15), $\det M(z,n) = 1$, and $m^+(z) + m^-(z) = 2\Delta(z)$.

By (2.141) and (7.45) the spectrum of H is given by

$$(7.48) \quad \sigma(H) = \{\lambda \in \mathbb{R} | |\Delta(\lambda)| \leq 1\} = \bigcup_{j=0}^{N-1}[E_{2j}, E_{2j+1}]$$

and it is purely absolutely continuous ($\sigma_p(H) = \sigma_{sc}(H) = \emptyset$).

The sets

$$(7.49) \quad \overline{\rho}_0 = (-\infty, E_0], \quad \overline{\rho}_j = [E_{2j-1}, E_{2j}], \quad \overline{\rho}_N = [E_{2N-1}, \infty),$$

$1 \leq j \leq N-1$, are called spectral gaps. In addition, since $g(z,n)$ and $\gamma^\beta(z,n)$ are Herglotz functions we infer (compare the argument in Section 8.1)

$$(7.50) \quad \lambda_j^\beta(n) \in \overline{\rho}_j, \quad 1 \leq j \leq N-1, \beta \in \mathbb{R} \cup \{\infty\},$$

and

$$(7.51) \quad \lambda_N^\beta(n) \in \overline{\rho}_N \text{ for } a(n)\beta < 0, \quad \lambda_N^\beta(n) \in \overline{\rho}_0 \text{ for } a(n)\beta > 0.$$

The most important case is $\beta = \infty$, $\lambda_j^\infty(n) = \mu_j(n)$ and the typical situation looks as follows.

The fact that $s_\beta(\lambda_j^\beta(n_0), n, n_0)$ satisfies the same boundary condition at n_0 and $n_0 + N$ implies that $s_\beta(\lambda_j^\beta(n_0), n, n_0)$ is a Floquet solution. Hence we can associate a sign $\sigma_j^\beta(n)$ with $\lambda_j^\beta(n)$ such that

$$(7.52) \quad M(n, \lambda_j^\beta(n)) \begin{pmatrix} \sin(\alpha) \\ -\cos(\alpha) \end{pmatrix} = m^{\sigma_j^\beta(n)}(\lambda_j^\beta(n)) \begin{pmatrix} \sin(\alpha) \\ -\cos(\alpha) \end{pmatrix}.$$

7.3. Polynomial identities

If $\beta = \infty$ we abbreviate $\sigma_j^\infty(n) = \sigma_j(n)$. Moreover, if $\lambda_j^\beta(n)$ does not coincide with one of the band edges E_j, then the corresponding Floquet solution is an eigenfunction of $H_{\sigma_j^\beta(n),n}^\beta$ and consequently

$$(7.53) \qquad \sigma(H_n^\beta) = \sigma(H) \cup \{\lambda_j^\beta(n)\}_{j=1}^{\tilde N}.$$

This characterization of the pairs $(\lambda_j^\beta(n), \sigma_j^\beta(n))$ will be useful later on because it does not rely on periodicity.

Remark 7.5. Another way to investigate periodic operators would be a (constant fiber) direct integral decomposition of H

$$(7.54) \qquad \ell^2(\mathbb{Z}) \cong \int_{[0,2\pi)}^\oplus \ell^2(n_0, n_0 + N + 1) \frac{d\theta}{2\pi}, \qquad H \cong \int_{[0,2\pi)}^\oplus \tilde H_{n_0}^\theta \frac{d\theta}{2\pi},$$

compare [**195**], Chapter XIII.16. However, since this approach will not reveal any new information for us, we will not investigate it further.

7.3. Polynomial identities

In the previous section we have seen that several objects exhibit a simple polynomial structure with respect to z. The present section further pursues this observation.

We first define the following polynomials

$$G(z,n) = \frac{A}{a(n)} s(z, n+N, n) = \frac{-A}{a(n-1)} c(z, n+N, n-1)$$

$$= \prod_{j=1}^{N-1} (z - \mu_j(n)),$$

$$H(z,n) = A\bigl(s(z, n+N+1, n) - c(z, n+N, n)\bigr),$$

$$(7.55) \qquad R(z) = 4A^2(\Delta(z)^2 - 1) = \prod_{j=0}^{2N-1} (z - E_j).$$

In terms of these polynomials the monodromy matrix reads

$$(7.56) \qquad M(z,n) = \frac{1}{A}\begin{pmatrix} A\Delta(z) - \tfrac{1}{2}H(z,n) & a(n)G(z,n) \\ -a(n)G(z,n+1) & A\Delta(z) + \tfrac{1}{2}H(z,n) \end{pmatrix}.$$

Using (7.56) and (7.52) we get

$$(7.57) \qquad H(n, \mu_j(n+1)) = -R^{1/2}(\hat\mu_j(n+1)), \qquad H(n, \mu_j(n)) = R^{1/2}(\hat\mu_j(n)),$$

where we have set $R^{1/2}(z) = 2A(\Delta(z)^2 - 1)^{1/2}$ and used the convenient abbreviation

$$(7.58) \qquad R^{1/2}(\hat\lambda_j^\beta(n)) = \sigma_j^\beta(n) R^{1/2}(\lambda_j^\beta(n)), \qquad \hat\lambda_j^\beta(n) = (\lambda_j^\beta(n), \sigma_j^\beta(n)).$$

Thus Newton's interpolation formula yields

$$H(z,n) = \sum_{j=1}^{N-1} R^{1/2}(\hat\mu_j(n)) \prod_{k\neq j}^{N-1} \frac{z - \mu_k(n)}{\mu_j(n) - \mu_k(n)} + (z - b(n)) G(z,n)$$

$$= -\sum_{j=1}^{N-1} R^{1/2}(\hat\mu_j(n+1)) \prod_{k\neq j}^{N-1} \frac{z - \mu_k(n+1)}{\mu_j(n+1) - \mu_k(n+1)}$$

$$(7.59) \qquad + (z - b(n+1)) G(z, n+1),$$

where the factor in front of $G(z,n)$, $G(z,n+1)$ follows after expanding both sides and considering the highest two coefficients. In addition, we obtain from (7.59)

$$H(z,n) + H(z,n-1) = 2(z-b(n))G(z,n),$$

(7.60) $$H(z,n) - H(z,n-1) = 2\sum_{j=1}^{N-1} R^{1/2}(\hat{\mu}_j(n)) \prod_{k\neq j}^{N-1} \frac{z-\mu_k(n)}{\mu_j(n)-\mu_k(n)}.$$

Next, from (7.56) and (7.60) we obtain

$$\frac{1}{A}\sum_{j=1}^{N-1} R^{1/2}(\hat{\mu}_j(n)) \prod_{k\neq j}^{N-1} \frac{z-\mu_k(n)}{\mu_j(n)-\mu_k(n)}$$

$$= c(z, n+N-1, n-1) - c(z, n+N, n)$$

(7.61) $$= s(z, n+N+1, n) - s(z, n+N, n-1).$$

Finally, we have

(7.62) $$\phi_\pm(z,n) = \frac{H(z,n) \pm R^{1/2}(z)}{2a(n)G(z,n)} = \frac{2a(n)G(z,n+1)}{H(z,n) \mp R^{1/2}(z)},$$

and thus

(7.63) $$4a(n)^2 G(z,n)G(z,n+1) = H(z,n)^2 - R(z),$$

which is equivalent to $\det M(z,n) = 1$.

Remark 7.6. In the case of closed gaps the factor

(7.64) $$Q(z) = \prod_{j\in\Gamma'}(z - E_{2j})$$

cancels from all equations since $\mu_j(n) = E_{2j-1} = E_{2j}$ if $j \in \Gamma'$. Here

(7.65) $\Gamma' = \{1 \leq j \leq N-1 | E_{2j-1} = E_{2j}\}$, $\Gamma = \{1 \leq j \leq N-1 | E_{2j-1} < E_{2j}\}$.

In particular, observe $\sqrt{z-E_{2j-1}}\sqrt{z-E_{2j}} = z-\mu_j(n)$ for $j \in \Gamma'$.

7.4. Two examples: period one and two

In this section we consider the two simplest cases when the period is one or two.

We start with period one. We assume $a(n) = \frac{1}{2}$, $b(n) = 0$ and consider them to be periodic with period N. By inspection, the Floquet functions ($n_0 = 0$) are given by

$$\psi_\pm(z,n) = u_\pm(z,n) = (z \pm R_2^{1/2}(z))^n,$$

(7.66) $$\phi_\pm(z) = z \pm R_2^{1/2}(z).$$

Hence the discriminant and the Floquet multipliers read

$$\Delta(z) = \frac{1}{2}\big(s(z,N+1) - s(z,N-1)\big) = \frac{1}{2}\big(m^+(z) + m^-(z)\big),$$

(7.67) $$m^\pm(z) = \Delta(z) \pm (\Delta(z)^2 - 1)^{1/2} = (z \pm R_2^{1/2}(z))^N.$$

7.4. Two examples: period one and two

Next we calculate

$$\begin{aligned}
\Delta(z)^2 - 1 &= \frac{1}{4}\big(m^+(z) - m^-(z)\big)^2 \\
&= \frac{1}{4}\big((z + R_2^{1/2}(z))^N - (z - R_2^{1/2}(z))^N\big)^2 \\
&= (z^2 - 1)s(z, N)^2.
\end{aligned}$$
(7.68)

Therefore we have

(7.69) $\quad E_0 = -1, \quad E_{2j-1} = E_{2j} = \mu_j = \lambda_j^\beta = -\cos(\frac{j\pi}{N}), \quad E_{2N-1} = 1,$

$1 \leq j \leq N - 1$, and all spectral gaps $\overline{\rho}_j$, $1 \leq j \leq N - 1$ are closed. We can now verify the first relation in (7.38) directly and from the second we obtain

(7.70) $$\lambda_N^\beta = -\frac{\beta + \beta^{-1}}{2}.$$

Which shows that $\lambda_N^\beta \in \overline{\rho}_N$ for $\beta < 0$ and $\lambda_N^\beta \in \overline{\rho}_0$ for $0 < \beta$.

Now let us consider the case of period $N = 2$. In this case $A = a(0)a(1)$ and $B = b(0) + b(1)$. Moreover, one computes

(7.71) $$\Delta(z) = \frac{(z - \frac{B}{2})^2 - C^2}{2A}, \quad C^2 = a(0)^2 + a(1)^2 + \frac{(b(0) - b(1))^2}{4}$$

and hence

$$\begin{aligned}
R(z) &= 4A^2(\Delta(z)^2 - 1) = ((z - \frac{B}{2})^2 - C^2)^2 - 4A^2 \\
&= (z - E_0)(z - E_1)(z - E_2)(z - E_3),
\end{aligned}$$
(7.72)

where

(7.73) $\quad E_{0,1} = \frac{B}{2} - \sqrt{C^2 \pm 2|A|}, \quad E_{2,3} = \frac{B}{2} + \sqrt{C^2 \pm 2|A|}.$

Conversely, we have $B = E_0 + E_2 = E_1 + E_3$, $C^2 = ((E_0 - E_2)^2 + (E_1 - E_3)^2)/8$, and $|A| = ((E_0 - E_2)^2 - (E_1 - E_3)^2)/16$. Our trace formulas (7.38) produce

(7.74) $\quad \mu_1(0) = b(1), \quad \mu_1(1) = b(0), \quad \lambda_1^\beta(n) = B - (\beta + \beta^{-1})a(n)$

and we observe $\Delta(\mu_1(0)) = \Delta(\mu_1(1)) = \frac{(\mu_1(0) - \mu_1(1))^2 - 4C^2}{8A} = \Delta(\mu_1)$. A short computation shows

(7.75) $\quad m^{\sigma_1(0)}(\mu_1) = -\frac{a(1)}{a(0)}, \quad m^{\sigma_1(1)}(\mu_1) = -\frac{a(0)}{a(1)},$

where $m^\pm(\mu_1) = m^\pm(\mu_1(0)) = m^\pm(\mu_1(1))$. Moreover, we have $\sigma_1(0) = -\sigma_1(1) = \pm\mathrm{sgn}(|a(0)| - |a(1)|)$ and $a(0) = a(1)$ implies $\Delta(\mu_1)^2 = 1$. Solving $m^\pm(\mu_1)$ for $a(n)$ shows

$$\begin{aligned}
a(0)^2 &= -A\, m^{-\sigma_1(0)}(\mu_1) = -A(\Delta(\mu_1) - \sigma_1(0)(\Delta(\mu_1)^2 - 1)^{1/2}), \\
a(1)^2 &= -A\, m^{\sigma_1(0)}(\mu_1) = -A(\Delta(\mu_1) + \sigma_1(0)(\Delta(\mu_1)^2 - 1)^{1/2}).
\end{aligned}$$
(7.76)

In particular, this shows that the coefficients $a(n), b(n)$ are expressible in terms of E_j, $0 \leq j \leq 3$, and (e.g.) $\mu_1(0), \sigma_1(0)$. This result will be generalized in Section 8.1.

7.5. Perturbations of periodic operators

In this section we are going to study short-range perturbations H of periodic operators H_p associated with sequences a, b satisfying $a(n) \to a_p(n)$ and $b(n) \to b_p(n)$ as $|n| \to \infty$. Our main hypothesis for this section reads

Hypothesis H. 7.7. Suppose a_p, b_p are given periodic sequences and H_p is the corresponding Jacobi operator. Let H be a perturbation of H_p such that

$$(7.77) \qquad \sum_{n \in \mathbb{Z}} |n(a(n) - a_p(n))| < \infty, \quad \sum_{n \in \mathbb{Z}} |n(b(n) - b_p(n))| < \infty.$$

We start with two preliminary lemmas.

Lemma 7.8. *Consider the* **Volterra sum equation**

$$(7.78) \qquad f(n) = g(n) + \sum_{m=n+1}^{\infty} K(n, m) f(m).$$

Suppose there is a sequence $\hat{K}(n, m)$ such that

$$(7.79) \quad |K(n,m)| \leq \hat{K}(n,m), \quad \hat{K}(n+1,m) \leq \hat{K}(n,m), \quad \hat{K}(n,.) \in \ell^1(0,\infty).$$

Then, for given $g \in \ell^\infty(0, \infty)$, there is a unique solution $f \in \ell^\infty(0, \infty)$, fulfilling the estimate

$$(7.80) \qquad |f(n)| \leq \Big(\sup_{m>n} |g(m)| \Big) \exp \Big(\sum_{m=n+1}^{\infty} \hat{K}(n,m) \Big).$$

If $g(n)$ and $K(n, m)$ depend continuously (resp. holomorphically) on a parameter and if $\hat{K}$ does not depend on this parameter, then the same is true for $f(n)$.

Proof. Using the standard iteration trick

$$(7.81) \qquad f_0(n) = g(n), \quad f_{j+1}(n) = \sum_{m=n+1}^{\infty} K(n,m) f_j(m),$$

we see that the solution is formally given by

$$(7.82) \qquad f(n) = \sum_{j=0}^{\infty} f_j(n).$$

To prove uniqueness and existence it remains to show that this iteration converges. We claim

$$(7.83) \qquad |f_j(n)| \leq \frac{\sup_{m>n} |g(m)|}{j!} \Big(\sum_{m=n+1}^{\infty} \hat{K}(n,m) \Big)^j, \quad j \geq 1,$$

7.5. Perturbations of periodic operators

which follows from

$$\begin{aligned}
|f_{j+1}(n)| &\leq \sum_{m=n+1}^{\infty} \hat{K}(n,m) \frac{\sup_{l>m} |g(l)|}{j!} \Big(\sum_{l=m+1}^{\infty} \hat{K}(m,l) \Big)^j \\
&\leq \frac{\sup_{m>n} |g(m)|}{j!} \sum_{m=n+1}^{\infty} \hat{K}(n,m) \Big(\sum_{l=m+1}^{\infty} \hat{K}(n,l) \Big)^j \\
&\leq \frac{\sup_{m>n} |g(m)|}{(j+1)!} \sum_{m=n+1}^{\infty} \Bigg(\Big(\sum_{l=m}^{\infty} \hat{K}(n,l) \Big)^{j+1} \\
&\qquad - \Big(\sum_{l=m+1}^{\infty} \hat{K}(n,l) \Big)^{j+1} \Bigg) \\
&= \frac{\sup_{m>n} |g(m)|}{(j+1)!} \Big(\sum_{l=n+1}^{\infty} \hat{K}(n,l) \Big)^{j+1},
\end{aligned}$$

where we have used ($s = \hat{K}(n,m) \geq 0$, $S = \sum_{\ell=m+1}^{\infty} \hat{K}(n,\ell) \geq 0$)

(7.84) $$(S+s)^{j+1} - S^{j+1} = sS^j \sum_{\ell=0}^{j} (1+\frac{s}{S})^\ell \geq (j+1)sS^j, \quad j \in \mathbb{N}.$$

This settles the iteration and the estimate (7.80). The rest follows from uniform convergence of the series (7.82). □

Remark 7.9. A similar result holds for equations of the type

(7.85) $$f(n) = g(n) + \sum_{m=-\infty}^{n-1} K(n,m) f(m).$$

Lemma 7.10. *Assume (H.7.7). Then there exist solutions $u_\pm(z,.)$, $z \in \mathbb{C}$, of $\tau u = zu$ satisfying*

(7.86) $$\lim_{n \to \infty} |m_p^\mp(z)^{n/N} (u_\pm(z,n) - u_{p,\pm}(z,n))| = 0,$$

where $u_{p,\pm}(z,.)$, $m_p^\pm(z)$, and N are the Floquet solutions, Floquet multipliers, and period of H_p, respectively (cf. (7.19)). In addition, $u_\pm(z,.)$ can be assumed continuous (resp. holomorphic) with respect to z whenever $u_{p,\pm}(z,.)$ are and they satisfy the equations

(7.87) $$u_\pm(z,n) = \frac{a_p(n - {0 \atop 1})}{a(n - {0 \atop 1})} u_{p,\pm}(z,n) \mp \sum_{m={n+1 \atop -\infty}}^{n-1} \frac{a_p(n - {0 \atop 1})}{a(n - {0 \atop 1})} K(z,n,m) u_\pm(z,m),$$

where

$$\begin{aligned}
K(z,n,m) &= \frac{((\tau - \tau_p) u_{p,-}(z))(m) u_{p,+}(z,n) - u_{p,-}(z,n)((\tau - \tau_p) u_{p,+}(z))(m)}{W_p(u_{p,-}(z), u_{p,+}(z))} \\
&= \frac{s_p(z,n,m+1)}{a_p(m+1)}(a(m) - a_p(m)) + \frac{s_p(z,n,m)}{a_p(m)}(b(m) - b_p(m)) \\
&\quad + \frac{s_p(z,n,m-1)}{a_p(m-1)}(a(m-1) - a_p(m-1))
\end{aligned}$$

(7.88)

($W_p(.,..)$ denotes the Wronskian formed with a_p rather than a).

Proof. We only prove the claim for $u_+(z,.)$, the one for $u_-(z,.)$ being similar. Using the transformation $\tilde{u}_+(z,.) = m_p^-(z)^{n/N} u_+(z,.)$ we get a sequence which is bounded near $+\infty$.

Suppose $\tilde{u}_+(z,.)$ satisfies (disregarding summability for a moment)

$$\tilde{u}_+(z,n) = \frac{a_p(n)}{a(n)} \tilde{u}_{p,+}(z,n)$$

(7.89)
$$+ \sum_{m=n+1}^{\infty} \frac{a_p(n)}{a(n)} m_p^-(z)^{(n-m)/N} K(z,n,m) \tilde{u}_+(z,m).$$

Then $u_+(z,.)$ fulfills $\tau u = zu$ and (7.86). Hence, if we can apply Lemma 7.8, we are done. To do this, we need an estimate for $K(z,n,m)$ or, equivalently, for the growth rate of the transfer matrix $\Phi(z,n,m)$. By (7.6) it suffices to show

(7.90)
$$\|(m_p^-(z) M_p(z,m))^n\| \leq const(z,m)\, n, \quad n \in \mathbb{N}.$$

Abbreviate $M(z) = m_p^-(z) M_p(z,m)$ and note that the eigenvalues of $M(z)$ are 1 and $m_p^-(z)^2$. After performing the unitary transformation $U(z) = (e(z), e^\perp(z))$, where $e(z)$ is the normalized eigenvector of $M(z)$ corresponding to the eigenvalue 1 and $e^\perp(z)$ is its orthogonal complement, we have

(7.91)
$$\tilde{M}(z) = U(z)^{-1} M(z) U(z) = \begin{pmatrix} 1 & \alpha(z) \\ 0 & m_p^-(z)^2 \end{pmatrix}.$$

Using

(7.92)
$$\tilde{M}(z)^n = \begin{pmatrix} 1 & \alpha(z) \sum_{j=0}^{n-1} m_p^-(z)^{2j} \\ 0 & m_p^-(z)^{2n} \end{pmatrix}, \quad n \in \mathbb{N},$$

and $|m_p^-(z)| \leq 1$ we see $\|\tilde{M}(z)^n\| \leq \max\{1, |\alpha(z)| n\} \leq (1+|\alpha(z)|)n$, which is the desired estimate. Since $e(z)$ and $\alpha(z)$ are continuous with respect to z, the constant in (7.90) can be chosen independent of z as long as z varies in compacts. □

Theorem 7.11. *Suppose (H.7.7) holds. Then we have $\sigma_{ess}(H) = \sigma(H_p)$, the point spectrum of H is finite and confined to the spectral gaps of H_p, that is, $\sigma_p(H) \subset \mathbb{R}\backslash\sigma(H_p)$. Furthermore, the essential spectrum of H_p is purely absolutely continuous.*

Proof. That $\sigma_{ess}(H) = \sigma_{ess}(H_p)$ follows from Lemma 3.9. To prove the remaining claims we use the solutions $u_\pm(\lambda,.)$ of $\tau u = \lambda u$ for $\lambda \in \sigma(H_p)$ found in Lemma 7.10. Since $u_\pm(\lambda,.)$, $\lambda \in \sigma(H_p)$ are bounded and do not vanish near $\pm\infty$, there are no eigenvalues in the essential spectrum of H and invoking Theorem 3.18 shows that the essential spectrum of H is purely absolutely continuous. Moreover, (7.86) with $\lambda = E_0$ implies that $H - E_0$ is non-oscillatory since we can assume (perhaps after flipping signs) $u_{p,\pm}(E_0,n) \geq \varepsilon > 0$, $n \in \mathbb{Z}$, and by Corollary 4.11 there are only finitely many eigenvalues below E_0. Similarly, (using Remark 4.12) there are only finitely many eigenvalues above E_{2N+1}. Applying Corollary 4.20 in each gap (E_{2j-1}, E_{2j}), $1 \leq j \leq N$, shows that the number of eigenvalues in each gap is finite as well. □

7.5. Perturbations of periodic operators

These results enable us to define what is known as **scattering theory** for the pair (H, H_p), where H is a Jacobi operator satisfying (H.7.7).

Since we are most of the time interested in the case $z \in \sigma(H_p)$, we shall normalize $u_{p,\pm}(\lambda, 0) = 1$ for $\lambda \in \sigma(H_p)$. In particular, note that we have $\overline{u_{p,\pm}(\lambda)} = u_{p,\mp}(\lambda)$, where the bar denotes complex conjugation. Since one computes

$$(7.93) \quad W(u_\pm(\lambda), \overline{u_\pm(\lambda)}) = W_p(u_{p,\pm}(\lambda), u_{p,\mp}(\lambda)) = \mp \frac{2\mathrm{i}\sin(q(\lambda)N)}{s_p(\lambda, N)}, \quad \lambda \in \sigma(H_p),$$

($s_p(\lambda, n)$ is the solution of $\tau_p u = zu$ corresponding to the initial condition $s_p(\lambda, 0) = 0$, $s_p(\lambda, 1) = 1$) we conclude that $u_\pm(\lambda)$, $\overline{u_\pm(\lambda)}$ are linearly independent for λ in the interior of $\sigma(H_p)$ (if two bands collide at E, the numerator and denominator of (7.93) both approach zero when $\lambda \to E$ and both have a nonzero limit). Hence we might set

$$(7.94) \quad u_\pm(\lambda, n) = \alpha(\lambda)\overline{u_\mp(\lambda, n)} + \beta_\mp(\lambda)u_\mp(\lambda, n), \quad \lambda \in \sigma(H_p),$$

where

$$(7.95) \quad \begin{aligned} \alpha(\lambda) &= \frac{W(u_\mp(\lambda), u_\pm(\lambda))}{W(u_\mp(\lambda), \overline{u_\mp(\lambda)})} = \frac{s_p(\lambda, N)}{2\mathrm{i}\sin(q(\lambda)N)} W(u_-(\lambda), u_+(\lambda)), \\ \beta_\pm(z) &= \frac{W(u_\mp(\lambda), \overline{u_\pm(\lambda)})}{W(u_\pm(\lambda), \overline{u_\pm(\lambda)})} = \pm \frac{s_p(\lambda, N)}{2\mathrm{i}\sin(q(\lambda)N)} W(u_\mp(\lambda), \overline{u_\pm(\lambda)}). \end{aligned}$$

The function $\alpha(\lambda)$ can be defined for all $\lambda \in \mathbb{C}\setminus\{E_{p,j}\}$. Moreover, the Plücker identity (2.169) with $f_1 = u_\mp$, $f_2 = u_\pm$, $f_3 = \overline{u_\mp}$, $f_4 = \overline{u_\pm}$, implies

$$(7.96) \quad |\alpha(\lambda)|^2 = 1 + |\beta_\pm(\lambda)|^2 \quad \text{and} \quad \overline{\beta_\pm(\lambda)} = -\beta_\mp(\lambda).$$

Using (7.87) one can also show

$$(7.97) \quad W(u_-(\lambda), u_+(\lambda)) = W_p(u_{p,-}(\lambda), u_{p,+}(\lambda)) + \sum_{n \in \mathbb{Z}} u_\pm(\lambda, n)((\tau - \tau_p)u_{p,\mp}(\lambda))(n)$$

and

$$(7.98) \quad W(u_\mp(\lambda), \overline{u_\pm(\lambda)}) = \mp \sum_{n \in \mathbb{Z}} u_\pm(\lambda, n)((\tau - \tau_p)\overline{u_{p,\pm}(\lambda)})(n).$$

We now define the scattering matrix

$$(7.99) \quad S(\lambda) = \begin{pmatrix} T(\lambda) & R_-(\lambda) \\ R_+(\lambda) & T(\lambda) \end{pmatrix}, \quad \lambda \in \sigma(H_p),$$

of the pair (H, H_p), where $T(\lambda) = \alpha(\lambda)^{-1}$ and $R_\pm(\lambda) = \alpha(\lambda)^{-1}\beta_\pm(\lambda)$. The matrix $S(\lambda)$ is easily seen to be unitary since by (7.96) we have $|T(\lambda)|^2 + |R_\pm(\lambda)|^2 = 1$ and $T(\lambda)\overline{R_+(\lambda)} = -\overline{T(\lambda)}R_-(\lambda)$.

The quantities $T(\lambda)$ and $R_\pm(\lambda)$ are called transmission and reflection coefficients respectively. The following equation further explains this notation:

$$(7.100)$$
$$T(\lambda)u_\pm(\lambda, n) = \begin{cases} T(\lambda)u_{p,\pm}(\lambda, n), & n \to \pm\infty \\ u_{p,\pm}(\lambda, n) + R_\mp(\lambda)u_{p,\mp}(\lambda, n), & n \to \mp\infty \end{cases}, \quad \lambda \in \sigma(H_p).$$

If we regard $u_{p,\pm}(\lambda, n)$ as incoming plain wave packet, then $T(\lambda)u_{p,\pm}(\lambda, n)$ and $R_\mp(\lambda)u_{p,\mp}(\lambda, n)$ are the transmitted and reflected packets respectively.

The quantities $T(\lambda)$ and $R_\pm(\lambda)$ can be expressed in terms of $\tilde{m}_\pm(z) = \tilde{m}_\pm(z, 0)$ as follows

$$T(\lambda) = \frac{\overline{u_\pm(\lambda, 0)}}{u_\mp(\lambda, 0)} \frac{2\mathrm{i}\mathrm{Im}(\tilde{m}_\pm(\lambda + \mathrm{i}0))}{\tilde{m}_-(\lambda + \mathrm{i}0) + \tilde{m}_+(\lambda)},$$

(7.101) $$R_\pm(\lambda) = -\frac{\overline{u_\pm(\lambda, 0)}}{u_\pm(\lambda, 0)} \frac{\tilde{m}_\mp(\lambda) + \overline{\tilde{m}_\pm(\lambda)}}{\tilde{m}_-(\lambda) + \tilde{m}_+(\lambda)}, \quad \lambda \in \sigma(H_p).$$

Here we have set $\tilde{m}_\pm(\lambda) = \lim_{\varepsilon \downarrow 0} \tilde{m}_\pm(\lambda + \mathrm{i}\varepsilon)$, $\lambda \in \mathbb{R}$ as usual. In addition, one verifies ($\lambda \in \sigma(H_p)$)

$$g(\lambda + \mathrm{i}0, n) = \frac{u_-(\lambda, n)u_+(\lambda, n)}{W(u_-(\lambda), u_+(\lambda))} = T(\lambda)\frac{s_p(\lambda, N)}{2\mathrm{i}\sin(q(\lambda)N)}u_-(\lambda, n)u_+(\lambda, n)$$

(7.102) $$= \frac{s_p(\lambda, N)}{2\mathrm{i}\sin(q(\lambda)N)}|u_\pm(\lambda, n)|^2\left(1 + R_\pm(\lambda)\frac{\overline{u_\pm(\lambda, n)}}{u_\pm(\lambda, n)}\right).$$

We can now give a first application of the trace formulas derived in Section 6.2. Denote by $E_0 \leq E_1 < E_2 \leq \cdots < E_{2M} \leq E_{2M+1}$ the band edges of H, where equality holds if and only if $E_{2j} = E_{2j+1}$ is an eigenvalue of H. That is,

(7.103) $$\sigma(H) = \bigcup_{j=0}^{M-1}[E_{2j}, E_{2j+1}].$$

Furthermore, define the number $\mu_j(n)$ associated with each spectral gap $\rho_j = (E_{2j-1}, E_{2j})$ by (compare Section 8.1 for further details)

(7.104) $$\mu_j(n) = \sup\{E_{2j-1}\} \cup \{\lambda \in \rho_j | g(\lambda, n) < 0\} \in \overline{\rho_j}, \quad 1 \leq j \leq M.$$

Then we infer

$$\xi(\lambda, n) = \frac{1}{2}\chi_{(E_0, E_\infty)}(\lambda) + \frac{1}{2}\sum_{j=1}^M \left(\chi_{(E_{2j-1}, \mu_j(n))}(\lambda) - \chi_{(\mu_j(n), E_{2j})}(\lambda)\right)$$

(7.105) $$+ \chi_{(E_\infty, \infty)}(\lambda) + \frac{1}{\pi}\arg\left(1 + R_\pm(\lambda)\frac{\overline{u_\pm(\lambda, n)}}{u_\pm(\lambda, n)}\right)\chi_{\sigma(H_p)}(\lambda)$$

since we have

(7.106) $$\xi(\lambda, n) = \frac{1}{2} + \frac{1}{\pi}\arg\left(1 + R_\pm(\lambda)\frac{\overline{u_\pm(\lambda, n)}}{u_\pm(\lambda, n)}\right), \quad \lambda \in \sigma(H_p).$$

Hence we obtain from (6.58)

$$b^{(\ell)}(n) = \frac{1}{2}\sum_{j=0}^{2M+1} E_j^\ell - \sum_{j=1}^{M-1}\mu_j(n)^\ell$$

(7.107) $$+ \frac{\ell}{\pi}\int_{\sigma(H_p)} \lambda^{\ell-1}\arg\left(1 + R_\pm(\lambda)\frac{\overline{u_\pm(\lambda, n)}}{u_\pm(\lambda, n)}\right)d\lambda.$$

Remark 7.12. If H is reflectionless, that is $R_\pm(\lambda) = 0$, then H can be obtained from H_p by inserting the corresponding number of eigenvalues using the double commutation method provided in Section 11.6 since this transformation preserves the reflectionless property (cf. Remark 11.22).

7.5. Perturbations of periodic operators

Finally, we will write down the eigenfunction expansions for this case. We set (cf. (2.123))

(7.108) $$\underline{U}(\lambda, n) = \begin{pmatrix} u_+(\lambda, n) \\ u_-(\lambda, n) \end{pmatrix} = U(\lambda) \begin{pmatrix} c(\lambda, n) \\ s(\lambda, n) \end{pmatrix},$$

where

(7.109) $$U(\lambda) = \begin{pmatrix} u_+(\lambda, 0) & -u_+(\lambda, 0)\tilde{m}_+(\lambda) \\ u_-(\lambda, 0) & u_-(\lambda, 0)\tilde{m}_-(\lambda) \end{pmatrix}.$$

By the considerations at the end of Section 2.5 this choice of basis will diagonalize the matrix measure in the eigenfunction expansion. A short calculation (using (7.93)) shows

(7.110) $$|u_\pm(\lambda, 0)|^2 = \frac{-\sin(q(\lambda)N)}{a(0)^2 s_p(\lambda, N)\mathrm{Im}(\tilde{m}_\pm(\lambda))}, \quad \lambda \in \sigma(H_p),$$

that $u_\pm(\lambda, n)$ are not correctly normalized. However, this can be taken into account easily. Transforming the spectral measure to this new basis yields (use (7.101))

(7.111) $$d\tilde{\rho}_{ac}(\lambda) = (U^{-1}(\lambda))^\top d\rho_{ac}(\lambda) \overline{U^{-1}(\lambda)}$$
$$= \frac{-s_p(\lambda, N)}{4\sin(q(\lambda)N)} |T(\lambda)|^2 \begin{pmatrix} 1 & 0 \\ 0 & 1 \end{pmatrix} \chi_{\sigma(H_p)}(\lambda) d\lambda.$$

For the pure point part we can choose (e.g.)

(7.112) $$d\tilde{\rho}_{pp}(\lambda) = \sum_{E_j \in \sigma_p(H)} \begin{pmatrix} \gamma_+(E_j) & 0 \\ 0 & 0 \end{pmatrix} d\Theta(\lambda - E_j),$$

where $\gamma_+(E_j)$ is the norming constant corresponding to the eigenvalue E_j, that is,

(7.113) $$\gamma_+(E_j) = \left(\sum_{n \in \mathbb{Z}} |u_+(E_j, n)|^2 \right)^{-1}$$

and $\Theta(\lambda) = 0$ for $\lambda \leq 0$ respectively $\Theta(\lambda) = 1$ for $\lambda > 0$.

Transforming (2.124) shows that $\underline{U}(\lambda, n)$ are orthogonal with respect to $d\tilde{\rho} = d\tilde{\rho}_{ac} + d\tilde{\rho}_{pp}$,

(7.114) $$\langle \underline{U}(\lambda, m), \underline{U}(\lambda, n) \rangle_{L^2} = \sum_{i,j=0}^{1} \int_{\mathbb{R}} U_i(\lambda, m) U_j(\lambda, n) d\tilde{\rho}_{i,j}(\lambda)$$
$$\equiv \int_{\mathbb{R}} \underline{U}(\lambda, m) \underline{U}(\lambda, n) d\tilde{\rho}(\lambda) = \delta_{m,n}.$$

Moreover, we obtain a unitary transformation $\tilde{U} : \ell^2(\mathbb{Z}) \to L^2(\mathbb{R}, d\tilde{\rho})$ defined by

(7.115) $$(\tilde{U}f)(\lambda) = \sum_{n \in \mathbb{Z}} f(n) \underline{U}(\lambda, n),$$
$$(\tilde{U}^{-1}\underline{F})(n) = \int_{\mathbb{R}} \underline{U}(\lambda, n) \underline{F}(\lambda) d\tilde{\rho}(\lambda),$$

which maps the operator H to the multiplication operator by λ,

(7.116) $$\tilde{U} H \tilde{U}^{-1} = \tilde{H},$$

where
(7.117) $$\tilde{H}\underline{F}(\lambda) = \lambda \underline{F}(\lambda), \quad \underline{F}(\lambda) \in L^2(\mathbb{R}, d\tilde{\rho}).$$

Chapter 8

Reflectionless Jacobi operators

In this chapter we are going to look at a class of operators for which the trace formulas of Section 6.2 become particularly simple. Based on this fact we are able to give a detailed spectral and inverse spectral analysis.

8.1. Spectral analysis and trace formulas

In this section we discuss the direct spectral problem for a certain class of reflectionless bounded Jacobi operators.

A Jacobi operator H is called **reflectionless** if for all $n \in \mathbb{Z}$,

(8.1) $$\xi(\lambda, n) = \frac{1}{2} \text{ for a.e. } \lambda \in \sigma_{ess}(H),$$

where $\xi(\lambda, n)$ is the xi-function of H introduced in (6.53).

For instance, periodic operators and operators with purely discrete spectrum are special cases of reflectionless operators. The following result further illustrates the reflectionless condition (8.1).

Lemma 8.1. Suppose $\Lambda \subset \sigma(H)$. Then the following conditions are equivalent.
(i). For all $n \in \mathbb{Z}$, $\xi(\lambda, n) = \frac{1}{2}$ for a.e. $\lambda \in \Lambda$.
(ii). For some $n_0 \in \mathbb{Z}$, $n_1 \in \mathbb{Z}\backslash\{n_0, n_0 + 1\}$,

$$\xi(\lambda, n_0) = \xi(\lambda, n_0 + 1) = \xi(\lambda, n_1) = \frac{1}{2} \text{ for a.e } \lambda \in \Lambda.$$

(iii). For some $n_0 \in \mathbb{Z}$,

$$\tilde{m}_+(\lambda + i0, n_0) = -\overline{\tilde{m}_-(\lambda + i0, n_0)} \text{ for a.e. } \lambda \in \Lambda,$$

or equivalently $a(n_0)^2 m_+(\lambda + i0, n_0) = -a(n_0 - 1)^2 \overline{m_-(\lambda + i0, n_0)} - \lambda + b(n_0)$.
(iv) For some $n_0 \in \mathbb{Z}$,

$$\text{Re}(g(\lambda, n_0)) = \text{Re}(h(\lambda, n_0)) = 0 \text{ for a.e. } \lambda \in \Lambda.$$

(v) For some $n_0 \in \mathbb{Z}$,
$$\tilde{m}_+^\beta(\lambda + i0, n_0) = -\overline{\tilde{m}_-^\beta(\lambda + i0, n_0)} \text{ for a.e. } \lambda \in \Lambda.$$

(vi). For some $n_0 \in \mathbb{Z}$ and $u_\pm(z, n)$ as in (2.20),
$$u_+(\lambda + i0, n_0) = \overline{u_-(\lambda + i0, n_0)} \text{ for a.e. } \lambda \in \Lambda.$$

Proof. Without restriction we choose $n_0 = 0$. Moreover, for a.e. $\lambda \in \Lambda$ we can assume that both $\tilde{m}_\pm(\lambda + i0)$ exist and that $|\tilde{m}_-(\lambda + i0)|^2 + |\tilde{m}_+(\lambda + i0)|^2 > 0$, $s(\lambda, n_1) c(\lambda, n_1) \neq 0$. Hence we fix such a λ and abbreviate $\tilde{m}_\pm = \tilde{m}_\pm(\lambda + i0)$. Using (2.20) we see that the requirement $\xi(\lambda, n) = \frac{1}{2}$, that is, $g(\lambda + i0, n) = -\overline{g(\lambda + i0, n)}$, is equivalent to

$$\text{Re}\Big(\overline{(\tilde{m}_- + \tilde{m}_+)}(c(\lambda, n)^2 + a(n)(\tilde{m}_- - \tilde{m}_+)s(\lambda, n)c(\lambda, n))$$
(8.2)
$$- a(n)^2(\tilde{m}_- \tilde{m}_+)s(\lambda, n)^2\Big) = 0.$$

Clearly (i) $\Rightarrow$ (ii). In order to prove that (ii) $\Rightarrow$ (iii) we first pick $n = 0$ in (8.2) yielding $\text{Re}(\tilde{m}_-) = -\text{Re}(\tilde{m}_+)$. Furthermore choosing $n = 1$ shows $\text{Im}(\tilde{m}_-) = \text{Im}(\tilde{m}_+)$ or $\text{Re}(\tilde{m}_-) = \text{Re}(\tilde{m}_+) = 0$. In the latter case, $\text{Re}(\tilde{m}_-) = \text{Re}(\tilde{m}_+) = 0$, we obtain for $n = n_1$ that $\text{Im}(\tilde{m}_- + \tilde{m}_+) \text{Im}(\tilde{m}_- - \tilde{m}_+) = 0$ and since $\text{Im}(\tilde{m}_\pm) > 0$ again $\text{Im}(\tilde{m}_-) = \text{Im}(\tilde{m}_+)$. This completes (ii) $\Rightarrow$ (iii). The case (iii) $\Rightarrow$ (i) is obvious from (8.2) and the rest is evident from (2.20) and (2.92). $\square$

The last condition (vi) implies vanishing of the *reflection coefficients* $R_\pm(\lambda)$ (cf. (7.99)), explaining the term reflectionless (see also (7.106)). Moreover, condition (iii) allows us to strengthen Theorem 2.29 for reflectionless Jacobi operators.

Lemma 8.2. *Suppose H is reflectionless and $\sigma_{ess}(H)$ is of positive Lebesgue measure, then H is uniquely determined by one of the Weyl $\tilde{m}$-functions $\tilde{m}_+(z, n_0)$ or $\tilde{m}_-(z, n_0)$. The result still holds if the reflectionless condition only holds for a set of positive Lebesgue measure.*

Proof. Combine Lemma 8.1 (iii) and Theorem B.8 (iii). $\square$

Since the spectrum of H is closed, it can be written as the complement of a countable union of disjoint open intervals, that is,

(8.3)
$$\sigma(H) = \Sigma = \mathbb{R} \setminus \bigcup_{j \in J_0 \cup \{\infty\}} \rho_j,$$

where $J \subseteq \mathbb{N}$, $J_0 = J \cup \{0\}$,

(8.4)
$$\rho_0 = (-\infty, E_0), \quad \rho_\infty = (E_\infty, \infty),$$
$$E_0 \leq E_{2j-1} < E_{2j} \leq E_\infty, \quad \rho_j = (E_{2j-1}, E_{2j}), \quad j \in J,$$
$$-\infty < E_0 < E_\infty < \infty, \quad \rho_j \cap \rho_k = \emptyset \text{ for } j \neq k.$$

We emphasize that the notation employed in (8.4) implies that $E_{2j} = E_{2k+1}$ for some $k \in J_0 \cup \{\infty\}$ whenever $E_{2j} \in \sigma_d(H)$.

Next, we turn to Dirichlet eigenvalues associated with each spectral gap ρ_j. Formally $\mu_j(n)$ is the zero of $g(z, n)$ in ρ_j. Unfortunately, this definition causes

8.1. Spectral analysis and trace formulas

trouble since μ_j could lie at a boundary point of ρ_j where $g(.,n)$ is no longer holomorphic. Hence we use the more sophisticated one

(8.5) $$\mu_j(n) = \sup\{E_{2j-1}\} \cup \{\lambda \in \rho_j | g(\lambda, n) < 0\} \in \overline{\rho_j}, \quad j \in J.$$

Strict monotonicity of $g(\lambda, n)$ with respect to $\lambda \in \rho_j$ (cf. (2.37)) then yields

(8.6) $$\begin{aligned} g(\lambda, n) < 0, & \quad \lambda \in (E_{2j-1}, \mu_j(n)), \\ g(\lambda, n) > 0, & \quad \lambda \in (\mu_j(n), E_{2j}), \end{aligned} \quad j \in J.$$

Moreover, let $\mu \in \mathbb{R} \backslash \sigma_{ess}(H)$, then $g(\mu, n) = 0$ implies $u_-(\mu, n) = 0$ or $u_+(\mu, n) = 0$ (if $\mu \in \sigma_d(H)$ we have $u_-(\mu, n) = u_+(\mu, n) = 0$). Thus $\mu \in \sigma_d(H_n^\infty)$ and hence we infer

(8.7) $$\sigma(H_n^\infty) = \sigma_{ess}(H) \cup \{\mu_j(n)\}_{j \in J}.$$

However, observe that $\mu_j(n)$ is not necessarily an eigenvalue of H_n^∞ unless $\mu_j(n) \notin \sigma_{ess}(H)$.

Equation (8.6) shows that $\xi(\lambda, n) = 1$, $\lambda \in (E_{2j-1}, \mu_j(n))$, and $\xi(\lambda, n) = 0$, $\lambda \in (\mu_j(n), E_{2j})$, $j \in J$. Thus we know $\xi(\lambda, n)$ for $\lambda \in \mathbb{R} \backslash \sigma_{ess}(H)$ and if we assume $\xi(\lambda, n) = 1/2$ for $\lambda \in \sigma_{ess}(H)$ we obtain (see also Section 6.2)

Lemma 8.3. *Let H be reflectionless. Then*

(8.8) $$\begin{aligned} g(z, n) &= \frac{-1}{z - E_\infty} \exp\left(\int_{E_0}^{E_\infty} \frac{\xi(\lambda, n) d\lambda}{\lambda - z}\right) \\ &= \frac{-1}{\sqrt{z - E_0}\sqrt{z - E_\infty}} \prod_{j \in J} \frac{z - \mu_j(n)}{\sqrt{z - E_{2j-1}}\sqrt{z - E_{2j}}}. \end{aligned}$$

In particular, denoting by $\chi_\Lambda(.)$ the characteristic function of the set $\Lambda \subset \mathbb{R}$, one can represent $\xi(\lambda, n)$ by

$$\begin{aligned} \xi(\lambda, n) &= \frac{1}{2}\Big(\chi_{(E_0, \infty)}(\lambda) + \chi_{(E_\infty, \infty)}(\lambda)\Big) \\ &\quad + \frac{1}{2}\sum_{j \in J}\Big(\chi_{(E_{2j-1}, \infty)}(\lambda) + \chi_{(E_{2j}, \infty)}(\lambda) - 2\chi_{(\mu_j(n), \infty)}(\lambda)\Big) \\ &= \frac{1}{2}\chi_{(E_0, E_\infty)}(\lambda) + \frac{1}{2}\sum_{j \in J}\Big(\chi_{(E_{2j-1}, \mu_j(n))}(\lambda) - \chi_{(\mu_j(n), E_{2j})}(\lambda)\Big) \end{aligned}$$

(8.9) $$+ \chi_{(E_\infty, \infty)}(\lambda) \quad \text{for a.e. } \lambda \in \mathbb{R}.$$

In the case of two gaps the situation is depicted below.

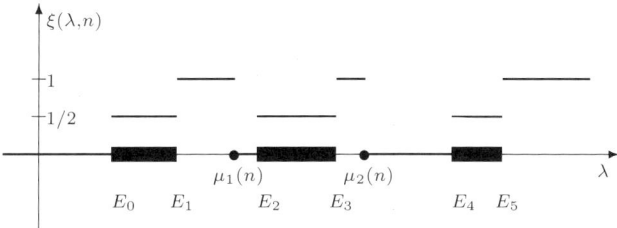

For later purpose we observe that evaluation of (6.58) shows

$$b^{(\ell)}(n) = \frac{1}{2}\Big(E_0^\ell + E_\infty^\ell + \sum_{j \in J}(E_{2j-1}^\ell + E_{2j}^\ell - 2\mu_j(n)^\ell)\Big) \tag{8.10}$$

and in the special case $\ell = 1$

$$b(n) = \frac{1}{2}\Big(E_0 + E_\infty + \sum_{j \in J}(E_{2j-1} + E_{2j} - 2\mu_j(n))\Big). \tag{8.11}$$

Now we consider the inverse spectral problem associated with reflectionless operators more closely. To reconstruct H we need (e.g.) $\tilde{m}_\pm(z,n)$. However, given E_j, $\mu_j(n)$ we only know $g(z,n)$ and thus only $a(n)^2(\tilde{m}_-(z,n)+\tilde{m}_+(z,n)) = -g(z,n)^{-1}$. Thus it suffices to split $-g(z,n)^{-1}$ into $a(n)^2\tilde{m}_+(z,n)$ and $a(n)^2\tilde{m}_-(z,n)$. Rather than treating $-g(z,n)^{-1}$ directly, our strategy is to consider the Herglotz representation

$$-g(z,n)^{-1} = z - b(n) + \int_\mathbb{R} \frac{d\tilde{\rho}_n(\lambda)}{\lambda - z} \tag{8.12}$$

and to seek the corresponding split up of the measure $d\tilde{\rho}_n = d\tilde{\rho}_{-,n} + d\tilde{\rho}_{+,n}$. For if we have $d\tilde{\rho}_{\pm,n}$, we also know (see (2.13))

$$a(n)^2 = \int_\mathbb{R} d\tilde{\rho}_{+,n}, \quad a(n-1)^2 = \int_\mathbb{R} d\tilde{\rho}_{-,n} \tag{8.13}$$

and

$$m_\pm(z,n) = a(n-{0 \atop 1})^{-2} \int_\mathbb{R} \frac{d\tilde{\rho}_{\pm,n}(\lambda)}{\lambda - z}. \tag{8.14}$$

Invoking Theorem 2.29 will then show that H is uniquely determined.

The process of splitting up $d\tilde{\rho}_n$ can be reduced to splitting up the pure point, absolutely continuous, and singularly continuous part, respectively.

We start with the pure point part and introduce

$$\sigma_p(H_n^\infty) = \{\tilde{\mu}_j(n)\}_{j \in \tilde{J}} = \{\mu \in \mathbb{R} | \tilde{\rho}_n(\{\mu\}) > 0\}, \quad \tilde{J} \subseteq \mathbb{N}. \tag{8.15}$$

Note that $\sigma_p(H_n^\infty)$ is known from $d\tilde{\rho}_n$. Even though the two sets $\{\tilde{\mu}_j(n)\}_{j \in \tilde{J}}$ and $\{\mu_j(n)\}_{j \in J}$ are closely related, unfortunately neither $\{\tilde{\mu}_j(n)\}_{j \in \tilde{J}} \subseteq \{\mu_j(n)\}_{j \in J}$ nor $\{\mu_j(n)\}_{j \in J} \subseteq \{\tilde{\mu}_j(n)\}_{j \in \tilde{J}}$ is true in general. We have observed before that μ_j might not be an eigenvalue if it lies at the boundary of its gap. On the other hand, an accumulation point of the μ_j's might be an eigenvalue (see Remark 8.11 (i) below).

Moreover, we introduce the numbers

$$\tilde{R}_j(n) = \lim_{\varepsilon \downarrow 0} i\varepsilon g(\tilde{\mu}_j(n) + i\varepsilon, n)^{-1} = \tilde{\rho}_n(\{\tilde{\mu}_j(n)\}) > 0, \tag{8.16}$$

and

$$\tilde{\sigma}_j(n) = \lim_{\varepsilon \downarrow 0} h(\tilde{\mu}_j(n) + i\varepsilon, n) = \frac{\tilde{\rho}_{+,n}(\{\tilde{\mu}_j(n)\}) - \tilde{\rho}_{-,n}(\{\tilde{\mu}_j(n)\})}{\tilde{\rho}_{+,n}(\{\tilde{\mu}_j(n)\}) + \tilde{\rho}_{-,n}(\{\tilde{\mu}_j(n)\})} \in [-1,1]. \tag{8.17}$$

Equivalently, we have (use (2.201))

$$\tilde{m}_\pm(z,n) = -\frac{1 \mp \tilde{\sigma}_j(n)}{2} \frac{\tilde{R}_j(n)a(n)^{-2}}{z - \tilde{\mu}_j(n)} + O(z - \tilde{\mu}_j(n))^0. \tag{8.18}$$

In particular, $\tilde{\sigma}_j(n)$ is either ± 1 (depending on whether $\tilde{\mu}_j$ is an eigenvalue of $H_{+,n}$ or $H_{-,n}$) or in $(-1,+1)$ (if $\tilde{\mu}_j$ is an eigenvalue of both $H_{\pm,n}$ and hence also of H).

8.1. Spectral analysis and trace formulas

The numbers $\tilde{R}_j(n)$ can be evaluated using (8.8)

$$\tilde{R}_j(n)^2 = (\tilde{\mu}_j(n) - E_0)(\tilde{\mu}_j(n) - E_\infty)(\tilde{\mu}_j(n) - E_{2j-1})(\tilde{\mu}_j(n) - E_{2j}) \times$$
$$\tag{8.19} \prod_{k \in J \setminus \{j\}} \frac{(\tilde{\mu}_j(n) - E_{2k-1})(\tilde{\mu}_j(n) - E_{2k})}{(\tilde{\mu}_j(n) - \mu_k(n))^2}.$$

If $\tilde{\mu}_j = \mu_k = E_{2k} = E_{2j-1}$ for some k (resp. $\tilde{\mu}_j = \mu_k = E_{2k-1} = E_{2j}$) the vanishing factors $(\tilde{\mu}_j - \mu_k)^2$ in the denominator and $(\tilde{\mu}_j - E_{2j-1})(\tilde{\mu}_j - E_{2k})$ (resp. $(\tilde{\mu}_j - E_{2k-1})(\tilde{\mu}_j - E_{2j})$) in the numerator have to be omitted.

By now we know how to split up the pure point and the absolutely continuous (which is being taken care of by the reflectionless condition) part of $d\tilde{\rho}_n$, that is,

$$\tag{8.20} d\tilde{\rho}_{n,\pm,pp}(\{\tilde{\mu}_j(n)\}) = \frac{1 \pm \tilde{\sigma}_j(n)}{2} d\tilde{\rho}_{n,pp}(\{\tilde{\mu}_j(n)\}), \quad d\tilde{\rho}_{n,\pm,ac} = \frac{1}{2} d\tilde{\rho}_{n,ac}.$$

Since we do not want to bother with the singularly continuous part, we will assume $\sigma_{sc}(H_n^\infty) = \emptyset$, that is, $\tilde{\rho}_{n,sc} = 0$, for simplicity. In this case the spectral data E_j, $j \in J \cup \{0, \infty\}$, plus $\mu_j(n_0)$, $j \in J$, plus $\tilde{\sigma}_j(n_0)$, $j \in \tilde{J}$, for one fixed $n_0 \in \mathbb{Z}$ are minimal and uniquely determine $a(n)^2$, $b(n)$. Without this assumption we would need to introduce $\sigma(n,.) : \sigma_{sc}(H_n^\infty) \to [-1,1]$ such that $d\tilde{\rho}_{n,\pm,sc}(\lambda) = \frac{1 \pm \sigma(n,\lambda)}{2} d\tilde{\rho}_{n,sc}(\lambda)$. The corresponding modifications are straightforward.

Now let us focus on the actual reconstruction of $a(n)^2$, $b(n)$ from given spectral data as above and present an explicit expression of $a(n)^2$, $b(n)$ in terms of the spectral data.

As a preparation we observe that $d\tilde{\rho}_{n,+} - d\tilde{\rho}_{n,-}$ consists only of a pure point part. In fact, due to the reflectionless condition, the absolutely continuous part cancels and the singularly continuous part is absent by assumption. As a consequence $a(n)^2(\tilde{m}_-(z,n) - \tilde{m}_+(z,n))$ is meromorphic and we obtain an interesting expression for $h(z,n)$ (cf. (2.201)).

Lemma 8.4. *Suppose H is reflectionless and $\sigma_{sc}(H_n^\infty) = \emptyset$. Then we have*

$$\tag{8.21} h(z,n) = g(z,n)\Big(z - b(n) + \sum_{j \in \tilde{J}} \frac{\tilde{\sigma}_j(n)\tilde{R}_j(n)}{z - \tilde{\mu}_j(n)}\Big).$$

Proof. Since the (signed) measure $d\tilde{\rho}_{n,+} - d\tilde{\rho}_{n,-}$ is pure point, the integral in the corresponding Herglotz representation can be easily computed. $\square$

Our idea for reconstructing $a(n)$, $b(n)$ is to express the moments of $m_\pm(z,n)$ in terms of the spectral data. Once this is done, the rest follows from (2.118). Since we already know $g(z,n)$ and $h(z,n)$ in terms of the spectral data, it suffices to relate these quantities to $m_\pm(z,n)$. But this is the content of (2.202) from which we infer

$$a(n)^2 m_+(z,n) \pm a(n-1)^2 m_-(z,n) = \mp z \pm b(n) - \begin{cases} \frac{1}{g(z,n)} \\ \frac{h(z,n)}{g(z,n)} \end{cases}$$

$$\tag{8.22} = -\sum_{j=0}^{\infty} \frac{c_{\pm,j}(n)}{z^{j+1}},$$

where the coefficients $c_{\pm,j}(n)$ are to be determined. From (8.21) we infer

(8.23) $$c_{-,\ell}(n) = \sum_{j\in\tilde{J}} \tilde{\sigma}_j(n)\tilde{R}_j(n)\tilde{\mu}_j(n)^\ell, \quad \ell \in \mathbb{N}_0.$$

Moreover, $c_{+,\ell}(n)$ can be determined from (cf. (6.56))

(8.24) $$\frac{1}{g(z,n)} = -z \exp\Big(-\sum_{\ell=1}^\infty \frac{b^{(\ell)}(n)}{\ell z^\ell}\Big),$$

which implies

$$c_{+,-2}(n) = 1,$$

(8.25) $$c_{+,\ell-2}(n) = \frac{1}{\ell}\sum_{j=1}^\ell c_{+,\ell-j-2}(n)b^{(j)}(n), \quad \ell \in \mathbb{N}.$$

Thus $c_{+,\ell}$ are expressed in terms of E_j, $\mu_j(n)$ (by (8.10)). Here $c_{+,-2}(n)$ and $c_{+,-1}(n)$ have been introduced for notational convenience only.

In particular, combining the case $\ell = 0$ with our previous results we obtain

(8.26) $$a(n - {}^0_1)^2 = \frac{b^{(2)}(n) - b(n)^2}{4} \pm \sum_{j\in\tilde{J}} \frac{\tilde{\sigma}_j(n)}{2} \tilde{R}_j(n).$$

Similarly, for $\ell = 1$,

(8.27) $$b(n \pm 1) = \frac{1}{a(n - {}^0_1)^2}\Big(\frac{2b^{(3)}(n) - 3b(n)b^{(2)}(n) + b(n)^3}{12} \\ \pm \sum_{j\in\tilde{J}} \frac{\tilde{\sigma}_j(n)}{2} \tilde{R}_j(n)\tilde{\mu}_j(n)\Big).$$

However, these formulas are only the tip of the iceberg. Combining

(8.28) $$c_{\pm,\ell}(n) = \int_\mathbb{R} \lambda^\ell \Big(a(n)^2 d\rho_{+,n}(\lambda) \pm a(n-1)^2 d\rho_{-,n}(\lambda)\Big) \\ = a(n)^2 m_{+,\ell}(n) \pm a(n-1)^2 m_{-,\ell}(n), \quad \ell \in \mathbb{N}_0,$$

with (2.118) we obtain the main result of this section.

Theorem 8.5. *Let H be a given bounded reflectionless Jacobi operator. Suppose the singular continuous spectrum of H_n^∞ is empty and the spectral data (E_j), (μ_j), and $(\tilde{\sigma}_j)$ corresponding to H (as above) are given for one fixed $n \in \mathbb{Z}$. Then the sequences a^2, b can be expressed explicitly in terms of the spectral data as follows*

(8.29) $$a(n \pm k - {}^0_1)^2 = \frac{C_{\pm,n}(k+1)C_{\pm,n}(k-1)}{C_{\pm,n}(k)^2}, \\ b(n \pm k) = \frac{D_{\pm,n}(k)}{C_{\pm,n}(k)} - \frac{D_{\pm,n}(k-1)}{C_{\pm,n}(k-1)}, \quad k \in \mathbb{N},$$

where $C_{\pm,n}(k)$ and $D_{\pm,n}(k)$ are defined as in (2.109) and (2.113) using $m_{\pm,\ell}(n) = \frac{c_{+,\ell}(n)\pm c_{-,\ell}(n)}{2a(n-{}^0_1)}$. The quantities $a(n)^2$, $a(n-1)^2$, and $c_{\pm,\ell}(n)$ have to be expressed in terms of the spectral data using (8.26), (8.25), and (8.23), (8.10), respectively.

In addition, we get the neat

8.1. Spectral analysis and trace formulas

Corollary 8.6. *Let H be a reflectionless Jacobi operator with spectrum consisting of only one band, that is $\sigma(H) = [E_0, E_\infty]$. Then the sequences $a(n)^2, b(n)$ are necessarily constant*

$$(8.30) \qquad a(n)^2 = \frac{(E_\infty - E_0)^2}{16}, \quad b(n) = \frac{E_0 + E_\infty}{2}.$$

If J is finite, that is, H has only finitely many spectral gaps, then $\{\tilde{\mu}_j(n)\}_{j \in \tilde{J}} \subseteq \{\mu_j(n)\}_{j \in J}$ and we can forget about the $\tilde{\mu}$'s. This case will be investigated more closely in Section 8.3.

Finally, we turn to general eigenvalues associated with H_n^β. Associated with each spectral gap ρ_j we set

$$(8.31) \qquad \lambda_j^\beta(n) = \sup\{E_{2j-1}\} \cup \{\lambda \in \rho_j | \gamma^\beta(\lambda, n) < 0\} \in \overline{\rho_j}, \quad j \in J.$$

The strict monotonicity of $\gamma^\beta(\lambda, n)$ with respect to $\lambda \in \rho_j, j \in J_0 \cup \{\infty\}$, that is,

$$(8.32) \qquad \frac{d}{d\lambda}\gamma^\beta(\lambda, n) = (1+\beta^2)\langle \delta_n^\beta, (H-\lambda)^{-2}\delta_n^\beta \rangle, \quad \lambda \in \rho_j,$$

then yields

$$(8.33) \qquad \begin{array}{l} \gamma^\beta(\lambda, n) < 0, \quad \lambda \in (E_{2j-1}, \lambda_j^\beta(n)), \\ \gamma^\beta(\lambda, n) > 0, \quad \lambda \in (\lambda_j^\beta(n), E_{2j}), \end{array} \quad j \in J.$$

Since $\gamma^\beta(\lambda, n)$ is positive (resp. negative) for $a(n)\beta < 0$ (resp. $a(n)\beta > 0$) as $|\lambda| \to \infty$, there must be an additional zero

$$(8.34) \qquad \lambda_\infty^\beta = \begin{cases} \sup\{E_\infty\} \cup \{\lambda \in \rho_\infty | \gamma^\beta(\lambda, n) < 0\}, & a(n)\beta < 0 \\ \sup\{\lambda \in \rho_0 | \gamma^\beta(\lambda, n) < 0\}, & a(n)\beta > 0 \end{cases}$$

for $\lambda \geq E_\infty$ (resp. $\lambda \leq E_0$). Summarizing, $\xi^\beta(\lambda, n)$ is given by

$$\xi^\beta(\lambda, n) = \frac{1}{2}\chi_{(E_0, E_\infty)}(\lambda) + \frac{1}{2}\sum_{j \in J}\left(\chi_{(E_{2j-1}, \lambda_j^\beta(n))}(\lambda) - \chi_{(\lambda_j^\beta(n), E_{2j})}(\lambda)\right)$$
$$(8.35) \qquad + \chi_{(E_\infty, \lambda_\infty^\beta)}(\lambda), \quad a(n)\beta < 0,$$

and

$$\xi^\beta(\lambda, n) = -\frac{1}{2}\chi_{(E_0, E_\infty)}(\lambda) + \frac{1}{2}\sum_{j \in J}\left(\chi_{(E_{2j-1}, \lambda_j^\beta(n))}(\lambda) - \chi_{(\lambda_j^\beta(n), E_{2j})}(\lambda)\right)$$
$$(8.36) \qquad - \chi_{(\lambda_\infty^\beta, E_0)}(\lambda), \quad a(n)\beta > 0.$$

Thus we have for $\beta \neq 0, \infty$,

$$(8.37) \qquad \gamma^\beta(z, n) = -\frac{\beta}{a(n)} \frac{z - \lambda_\infty^\beta(n)}{\sqrt{z - E_0}\sqrt{z - E_\infty}} \prod_{j \in J} \frac{z - \lambda_j^\beta(n)}{\sqrt{z - E_{2j-1}}\sqrt{z - E_{2j}}},$$

and we remark that the numbers $\lambda_j^\beta(n)$ are related to the spectrum of H_n^β as follows

$$(8.38) \qquad \sigma(H_n^\beta) = \sigma_{ess}(H) \cup \{\lambda_j^\beta(n)\}_{j \in J \cup \{\infty\}}.$$

Again observe that $\lambda_j^\beta(n)$ is not necessarily an eigenvalue of H_n^β unless $\lambda_j^\beta(n) \notin \sigma_{ess}(H)$.

Evaluation of (6.58) shows

$$b^{\beta,(\ell)}(n) = \frac{-\beta}{2a(n)}\Big(E_0^{\ell+1} + E_\infty^{\ell+1} - 2\lambda_\infty^\beta(n)^{\ell+1}$$
(8.39)
$$+ \sum_{j\in J}(E_{2j-1}^{\ell+1} + E_{2j}^{\ell+1} - 2\lambda_j^\beta(n)^{\ell+1})\Big)$$

and in the special case $\ell = 0$,

(8.40) $\quad a(n) = \dfrac{1}{2(\beta+\beta^{-1})}\Big(E_0 + E_\infty - 2\lambda_\infty^\beta(n) + \sum_{j\in J}(E_{2j-1} + E_{2j} - 2\lambda_j^\beta(n))\Big).$

8.2. Isospectral operators

In this section we will use the results of the previous section to characterize the isospectral class of certain reflectionless Jacobi operators.

Before we can do this, we need a condition which ensures $\sigma_{sc}(H_n^\infty) = \emptyset$.

Hypothesis H. 8.7. Suppose Σ is infinite and the set of all accumulation points of the set of band edges $\{E_{2j-1}, E_{2j}\}_{j\in J}$ is countable.

We will denote the set of all accumulation points of the set of band edges by

(8.41) $\qquad\qquad\qquad \{\hat{E}_j\}_{j\in \hat{J}}, \qquad \hat{J} \subseteq \mathbb{N}.$

Let us first note some consequences of this hypothesis.

Lemma 8.8. *Condition (H.8.7) holds if and only if $\sigma(H)$ is a countable union of disjoint closed intervals (which may degenerate to points).*

Explicitly, we have

(8.42) $\qquad\qquad \sigma(H) = \Big(\bigcup_{j\in J_0} \Sigma_j\Big) \cup \Big(\bigcup_{j\in \hat{J}} \hat{\Sigma}_j\Big),$

where

(8.43) $\qquad \Sigma_j = [E_{2j}, E_{2j}^{(r)}], \quad j \in J_0, \quad \hat{\Sigma}_j = [\hat{E}_j, \hat{E}_j^{(r)}], \quad j \in \hat{J},$

with

(8.44) $\qquad x^{(r)} = \inf\{E_j | x < E_j \text{ or } j = \infty\} \in \{E_j\}_{j\in J\cup\{\infty\}} \cup \{\hat{E}_j\}_{j\in \hat{J}}.$

Proof. Denote the right hand side of (8.42) by Σ. Since $x^{(r)} = \inf\{\lambda \in \rho(H) | x < \lambda\}$ we see $\Sigma \subseteq \sigma(H)$. To show $\mathbb{R}\backslash\Sigma \subseteq \rho(H)$ pick $x \notin \Sigma$. There are three cases $x^{(r)} = E_{2j}$, $x^{(r)} = E_{2j-1}$, or $x^{(r)} = \hat{E}_j$. In the first case we have $x \in (E_{2j-1}, E_{2j}) \subset \rho(H)$ which is what we want. In the second case we set $E = \sup\{E_{2k} | E_{2k} \leq E_{2j-1}\}$. By minimality of $x^{(r)} = E_{2j-1}$ there cannot be an E_{2k} in $[x, E_{2j-1})$ and we either have $E_{2j-1} = E_{2k}$, reducing it to the first case, or $E \leq x$ and hence $x \in [E, E_{2j-1}] \subseteq \Sigma$ contradicting case two. Case three is similar to case two. Thus $\sigma(H) = \Sigma$. The intervals in (8.42) are disjoint unless $E_{2j} = \hat{E}_k$ in which case the corresponding intervals are equal. Hence after throwing out the superfluous intervals we obtain a disjoint union.

Conversely, let $\sigma(H) = \cup_{j\in \tilde{J}}[\tilde{E}_{2j}, \tilde{E}_{2j+1}]$ be a countable union of disjoint closed intervals. Then we have $\{E_j\} \subseteq \{\tilde{E}_j\}$. and the claim follows since $\{\tilde{E}_j\}$ is closed. □

8.2. Isospectral operators

Since the spectral measure of $-g(z,n)^{-1}$ is purely absolutely continuous on the interior of each of the intervals $\Sigma_j, \hat{\Sigma}_j$ the singularly continuous part must be zero (since it is supported on a countable set). The considerations of the previous section indicate that we can choose a number $\mu_j(n)$ for each spectral gap and a number $\tilde{\sigma}_j$ for each Dirichlet eigenvalue. Since the set of all Dirichlet eigenvalues is contained in the set $\{\mu_j(n)\}_{j \in J} \cup \{\hat{E}_j\}_{j \in \hat{J}}$ we will specify $\sigma_j(n), \hat{\sigma}_j(n)$ for each $\mu_j(n), \hat{E}_j, j \in J, \hat{J}$, respectively. However, the previous section shows that $\sigma_j(n)$ (resp. $\hat{\sigma}_j(n)$) is not needed if $\mu_j(n)$ (resp. $\hat{E}_j$) is not in $\sigma_p(H_n^\infty)$, we will get rid of all superfluous σ's by an equivalence relation later on.

We first introduce the following hypothesis.

Hypothesis H.8.9. Suppose the set $\left\{ \left((\mu_j, \sigma_j)\right)_{j \in J}, (\hat{\sigma}_j)_{j \in \hat{J}} \right\}$ satisfies the following requirements.
(i). $(\mu_j, \sigma_j) \in [E_{2j-1}, E_{2j}] \times [-1, 1], j \in J$, and $\hat{\sigma}_j \in [-1, 1], j \in \hat{J}$.
(ii). If $\mu_j \in (E_{2j-1}, E_{2j})$ then $\sigma_j \in \{\pm 1\}$.
(iii). If $\mu_j = E_{2j-1} = E_{2k}$ (resp. $\mu_j = E_{2j} = E_{2k-1}$) for some $j, k \in J$ then $(\mu_j, \sigma_j) = (\mu_k, \sigma_k)$ and $\sigma_j \in (-1, 1)$.
(iv). If $\mu_j = \hat{E}_k$ for some $j \in J, k \in \hat{J}$ we have $\sigma_j = \hat{\sigma}_k$.
(v). For both $\pm$, at least one of the sets $\{j \in J| \pm \sigma_j < 1\}$ or $\{j \in \hat{J}| \pm \hat{\sigma}_j < 1\}$ is infinite.

Given Hypothesis (H.8.9) we define the set

$$(8.45) \qquad \mathcal{D}_0(\Sigma) = \left\{ \left\{ \left((\mu_j, \sigma_j)\right)_{j \in J}, (\hat{\sigma}_j)_{j \in \hat{J}} \right\} | (\text{H.8.9}) \text{ holds} \right\}.$$

Furthermore, for each element in $\mathcal{D}_0(\Sigma)$ we can define $g(z)$ as in (8.8) and $R_j, \hat{R}_j$, $j \in J, \hat{J}$ as in (8.16) with $\tilde{\mu}_j(n)$ replaced by $\mu_j, \hat{E}_j$, respectively. We will set

$$(8.46) \qquad \left\{ \left((\mu_j, \sigma_{1,j})\right)_{j \in J}, (\hat{\sigma}_{1,j})_{j \in \hat{J}} \right\} \cong \left\{ \left((\mu_j, \sigma_{2,j})\right)_{j \in J}, (\hat{\sigma}_{2,j})_{j \in \hat{J}} \right\}$$

if $\sigma_{1,j} = \sigma_{2,j}$ for all $j \in \{j \in J | R_j > 0\}$ and if $\hat{\sigma}_{1,j} = \hat{\sigma}_{2,j}$ for all $j \in \{j \in \hat{J} | \hat{R}_j > 0\}$. The set $\mathcal{D}_0(\Sigma)$ modulo this equivalence relation will be denoted by $\mathcal{D}(\Sigma)$.

For any set Σ satisfying (H.8.7) the isospectral set of reflectionless Jacobi operators H in $\ell^2(\mathbb{Z})$ is denoted by

$$(8.47) \quad \text{Iso}_R(\Sigma) = \{\text{reflectionless Jacobi operators } H \text{ on } \ell^2(\mathbb{Z}) | \sigma(H) = \Sigma, a > 0\}.$$

With each $H \in \text{Iso}_R(\Sigma)$ we can associate an element in $\mathcal{D}(\Sigma)$ by fixing a base point $n \in \mathbb{Z}$. In fact, defining $\mu_j(n)$ as in (8.5) and $\sigma_j(n), \hat{\sigma}_j(n)$ as in (8.17) with $\tilde{\mu}_j(n)$ replaced by $\mu_j, \hat{E}_j$, respectively, we obtain a mapping

$$(8.48) \qquad\qquad I_n : \text{Iso}_R(\Sigma) \to \mathcal{D}(\Sigma).$$

Theorem 8.10. *Suppose Σ satisfies (H.8.7) and fix $n \in \mathbb{Z}$. Then the map I_n is a bijection.*

Proof. It suffices to show that I_n is surjective (since Dirichlet data uniquely determine an operator). Let an element of $\mathcal{D}(\Sigma)$ be given. Define $g(z, n)$, $d\tilde{\rho}_n$, $R_j(n)$, $\hat{R}_j(n)$ as in (8.8), (8.12), (8.13), respectively. Since $d\tilde{\rho}_{n,pp}$ is supported on $\{\mu_j\}_{j\in J} \cup \{\hat{E}_j\}_{j\in \hat{J}}$ (as noted before), we can split it up according to

$$d\tilde{\rho}_{n,\pm,pp}(\{\mu_j\}) = \frac{1 \pm \sigma_j}{2} d\tilde{\rho}_{n,pp}(\{\mu_j\}),$$

(8.49)
$$d\tilde{\rho}_{n,\pm,pp}(\{\hat{E}_j(n)\}) = \frac{1 \pm \hat{\sigma}_j}{2} d\tilde{\rho}_{n,pp}(\{\hat{E}_j\}).$$

Similarly, the absolutely continuous part splits up according to $d\tilde{\rho}_{n,\pm,ac} = \frac{1}{2} d\tilde{\rho}_{n,ac}$. Due to (H.8.9) this splitting is well-defined and by virtue of (8.14) and Theorem 2.13 this defines an operator H. It remains to show $\sigma(H) = \Sigma$. By Remark 3.5(ii) it suffices to consider all points μ_j for which both $\tilde{m}_\pm(z,n)$ have a pole (i.e., $\sigma_j \in (-1,1)$ and $R_j > 0$). Near such a point we infer using (8.18)

(8.50) $$g(z, n+1) = -(1-\sigma_j^2)\frac{R_j}{z-\mu_j} + O(z-\mu_j)^0$$

showing $\mu_j \in \sigma(H)$. This proves $\sigma(H) = \Sigma$. □

We conclude this section with a simple example illustrating an explicit construction to the effect that an accumulation point of eigenvalues of H may or may not be an eigenvalue of H.

Remark 8.11. (i). Suppose H is reflectionless and has pure point spectrum only. Pick an accumulation point $\hat{E}$ of eigenvalues and define

(8.51) $$\hat{R} = \lim_{\varepsilon \downarrow 0} i\varepsilon g(\hat{E} + i\varepsilon, 0)^{-1}$$

and

(8.52) $$g_\delta(z,0) = -\left(-g(z,0)^{-1} - (\delta - \hat{R})(z-\hat{E})^{-1}\right)^{-1}, \quad \delta \geq 0.$$

Then g_δ is a Herglotz function corresponding to a pure point measure. Computing the zeros $\mu_{\delta,j}$ of $g_\delta(z,0)$ and choosing $\sigma_{\delta,j}, \hat{\sigma}_{\delta,j} \in [-1,1]$ according to our requirements yields a corresponding Jacobi operator H_δ by Theorem 8.10. Since

(8.53) $$\hat{R}_\delta = \lim_{\varepsilon \downarrow 0} i\varepsilon g_\delta(\tilde{E} + i\varepsilon, 0)^{-1} = \delta,$$

one obtains the following case distinctions ($\hat{\sigma}$ the sigma corresponding to $\hat{E}$).

(1). $\delta = 0$, then $\tilde{E} \notin \sigma_p(H_{\delta,\pm})$.
(2). $\delta > 0$, $\hat{\sigma} \in \{\pm 1\}$, then $\tilde{E} \in \sigma_p(H_{\delta,\hat{\sigma}})$, $\tilde{E}_{j_0} \notin \sigma_p(H_\delta)$.
(3). $\delta > 0$, $\hat{\sigma} \in (-1,1)$, then $\tilde{E} \in \sigma_p(H_{\delta,\pm}) \cap \sigma_p(H_\delta)$.

(ii). Let us further comment on (H.8.9). If (ii) should not hold, then μ_j would be an eigenvalue of the operator obtained from I_n^{-1} (by (8.50)). Similarly, if we drop condition (iii), then $E_{2j} = \mu_j$ might not be an eigenvalue.

8.3. The finite-gap case

In this section we will prove some additional results in the case of reflectionless Jacobi operators with finitely many gaps. This class is closely connected to the Toda hierarchy (to be studied in Chapter 12) and hence deserves special attention. Given $(E_j)_{j=0}^{2r+1}$ and $((\mu_j(n_0), \sigma_j(n_0)))_{j=1}^{r}$ for one fixed n_0 we will reconstruct the sequences a, b recursively. We will derive all results without using information from the previous two sections.

8.3. The finite-gap case

Our approach will be modeled after the information gained in the previous chapter about periodic operators (in particular Section 7.3). However, note that for periodic sequences (with no closed gaps) $(E_j)_{j=0}^{2r+1}$ is already too much $((E_j^\theta)_{j=1}^{r+1}$ for some θ plus A would be enough). If H has no eigenvalues, that is $E_{2j} < E_{2j+1}$, the constructed sequences will turn out to be quasi-periodic and to be expressible in terms of Riemann theta functions. This case will be investigated more closely in Section 9.2.

Let $r \in \mathbb{N}_0$ and

$$(8.54) \qquad E_0 \leq E_1 < E_2 \leq E_3 \leq E_4 \leq \cdots < E_{2r-1} \leq E_{2r} < E_{2r+1}$$

be given real numbers. Here we have excluded the case $E_{2j} = E_{2j+1}$ since the corresponding factors would cancel from all equations. In addition, we will assume that $E_{2j} < E_{2j+1}$ for at least one j.

Define $R_{2r+2}^{1/2}(z)$ as

$$(8.55) \qquad R_{2r+2}^{1/2}(z) = -\prod_{j=0}^{2r+1} \sqrt{z - E_j},$$

where $\sqrt{\cdot}$ is defined as in (1.116). It is most natural to consider $R_{2r+2}^{1/2}(z)$ as a function on the associated Riemann surface M which is the same as the Riemann surface of

$$(8.56) \qquad \tilde{R}_{2g+2}^{1/2}(z) = -\prod_{j \in \Gamma} \sqrt{z - E_{2j}}\sqrt{z - E_{2j+1}},$$

where $\Gamma = \{j | 0 \leq j \leq r,\ E_{2j} < E_{2j+1}\}$, $g = |\Gamma| - 1 \geq 0$. As a set, M is given by all pairs

$$(8.57) \qquad p = (z, \pm\tilde{R}_{2g+2}^{1/2}(z)), \quad z \in \mathbb{C},$$

plus two points at infinity, $\infty_\pm$. We will denote by π the projection on M and by $\tilde{R}_{2g+2}^{1/2}$ the evaluation map, that is,

$$(8.58) \qquad \pi(p) = z, \quad \tilde{R}_{2g+2}^{1/2}(p) = \pm\tilde{R}_{2g+2}^{1/2}(z), \quad p = (z, \pm\tilde{R}_{2g+2}^{1/2}(z)).$$

The points $\{(E_{2j}, 0), (E_{2j+1}, 0) | j \in \Gamma\} \subseteq M$ are called branch points and g is called the genus of M. In addition, we define

$$(8.59) \qquad R_{2r+2}^{1/2}(p) = \tilde{R}_{2g+2}^{1/2}(p) \prod_{j \notin \Gamma}(\pi(p) - E_{2j})$$

and $p^* = (z, \pm\tilde{R}_{2g+2}^{1/2}(z))^* = (z, \mp\tilde{R}_{2g+2}^{1/2}(z))$, $\infty_\pm^* = \infty_\mp$. Clearly, any polynomial $P(z)$ gives rise to a function $P(p) = P(\pi(p))$ on M. For additional information on Riemann surfaces we refer to Section 9.1 and Appendix A.

To set the stage, let

$$(8.60) \qquad (\hat{\mu}_j(n_0))_{j=1}^r = (\mu_j(n_0), \sigma_j(n_0))_{j=1}^r$$

be a list of pairs satisfying (cf. (H.8.9))

Hypothesis H. 8.12. Suppose $\hat{\mu}_j(n_0)$, $1 \leq j \leq r$, satisfy
(i).

(8.61) $\quad \begin{aligned} &\mu_j(n_0) \in (E_{2j-1}, E_{2j}) \text{ and } \sigma_j(n_0) \in \{\pm 1\} \text{ or} \\ &\mu_j(n_0) \in \{E_{2j-1}, E_{2j}\} \text{ and } \sigma_j(n_0) \in (-1, 1). \end{aligned}$

If $E_{2j-2} < E_{2j-1} = \mu_j(n_0)$ (resp. $E_{2j} < E_{2j+1} = \mu_j(n_0)$), then $\sigma_j(n_0)$ is superfluous and we will set $\sigma_j(n_0) = 0$ in this case.
(ii).

(8.62) $\quad \begin{aligned} E_{2j-2} = E_{2j-1} = \mu_j(n_0) &\Rightarrow \hat{\mu}_{j-1}(n_0) = \hat{\mu}_j(n_0) \\ E_{2j} = E_{2j+1} = \mu_j(n_0) &\Rightarrow \hat{\mu}_{j+1}(n_0) = \hat{\mu}_j(n_0) \end{aligned}$,

for $1 \leq j \leq r$. In particular, $\mu_1(n_0) \neq E_0$ and $\mu_r(n_0) \neq E_{2r+1}$.

In the notation of the previous section we have

(8.63) $$\mathcal{D}(\Sigma) = \{(\hat{\mu}_j(n_0))_{j=1}^r | (\text{H.8.12}) \text{ holds}\}$$

with

(8.64) $$\Sigma = \bigcup_{j=0}^r [E_{2j}, E_{2j+1}]$$

and Theorem 8.10 is valid.

We require (H.8.12) (ii) without loss of generality since otherwise the construction below would give $\mu_j(n) = \mu_j(n_0)$ for all n and the factor $z - \mu_j(n)$ would cancel with the factor $\sqrt{z - E_{2j-2}}\sqrt{z - E_{2j-1}}$ (resp. $\sqrt{z - E_{2j}}\sqrt{z - E_{2j+1}}$) from all equations. For convenience we set

(8.65) $\quad R_{2r+2}^{1/2}(\hat{\mu}_j(n_0)) = \sigma_j(n_0) R_{2r+2}^{1/2}(\mu_j(n_0)), \quad R_{2r+2}(z) = \left(R_{2r+2}^{1/2}(z)\right)^2.$

To begin with, we define the polynomial associated with the Dirichlet eigenvalues

(8.66) $$G_r(z, n_0) = \prod_{j=1}^r (z - \mu_j(n_0))$$

and the residues

(8.67) $$R_j(n_0) = \lim_{z \to \mu_j(n_0)} (z - \mu_j(n_0)) \frac{R_{2r+2}^{1/2}(z)}{G_r(z, n_0)}.$$

If $\mu_j(n_0) \neq \mu_{j-1}(n_0), \mu_{j+1}(n_0)$ we have

(8.68) $$R_j(n_0) = \frac{R_{2r+2}^{1/2}(\mu_j(n_0))}{\prod_{k \neq j}^r (\mu_j(n_0) - \mu_k(n_0))} \geq 0.$$

If $\mu_j(n_0) = \mu_{j-1}(n_0)$ or $\mu_j(n_0) = \mu_{j+1}(n_0)$ the vanishing factors in the numerator and denominator have to be omitted. For notational convenience we set in addition

(8.69) $\quad \hat{R}_j(n_0) = \begin{cases} \sigma_j(n_0) R_j(n_0), & \mu_j(n_0) \neq \mu_{j-1}(n_0), \mu_{j+1}(n_0) \\ \frac{\sigma_j(n_0)}{2} R_j(n_0), & \mu_j(n_0) = \mu_{j-1}(n_0) \text{ or } \mu_{j+1}(n_0) \end{cases}.$

The factor $1/2$ in the case where two Dirichlet eigenvalues coincide will ensure that we do not count the corresponding residue twice.

8.3. The finite-gap case

Using this notation we define a second polynomial

$$H_{r+1}(z, n_0) = \sum_{j=1}^{r} \hat{R}_j(n_0) \prod_{\substack{k=1 \\ k \neq j}}^{r} \big(z - \mu_k(n_0)\big) + (z - b(n_0))G_r(z, n_0)$$

(8.70)
$$= G_r(z, n_0)\Big(z - b(n_0) + \sum_{j=1}^{r} \frac{\hat{R}_j(n_0)}{z - \mu_j(n_0)}\Big),$$

where

(8.71)
$$b(n_0) = \frac{1}{2}\sum_{j=0}^{2r+1} E_j - \sum_{j=1}^{r} \mu_j(n_0).$$

This definition implies

(8.72)
$$\lim_{z \to \mu_j(n_0)} \frac{H_{r+1}(z, n_0)}{R_{2r+2}^{1/2}(z)} = \sigma_j(n_0), \quad 1 \leq j \leq r.$$

Our next task is to define $\hat{\mu}_j(n_0 + 1)$. This will be done using a procedure motivated by the approach of Section 2.7. The last equation (8.72) shows that we can define a polynomial $G_r(z, n_0 + 1)$ by (compare (2.187))

(8.73) $\quad 4a(n_0)^2 G_r(z, n_0) G_r(z, n_0 + 1) = H_{r+1}(z, n_0)^2 - R_{2r+2}(z).$

As we want the highest coefficient of $G_r(z, n_0 + 1)$ to be one, we have introduced a nonzero constant $a(n_0)^2 \neq 0$.

First of all, note that $G_r(z, n_0 + 1)$ does not vanish identically (H_{r+1}^2 is non-negative and R_{2r+2} is not since $E_{2j} < E_{2j+1}$ for at least one j) and that it is a polynomial of degree at most r (since the highest two powers of z on the right cancel due to the definition of $b(n_0)$).

We claim that there are at least two zeros of $G_r(., n_0) G_r(., n_0+1)$ in the interval $[E_{2j-1}, E_{2j}]$. Since

$$4a(n_0)^2 G_r(E_{2j-1}, n_0) G_r(E_{2j-1}, n_0 + 1) \geq H_{r+1}(E_{2j-1}, n_0)^2 \geq 0 \text{ and}$$
(8.74) $\quad 4a(n_0)^2 G_r(E_{2j}, n_0) G_r(E_{2j}, n_0 + 1) \geq H_{r+1}(E_{2j}, n_0)^2 \geq 0,$

this is clear if $\mu_j(n_0) \in (E_{2j-1}, E_{2j})$. If $\mu_j(n_0) \in \{E_{2j-1}, E_{2j}\}$, say $\mu_j(n_0) = E_{2j-1}$, we have to consider two cases. First, let $E_{2j-2} < E_{2j-1}$, then $R_{2r+2}(.)$ has a first order zero near E_{2j-1} and hence $H_{r+1}(z, n_0)^2 - R_{2r+2}(z) < 0$ for $z \in (E_{2j-1}, E_{2j-1} + \varepsilon)$ and some $\varepsilon > 0$ sufficiently small. Implying one zero in (E_{2j-1}, E_{2j}). Second, let $E_{2j-2} = E_{2j-1}$, then the same conclusion applies because of (8.72) and $\sigma_j(n_0) \in (-1, 1)$.

Since $G_r(z, n_0 + 1)$ is of degree at most r we conclude that it must be of the form

(8.75)
$$G_r(z, n_0 + 1) = \prod_{j=1}^{r} \big(z - \mu_j(n_0 + 1)\big),$$

with

(8.76)
$$\mu_j(n_0 + 1) \in [E_{2j-1}, E_{2j}].$$

Observe that $E_{2j-2} = E_{2j-1} = \mu_j(n_0)$ (resp. $E_{2j} = E_{2j+1} = \mu_j(n_0)$) implies $E_{2j-1} \neq \mu_j(n_0 + 1)$ (resp. $E_{2j} \neq \mu_j(n_0 + 1)$). Moreover, if $E_{2j-2} = E_{2j-1} = $

$\mu_j(n_0+1)$, then $\mu_j(n_0+1) = \mu_{j-1}(n_0+1)$ and if $E_{2j} = E_{2j+1} = \mu_j(n_0+1)$, then $\mu_j(n_0+1) = \mu_{j+1}(n_0+1)$ (since the left-hand side of (8.73) has a double zero in this case). In addition, $\mu_1(n_0+1) \neq E_0$ (resp. $\mu_r(n_0+1) \neq E_{2r+1}$) since otherwise we must have $E_0 = E_1$ implying that the left-hand side of (8.73) has a single zero and the right-hand side a double zero. It remains to define

$$(8.77) \qquad \sigma_j(n_0+1) = -\lim_{z \to \mu_j(n_0+1)} \frac{H_{r+1}(z, n_0)}{R_{2r+2}^{1/2}(z)}, \quad 1 \leq j \leq r.$$

Clearly, if $H_{r+1}(\mu_j(n_0+1), n_0) \neq 0$ we have $\sigma_j(n_0+1) \in \{\pm 1\}$ by (8.73). If $H_{r+1}(\mu_j(n_0+1), n_0) = 0$ we have $\mu_j(n_0+1) \in \{E_{2j-1}, E_{2j}\}$, say E_{2j-1}. Then, if $E_{2j-2} < E_{2j-1}$ we get $\sigma_j(n_0+1) = 0$ and if $E_{2j-2} = E_{2j-1}$ we obtain $\sigma_j(n_0+1) \in (-1,1)$ since (8.73) is negative near E_{2j-1}.

Thus we have constructed the list $(\hat{\mu}_j(n_0+1))_{j=1}^r$ from the list $(\hat{\mu}_j(n_0))_{j=1}^r$ such that $(\hat{\mu}_j(n_0+1))_{j=1}^r$ satisfies (H.8.12).

The coefficient $a(n_0)^2$ can be calculated by expanding both sides of (8.73) (using (8.66), (8.70), and (8.55)). Comparing the coefficient of z^{2g} shows

$$(8.78) \qquad a(n_0)^2 = \frac{1}{2}\sum_{j=1}^r \hat{R}_j(n_0) + \frac{b^{(2)}(n_0) - b(n_0)^2}{4} > 0,$$

where

$$(8.79) \qquad b^{(2)}(n) = \frac{1}{2}\sum_{j=0}^{2r+1} E_j^2 - \sum_{j=1}^r \mu_j(n)^2.$$

That $a(n_0)^2$ is positive follows from $a(n_0)^2 G_r(E_{2r+1}, n_0) G_r(E_{2r+1}, n_0+1) \geq 0$. In addition, we obtain

$$(8.80) \qquad \begin{aligned} H_{r+1}(z, n_0) = & -\sum_{j=1}^r \hat{R}_j(n_0+1) \prod_{k \neq j}^r (z - \mu_k(n_0+1)) \\ & + (z - b(n_0+1)) G_r(z, n_0+1), \end{aligned}$$

implying

$$(8.81) \qquad a(n_0)^2 = -\frac{1}{2}\sum_{j=1}^r \hat{R}_j(n_0+1) + \frac{b^{(2)}(n_0+1) - b(n_0+1)^2}{4} > 0,$$

and

$$(8.82) \qquad H_{r+1}(z, n_0+1) = -H_{r+1}(z, n_0) + 2(z - b(n_0+1)) G_r(z, n_0+1).$$

As (8.70) is symmetric in $G_r(z, n_0+1)$ and $G_r(z, n_0)$ we can even reverse this process using (8.80) as definition for H_{r+1} (which only differs from (8.70) by the sign in front of $R_{2r+2}^{1/2}(.)$) and defining $\sigma_j(n_0)$ with reversed sign, that is, $\hat{\mu}_j(n_0) = (\mu(n_0), \lim_{z \to \mu_j(n_0)} H_{r+1}(z, n_0)/R_{2r+2}^{1/2}(z))$. Thus given $(\hat{\mu}_j(n_0))_{j=1}^r$ we can recursively define the set $(\hat{\mu}_j(n))_{j=1}^r$ for all $n \in \mathbb{Z}$. In summary we have proven the following theorem.

Theorem 8.13. *Given the list $(\hat{\mu}_j(n_0))_{j=1}^r$ we can define the list $(\hat{\mu}_j(n))_{j=1}^r$ for all $n \in \mathbb{Z}$, which again fulfills (H.8.12). In addition, we get two sequences $a(n)$, $b(n)$*

8.3. The finite-gap case

defined as

$$a(n)^2 = \frac{1}{2}\sum_{j=1}^{r}\hat{R}_j(n) + \frac{1}{8}\sum_{j=0}^{2r+1}E_j^2 - \frac{1}{4}\sum_{j=1}^{r}\mu_j(n)^2 - \frac{b(n)^2}{4} > 0,$$

(8.83) $$b(n) = \frac{1}{2}\sum_{j=0}^{2r+1}E_j - \sum_{j=1}^{r}\mu_j(n),$$

(cf (8.69)). The sign of $a(n)$ can be chosen freely. The sequences $a(n), b(n)$ are real and bounded.

Proof. It remains to show boundedness of $a(n)$ and $b(n)$. This is clear for $b(n)$ by virtue of (8.83). Combining (8.78) and (8.81)

(8.84) $$a(n)^2 + a(n-1)^2 = \frac{b^{(2)}(n) - b(n)^2}{2}$$

we see that $a(n)$ is bounded as well. □

Remark 8.14. (i). The question when $a(n)^2$, $b(n)$ are periodic will be answered in Theorem 9.6.
(ii). Suppose $\mu_{j-1} \neq \mu_j \neq \mu_{j+1}$ for a moment. Using the residue theorem we may write

(8.85) $$\mu_j(n+1) + \mu_j(n) = \frac{1}{2\pi i}\int_{\Gamma_j} z \frac{2H_{r+1}(z,n)H'_{r+1}(z,n) - R'_{2r+2}(z)}{H_{r+1}(z,n)^2 - R_{2r+2}(z)} dz,$$

(here the prime denotes the derivative with respect to z) where Γ_j is a closed path in the complex plane encircling (once) the interval $[E_{2j-1}, E_{2j}]$ counter-clock-wise. Further we may also write

(8.86) $$\mu_j(n+1) - \mu_j(n) = \frac{1}{2\pi i}\int_{\gamma_j} z\, d\ln\frac{G_r(z,n+1)}{G_r(z,n)}.$$

Let me point out another interesting observation. The procedure for calculating $\mu_j(n)$ from $\mu_j(n_0)$ with $n > n_0$ differs from that with $n < n_0$ only by two signs! One in the definition of H_{r+1} and one in the definition of $\hat{\mu}_j$. These cancel in each intermediate step but not in the first and not in the last. This means calculating $\hat{\mu}_j(n_0 + k)$ from $\hat{\mu}_j(n_0)$ yields the same result as calculating $\hat{\mu}_j(n_0 - k)$ from $\hat{\mu}_j(n_0)^*$ after switching signs in the last result. Here $\hat{\mu}_j(n_0)^* = (\mu_j(n_0), \sigma_j(n_0))^* = (\mu_j(n_0), -\sigma_j(n_0))$.

Lemma 8.15. *Denote the sequences calculated from $\{\hat{\mu}_j(n_0)^*\}_{j=1}^r$ by a_R, b_R. Then we have $a_R(n_0 + k) = a(n_0 - k - 1)$, $b_R(n_0 + k) = b(n_0 - k)$ (cf. Lemma 1.7).*

In addition, if all $\mu_j(n_0)$ coincide with band edges, that is if $\{\mu_j(n_0)\}_{j=1}^r \subset \{E_{2j-1}, E_{2j}\}_{j=1}^r$, then we have $\hat{\mu}_j(n_0 + k) = \hat{\mu}_j(n_0 - k)^*$, $1 \leq j \leq r$, and hence $a(n_0 + k) = a(n_0 - k - 1)$, $b(n_0 + k) = b(n_0 - k)$.

Proof. Our observation says in formulas that $\hat{\mu}_{R,j}(n_0 + k) = \hat{\mu}_j(n_0 - k)^*$. From this the result for b_R is clear from (8.83). For the corresponding result for a_R use (8.81) and (8.81). □

Our next task is to investigate the spectrum of the Jacobi operator H associated with the sequences a, b. To do this we need solutions of the Jacobi and Riccati equations first.

Using our polynomials we define two meromorphic functions on M,

$$(8.87) \quad \phi(p,n) = \frac{H_{r+1}(p,n) + R_{2r+2}^{1/2}(p)}{2a(n)G_r(p,n)} = \frac{2a(n)G_r(p,n+1)}{H_{r+1}(p,n) - R_{2r+2}^{1/2}(p)}$$

and, the **Baker-Akhiezer function**,

$$(8.88) \quad \psi(p,n,n_0) = \prod_{j=n_0}^{n-1}{}^* \phi(p,j).$$

Notice

$$(8.89) \quad \phi(p,n)^2 = \frac{G_r(p,n+1)}{G_r(p,n)} \frac{H_{r+1}(p,n) + R_{2r+2}^{1/2}(p)}{H_{r+1}(p,n) - R_{2r+2}^{1/2}(p)}$$

and hence

$$(8.90) \quad \psi(p,n,n_0)^2 = \frac{G_r(p,n)}{G_r(p,n_0)} \prod_{j=n_0}^{n-1}{}^* \frac{H_{r+1}(p,j) + R_{2r+2}^{1/2}(p)}{H_{r+1}(p,j) - R_{2r+2}^{1/2}(p)}.$$

Theorem 8.16. *The function $\phi(p,n)$ fulfills*

$$(8.91) \quad a(n)\phi(p,n) + \frac{a(n-1)}{\phi(p,n-1)} = \pi(p) - b(n),$$

and hence $\psi(p,n,n_0)$ satisfies

$$(8.92) \quad \begin{aligned} a(n)\psi(p,n+1,n_0) &+ a(n-1)\psi(p,n-1,n_0) + b(n)\psi(p,n,n_0) \\ &= \pi(p)\psi(p,n,n_0). \end{aligned}$$

Proof. A straightforward calculation (using (8.82)) yields

$$(8.93) \quad \begin{aligned} a(n)\phi(p,n) + \frac{a(n-1)}{\phi(p,n-1)} &= \frac{H_{r+1}(p,n) + R_{2r+2}^{1/2}(p)}{2G_r(p,n)} \\ &+ \frac{H_{r+1}(p,n-1) - R_{2r+2}^{1/2}(p)}{2G_r(p,n)} = \pi(p) - b(n), \end{aligned}$$

and the rest follows from

$$(8.94) \quad \phi(p,n) = \frac{\psi(p,n+1,n_0)}{\psi(p,n,n_0)}.$$

$\square$

With $\psi_\pm, \phi_\pm$ we mean the chart expression of ψ, ϕ in the charts $(\Pi_\pm, \pi)$ (cf. Appendix A.7), that is,

$$(8.95) \quad \phi_\pm(z,n) = \frac{H_{r+1}(z,n) \pm R_{2r+2}^{1/2}(z)}{2a(n)G_r(z,n)}, \text{ etc..}$$

Now Lemma 6.7 together with $\phi_\pm(z,n) = (a(n)/z)^{\pm 1}(1 + O(z^{-1}))$ implies

8.3. The finite-gap case

Theorem 8.17. *The functions $\psi_\pm(z,n,n_0)$ are square summable near $\pm\infty$ and the Green function of H is given by (recall the notation from (1.100))*

$$
\begin{aligned}
g(z,n) &= \frac{G_r(z,n)}{R_{2r+2}^{1/2}(z)}, & z \in \rho(H), \\
h(z,n) &= \frac{H_{r+1}(z,n)}{R_{2r+2}^{1/2}(z)}, & z \in \rho(H).
\end{aligned}
$$
(8.96)

Hence the spectrum of H reads

$$\sigma(H) = \bigcup_{j=0}^{r} [E_{2j}, E_{2j+1}],$$

(8.97) $\quad \sigma_{ac}(H) = \bigcup_{j\in\Gamma} [E_j, E_{j+1}], \quad \sigma_{sc}(H) = \emptyset, \quad \sigma_p(H) = \{E_j | j \notin \Gamma\}.$

In addition, if $\mu_j(n) \neq \mu_j(n_0)$, we have $\psi_{\sigma_j(n)}(\mu_j(n), n, n_0) = 0$ if $\sigma_j(n)^2 = 1$ and $\psi_-(\mu_j(n), n, n_0) = \psi_+(\mu_j(n), n, n_0) = 0$ if $\sigma_j(n)^2 < 1$.

The facts concerning the spectrum of H can be read off from the following lemma:

Lemma 8.18. *The Weyl M-matrix is given by*

(8.98) $\quad M(z) = \frac{1}{2a(0)R_{2r+2}^{1/2}(z)} \begin{pmatrix} 2a(0)G_r(z,0) & H_{r+1}(z,0) \\ H_{r+1}(z,0) & 2a(0)G_r(z,1) \end{pmatrix},$

and the associated matrix valued measure $d\rho$ by

$$
\begin{aligned}
d\rho_{0,0}(\lambda) &= \frac{G_r(\lambda,0)}{\pi i R_{2r+2}^{1/2}(\lambda)} \chi_{\sigma_{ac}(H)}(\lambda) d\lambda + \sum_{E_j \in \sigma_p(H)} P(E_j, 0, 0) d\Theta(\lambda - E_j), \\
d\rho_{0,1}(\lambda) &= \frac{H_{r+1}(\lambda,0)}{2\pi i a(0) R_{2r+2}^{1/2}(\lambda)} \chi_{\sigma_{ac}(H)}(\lambda) d\lambda + \sum_{E_j \in \sigma_p(H)} P(E_j, 0, 1) d\Theta(\lambda - E_j), \\
d\rho_{1,1}(\lambda) &= \frac{G_r(\lambda,1)}{\pi i R_{2r+2}^{1/2}(\lambda)} \chi_{\sigma_{ac}(H)}(\lambda) d\lambda + \sum_{E_j \in \sigma_p(H)} P(E_j, 1, 1) d\Theta(\lambda - E_j),
\end{aligned}
$$
(8.99)

where $-P(E_j, n, m)$ is the residue of $G(z, n, m)$ at $z = E_j$ (cf. (2.34)). Here $\Theta(\lambda) = 0$ for $\lambda \leq 0$ and $\Theta(\lambda) = 1$ for $\lambda > 0$.

Proof. The M-matrix easily follows from (8.96) and $d\rho_{0,0}$, $d\rho_{0,1}$, $d\rho_{1,1}$ are immediate from (B.31) and (B.41) (cf. Remark 3.5(i)). □

We remark that $\sigma_j(n_0)$ in the case of $\mu_j(n) = E_{2j-2} = E_{2j-1}$ can be alternatively interpreted in terms of Weyl m-functions

(8.100) $\quad \dfrac{a(n-1)^2}{a(n)^2} \lim_{z \to \mu_j(n)} \dfrac{m_{+,n}(z)}{m_{-,n}(z)} = \lim_{z \to \mu_j(n)} \dfrac{\phi_+(z,n)}{\phi_-(z,n)} = \dfrac{1+\sigma_j(n)}{1-\sigma_j(n)} > 0.$

Last, we collect some useful relations. First note

$$a(n)^2 G_r(z, n+1) - a(n-1)^2 G_r(z, n-1) + (z-b(n))^2 G_r(z,n)$$
(8.101) $\quad = (z - b(n)) H_{r+1}(z, n).$

Further, from (8.87) we get

$$\phi(p,n)\phi(p^*,n) = \frac{G_r(p,n+1)}{G_r(p,n)}, \tag{8.102}$$

and hence

$$\psi(p,n,n_0)\psi(p^*,n,n_0) = \frac{G_r(p,n)}{G_r(p,n_0)}. \tag{8.103}$$

Moreover,

$$\begin{aligned}\phi(p,n) - \phi(p^*,n) &= \frac{R_{2r+2}^{1/2}(p)}{a(n)G_r(p,n)}, \\ \phi(p,n) + \phi(p^*,n) &= \frac{H_{r+1}(p,n)}{a(n)G_r(p,n)}.\end{aligned} \tag{8.104}$$

8.4. Further spectral interpretation

In this section we will try to relate $(\hat{\mu}_j(n))_{j=1}^r$ to the spectra of the operators H_n^∞ associated with a, b. In fact we will even be a bit more general.

We first define a polynomial $K_{r+1}^\beta(z,n)$ for $\beta \in \mathbb{R}$ by

$$\begin{aligned}\phi(p,n) + \beta &= \frac{H_{r+1}(p,n) + 2a(n)\beta G_r(p,n) + R_{2r+2}^{1/2}(p)}{2a(n)G_r(p,n)} \\ &= \frac{2a(n)K_{r+1}^\beta(p,n)}{H_{r+1}(p,n) + 2a(n)\beta G_r(p,n) - R_{2r+2}^{1/2}(p)}.\end{aligned} \tag{8.105}$$

Thus we have

$$\begin{aligned}K_{r+1}^\beta(z,n) &= G_r(z,n+1) + \frac{\beta}{a(n)}H_{r+1}(z,n) + \beta^2 G_r(z,n) \\ &= \frac{\beta}{a(n)}\prod_{j=1}^{r+1}(z - \lambda_j^\beta(n)), \quad \beta \in \mathbb{R}\backslash\{0\},\end{aligned} \tag{8.106}$$

where $(\lambda_j^\beta(n))_{j=1}^{\tilde{r}}$, with $\tilde{r} = r$ if $\beta = 0$ and $\tilde{r} = r+1$ otherwise, are the zeros of $K_{r+1}^\beta(z,n)$. Since we have $K_{r+1}^\beta(z,n) \to G_r(z,n+1)$ as $\beta \to 0$ we conclude $\lambda_j^\beta(n) = \mu_j(n+1) + O(\beta)$ for $1 \leq j \leq r$ and $\lambda_{r+1}^\beta(n) = a(n)\beta^{-1} + b(n+1) + O(\beta)$. Also note that the analogue of (8.96) reads ($z \in \rho(H)$)

$$\begin{aligned}\frac{K_{r+1}^\beta(z,n)}{R_{2r+2}^{1/2}(z)} &= \frac{(\psi_+(z,n+1) + \beta\psi_+(z,n))(\psi_-(z,n+1) + \beta\psi_-(z,n))}{W(\psi_-(z),\psi_+(z))} \\ &= \gamma^\beta(z,n),\end{aligned} \tag{8.107}$$

where we have omitted the dependence of $\psi_\pm$ on n_0 (which cancels).

As in the last section we can use

$$(H_{r+1}(z,n) + 2a(n)\beta G_r(z,n))^2 - R_{2r+2}(z) = K_{r+1}^\beta(z,n)G_r(z,n) \tag{8.108}$$

to show

$$\lambda_j^\beta(n) \in [E_{2j-1}, E_{2j}], \quad 1 \leq j \leq r, \tag{8.109}$$

8.4. Further spectral interpretation

and ($\beta \in \mathbb{R}\backslash\{0\}$)

(8.110) $\quad \lambda^\beta_{r+1}(n) \geq E_{2r+1} \quad$ for $a(n)\beta < 0, \qquad \lambda^\beta_{r+1}(n) \leq E_0 \quad$ for $a(n)\beta > 0$,

(whether $\lambda^\beta_{r+1}(n) \geq E_{2r+1}$ or $\lambda^\beta_{r+1}(n) \leq E_0$ follows from the behavior as $\beta \to 0$ together with a continuity argument). Evaluating (8.108) at $z = \lambda^\beta_j(n)$ yields

(8.111) $\quad R_{2r+2}(\lambda^\beta_j(n)) = \bigl(H_{r+1}(\lambda^\beta_j(n), n) + 2a(n)\beta G_r(\lambda^\beta_j(n), n)\bigr)^2$,

and hence we define

(8.112) $\quad \sigma^\beta_j(n) = -\lim_{z \to \lambda^\beta_j(n)} \dfrac{H_{r+1}(z, n) + 2a(n)\beta G_r(z, n)}{R^{1/2}_{2r+2}(z)}$.

This implies (use (8.106))

$$R^{1/2}_{2r+2}(\hat{\lambda}^\beta_j(n)) = -H_{r+1}(\lambda^\beta_j(n), n) - 2a(n)\beta G_r(\lambda^\beta_j(n), n)$$
(8.113)
$$= H_{r+1}(\lambda^\beta_j(n), n) + 2a(n)\beta^{-1} G_r(\lambda^\beta_j(n), n+1).$$

For the sake of completeness we also state the trace formulas in the finite-gap case.

Lemma 8.19. *The zeros $\lambda^\beta_j(n)$ of $K^\beta_{r+1}(z, n)$ fulfill for $\beta \in \mathbb{R}$ (cf. (6.64))*

(8.114) $\quad b^{\beta,(\ell)}(n) = \dfrac{\beta}{2a(n)}\Bigl(\sum_{j=0}^{2r+1} E^{\ell+1}_j - 2\sum_{j=1}^{r+1} \lambda^\beta_j(n)^{\ell+1}\Bigr)$.

Especially for $\ell = 0$

(8.115) $\quad a(n) = \dfrac{1}{2(\beta + \beta^{-1})}\Bigl(\sum_{j=0}^{2r+1} E_j - 2\sum_{j=1}^{r+1} \lambda^\beta_j(n)\Bigr)$.

We further compute

(8.116) $\quad (\phi(p, n) + \beta)(\phi(n, p^*) + \beta) = \dfrac{K^\beta_{r+1}(p, n)}{G_r(p, n)}$,

and

$$a(n+1)^2 K^\beta_{r+1}(z, n+1) = a(n)^2 G_r(z, n) + b_\beta(n+1) H_{r+1}(z, n)$$
(8.117)
$$+ b_\beta(n+1)^2 G_r(z, n+1),$$

where $b_\beta(n) = b(n) - \beta a(n)$.

The next theorem characterizes the spectrum of H^β_n.

Theorem 8.20. *Suppose $\beta \in \mathbb{R} \cup \{\infty\}$. Then*

$$\sigma(H^\beta_n) = \sigma(H) \cup \{\lambda^\beta_j(n)\}^{\tilde{r}}_{j=1},$$
(8.118) $\quad \sigma_{ac}(H^\beta_n) = \sigma_{ac}(H), \quad \sigma_{sc}(H^\beta_n) = \emptyset$,

where $\tilde{r} = r$ for $\beta = 0, \infty$ and $\tilde{r} = r+1$ otherwise. Whether $\lambda^\beta_j(n)$ is an eigenvalue of $H^\beta_{+,n}$ or $H^\beta_{-,n}$ depends on whether $\sigma^\beta_j(n) = +1$ or $\sigma^\beta_j(n) = -1$, respectively. If $\sigma^\beta_j(n) \in (-1, 1)$, then $\lambda^\beta_j(n_0)$ is an eigenvalue of both $H^\beta_{\pm,n_0}$ if $\lambda^\beta_j(n) \in \sigma_p(H)$ and no eigenvalue for both otherwise. Moreover, for all $\beta_{1,2} \in \mathbb{R} \cup \{\infty\}$ we have $\hat{\lambda}^{\beta_1}_j(n) \neq \hat{\lambda}^{\beta_2}_j(n)$ provided $\beta_1 \neq \beta_2$.

Proof. The claims concerning the spectra are immediate by looking at Weyl m-functions and applying Lemma B.7 and the following considerations. It remains to show $\hat{\lambda}_j^{\beta_1}(n) \neq \hat{\lambda}_j^{\beta_2}(n)$ provided $\beta_1 \neq \beta_2$. Suppose the contrary. Then we have $\beta_1 = \phi(\hat{\lambda}_j^{\beta_2}(n), n) = \phi(\hat{\lambda}_j^{\beta_1}(n), n) = \beta_2$ a contradiction. Nevertheless we remark that the case $\lambda_j^{\beta_1}(n) = \lambda_j^{\beta_2}(n)$ (but $\sigma_j^{\beta_1}(n) \neq \sigma_j^{\beta_2}(n)$) can occur. $\square$

Next, we will investigate how $\lambda_j^\beta(n)$ varies with β. Therefore we use (8.106) to calculate

(8.119)
$$\frac{\partial}{\partial \beta} K_{r+1}^\beta(z, n) \Big|_{z = \lambda_j^\beta(n)}$$

in two ways

$$\frac{-\beta}{a(n)} \prod_{k \neq j}^{r+1} \left(\lambda_j^\beta(n) - \lambda_k^\beta(n)\right) \frac{\partial}{\partial \beta} \lambda_j^\beta(n)$$

(8.120)
$$= \frac{1}{a(n)} \left(H_{r+1}(\lambda_j^\beta(n), n) + 2a(n)\beta G_r(\lambda_j^\beta(n), n) \right).$$

This yields

Lemma 8.21. *The variation of the zeros $\lambda_j^\beta(n)$ of $K_{r+1}^\beta(z, n)$ with respect to $\beta \in \mathbb{R} \backslash \{0\}$ is described by*

(8.121)
$$\frac{\partial}{\partial \beta} \lambda_j^\beta(n) = \frac{R_{2r+2}^{1/2}(\hat{\lambda}_j^\beta(n))}{\beta \prod_{k \neq j}^{r+1} (\lambda_j^\beta(n) - \lambda_k^\beta(n))}, \quad 1 \leq j \leq r+1.$$

If $\lambda_j^\beta(n) = \lambda_{j+1}^\beta(n)$ the vanishing factors in the numerator and denominator have to be omitted.

Finally, we note that the connection with (8.23) is given by

(8.122)
$$c_{-,\ell}(n) = \sum_{j=1}^r \mu_j(n)^\ell \hat{R}_j(n), \quad \ell \in \mathbb{N}.$$

Chapter 9

Quasi-periodic Jacobi operators and Riemann theta functions

In the previous chapter we have seen how Riemann surfaces arise naturally in the investigation of reflectionless finite-gap sequences. However, there we have only used them as a convenient abbreviation. In this chapter we want to use Riemann surfaces as an independent tool which will help us to gain further insight. For simplicity we will only consider the case of nonsingular surfaces, that is, $r = g$ in (8.59) (i.e., $\tilde{R}_{2g+2}(z) = R_{2r+2}(z)$). The corresponding sequences will turn out to be expressible in terms of Riemann theta functions and are hence quasi-periodic.

The present chapter is entirely independent of Chapter 8 in the sense that one could take Theorem 9.2 as a definition for the Baker-Akhiezer function. But without our knowledge from Chapter 7 and 8 there is no reason why this object should be of any interest. Nevertheless we will show how all major results from Chapter 8 can be obtained using only methods from this chapter.

9.1. Riemann surfaces

The purpose of this section is mainly to clarify notation. For further information and references we refer the reader to Appendix A.

We consider the Riemann surface M associated with the following function

$$(9.1) \qquad R_{2g+2}^{1/2}(z), \qquad R_{2g+2}(z) = \prod_{j=0}^{2g+1} (z - E_j), \qquad E_0 < E_1 < \cdots < E_{2g+1},$$

$g \in \mathbb{N}$. M is a compact, hyperelliptic Riemann surface of genus g. Again we will choose $R_{2g+2}^{1/2}(z)$ as the fixed branch

$$(9.2) \qquad R_{2g+2}^{1/2}(z) = -\prod_{j=0}^{2g+1} \sqrt{z - E_j},$$

where $\sqrt{.}$ is defied as in (1.116).

A point of M is denoted by

(9.3) $$p = (z, \pm R_{2g+2}^{1/2}(z)), \quad z \in \mathbb{C}, \quad \text{or} \quad p = \infty_\pm$$

and M comes with three holomorphic maps on it. The involution map

(9.4) $$\begin{aligned} *: \quad M &\to M \\ (z, \pm R_{2g+2}^{1/2}(z)) &\mapsto (z, \pm R_{2g+2}^{1/2}(z))^* = (z, \mp R_{2g+2}^{1/2}(z)) \ , \\ \infty_\pm &\mapsto \infty_\mp \end{aligned}$$

the projection

(9.5) $$\begin{aligned} \pi: \quad M &\to \mathbb{C} \cup \{\infty\} \\ (z, \pm R_{2g+2}^{1/2}(z)) &\mapsto z \\ \infty_\pm &\mapsto \infty \end{aligned} \ ,$$

and the evaluation map

(9.6) $$\begin{aligned} R_{2g+2}^{1/2}: \quad M &\to \mathbb{C} \cup \{\infty\} \\ (z, \pm R_{2g+2}^{1/2}(z)) &\mapsto \pm R_{2g+2}^{1/2}(z) \ . \\ \infty_\pm &\mapsto \infty \end{aligned}$$

The sets $\Pi_\pm = \{(z, \pm R_{2g+2}^{1/2}(z)) | z \in \mathbb{C} \backslash \sigma(H)\}$ are called upper, lower sheet, respectively.

Clearly, the eigenvalues $\hat{\lambda}_j^\beta(n)$ can be viewed as points on our Riemann surface

(9.7) $$\begin{aligned} \hat{\lambda}_j^\beta(n) &= (\lambda_j^\beta(n), \sigma_j^\beta(n) R_{2g+2}^{1/2}(\lambda_j^\beta(n))) \in M, \\ R_{2g+2}^{1/2}(\hat{\lambda}_j^\beta(n)) &= \sigma_j^\beta(n) R_{2g+2}^{1/2}(\lambda_j^\beta(n)). \end{aligned}$$

The divisor associated with the eigenvalues $\hat{\lambda}_j^\beta(n)$ is defined by

(9.8) $$\mathcal{D}_{\underline{\hat{\lambda}}^\beta(n)}(\hat{\lambda}_j^\beta(n)) = 1, \quad 1 \le j \le \tilde{g},$$

and 0 otherwise, where $\tilde{g} = g+1$ if $\beta \in \mathbb{R} \backslash \{0\}$ and $\tilde{g} = g$ if $\beta \in \{0, \infty\}$. We have

(9.9) $$\deg(\mathcal{D}_{\underline{\hat{\lambda}}^\beta(n)}) = \tilde{g}$$

and since $\sigma_p(H_n^\beta)$ is simple, $\mathcal{D}_{\underline{\hat{\lambda}}^\beta(n)}$ is nonspecial,

(9.10) $$i(\mathcal{D}_{\underline{\hat{\lambda}}^\beta(n)}) = 0,$$

by Lemma A.20. We set $\mathcal{D}_{\underline{\hat{\lambda}}^\beta} = \mathcal{D}_{\underline{\hat{\lambda}}^\beta(0)}$.

We proceed with considering the function $\phi(p, n) - \beta$ with $\beta \in \mathbb{R} \backslash \{0\}$. Equation (8.105) shows that there is a simple pole at ∞_- and that all further poles are also simple and can only be at $\hat{\mu}_j(n)$, $1 \le j \le g$. A closer investigation using (8.105), (8.113) shows that the divisor of $\phi(p, n) + \beta$ is given by

(9.11) $$(\phi(p, n) + \beta) = \mathcal{D}_{\underline{\hat{\lambda}}^\beta(n)} - \mathcal{D}_{\underline{\hat{\mu}}(n)} - \mathcal{D}_{\infty_-}, \quad \beta \in \mathbb{R} \backslash \{0\}.$$

If we want to include the case $\beta = 0$ we have to add an 'additional' eigenvalue at ∞_+ (the eigenvalue $\lambda_{g+1}^\beta(n)$ disappears at infinity)

(9.12) $$(\phi(p, n)) = \mathcal{D}_{\underline{\hat{\mu}}(n+1)} - \mathcal{D}_{\underline{\hat{\mu}}(n)} - \mathcal{D}_{\infty_-} + \mathcal{D}_{\infty_+}.$$

In principal, some of the poles and zeros might still cancel, but Theorem 8.20 shows that this does not happen.

9.2. Solutions in terms of theta functions

By taking Abel's map $\underline{\alpha}_{p_0}$ on both sides we get ($\beta \in \mathbb{R}\backslash\{0\}$)

$$\underline{\alpha}_{p_0}(\mathcal{D}_{\hat{\underline{\lambda}}^\beta(n)}) = \underline{\alpha}_{p_0}(\mathcal{D}_{\hat{\underline{\mu}}(n)} + \mathcal{D}_{\infty_-}),$$

(9.13) $$\underline{\alpha}_{p_0}(\mathcal{D}_{\hat{\underline{\mu}}(n+1)} + \mathcal{D}_{\infty_+}) = \underline{\alpha}_{p_0}(\mathcal{D}_{\hat{\underline{\mu}}(n)} + \mathcal{D}_{\infty_-}).$$

This shows that $\underline{\alpha}_{p_0}(\mathcal{D}_{\hat{\underline{\lambda}}^\beta(n)})$ is independent of $\beta \in \mathbb{R}\backslash\{0\}$ which can also be shown using (8.121) (compare (13.52)).

Combining the last two equations yields for arbitrary β

(9.14) $$\underline{\alpha}_{p_0}(\mathcal{D}_{\hat{\underline{\lambda}}^\beta(n+1)} + \mathcal{D}_{\infty_+}) = \underline{\alpha}_{p_0}(\mathcal{D}_{\hat{\underline{\lambda}}^\beta(n)} + \mathcal{D}_{\infty_-}).$$

A result we will soon investigate further.

Next, we are interested in the poles and zeros of the Baker-Akhiezer function. Because of (8.88) the divisor of $\psi(p,n) = \psi(p,n,0)$ reads

(9.15) $$(\psi(p,n)) = \mathcal{D}_{\hat{\underline{\mu}}(n)} - \mathcal{D}_{\hat{\underline{\mu}}} + n(\mathcal{D}_{\infty_+} - \mathcal{D}_{\infty_-})$$

and by taking Abel's map on both sides we obtain

(9.16) $$\underline{\alpha}_{p_0}(\mathcal{D}_{\hat{\underline{\mu}}(n)}) = \underline{\alpha}_{p_0}(\mathcal{D}_{\hat{\underline{\mu}}}) - n\underline{A}_{\infty_-}(\infty_+).$$

9.2. Solutions in terms of theta functions

We first fix the representatives of a canonical homology basis $\{a_j, b_j\}_{j=1}^g$ on M. We require a_j to surround the points E_{2j-1}, E_{2j} (changing sheets twice) and b_j to surround E_0, E_{2j-1} on the upper sheet. The corresponding canonical basis $\{\zeta_j\}_{j=1}^g$ for $\mathcal{H}^1(M)$ (the space of holomorphic differentials) has the following representation

(9.17) $$\underline{\zeta} = \sum_{j=1}^g \underline{c}(j) \frac{\pi^{j-1} d\pi}{R_{2g+2}^{1/2}}$$

and we have

(9.18) $$\int_{a_j} \zeta_k = \delta_{j,k}, \qquad \int_{b_j} \zeta_k = \tau_{j,k}.$$

To every compact Riemann surface of genus $g \in \mathbb{N}$ belongs a Riemann theta function (cf. Sections A.5 and A.6)

(9.19) $$\theta(\underline{z}) = \sum_{\underline{m} \in \mathbb{Z}^g} \exp 2\pi i \left(\langle \underline{m}, \underline{z}\rangle + \frac{\langle \underline{m}, \underline{\tau}\,\underline{m}\rangle}{2}\right), \qquad \underline{z} \in \mathbb{C}^g,$$

with $\langle \hat{\underline{z}}, \underline{z}\rangle = \sum_{j=1}^g \hat{z}_j z_j$ and $\underline{\tau}$ the matrix of b-periods of the differentials (9.17). It has the following fundamental properties

$$\theta(-\underline{z}) = \theta(\underline{z}),$$

(9.20) $$\theta(\underline{z} + \underline{m} + \underline{\tau}\,\underline{n}) = \exp 2\pi i \left(-\langle \underline{n}, \underline{z}\rangle - \frac{\langle \underline{n}, \underline{\tau}\,\underline{n}\rangle}{2}\right) \theta(\underline{z}), \qquad \underline{n}, \underline{m} \in \mathbb{Z}^g.$$

We introduce

(9.21) $$\underline{z}(p,n) = \underline{\hat{A}}_{p_0}(p) - \underline{\hat{\alpha}}_{p_0}(\mathcal{D}_{\hat{\underline{\mu}}(n)}) - \underline{\hat{\Xi}}_{p_0} \in \mathbb{C}^g, \quad \underline{z}(n) = \underline{z}(\infty_+, n),$$

where $\underline{\Xi}_{p_0}$ is the vector of Riemann constants

(9.22) $$\hat{\Xi}_{p_0,j} = \frac{1 - \sum_{k=1}^g \tau_{j,k}}{2}, \quad p_0 = (E_0, 0),$$

and $\underline{A}_{p_0}$ ($\underline{\alpha}_{p_0}$) is Abel's map (for divisors). The hat indicates that we regard it as a (single-valued) map from $\hat{M}$ (the fundamental polygon associated with M) to $\mathbb{C}^g$. $\underline{z}(p,n)$ is independent of the base point p_0 and we note that $\underline{z}(\infty_+, n) = \underline{z}(\infty_-, n+1)$ mod $\mathbb{Z}^g$. We recall that the function $\theta(\underline{z}(p,n))$ has zeros precisely at the points $\hat{\mu}_j(n)$, $1 \leq j \leq g$, that is,

(9.23) $$\theta(\underline{z}(p,n)) = 0 \quad \Leftrightarrow \quad p \in \{\hat{\mu}_j(n)\}_{j=1}^g.$$

Since divisors of the type $\mathcal{D}_{\underline{\hat{\mu}}}$ with $\mu_j \in [E_{2j-1}, E_{2j}]$ are of special importance to us, we want to tell whether a given divisor $\mathcal{D} \in M_g$ (M_g, the g-th symmetric power of M) is of the above type or not by looking at $\underline{\alpha}_{p_0}(\mathcal{D})$.

Lemma 9.1. *Introduce two manifolds*

(9.24) $$\begin{aligned} M^D &= \otimes_{j=1}^g \pi^{-1}([E_{2j-1}, E_{2j}]) \simeq \otimes_1^g S^1, \\ J^D &= \{[\underline{A}] \in J(M) | \, [\mathrm{i}\,\mathrm{Im}(\underline{A} + \hat{\underline{\Xi}}_{p_0})] = [\underline{0}]\} \simeq \otimes_1^g S^1, \end{aligned}$$

where [.] denotes the equivalence classes in the Jacobian variety $J(M)$. Then the mapping

(9.25) $$\begin{aligned} \underline{\alpha}_{p_0} : \quad M^D &\to J^D \\ \underline{\hat{\mu}} &\mapsto \underline{\alpha}_{p_0}(\mathcal{D}_{\underline{\hat{\mu}}}) \end{aligned}$$

is an isomorphism.

Proof. Let $\mu_j \in [E_{2j-1}, E_{2j}]$ and observe

(9.26) $$\mathrm{i}\,\mathrm{Im}\left(\hat{A}_{p_0,k}(\hat{\mu}_j)\right) = \frac{\tau_{j,k}}{2}.$$

This can be shown as in Section A.7 by taking all integrals as limits along the real axis (we may intersect some b-cycles though we are required to stay in $\hat{M}$, but we only add a-cycles which are real). Thus $\underline{\alpha}_{p_0}(M^D) \subseteq J^D$. Since divisors in M^D are nonspecial by Lemma A.20 we see that $\underline{\alpha}_{p_0}$ is an immersion by (A.86) and injective by Lemma A.9. Moreover, $\underline{\alpha}_{p_0}(M^D)$ is open (since holomorphic maps are open) and compact (since M^D is). So $\underline{\alpha}_{p_0}(M^D)$ is both open and closed, and hence must be the whole of J^D by connectedness. $\square$

As a consequence we note that when n changes continuously to $n+1$, $\mu_j(n)$ changes continuously to $\mu_j(n+1)$, always remaining in its gap. Thus we obtain

(9.27) $$\theta(\underline{z}(p,m))\theta(\underline{z}(p,n)) > 0$$

for $\pi(p) \in (-\infty, E_0) \cup (E_{2g+1}, \infty) \cup \{\infty\}$ and $m, n \in \mathbb{Z}$. Moreover, since the set of all possible Dirichlet eigenvalues M^D is compact, we infer that there are positive constants $C_1 \leq C_2$ such that

(9.28) $$C_1 \leq |\theta(\underline{z}(n))| \leq C_2.$$

Finally, we recall $\omega_{\infty_+, \infty_-}$, the normalized abelian differential of the third kind with simple poles at $\infty_\pm$ and residues ± 1, respectively (cf. (A.20)). We can make the following ansatz

(9.29) $$\omega_{\infty_+, \infty_-} = \frac{\prod_{j=1}^g (\pi - \lambda_j)}{R_{2g+2}^{1/2}} d\pi,$$

9.2. Solutions in terms of theta functions

where the constants λ_j have to be determined from the normalization

$$(9.30) \qquad \int_{a_j} \omega_{\infty_+,\infty_-} = 2 \int_{E_{2j-1}}^{E_{2j}} \frac{\prod_{k=1}^{g}(z-\lambda_k)}{R_{2g+2}^{1/2}(z)} dz = 0,$$

which shows $\lambda_j \in (E_{2j-1}, E_{2j})$. With these preparations we are now able to express the Baker-Akhiezer function $\psi(p,n)$ in terms of theta functions.

Theorem 9.2. *The Baker-Akhiezer function ψ is given by*

$$(9.31) \qquad \psi(p,n) = C(n,0) \frac{\theta(\underline{z}(p,n))}{\theta(\underline{z}(p,0))} \exp\left(n \int_{p_0}^{p} \hat{\omega}_{\infty_+,\infty_-}\right),$$

and ϕ is given by

$$(9.32) \qquad \phi(p,n) = C(n) \frac{\theta(\underline{z}(p,n+1))}{\theta(\underline{z}(p,n))} \exp\left(\int_{p_0}^{p} \hat{\omega}_{\infty_+,\infty_-}\right).$$

The hat indicates that we require the path of integration to lie in $\hat{M}$. The base point p_0 has been chosen to be $(E_0, 0)$ (any other branch point would do as well). The constants $C(n)$ and $C(n, n_0)$ are real and read

$$(9.33) \qquad C(n)^2 = \frac{\theta(\underline{z}(n-1))}{\theta(\underline{z}(n+1))}, \quad C(n,m) = \prod_{j=m}^{n-1}{}^{*} C(j).$$

Proof. The function ψ is at least well-defined on $\hat{M}$ since changing the path in $\hat{M}$ can only change the values of the integral in the exponential by $2\pi i$. But by

$$(9.34) \qquad \int_{a_j} \omega_{\infty_+,\infty_-} = 0,$$

the b-periods are given by (cf. (A.21))

$$(9.35) \qquad \int_{b_j} \omega_{\infty_+,\infty_-} = 2\pi i \hat{A}_{\infty_-,j}(\infty_+),$$

which shows together with (9.20) that it extends to a meromorphic function on M. Since ψ is fixed by its poles and zeros up to a constant (depending on n) it remains to determine this constant. This can be done using $\psi(\infty_+, n)\psi(\infty_-, n) = 1$, which follows from (8.103). From Section A.7 we recall $\hat{A}_{p_0}(\infty_+) = -\hat{A}_{p_0}(\infty_-) \mod \mathbb{Z}^g$ and similarly we obtain $\int_{p_0}^{p} \hat{\omega}_{\infty_+,\infty_-} = -\int_{p_0}^{p^*} \hat{\omega}_{\infty_+,\infty_-}$ for $\pi(p) < E_0$. This justifies our choice for $C(n,0)^2$. By (9.27), $C(n,m)^2$ is positive and the formula for ϕ now follows from $\phi(p,n) = \psi(n+1,p)/\psi(n,p)$. $\square$

The sign of $C(n)$ is related to that of $a(n)$ as will be shown in Theorem 9.4 below. Next, let us prove some basic properties of ϕ and ψ.

Lemma 9.3. *Set $\sigma(H) = \bigcup_{j=0}^{g}[E_{2j}, E_{2j+1}]$. The functions ϕ and ψ have the following properties*
(i). ϕ and hence ψ is real for $\pi(p) \in \mathbb{R} \backslash \sigma(H)$.
(ii). Let $z \in \mathbb{R}$, then we have

$$(9.36) \qquad |\psi_{\pm}(z,n)| \leq M k(z)^{\pm n}, \quad M > 0,$$

where

$$(9.37) \quad k(z) = \exp\left(\text{Re}\left(\int_{p_0}^{p} \hat{\omega}_{\infty_+, \infty_-}\right)\right) > 0, \quad p = (z, R_{2g+2}^{1/2}(z))$$

and $k(z) = 1$ for $z \in \sigma(H)$ and $k(z) < 1$ for $z \in \mathbb{R}\backslash\sigma(H)$. Furthermore, $k(z)$ has minima precisely at the points λ_j, $1 \leq j \leq g$, and tends to 0 as $|z| \to \infty$.

Proof. We begin with realvaluedness of ϕ (which implies realvaluedness of ψ). Let $\hat{\mu}_k$ be an arbitrary point with $\pi(\hat{\mu}_k) \in [E_{2k-1}, E_{2k}]$. Proceeding as usual we get from (9.35) for the differential of the third kind

$$(9.38) \quad \text{i}\,\text{Im}\left(\int_{p_0}^{\hat{\mu}_k} \hat{\omega}_{\infty_+, \infty_-}\right) = -\pi\text{i}\hat{A}_{\infty_-, k}(\infty_+)$$

and, similarly, $\text{i}\,\text{Im}(z_j(\hat{\mu}_k, n)) = \frac{\tau_{j,k}}{2}$ (cf. (9.26)). Now we recall theta functions with integer characteristics (cf. [**81**], VI.1.5)

$$(9.39) \quad \theta\begin{bmatrix}\underline{n}\\\underline{\hat{n}}\end{bmatrix}(\underline{z}) = \sum_{\underline{m}\in\mathbb{Z}^g} \exp 2\pi\text{i}\left(\langle\underline{m}+\tfrac{1}{2}\underline{\hat{n}}, \underline{z}+\tfrac{1}{2}\underline{n}\rangle + \tfrac{1}{2}\langle\underline{m}+\tfrac{1}{2}\underline{\hat{n}}, \underline{\tau}(\underline{m}+\tfrac{1}{2}\underline{\hat{n}})\rangle\right),$$

$\underline{z} \in \mathbb{C}^g$, which are clearly real for $\underline{z} \in \mathbb{R}$. Inserting

$$(9.40) \quad \begin{aligned}\theta(\underline{z}(\hat{\mu}_k, n)) &= \exp 2\pi\text{i}\left(\tfrac{1}{8}\langle\underline{\delta}(k), \underline{\tau}\,\underline{\delta}(k)\rangle + \tfrac{1}{2}\langle\underline{\delta}(k), \text{Re}(\underline{z}(\hat{\mu}_k, n))\rangle\right) \\ &\quad \times \theta\begin{bmatrix}\underline{0}\\\underline{\delta}(k)\end{bmatrix}(\text{Re}(\underline{z}(\hat{\mu}_k, n)))\end{aligned}$$

into ϕ shows that the complex factors cancel since $C(n) \in \mathbb{R}$ and $z_j(\hat{\mu}_k, n+1) = z_j(\hat{\mu}_k, n) + \hat{A}_{\infty_-, j}(\infty_+)$. The argument for $\mu > E_{2g+1}$ and $\mu < E_0$ is even easier since $\text{Im}(\underline{z}(\hat{\mu}, n)) = \text{Im}(\int_{p_0}^{\hat{\mu}} \hat{\omega}_{\infty_+, \infty_-}) = 0$ in this case.

Now we turn to the second claim. Since the theta functions are quasi-periodic we only have to investigate the exponential factor. For the asymptotic behavior we need to know $\text{Re}(\int_{p_0}^{p} \hat{\omega}_{\infty_+, \infty_-})$. We first assume $p \in \Pi_+$, $\pi(p) = \lambda \in \mathbb{R}$, and choose as integration path the lift of the straight line from $E_0 + \text{i}\varepsilon$ to $\lambda + \text{i}\varepsilon$ and take the limit $\varepsilon \downarrow 0$ (we may intersect as many b-cycles as we like since all a-periods are zero). Hence

$$(9.41) \quad \int_{p_0}^{p} \hat{\omega}_{\infty_+, \infty_-} = \int_{E_0}^{\lambda} \frac{\prod_{j=1}^{g}(\pi - \lambda_j)d\pi}{R_{2g+2}^{1/2}}.$$

Clearly, (9.41) is negative for $\lambda < E_0$. For $\lambda \in [E_1, E_2]$, (9.41) is purely imaginary and hence the real part is 0. At E_2 our real part starts to decrease from zero until it hits its minimum at λ_1. Then it increases until it reaches 0, which must be at E_3 by (9.30). Proceeding like this we get the desired behavior of $k(z)$. For $p \in \Pi_-$ we have to change the sign of $\text{Re}(\int_{p_0}^{p} \omega_{\infty_+, \infty_-})$. $\square$

Next we are interested in expanding ϕ near $\infty_\pm$ (in the coordinates $(\Pi_\pm, z)$ induced by the projection on the upper/lower sheet $\Pi_\pm$). We start with

$$(9.42) \quad \exp\left(\int_{p_0}^{p} \hat{\omega}_{\infty_+, \infty_-}\right) = -\left(\frac{\tilde{a}}{z}\right)^{\pm 1}\left(1 + \frac{\tilde{b}}{z} + \frac{\tilde{c}}{z^2} + O(\frac{1}{z^3})\right)^{\pm 1}, \quad p \in \Pi_\pm,$$

9.2. Solutions in terms of theta functions

where $\tilde{a}, \tilde{b}, \tilde{c} \in \mathbb{R}$ are constants depending only on the Riemann surface (i.e., on E_j). We have

$$\tilde{a} = \lim_{\lambda \to \infty} \frac{1}{\lambda} \exp\Big(\int_{-\lambda}^{E_0} \frac{\prod_{j=1}^{g}(z - \lambda_j)}{R_{2g+2}^{1/2}(z)} dz\Big) > 0 \tag{9.43}$$

and for $\tilde{b}$ we get from (9.30)

$$\tilde{b} = \frac{1}{2} \sum_{j=0}^{2g+1} E_j - \sum_{j=1}^{g} \lambda_j. \tag{9.44}$$

For the theta functions we get near ∞_+ (cf. (A.119) and don't confuse z and $\underline{z}$)

$$\begin{aligned}\frac{\theta(\underline{z}(p, n+1))}{\theta(\underline{z}(p, n))} = \frac{\theta(\underline{z}(n+1))}{\theta(\underline{z}(n))} &\Big(1 + \frac{1}{z} \sum_{j=1}^{g} B_j(n+1) \\&+ \frac{1}{2z^2}\Big(\Big(\sum_{j=1}^{g} B_j(n+1)\Big)^2 \\&+ \sum_{j=1}^{g}\Big(\sum_{k=0}^{2g+1} \frac{E_k}{2} + \frac{c_j(g-1)}{c_j(g)}\Big) B_j(n+1) \\&+ \sum_{j=1}^{g}\sum_{k=1}^{g} BB_{j,k}(n+1)\Big) + O(\frac{1}{z^3})\Big),\end{aligned} \tag{9.45}$$

with

$$\begin{aligned}B_j(n) &= c_j(g) \frac{\partial}{\partial w_j} \ln\Big(\frac{\theta(\underline{w} + \underline{z}(n))}{\theta(\underline{w} + \underline{z}(n-1))}\Big)\Big|_{\underline{w}=0}, \\BB_{j,k}(n) &= c_j(g) c_k(g) \frac{\partial^2}{\partial w_j \partial w_k} \ln\Big(\frac{\theta(\underline{w} + \underline{z}(n))}{\theta(\underline{w} + \underline{z}(n-1))}\Big)\Big|_{\underline{w}=0}.\end{aligned} \tag{9.46}$$

Near ∞_- one has to make the substitutions $\underline{c}(n) \to -\underline{c}(n)$ and $n \to n-1$. The constants $\underline{c}(g)$ come from (9.17). In summary,

Theorem 9.4. *The functions $\phi_{\pm}$ have the following expansions*

$$\phi_{\pm}(z, n) = \Big(\frac{a(n)}{z}\Big)^{\pm 1}\Big(1 \pm \frac{b(n + {}^{1}_{0})}{z} + O(\frac{1}{z^2})\Big), \tag{9.47}$$

where the sequences $a(n)$, $b(n)$ are given by

$$\begin{aligned}a(n)^2 &= \tilde{a}^2 \frac{\theta(\underline{z}(n+1))\theta(\underline{z}(n-1))}{\theta(\underline{z}(n))^2}, \\b(n) &= \tilde{b} + \sum_{j=1}^{g} c_j(g) \frac{\partial}{\partial w_j} \ln\Big(\frac{\theta(\underline{w} + \underline{z}(n))}{\theta(\underline{w} + \underline{z}(n-1))}\Big)\Big|_{\underline{w}=0}.\end{aligned} \tag{9.48}$$

The sings of $a(n)$ and $C(n)$ must be opposite

$$a(n) = -\tilde{a}\, C(n) \frac{\theta(\underline{z}(n+1))}{\theta(\underline{z}(n))}. \tag{9.49}$$

The sequences $a(n), b(n)$ are obviously quasi-periodic with g periods (i.e., their Fourier exponents possess a g-term integral basis).

We remark that we have

(9.50) $\{\mu_j(n_0)\}_{j=1}^g \subset \{E_{2j-1}, E_{2j}\}_{j=1}^g \;\Leftrightarrow\; [\underline{z}(p, n_0 + k)] = -[\underline{z}(p^*, n_0 - k)].$

Inserting this into (9.48) yields again $a(n_0+k) = a(n_0-k-1)$, $b(n_0+k) = b(n_0-k)$ (compare Lemma 8.15).

The expansion coefficients $\tilde{a}, \tilde{b}$ can be interpreted as averages of our coefficients $a(n)$ and $b(n)$ as follows.

Lemma 9.5. *Suppose $a(n) > 0$ and $\pi(p) \in (-\infty, E_0)$, then*

(9.51) $$\lim_{N \to \infty} \sqrt[N]{\psi(p, N)} = -\exp\left(\int_{p_0}^p \hat{\omega}_{\infty_+, \infty_-}\right).$$

In particular, we obtain

(9.52) $$\tilde{a} = \lim_{N \to \infty} \sqrt[N]{\prod_{j=1}^N a(j)}, \quad \tilde{b} = \lim_{N \to \infty} \frac{1}{N} \sum_{j=1}^N b(j),$$

that is, $\tilde{a}$ is the geometric mean of the a's and $\tilde{b}$ is the arithmetic mean of the b's.

Proof. The first claim follows from $\lim_{N \to \infty} \sqrt[N]{|\theta(\underline{z}(N))|} = 1$ (cf. (9.28)). The second follows by comparing asymptotic coefficients with (9.47) respectively (6.27). $\square$

Now we want to know when our sequences are periodic.

Theorem 9.6. *A necessary and sufficient condition for $a(n), b(n)$ to be periodic is that $R_{2g+2}(z)$ is of the form*

(9.53) $$\frac{1}{4A^2} R_{2g+2}(z) Q(z)^2 = \Delta(z)^2 - 1,$$

where $Q(z)$ and $\Delta(z)$ are polynomials, $Q(z)$ has leading coefficient one, and $A^2 > 0$ (compare (7.64)). The period N is given by

(9.54) $$N = \deg(Q) + g + 1.$$

Proof. Necessity has been shown in Chapter 7 (see (7.55) and Remark 7.6). For the remaining direction note that in the above case $\omega_{\infty_+, \infty_-}$ is given by

(9.55) $$\omega_{\infty_+, \infty_-} = \frac{1}{N} \frac{2A\Delta'\, d\pi}{R_{2g+2}^{1/2} Q},$$

where $N = \deg(Q) + g + 1$ and the prime denotes the derivative with respect to z. We only have to check that the a-periods are zero

(9.56) $$\int_{a_j} \omega_{\infty_+, \infty_-} = \frac{2}{N} \int_{E_{2j-1}}^{E_{2j}} \frac{\Delta'(z)\, dz}{(\Delta(z)^2 - 1)^{1/2}}$$
$$= \frac{2}{N} \ln|\Delta(z) + (\Delta(z)^2 - 1)^{1/2}|\Big|_{E_{2j-1}}^{E_{2j}} = 0.$$

9.2. Solutions in terms of theta functions

For the b-periods we get

$$\int_{b_j} \omega_{\infty_+,\infty_-} = \frac{-2}{N} \sum_{k=0}^{j-1} \int_{E_{2k}}^{E_{2k+1}} \frac{\Delta'(z)\,dz}{(\Delta(z)^2 - 1)^{1/2}}$$

$$= \frac{-2\mathrm{i}}{N} \sum_{k=0}^{j-1} (-1)^{g-k} \int_{E_{2k}}^{E_{2k+1}} \frac{\Delta'(z)\,dz}{\sqrt{1 - \Delta(z)^2}}$$

(9.57) $$= \frac{-2\mathrm{i}}{N} \sum_{k=0}^{j-1} (-1)^{g-k} \arcsin(\Delta(z))\Big|_{E_{2k}}^{E_{2k+1}} = -\frac{2\pi\mathrm{i}}{N} j.$$

Which proves the assertion, since we infer $2N\underline{\hat{A}}_{p_0}(\infty_+) \in \mathbb{Z}^g$. □

Further we obtain in the periodic case for the Floquet multipliers

(9.58) $$m(p) = \psi(p, N) = (-1)^N \mathrm{sgn}(A) \exp\left(N \int_{p_0}^{p} \hat{\omega}_{\infty_+,\infty_-}\right)$$

and $\tilde{a} = \sqrt[N]{|A|}$, $\tilde{b} = \frac{B}{N}$.

Next we will show that the Baker-Akhiezer function is even fixed by its (finite) poles and its behavior at infinity

Lemma 9.7. *Let $\mathcal{D}_{\underline{\hat{\mu}}} \in M_g$ be nonspecial (i.e., $i(\mathcal{D}_{\underline{\hat{\mu}}}) = 0$). Then there is a meromorphic function $\psi(n,.)$ on M fulfilling*

(9.59) $$(\psi(n)) \geq -\mathcal{D}_{\underline{\hat{\mu}}} + n(\mathcal{D}_{\infty_+} - \mathcal{D}_{\infty_-}).$$

Moreover, $\psi(n,.)$ is unique up to a multiple depending only on n and the divisor $\mathcal{D}_{\mathrm{zer}}(n) \in M_g$ defined by

(9.60) $$(\psi(n)) = \mathcal{D}_{\mathrm{zer}}(n) - \mathcal{D}_{\underline{\hat{\mu}}} + n(\mathcal{D}_{\infty_+} - \mathcal{D}_{\infty_-}),$$

is also nonspecial.

Proof. Abbreviate $\mathcal{D}_n = \mathcal{D}_{\underline{\hat{\mu}}} - n(\mathcal{D}_{\infty_+} - \mathcal{D}_{\infty_-})$. The Riemann-Roch theorem (cf. Theorem (A.2)) implies existence of at least one Baker-Akhiezer function since $r(-\mathcal{D}_n) = 1 + i(\mathcal{D}_{\underline{\hat{\mu}}} - n(\mathcal{D}_{\infty_+} - \mathcal{D}_{\infty_-})) \geq 1$. Moreover, since there is a function satisfying (9.60), we obtain $i(\mathcal{D}_{\mathrm{zer}}(n)) = i(\mathcal{D}_n) = r(-\mathcal{D}_n) - 1$ by (A.39) and Riemann-Roch.

There is nothing to prove for $n = 0$, since $\mathcal{L}(-\mathcal{D}_{\underline{\hat{\mu}}}) = \mathbb{C}$ and $\mathcal{D}_{\mathrm{zer}}(0) = \mathcal{D}_{\underline{\hat{\mu}}}$. Next let us consider $n = 1$. From $\mathcal{L}(-\mathcal{D}_{\underline{\hat{\mu}}}) = \mathbb{C}$ we see $\mathcal{L}(-\mathcal{D}_{\underline{\hat{\mu}}} + \mathcal{D}_{\infty_-}) = \{0\}$ and hence $r(-\mathcal{D}_1) \leq 1$ (otherwise, if $r(-\mathcal{D}_1) > 1$, a proper linear combination of two elements would lie in $\mathcal{L}(-\mathcal{D}_{\underline{\hat{\mu}}} + \mathcal{D}_{\infty_-})$, a contradiction). Now suppose the claim holds for $n > 1$. Then it also holds for $n+1$ since $r(-\mathcal{D}_{n+1}) = r(-\mathcal{D}_{n+1} - (\psi(n))) = r(-\mathcal{D}_{\mathrm{zer}}(n) + (\mathcal{D}_{\infty_+} - \mathcal{D}_{\infty_-})) = 1$ again by (A.39). So by induction the claim holds for $n \in \mathbb{N}$. Similarly, one can show that it holds for $n \in -\mathbb{N}$. □

Now using (9.47) it is not hard to show that (near $\infty_\pm$)

(9.61) $$a(n)\phi(p,n) + \frac{a(n-1)}{\phi(p,n-1)} + b(n) - \pi(p) = O\left(\frac{1}{z}\right)$$

or equivalently

(9.62) $$a(n)\psi(p, n+1) + a(n-1)\psi(p, n-1) - (\pi(p) - b(n))\psi(p, n) = z^{\mp n} O(\frac{1}{z}).$$

The function on the left fulfills the requirements for a Baker-Akhiezer function (9.59) and $\mathcal{D}_{\infty_+} + \mathcal{D}_{\infty_-}$ is part of $\mathcal{D}_{\text{zer}}(n)$. Since $\mathcal{D}_{\text{zer}}(n)$ is nonspecial, this is impossible (by Lemma A.20) unless the left hand side is zero. This is clearly an independent proof that $\psi(p, n)$ satisfies $\tau \psi(p, n) = \pi(p) \psi(p, n)$.

Remark 9.8. From (9.61) we also obtain a different representation of $b(n)$. If we choose $p = p_0$ it reads

(9.63) $$b(n) = E_0 + \tilde{a} \frac{\theta(\underline{z}(n-1))}{\theta(\underline{z}(n))} \frac{\theta(\underline{z}(p_0, n+1))}{\theta(\underline{z}(p_0, n))} + \tilde{a} \frac{\theta(\underline{z}(n))}{\theta(\underline{z}(n-1))} \frac{\theta(\underline{z}(p_0, n-1))}{\theta(\underline{z}(p_0, n))}.$$

We note that (8.101)–(8.104) can be easily verified independently by considering poles, zeros, and the behavior at $\infty_\pm$. In particular, by expanding (8.104) around $\infty_\pm$ and comparing the second term in the expansion we get our trace relation

(9.64) $$b(n) = \frac{1}{2} \sum_{j=0}^{2g+1} E_j - \sum_{j=1}^{g} \mu_j(n).$$

Upon evaluating the integral

(9.65) $$I = \frac{1}{2\pi i} \oint_{\partial \hat{M}} \pi(.) d \ln(\theta(\underline{z}(., n))),$$

we get from the residue theorem

(9.66) $$I = \sum_{j=1}^{g} \mu_j(n) + \sum_{p \in \{\infty_\pm\}} \text{res}_p \big(\pi(.) d \ln(\theta(\underline{z}(., n))) \big),$$

and by direct calculation (using (A.83))

(9.67) $$I = \sum_{j=1}^{g} \int_{a_j} \pi \zeta_j.$$

Which yields the relation

(9.68) $$b(n) = \frac{1}{2} \sum_{j=0}^{2g+1} E_j - \sum_{j=1}^{g} \int_{a_j} \pi \zeta_j + \sum_{j=1}^{g} c_j(g) \frac{\partial}{\partial w_j} \ln \left(\frac{\theta(\underline{w} + \underline{z}(n))}{\theta(\underline{w} + \underline{z}(n-1))} \right) \Big|_{\underline{w}=0},$$

and thus

(9.69) $$\sum_{j=1}^{g} \int_{a_j} \pi \zeta_j = \sum_{j=1}^{g} \lambda_j.$$

9.3. The elliptic case, genus one

In this section we will show how the formulas of the last section can be expressed in terms of Jacobi's elliptic functions if $g = 1$. All integrals are evaluated using the tables in [**38**].

9.3. The elliptic case, genus one

Suppose $E_0 < E_1 < E_2 < E_3$ are given and $R_4^{1/2}(z)$ is defined as usual. We introduce

$$(9.70) \quad k = \sqrt{\frac{E_2 - E_1}{E_3 - E_1}\frac{E_3 - E_0}{E_2 - E_0}} \in (0,1), \quad k' = \sqrt{\frac{E_3 - E_2}{E_3 - E_1}\frac{E_1 - E_0}{E_2 - E_0}} \in (0,1),$$

such that $k^2 + k'^2 = 1$, and

$$(9.71) \quad u(z) = \sqrt{\frac{E_3 - E_1}{E_3 - E_0}\frac{E_0 - z}{E_1 - z}}, \quad C = \frac{2}{\sqrt{(E_3 - E_1)(E_2 - E_0)}}.$$

Furthermore, we will use the following notation for Jacobi's elliptic integrals of the first

$$(9.72) \quad \int_0^z \frac{dx}{\sqrt{(1-x^2)(1-k^2 x^2)}} = F(z,k),$$

second

$$(9.73) \quad \int_0^z \sqrt{\frac{1-x^2}{1-k^2 x^2}}\, dx = E(z,k),$$

and third kind

$$(9.74) \quad \int_0^z \frac{dx}{(1-\alpha^2 x^2)\sqrt{(1-x^2)(1-k^2 x^2)}} = \Pi(z,\alpha^2,k).$$

We set $F(1,k) = K(k)$, $E(1,k) = E(k)$, and $\Pi(1,\alpha^2,k) = \Pi(\alpha^2,k)$, where $k \in (0,1)$, $\alpha^2 \in \mathbb{R}$ and all roots are assumed to be positive for $x \in (0,1)$. We remark that $\Pi(z,\alpha^2,k)$ has simple poles at $z^2 = \alpha^{-2}$ and that $E(z,k)$ has a simple pole at ∞.

First we determine ζ, the abelian differential of the first kind (in the charts $(\Pi_\pm, z)$).

$$(9.75) \quad \zeta = c(1)\frac{d\pi}{R_4^{1/2}} = c(1)\frac{dz}{\pm R_4^{1/2}(z)}.$$

The normalization ([**38**], 254.00)

$$(9.76) \quad \int_{a_1} \zeta = 2c(1)\int_{E_1}^{E_2} \frac{dz}{R_4^{1/2}(z)} = 2c(1)CK(k) = 1$$

yields

$$(9.77) \quad c(1) = \frac{1}{2CK(k)}.$$

In addition, we obtain ([**38**], 252.00)

$$(9.78) \quad \tau_{1,1} = \int_{b_1} \zeta = -2c(1)\int_{E_0}^{E_1} \frac{dx}{R_4^{1/2}(x)} = 2ic(1)CK(k') = i\frac{K(k')}{K(k)}.$$

Next we are interested in Abel's map A_{p_0}, $p_0 = (E_0, 0)$. Let $p = (z, \pm R_4^{1/2}(z))$, then we have ([**38**], 252.00)

$$(9.79) \quad A_{p_0}(p) = [\int_{p_0}^p \zeta] = [\pm c(1)\int_{E_0}^z \frac{dx}{R_4^{1/2}(x)}] = [\pm\frac{F(u(z),k)}{2K(k)}]$$

and especially for $p = \infty_+$

$$A_{p_0}(\infty_+) = [\frac{F(\sqrt{\frac{E_3-E_1}{E_3-E_0}}, k)}{2K(k)}]. \tag{9.80}$$

The vector of Riemann constants is independent of the base point p_0 and clearly given by

$$\Xi = [\frac{1 - \tau_{1,1}}{2}]. \tag{9.81}$$

Now we turn to the Dirichlet eigenvalue $\mu_1(n)$. We recall

$$A_{p_0}(\hat{\mu}_1(n)) = A_{p_0}(\hat{\mu}_1(0)) - 2n A_{p_0}(\infty_+). \tag{9.82}$$

If we set $A_{p_0}(\hat{\mu}_1(0)) = [\delta + \frac{\tau_{1,1}}{2}]$, where $\delta \in [0, 1)$, we obtain

$$A_{p_0}(\hat{\mu}_1(n)) = [\frac{\tau_{1,1}}{2} + \sigma_1(n) \int_{E_1}^{\mu_1(n)} \frac{c(1)dx}{R_4^{1/2}(x)}] = [\delta + \frac{\tau_{1,1}}{2} - n\frac{F(\sqrt{\frac{E_3-E_1}{E_3-E_0}}, k)}{K(k)}], \tag{9.83}$$

with $\hat{\mu}_1(n) = (\mu_1(n), \sigma_1(n) R_4^{1/2}(\mu_1(n)))$. Evaluation of the integral yields ([**38**], 254.00)

$$\int_{E_1}^{\mu_1(n)} \frac{c(1)dx}{R_4^{1/2}(x)} = \frac{F(\sqrt{\frac{E_2-E_0}{E_2-E_1}\frac{\mu_1(n)-E_1}{\mu_1(n)-E_0}}, k)}{2K(k)} \tag{9.84}$$

and hence we infer

$$\mu_1(n) = E_1 \frac{1 - \frac{E_2-E_1}{E_2-E_0}\frac{E_0}{E_1}\operatorname{sn}^2(2K(k)\delta - 2nF(\sqrt{\frac{E_3-E_1}{E_3-E_0}}, k))}{1 - \frac{E_2-E_1}{E_2-E_0}\operatorname{sn}^2(2K(k)\delta - 2nF(\sqrt{\frac{E_3-E_1}{E_3-E_0}}, k))}, \tag{9.85}$$

where $\operatorname{sn}(z)$ is Jacobi's elliptic function (i.e., the inverse of $F(z, k)$). The sheet can be read off from the sign of $\operatorname{sn}(2K(k)\delta - 2nF(\dots))$.

Now we search a more explicit expression for the differential $\omega_{\infty_-,\infty_+}$. We first determine the constant λ_1 from ([**38**], 254.00, 254.10, 340.01)

$$\int_{a_1} \omega_{\infty_-,\infty_+} = 2 \int_{E_1}^{E_2} \frac{x - \lambda_1}{R_4^{1/2}(x)} dx$$
$$= 2C\Big((E_1 - E_0)\Pi(\frac{E_2-E_1}{E_2-E_0}, k) + (E_0 - \lambda_1)K(k)\Big) = 0, \tag{9.86}$$

which implies

$$\lambda_1 = E_0 + \frac{(E_1 - E_0)}{K(k)} \Pi(\frac{E_2-E_1}{E_2-E_0}, k). \tag{9.87}$$

The b-period is given by (cf. (9.35))

$$\int_{b_1} \omega_{\infty_-,\infty_+} = -2 \int_{E_0}^{E_1} \frac{x - \lambda_1}{R_4^{1/2}(x)} dx = -2\pi\mathrm{i}\Big(1 - \frac{F(\sqrt{\frac{E_3-E_1}{E_3-E_0}}, k)}{K(k)}\Big). \tag{9.88}$$

And for $p = (z, \pm R_4^{1/2}(z))$ near p_0 we have ([**38**], 251.00, 251.10, 340.01)

$$\int_{p_0}^{p} \omega_{\infty_-,\infty_+} = \pm \int_{E_0}^{z} \frac{x - \lambda_1}{R_4^{1/2}(x)} dx$$

$$= \pm C(E_1 - E_0)\Big(\big(1 - \frac{\Pi(\frac{E_2-E_1}{E_2-E_0}, k)}{K(k)}\big)F(u(z), k)$$

(9.89)
$$- \Pi\big(u(z), \frac{E_3 - E_0}{E_3 - E_1}, k\big)\Big).$$

To complete this section we recall that Riemann's theta function can be expressed as a Jacobi theta function, namely

(9.90) $$\theta(z) = \vartheta_3(z) = \sum_{n \in \mathbb{Z}} e^{\pi i (2nz + \tau_{1,1} n^2)}.$$

9.4. Some illustrations of the Riemann-Roch theorem

In this section we give some illustrations of the Riemann-Roch theorem (Theorem A.2) in connection with hyperelliptic Riemann surfaces not branched at infinity and obtain a basis for the vector space $\mathcal{L}(-n\mathcal{D}_{-\infty} - \mathcal{D}_{\hat{\mu}})$, $n \in \mathbb{N}_0$.

Theorem 9.9. *Assume $\mathcal{D}_{\hat{\mu}} \geq 0$ to be nonspecial (i.e., $i(\mathcal{D}_{\hat{\mu}}) = 0$) and of degree g. For $n \in \mathbb{N}_0$, a basis for the vector space $\mathcal{L}(-n\mathcal{D}_{-\infty} - \mathcal{D}_{\hat{\mu}})$ is given by,*

(9.91) $$\Big\{\prod_{\ell=0}^{j-1} {}^* \phi(.,\ell)\Big\}_{j=0}^{n},$$

or equivalently, by

(9.92) $$\mathcal{L}(-n\mathcal{D}_{-\infty} - \mathcal{D}_{\hat{\mu}}) = \mathrm{span}\,\{\psi(.,j)\}_{j=0}^{n}.$$

Proof. Since the above functions are linearly independent elements of the vector space $\mathcal{L}(-n\mathcal{D}_{-\infty} - \mathcal{D}_{\hat{\mu}})$ it remains to show that they are maximal. From $i(\mathcal{D}_{\hat{\mu}}) = 0 = i(n\mathcal{D}_{-\infty} + \mathcal{D}_{\hat{\mu}})$ and the Riemann-Roch theorem we obtain

(9.93) $$r(-n\mathcal{D}_{\infty} - \mathcal{D}_{\hat{\mu}}) = 1 + n,$$

which proves the theorem. □

Chapter 10

Scattering theory

In Section 7.5 we have considered perturbations of periodic Jacobi operators, that is, we have started developing **scattering theory** with periodic background. In this chapter we will investigate sequences $a(n) > 0, b(n)$ which look asymptotically like the free ones (compare Section 1.3).

10.1. Transformation operators

To begin with, we state the main hypothesis which will be assumed to hold throughout this chapter.

Hypothesis H. 10.1. Suppose $a, b \in \ell(\mathbb{Z}, \mathbb{R})$ satisfy $a > 0$ and

$$(10.1) \qquad n(1 - 2a(n)) \in \ell^1(\mathbb{Z}), \qquad n\, b(n) \in \ell^1(\mathbb{Z}).$$

In addition, it will turn out convenient to introduce

$$(10.2) \qquad A_-(n) = \prod_{m=-\infty}^{n-1} 2a(m), \quad A_+(n) = \prod_{m=n}^{\infty} 2a(m), \quad A = A_-(n)A_+(n).$$

From Section 1.3 we know that it is better to consider the free problem in terms of $k = z - \sqrt{z-1}\sqrt{z+1}$ rather than in terms of the spectral parameter z itself.

By virtue of Lemma 7.10 we infer (upon choosing $a_p(n) = 1/2$ and $b_p(n) = 0$) the existence of solutions $f_\pm(k, n)$ of $(\tau - (k + k^{-1})/2)f = 0$ which asymptotically look like the free ones, that is,

$$(10.3) \qquad f_\pm(k, n) = k^{\pm n}(1 + o(1)) \quad \text{as } n \to \pm\infty,$$

respectively, $\lim_{n \to \pm\infty} f_\pm(k, n) k^{\mp n} = 1$.

They are called **Jost solutions** and their properties are summarized in the following theorem.

Theorem 10.2. *Let $a(n), b(n)$ satisfy (H.10.1), then there exist Jost solutions $f_\pm(k, n)$ of*

$$(10.4) \qquad \tau f_\pm(k, n) = \frac{k + k^{-1}}{2} f_\pm(k, n), \quad 0 < |k| \leq 1,$$

167

fulfilling

(10.5) $$\lim_{n \to \pm\infty} f_\pm(k,n) k^{\mp n} = 1.$$

Moreover $\tilde{f}(k,n) = A_\pm(n) k^{\mp n} f_\pm(k,n)$ *is holomorphic in the domain* $|k| < 1$, *continuous for* $|k| \leq 1$, *and satisfies* $\tilde{f}(0,n) = 1$.

Proof. For the case $k \neq 0$ everything follows from Lemma 7.10. To show that the singularity of $\tilde{f}(k,n)$ at $k = 0$ is removable, one needs to investigate the limit $k \to 0$ in (7.88). To do this, use $a_p(n) = 1/2$, $b_p(n) = 0$, $m_p^\pm(z) = k^{\pm 1}$, $s_p(z,n,m) = \frac{k}{k^2-1}(k^{n-m} - k^{m-n})$ (see Section 1.3) and compute the limit $\lim_{k \to 0} k^{n-m} K(z,n,m) = 2a(m) - 1$. □

The previous theorem shows that we may set

(10.6) $$f_+(k,n) = \frac{k^n}{A_+(n)}\Big(1 + \sum_{j=1}^\infty K_{+,j}(n) k^j\Big), \quad |k| \leq 1,$$

where $K_{+,j}(n)$ are the Fourier coefficients of $\tilde{f}_+(k,n)$ given by

(10.7) $$K_{+,j}(n) = \frac{1}{2\pi\mathrm{i}} \int_{|k|=1} \tilde{f}_+(k,n) k^{-j} \frac{dk}{k}, \quad j \in \mathbb{N}.$$

Taking the limit $n \to \infty$ in this formula (using the estimate (7.80) and dominated convergence) shows $\lim_{n \to \infty} K_{+,j}(n) = 0$, $j \in \mathbb{N}$.

Inserting (10.6) into (10.4) (set $K_0^+(n) = 1$, $K_j^+(n) = 0$, $j < 0$) yields

(10.8) $$\begin{aligned} K_{+,j+1}(n) - K_{+,j+1}(n-1) &= 2b(n) K_{+,j}(n) - K_{+,j-1}(n) \\ &\quad + 4a(n)^2 K_{+,j-1}(n+1). \end{aligned}$$

Summing up (with respect to n) using $\lim_{n \to \infty} K_{+,j}(n) = 0$, $j \in \mathbb{N}$, we obtain

$$K_{+,1}(n) = -\sum_{m=n+1}^\infty 2b(m),$$

$$K_{+,2}(n) = -\sum_{m=n+1}^\infty \Big(2b(m) K_{+,1}(m) + (4a(m)^2 - 1)\Big),$$

(10.9) $$\begin{aligned} K_{+,j+1}(n) = K_{+,j-1}(n+1) &- \sum_{m=n+1}^\infty \Big(2b(m) K_{+,j}(m) \\ &+ (4a(m)^2 - 1) K_{+,j-1}(m+1)\Big). \end{aligned}$$

A straightforward induction argument shows

(10.10) $$|K_{+,j}(n)| \leq D_{+,j}(n) C_+(n + [\![\tfrac{j}{2}]\!] + 1), \quad j \in \mathbb{N},$$

where

(10.11) $$C_+(n) = \sum_{m=n}^\infty c(m), \quad D_{+,m}(n) = \prod_{j=1}^{m-1} \Big(1 + C_+(n+j)\Big),$$

10.1. Transformation operators

and $c(n) = 2|b(n)| + |4a(n)^2 - 1|$. It is important to note that $C_+ \in \ell^1_+(\mathbb{Z})$. In fact, summation by parts shows

(10.12) $$\sum_{m=n}^{\infty} m\, c(m) = (n-1)C_+(n) + \sum_{m=n}^{\infty} C_+(m)$$

since

(10.13) $$\lim_{n \to \infty} n\, C_+(n+1) \leq \lim_{n \to \infty} \sum_{m=n+1}^{\infty} m\, c(m) = 0.$$

Moreover, $1 \leq D_{+,m}(n) \leq D_+(n) = \lim_{j \to \infty} D_{+,j}(n)$ and $\lim_{n \to \infty} D_+(n) = 1$.

This gives also an estimate for

$$|K_{+,j+1}(n) - K_{+,j+1}(n-1)| \leq 2|b(n)||K_{+,j}(n)| + |4a(n)^2 - 1||K_{+,j-1}(n)|$$
$$+ |K_{+,j-1}(n+1) - K_{+,j-1}(n)|$$

(10.14) $$\leq c(n)D_+(n)C_+(n + [\![\frac{j}{2}]\!] + 1) + |K_{+,j-1}(n+1) - K_{+,j-1}(n)|.$$

Using induction we obtain

(10.15)
$$|K_{+,j}(n) - K_{+,j}(n-1)| \leq D_+(n)C_+(n)C_+(n + [\![\frac{j-1}{2}]\!] + 1) + c(n + [\![\frac{j}{2}]\!]).$$

In addition, the estimate (10.10) for $K_{+,j}(n)$ gives us an estimate on the rate of convergence in (10.3)

(10.16) $$|f_+(k,n)A(n)k^{-n} - 1| \leq \begin{cases} \dfrac{|k|}{1-|k|} D_+(n)C_+(n+1), & |k| < 1 \\ 2|k|D_+(n) \sum_{m=n}^{\infty} C_+(m), & |k| \leq 1 \end{cases}.$$

Moreover, since $K_{+,j}(n) \leq C_\infty(1 + C_\infty)^{j-1}$, $C_\infty = \lim_{n \to -\infty} C_+(n)$, we can take the limits ($K_0 = 1$)

$$K_1 = \lim_{n \to -\infty} K_{+,1}(n) = -2 \sum_{m \in \mathbb{Z}} b(m),$$

$$K_2 = \lim_{n \to -\infty} K_{+,2}(n) = -\sum_{m \in \mathbb{Z}} \Big(2b(m)K_{+,1}(m) + (4a(m)^2 - 1)\Big),$$

$$K_{j+1} = \lim_{n \to -\infty} K_{+,j+1}(n) = K_{j-1} - \sum_{m \in \mathbb{Z}} \Big(2b(m)K_{+,j}(m)$$

(10.17) $$+ (4a(m)^2 - 1)K_{+,j-1}(m+1)\Big).$$

(It will become clear later why we have dropped the $+$ subscript.)

Associated with $K_+(n, n+j) = K_{+,j}(n)$ is the operator

(10.18) $$(\mathcal{K}_+ f)(n) = \frac{1}{A_+(n)}\Big(f(n) + \sum_{m=n+1}^{\infty} K_+(n,m)f(m)\Big), \quad f \in \ell^\infty_+(\mathbb{Z}),$$

which acts as a **transformation operator** for the pair τ_0, τ.

Theorem 10.3. *Let τ_0 be the free Jacobi difference expression. Then we have*

(10.19) $$\tau \mathcal{K}_+ f = \mathcal{K}_+ \tau_0 f, \quad f \in \ell^\infty_+(\mathbb{Z}).$$

Proof. A straightforward calculation shows that $\tau \mathcal{K}_+ = \mathcal{K}_+ \tau_0$ is equivalent to (10.8). □

By our estimate (10.10), $\mathcal{K}_+$ maps $\ell_+^p(\mathbb{Z})$, $p \in [1, \infty) \cup \{\infty\}$, into itself. Moreover, by Lemma 7.8, $\mathcal{K}_+$ is an automorphism of $\ell_+^\infty(\mathbb{Z})$. Considering $f \in \ell_0(\mathbb{Z})$ we see that its inverse $\mathcal{L}_+$ is given by

$$(10.20) \quad (\mathcal{L}_+ f)(n) = A_+(n)\Big(f(n) + \sum_{m=n+1}^\infty \frac{L_+(n,m)}{\prod_{j=n}^{m-1} 2a(j)} f(m)\Big), \quad f \in \ell_+^\infty(\mathbb{Z}),$$

with

$$L_+(n,m) = -K_+(n,m) - \sum_{\ell=n+1}^{m-1} K_+(n,\ell) L_+(\ell,m)$$

$$= -K_+(n,m) - \sum_{\ell=n+1}^{m-1} L_+(n,\ell) K_+(\ell,m), \quad m > n.$$

From $\mathcal{L}_+ \tau = \tau_0 \mathcal{L}_+$ we infer that the coefficients $L_{+,j}(n) = L_+(n, n+j)$ satisfy

$$(10.21) \quad \begin{aligned} L_{+,j+1}(n) - L_{+,j+1}(n-1) &= -2b(n+j) L_{+,j}(n) - 4a(n+j-1)^2 L_{+,j-1}(n) \\ &\quad + L_{+,j-1}(n+1) \end{aligned}$$

as well as an estimate of type (10.10). In particular, $\mathcal{K}_+$ is an automorphism of $\ell_+^p(\mathbb{Z})$, $p \in [1, \infty) \cup \{\infty\}$.

Similarly, we set

$$(10.22) \quad f_-(k,n) = \frac{k^{-n}}{A_-(n)}\Big(1 + \sum_{j=1}^\infty K_{-,j}(n) k^j\Big), \quad |k| \leq 1,$$

which yields after inserting it into (1.19) (set $K_{-,0}(n) = 1$, $K_{-,j}(n) = 0$, $j < 0$)

$$(10.23) \quad \begin{aligned} K_{-,j+1}(n+1) - K_{-,j+1}(n) &= -2b(n) K_{-,j}(n) + K_{-,j-1}(n) \\ &\quad - 4a(n-1)^2 K_{-,j-1}(n-1). \end{aligned}$$

As before we obtain ($\lim_{n \to -\infty} K_{-,j}(n) = 0$, $j \in \mathbb{N}$)

$$K_{-,1}(n) = -\sum_{m=-\infty}^{n-1} 2b(m),$$

$$K_{-,2}(n) = -\sum_{m=-\infty}^{n-1} 2b(m) K_{-,1}(m) - (4a(m-1)^2 - 1),$$

$$K_{-,j+1}(n) = K_{-,j-1}(n) - \sum_{m=-\infty}^{n-1} \Big(2b(m) K_{-,j}(m)$$

$$(10.24) \qquad\qquad - (4a(m-1)^2 - 1) K_{-,j-1}(m-1)\Big).$$

Furthermore, $K_{-,j}(n) = K_-(n, n-j)$ satisfies an estimate of type (10.10) and we can define an associated operator

$$(10.25) \quad (\mathcal{K}_- f)(n) = \frac{1}{A_-(n)}\Big(f(n) + \sum_{m=-\infty}^{n-1} K_-(n,m) f(m)\Big), \quad f \in \ell_-^\infty(\mathbb{Z}),$$

10.2. The scattering matrix

with inverse $\mathcal{L}_-$.

10.2. The scattering matrix

Our next objective is to introduce the scattering matrix. For $|k| = 1, k^2 \neq 1$, the Jost solutions $f_\pm(k, n)$, $f_\pm(k^{-1}, n)$ are linearly independent since we have

(10.26) $$W(f_\pm(k), f_\pm(k^{-1})) = \pm \frac{1 - k^2}{2k}.$$

In particular, we may set

(10.27) $$f_\pm(k, n) = \alpha(k) f_\mp(k^{-1}, n) + \beta_\mp(k) f_\mp(k, n), \quad |k| = 1, k^2 \neq 1,$$

where

(10.28) $$\alpha(k) = \frac{W(f_\mp(k), f_\pm(k))}{W(f_\mp(k), f_\mp(k^{-1}))} = \frac{2k}{1 - k^2} W(f_+(k), f_-(k)),$$

$$\beta_\pm(k) = \frac{W(f_\mp(k), f_\pm(k^{-1}))}{W(f_\pm(k), f_\pm(k^{-1}))} = \pm \frac{2k}{1 - k^2} W(f_\mp(k), f_\pm(k^{-1})).$$

Note that we have

(10.29) $$\overline{\alpha(k)} = \alpha(\overline{k}), \quad \overline{\beta_\pm(k)} = \beta_\pm(\overline{k}), \quad \beta_+(k^{-1}) = -\beta_-(k),$$

and $\overline{k} = k^{-1}$ for $|k| = 1$. Moreover, the Plücker identity (2.169) with $f_1 = f_\mp$, $f_2 = f_\pm$, $f_3 = \overline{f_\mp}$, $f_4 = \overline{f_\pm}$, implies

(10.30) $$|\alpha(k)|^2 = 1 + |\beta_\pm(k)|^2, \quad |\beta_+(k)| = |\beta_-(k)|.$$

The function $\alpha(k)$ can be defined for all $|k| \leq 1$. Moreover, the power series of $\alpha(k)$ follows formally from

(10.31) $$\alpha(k) = \frac{2k}{1 - k^2} \lim_{n \to \pm\infty} a(n)\big(f_+(k, n)f_-(k, n+1) - f_+(k, n+1)f_-(k, n)\big)$$
$$= \frac{1}{A} \lim_{n \to \pm\infty} \tilde{f}_\mp(k, n) = \frac{1}{A} \sum_{j=0}^\infty K_j k^j,$$

where $K_j = \lim_{n \to \mp\infty} K_{\pm,j}(n)$ from (10.17). To make this precise we need to show that we can interchange limit and summation in the last line. This follows from $|K_{\pm,j}(n)| \leq C_\infty(1 + C_\infty)^{j-1}$ at least for $|k| < (1 + C_\infty)^{-1}$. Hence it follows for all $|k| < 1$ since $\alpha(k)$ is holomorphic for $|k| < 1$.

The function $\alpha(k)$ has even more interesting properties. To reveal its connection with Krein's spectral shift theory, we compute its derivative. By virtue of (2.29) we obtain

(10.32) $$\frac{d}{dk} W(f_+(k), f_-(k)) = W_m(f_+(k), \frac{d}{dk} f_-(k)) + W_n(\frac{d}{dk} f_+(k), f_-(k))$$
$$+ \frac{1 - k^2}{2k^2} \sum_{j=m+1}^n f_+(k, j) f_-(k, j).$$

From (10.6), (10.22) we get $\frac{d}{dk} f_\pm(k,n) = \pm \frac{n}{k} f_\pm(n,k) + k^{\pm n} o(1)$, $n \to \pm\infty$, and $f_\pm(k,n) = k^{\mp n}(\alpha(k) + o(1))$, $n \to \mp\infty$, respectively. Hence we see

$$W_m(f_+(k), \frac{d}{dk} f_-(k)) = -m \frac{1-k^2}{2k^2} \alpha(k) - \frac{1}{2} \alpha(k) + o(1), \quad m \to +\infty,$$

$$(10.33) \quad W_n(\frac{d}{dk} f_+(k), f_-(k)) = n \frac{1-k^2}{2k^2} \alpha(k) - \frac{1}{2k^2} \alpha(k) + o(1), \quad n \to -\infty$$

and using this to evaluate the limits $m \to -\infty$, $n \to \infty$ shows

$$(10.34) \quad \frac{d}{dk} \alpha(k) = \frac{1}{k} \sum_{n \in \mathbb{Z}} \big(f_+(k,n) f_-(k,n) - \alpha(k) \big).$$

Next, observe that, due to (H.10.1), $H - H_0$ is trace class. Indeed, the multiplication operators by $b(n)$, $2a(n) - 1$ are trace class and trace class operators form an ideal in the Banach space of bounded linear operators. So we can rewrite this equation in the form

$$-\frac{d}{dz} \ln \alpha(k(z)) = \sum_{j \in \mathbb{Z}} \big(G(z, n, n) - G_0(z, n, n) \big)$$

$$(10.35) \quad \qquad = \operatorname{tr}\big((H - z)^{-1} - (H_0 - z)^{-1} \big),$$

that is, $A\alpha(k)$ (since $A\alpha(0) = 1$) is the **perturbation determinant** of the pair H, H_0. This implies ([155], Theorem 1)

$$(10.36) \quad \alpha(k(z)) = \frac{1}{A} \exp \Big(\int_\mathbb{R} \frac{\xi_\alpha(\lambda) d\lambda}{\lambda - z} \Big),$$

where

$$(10.37) \quad \xi_\alpha(\lambda) = \frac{1}{\pi} \lim_{\varepsilon \downarrow 0} \arg \alpha(k(\lambda + i\varepsilon)) \in [-1, 1]$$

is of compact support. However, the reader should be warned that $\alpha(k(z))$ is *not* Herglotz since we do *not* have $0 \le \xi_\alpha(\lambda) \le 1$ in general.

In addition, Krein's result implies

$$(10.38) \quad \operatorname{tr}\big(H^\ell - (H_0)^\ell \big) = \ell \int_\mathbb{R} \lambda^{\ell-1} \xi_\alpha(\lambda) d\lambda.$$

The spectrum of H can be easily characterized upon specializing to the case $a_p(n) = 1/2$, $b_p(n) = 0$ in Theorem 7.11.

Theorem 10.4. *Suppose (H.10.1). Then we have*

$$\sigma_{ess}(H) = \sigma_{ac}(H) = [-1, 1],$$

$$(10.39) \quad \sigma_p(H) = \{\lambda = \frac{k + k^{-1}}{2} \mid |\lambda| > 1, \alpha(k) = 0\}.$$

Moreover, the essential spectrum of H is purely absolutely continuous and the point spectrum of H is finite.

Hence we can denote the number of eigenvalues by $N \in \mathbb{N}$ and set

$$(10.40) \quad \sigma_p(H) = \{\lambda_j = \frac{k_j + k_j^{-1}}{2}\}_{1 \le j \le N}.$$

Note that because of (10.30) all zeros of of $\alpha(k)$ lie inside the unit circle. Moreover, there can be no non-real zeros (otherwise H would have a non-real eigenvalue).

10.2. The scattering matrix

(Note also $a(0) = A^{-1} > 0$ by (10.31).) Using Theorem 3.20 we can even provide a bound on the number of eigenvalues.

Theorem 10.5. *Suppose (H.10.1). Then*

$$\dim \operatorname{Ran} P_{(\pm 1, \pm \infty)}(H) \leq 1 + \left(\sum_{n \in \mathbb{Z}} c_{\pm}(n) \right)^2, \tag{10.41}$$

where

$$c_{+}(n) = \sqrt{n(n+1)}|1 - 2a(n)| + |n| \max(\mp b(n), 0). \tag{10.42}$$

Proof. We only prove the first bound, the second being similar. Upon replacing $b(n)$ by $\min(b(n), 0)$ we can assume $b(n) \leq 0$ (cf. [195], Theorem XIII.10). Our goal is to apply Theorem 3.20. Since the limit of $G_0(\lambda, n, m)$ as $\lambda \to -1$ does not exist, we will consider $H_{\pm}$ and use

$$\begin{aligned}\dim \operatorname{Ran} P_{(-\infty,-1)}(H) &\leq 1 + \dim \operatorname{Ran} P_{(-\infty,-1)}(H_{+}) \\ &\quad + \dim \operatorname{Ran} P_{(-\infty,-1)}(H_{-}).\end{aligned} \tag{10.43}$$

Where the last claim follows from the fact that between two eigenvalues of H, there can be at most one Dirichlet eigenvalue (see Lemma 3.7). Hence we need to consider $H_{0,+}$. From ($0 < m \leq n$, $\lambda < -1$)

$$G_{0,+}(\lambda, n, m) = -2k \frac{1 - k^{2m}}{1 - k^2} k^{n-m} \tag{10.44}$$

($k = \lambda - \sqrt{\lambda^2 - 1} \in (-1, 0)$) we infer

$$|G_{0,+}(\lambda, n, m)| \leq 2 \min(m, n) \leq 2\sqrt{mn}, \quad \lambda \leq -1. \tag{10.45}$$

Evaluating Theorem 3.20 as in Corollary 3.21 gives

$$\dim \operatorname{Ran} P_{(-\infty,-1)}(H_{+}) \leq \left(\sum_{n=1}^{\infty} c_{+}(n) \right)^2. \tag{10.46}$$

The bound for H_{-} follows from reflection. $\square$

We now define the **scattering matrix**

$$S(k) = \begin{pmatrix} T(k) & R_{-}(k) \\ R_{+}(k) & T(k) \end{pmatrix}, \quad |k| = 1, \tag{10.47}$$

where $T(k) = \alpha(k)^{-1}$ and $R_{\pm}(k) = \alpha(k)^{-1} \beta_{\pm}(k)$. Equations (10.29) and (10.30) imply

Lemma 10.6. *The scattering matrix $S(k)$ is unitary and its determinant reads*

$$\det S(k) = \frac{T(k)}{T(\overline{k})}. \tag{10.48}$$

The coefficients $T(k), R_{\pm}(k)$ are continuous except at possibly $k = \pm 1$ and fulfill

$$|T(k)|^2 + |R_{\pm}(k)|^2 = 1, \quad T(k)\overline{R_{+}(\overline{k})} + \overline{T(\overline{k})}R_{-}(k) = 0, \tag{10.49}$$

and $\overline{T(k)} = T(\overline{k})$, $\overline{R_{\pm}(k)} = R_{\pm}(\overline{k})$.

The quantity $T(k)$ is called transmission coefficient and $R_\pm(k)$ are called reflection coefficients with respect to left respectively right incidence. We remark

(10.50) $$|T(k)|^2 = 1 - |R_\pm(k)|^2 \geq C^2|1 - k^2|^2,$$

where we can choose $C = \max_{|k|=1} |W(f_+(k), f_-(k))|/2$. In particular, $\ln(1 - |R_\pm(k)|^2)$ is integrable over the unit circle.

We proceed with a closer investigation of $T(k)$. Using the Laurent expansion for $\alpha(k)$ one computes

(10.51) $$T(k) = A\Big(1 + k \sum_{m \in \mathbb{Z}} 2b(m) + O(k^2)\Big).$$

The only poles of $T(k)$ are at k_j, where $\lambda_j = (k_j + k_j^{-1})/2 \in \sigma_p(H)$. The residue can be computed from (10.34) which yields (cf. (2.33))

(10.52) $$\operatorname{res}_{k_j} T(k) = -\frac{k_j \gamma_{+,j}}{\mu_j} = -k_j \mu_j \gamma_{-,j},$$

where

(10.53) $$\frac{1}{\gamma_{\pm,j}} = \sum_{m \in \mathbb{Z}} |f_\pm(m, k_j)|^2, \quad f_-(n, k_j) = \mu_j f_+(n, k_j).$$

The constants $\gamma_{\pm,j}$ are usually referred to as **norming constants**. The sets

(10.54) $$S_\pm(H) = \{R_\pm(k), |k| = 1;\ (k_j, \gamma_{\pm,j}),\ 1 \leq j \leq N\}$$

are called left/right **scattering data** for H, respectively. We will show that they determine the sequences a, b uniquely in the following sections. As a first step toward this result we note

Lemma 10.7. *Either one of the sets $S_\pm(H)$ determines the other and $T(k)$ via (Poisson-Jensen formula)*

(10.55) $$T(k) = \Big(\prod_{j=1}^{N} \frac{|k_j|(k - k_j^{-1})}{k - k_j}\Big) \exp\Big(\frac{1}{4\pi\mathrm{i}} \int_{|\kappa|=1} \ln(1 - |R_\pm(\kappa)|^2) \frac{\kappa + k}{\kappa - k} \frac{d\kappa}{\kappa}\Big),$$

(for $|k| < 1$) and

(10.56) $$\frac{R_-(k)}{R_+(\overline{k})} = -\frac{T(k)}{T(\overline{k})}, \quad \gamma_{+,j} \gamma_{-,j} = \Big(\frac{\operatorname{res}_{k_j} T(k)}{k_j}\Big)^2.$$

Proof. It suffices to prove the formula for $T(k)$ since evaluating the residues provides μ_j and thus $\gamma_{\mp,j} = \mu_j^{\mp 2} \gamma_{\pm,j}$. For this purpose consider

(10.57) $$t(k) = \prod_{j=1}^{N} \frac{k - k_j}{kk_j - 1} T(k)$$

which is holomorphic for $|k| < 1$ and has neither poles nor zeros inside the unit disk. In addition, $t(k)$ is continuous for $|k| \leq 1$ and has at most simple zeros at $k = \pm 1$. Applying the residue theorem in a circle with radius $r < 1$ we obtain

$$\frac{1}{2\pi\mathrm{i}} \int_{|\kappa|=r} \ln(t(\kappa)) \frac{\kappa + k}{\kappa - k} \frac{d\kappa}{\kappa} = 2\ln(t(k)) - \ln(t(0)),$$

(10.58) $$\frac{1}{2\pi\mathrm{i}} \int_{|\kappa|=r^{-1}} \ln(t(\kappa^{-1})) \frac{\kappa + k}{\kappa - k} \frac{d\kappa}{\kappa} = \ln(t(0)).$$

Letting $r \to 1$ (dominated convergence), using $\overline{t(k)} = t(k^{-1})$, $|k| = 1$, and adding up the resulting equations yields

(10.59) $$\frac{1}{2\pi\mathrm{i}} \int_{|\kappa|=1} \ln(|t(\kappa)|^2) \frac{\kappa + k}{\kappa - k} \frac{d\kappa}{\kappa} = 2\ln(t(k)) \mod 2\pi\mathrm{i},$$

where $\ln(|t(\kappa)|^2) \in \mathbb{R}$. Thus we conclude

(10.60) $$t(k)^2 = \exp\left(\frac{1}{2\pi\mathrm{i}} \int_{|\kappa|=1} \ln(|t(\kappa)|^2) \frac{\kappa + k}{\kappa - k} \frac{d\kappa}{\kappa}\right).$$

This proves the claim upon observing $|t(k)| = |T(k)|$, $|k| = 1$, and $T(0) > 0$. □

Comparing coefficients in the asymptotic expansions of (10.55) and (10.36) gives another trace formula

(10.61) $$\tilde{K}_m = \frac{-1}{\pi} \int_0^\pi \ln(1 - |R_\pm(\mathrm{e}^{\mathrm{i}\varphi}, t)|^2) \cos(m \mathrm{e}^{\mathrm{i}\varphi}) d\varphi + \sum_{j=1}^N \frac{k_j^m - k_j^{-m}}{m},$$

where $\tilde{K}_m = K_m - \sum_{j=1}^{m-1} \frac{j}{m} \tilde{K}_{m-j} K_j$ are the expansion coefficients of

(10.62) $$\ln \alpha(k) = -\ln A + \sum_{j=1}^\infty \tilde{K}_j k^j.$$

Moreover, expanding $\ln \alpha(k(z))$ one can express the traces $\mathrm{tr}(H^\ell - (H_0)^\ell)$ in terms of the coefficients K_m, for instance,

$$\mathrm{tr}\Big(H - (H_0)\Big) = -\frac{1}{2} K_1,$$
$$\mathrm{tr}\Big(H^2 - (H_0)^2\Big) = -\frac{1}{16}(2K_2 + K_1^2),$$

(10.63) etc. .

10.3. The Gel'fand-Levitan-Marchenko equations

Now we want to set up a procedure which allows the reconstruction of H from its scattering data $S_\pm(H)$. This will be done by deriving an equation for $K_\pm(n, m)$. In addition, the decay condition (H.10.1) is reflected by the estimates (10.10) and (10.15). These turn out to be equivalent to the corresponding estimates for $K_\pm(n, m)$ and will lead back to (H.10.1) when reconstructing H from $S_\pm(H)$.

Since $K_+(n, m)$ are essentially the Fourier coefficients of $f_+(k, n)$ we compute the Fourier coefficients of (10.27). Therefore we multiply

(10.64) $$T(k) f_-(k, n) = \Big(f_+(k^{-1}, n) + R_+(k) f_+(k, n)\Big)$$

by $(2\pi\mathrm{i})^{-1} A_+(n) k^{n+j-1}$ ($j \in \mathbb{N}_0$) and integrate around the unit circle. We first evaluate the right hand side

$$\frac{A_+(n)}{2\pi\mathrm{i}} \int_{|k|=1} f_+(k^{-1}, n) k^{n+j} \frac{dk}{k} = K_+(n, n + j),$$

$$\frac{A_+(n)}{2\pi\mathrm{i}} \int_{|k|=1} R_+(k) f_+(k, n) k^{n+j} \frac{dk}{k} = \sum_{m=0}^\infty K_+(n, n + m) \tilde{F}^+(2n + m + j),$$

(10.65)

where

(10.66) $$\tilde{F}^+(n) = \frac{1}{2\pi i} \int_{|k|=1} R_+(k) k^n \frac{dk}{k} \in \ell^2(\mathbb{Z}, \mathbb{R})$$

are the Fourier coefficients of $R_+(k^{-1})$. By Parseval's identity we have

(10.67) $$\sum_{n \in \mathbb{Z}} \tilde{F}^+(n)^2 = \frac{1}{2\pi i} \int_{|k|=1} |R_+(k)|^2 \frac{dk}{k} < 1.$$

Next we evaluate the left hand side using the residue theorem (take a contour inside the unit disk enclosing all poles and let this contour approach the unit circle)

$$\frac{A_+(n)}{2\pi i} \int_{|k|=1} T(k) f_-(k,n) k^{n+j} \frac{dk}{k}$$

$$= -A_+(n) \sum_{\ell=1}^{N} \frac{k_\ell \gamma_{+,\ell}}{\mu_\ell} f_-(k_\ell, n) k_\ell^{n+j-1} + A_+(n) \frac{T(0)}{A_-(n)} \delta_0(j)$$

$$= -A_+(n) \sum_{\ell=1}^{N} \gamma_{+,\ell} f_+(k_\ell, n) k_\ell^{n+j} + A_+(n)^2 \delta_0(j)$$

(10.68) $$= -\sum_{m=0}^{\infty} K_+(n, n+m) \sum_{\ell=1}^{N} \gamma_{+,\ell} k_\ell^{2n+m+j} + A_+(n)^2 \delta_0(j).$$

Defining the **Gel'fand-Levitan-Marchenko operator**

(10.69) $$\mathcal{F}_n^+ f(j) = \sum_{m=0}^{\infty} F^+(2n + m + j) f(m), \quad f \in \ell^2(\mathbb{N}_0),$$

where

(10.70) $$F^+(n) = \tilde{F}^+(n) + \sum_{\ell=1}^{N} \gamma_{+,\ell} k_\ell^n,$$

leads us to the **Gel'fand-Levitan-Marchenko equation**

(10.71) $$(\mathbb{1} + \mathcal{F}_n^+) K_+(n, n + \cdot) = A_+(n)^2 \delta_0.$$

Similarly, multiplying

(10.72) $$T(k) f_+(k, n) = \left(f_-(k^{-1}, n) + R_-(k) f_-(k, n) \right)$$

by $(2\pi i)^{-1} A_-(n) k^{-n+j-1}$ ($j \in \mathbb{N}_0$) and integrating around the unit circle we get for $K_-(n, m)$

(10.73) $$(\mathbb{1} + \mathcal{F}_n^-) K_-(n, n + \cdot) = A_-(n)^2 \delta_0,$$

where

(10.74) $$\mathcal{F}_n^- f(j) = \sum_{m=0}^{\infty} F^-(2n - m - j) f(m), \quad f \in \ell^2(\mathbb{N}_0),$$

and

(10.75) $$F^-(n) = \frac{1}{2\pi i} \int_{|k|=1} R_-(k) k^{-n} \frac{dk}{k} + \sum_{\ell=1}^{N} \gamma_{-,\ell} k_\ell^{-n}.$$

10.3. The Gel'fand-Levitan-Marchenko equations

Our next objective is to consider the Gel'fand-Levitan-Marchenko operator $\mathcal{F}_n^+$ a little closer. The structure of the Gel'fand-Levitan-Marchenko equation

$$K_+(n, n+j) + F^+(2n+j) + \sum_{\ell=1}^{\infty} F^+(2n+\ell+j) K_+(n, n+\ell)$$
(10.76)
$$= A_+(n) \delta_0(j)$$

suggests that the estimates (10.10) and (10.15) for $K_+(n, n+j)$ should imply similar estimates for $F^+(2n+j)$. This will be shown first.

As a warmup we prove a discrete Gronwall-type inequality.

Lemma 10.8. *Suppose $C_1, C_2, f \in \ell_+^\infty(\mathbb{Z}, [0, \infty))$, $C_3 \in \ell_+^1(\mathbb{Z}, [0, \infty))$, $C_1 C_3 < 1$, and*

(10.77)
$$f(n) \le C_1(n) \Big(C_2(n) + \sum_{j=n}^{\infty} C_3(j) f(j) \Big).$$

Then we have

(10.78)
$$f(n) \le C_1(n) \Big(C_2(n) + \sum_{j=n}^{\infty} \frac{C_1(j) C_2(j) C_3(j)}{\prod_{\ell=n}^{j} (1 - C_1(\ell) C_3(\ell))} \Big).$$

Proof. Using $g(n) = \sum_{j=n}^{\infty} C_3(j) f(j)$ one computes

(10.79)
$$g(n) - g(n+1) \le C_3(n) C_1(n) (C_2(n) + g(n))$$

and hence (solving for $g(n)$)

(10.80)
$$g(n) \le \sum_{j=n}^{\infty} \frac{C_1(j) C_2(j) C_3(j)}{\prod_{\ell=n}^{j} (1 - C_1(\ell) C_3(\ell))}.$$

The rest follows from $f(n) \le C_1(n)(C_2(n) + g(n))$. □

Solving the Gel'fand-Levitan-Marchenko equation (10.76) for $F_j(n) = |F^+(2n+j)|$, $j = 1, 2$, and using the estimate (10.10) we obtain

$$F(n) \le D_+(n) \Big(2 C_+(n+1) + \sum_{\ell=1}^{\infty} \big(|F^+(2n+\ell+1)| + |F^+(2n+\ell+2)| \big) $$
$$\times C_+(n + [\tfrac{\ell}{2}] + 1) \Big)$$
$$\le D_+(n) \Big(2 C_+(n+1) + \sum_{\ell=0}^{\infty} |F^+(2n+\ell+2)| C_+(n+k+1)$$
$$+ 2 \sum_{\ell=1}^{\infty} |F^+(2n+\ell+1)| C_+(n+k) + \sum_{\ell=1}^{\infty} |F^+(2n+\ell+2)| C_+(n+k) \Big)$$

(10.81)
$$\le 2 D_+(n) \Big(C_+(n+1) + \sum_{\ell=n}^{\infty} C_+(\ell) F(\ell) \Big),$$

where $F(n) = F_1(n) + F_2(n)$. And hence by the lemma

$$|F^+(2n+1)| + |F^+(2n+2)| \leq 2D_+(n)C_+(n+1)$$
(10.82)
$$\times \Big(1 + \sum_{j=n}^{\infty} \frac{2D_+(j)C_+(j)}{\prod_{\ell=n}^{j}(1 - 2D_+(\ell)C_+(\ell))}\Big).$$

for n large enough. In summary we have

(10.83) $$|F^+(2n+j)| \leq O(1)C_+(n + [\![\tfrac{j}{2}]\!] + 1),$$

where $O(1)$ are terms of order zero as $n \to \infty$. Next we want to derive an estimate for $\partial F^+(2 \cdot + j)$, $j = 1, 2$. Set $G_j(n) = |F^+(2n+j) - F^+(2n+j+2)|$, $j = 1, 2$, then

$$G_1(n) \leq |K_{+,1}(n) - K_{1,+}(n+1)|$$
$$+ \sum_{j=1}^{\infty} |F^+(2n+j+1)K_{+,j}(n) - F^+(2n+j+3)K_{+,j}(n+1)|$$
$$\leq c(n+1) + \sum_{j=1}^{\infty} |F^+(2n+j+1)||K_{+,j}(n) - K_{+,j}(n+1)|$$
$$+ \sum_{j=1}^{\infty} |F^+(2n+j+1) - F^+(2n+j+3)||K_{+,j}(n+1)|$$
$$\leq c(n+1)$$
$$+ O(1)\sum_{j=1}^{\infty} C_+(n + [\![\tfrac{j}{2}]\!] + 1)\big(C_+(n+1)C_+(n + [\![\tfrac{j+1}{2}]\!]) + c(n + [\![\tfrac{j}{2}]\!])\big)$$
$$+ O(1)\sum_{j=1}^{\infty} |F^+(2n+j+1) - F^+(2n+j+3)|C_+(n + [\![\tfrac{j}{2}]\!] + 2)$$

(10.84) $$\leq O(1)\Big(\tilde{c}(n) + \sum_{j=n}^{\infty}(G_1(j) + G_2(j))C_+(j+1)\Big),$$

where $\tilde{c}(n) = c(n+1) + C_+(n)C_+(n+1)$. Similarly, we obtain

(10.85) $$G_2(n) \leq O(1)\Big(\tilde{c}(n) + \sum_{j=n}^{\infty}(G_1(j) + G_2(j))C_+(j+1)\Big).$$

Adding both equations gives an estimate for $G(n) = G_1(n) + G_2(n)$ and again by the lemma

(10.86) $$G(n) \leq O(1)\Big(\tilde{c}(n) + \sum_{j=n}^{\infty} \tilde{c}(j)C_+(j+1)\Big).$$

As a consequence we infer

(10.87) $$nG(n) \leq O(1)\Big(n\tilde{c}(n) + C_+(n)\sum_{j=n}^{\infty} j\tilde{c}(j)\Big) \in \ell^1(\mathbb{N}).$$

Similar results hold for F^-. Let us summarize the conditions satisfied by $S_\pm(H)$.

10.3. The Gel'fand-Levitan-Marchenko equations

Hypothesis H. 10.9. The scattering data

(10.88) $$S_\pm(H) = \{R_\pm(k), |k| = 1;\ k_j, \gamma_{\pm,j},\ 1 \le j \le N\}$$

satisfy the following conditions.

(i). The consistency condition

(10.89) $$\frac{R_-(k)}{\overline{R_+(\bar{k})}} = -\frac{T(k)}{\overline{T(\bar{k})}}, \quad \gamma_{+,j}\gamma_{-,j} = \left(\frac{\mathrm{res}_{k_j} T(k)}{k_j}\right)^2,$$

where $T(k)$ is defined using (10.55).

(ii). The reflection coefficients $R_\pm(k)$ are continuous except possibly at $k = \pm 1$ and fulfill

(10.90) $$\overline{R_\pm(k)} = R_\pm(\bar{k}), \qquad C^2|1-k^2|^2 \le 1 - |R_\pm(k)|^2$$

for some $C > 0$. The Fourier coefficients $\tilde{F}^\pm$ of $R_\pm(k^{\mp 1})$ satisfy

(10.91) $$\sum_{j=1}^\infty j|\tilde{F}^\pm(\pm j) - \tilde{F}^\pm(\pm j \pm 2)| < \infty.$$

(iii). The values $k_j \in (-1,0) \cup (0,1)$, $1 \le j \le N$, are distinct and the norming constants $\gamma_{\pm,j}$, $1 \le j \le N$, are positive.

Finally, we collect some properties of the operator $\mathcal{F}_n^+$.

Theorem 10.10. *Fix $n \in \mathbb{Z}$ and consider $\mathcal{F}_n^+ : \ell^p(\mathbb{N}_0) \to \ell^p(\mathbb{N}_0)$, $p \in [1,\infty) \cup \{\infty\}$, as above. Then $\mathcal{F}_n^+$ is a compact operator with norm*

(10.92) $$\|\mathcal{F}_n^+\|_p \le \sum_{m=2n}^\infty |F^+(m)|.$$

For $p = 2$, $\mathcal{F}_n^+$ is a self-adjoint trace class operator satisfying

(10.93) $$\mathbb{1} + \mathcal{F}_n^+ \ge \varepsilon_n > 0, \quad \lim_{n \to \infty} \varepsilon_n = 1.$$

The trace of $\mathcal{F}_n^+$ is given by

(10.94) $$\mathrm{tr}(\mathcal{F}_n^+) = \sum_{j=0}^\infty F^+(2n+2j) + \sum_{\ell=1}^N \gamma_{+,\ell} \frac{k_\ell^{2n}}{1-k_\ell}.$$

Proof. The inequality for the norm of $\mathcal{F}_n^+$ is a special case of Young's inequality. Moreover, cutting off F^+ after finitely many terms gives a sequences of finite rank operators which converge to $\mathcal{F}_n^+$ in norm; hence $\mathcal{F}_n^+$ is compact.

Let $f \in \ell^2(\mathbb{N}_0)$ and abbreviate $\hat{f}(k) = \sum_{j=0}^\infty f(j) k^j$. Setting $f(j) = 0$ for $j < 0$ we obtain

(10.95) $$\sum_{j=0}^\infty \overline{f(j)} \mathcal{F}_n^+ f(j) = \frac{1}{2\pi \mathrm{i}} \int_{|k|=1} R_+(k) k^{2n} |\hat{f}(k)|^2 \frac{dk}{k} + \sum_{\ell=1}^N \gamma_{+,\ell} k_\ell^{2n} |\hat{f}(k_\ell)|^2$$

from the convolution formula. Since $\overline{R_+(k)} = R_+(\bar{k})$ the integral over the imaginary part vanishes and the real part can be replaced by

$$\mathrm{Re}(R_+(k) k^{2n}) = \frac{1}{2}\left(|1 + R_+(k) k^{2n}|^2 - 1 - |R_+(k) k^{2n}|^2\right)$$

(10.96) $$= \frac{1}{2}\left(|1 + R_+(k) k^{2n}|^2 + |T(k)|^2\right) - 1$$

(remember $|T(k)|^2 + |R_+(k)k^{2n}|^2 = 1$). This eventually yields the identity

$$\sum_{j=0}^{\infty} \overline{f(j)}(\mathbb{1} + \mathcal{F}_n^+)f(j) = \sum_{\ell=1}^{N} \gamma_{+,\ell} k_\ell^{2n} |\hat{f}(k_\ell)|^2$$

(10.97)
$$+ \frac{1}{4\pi\mathrm{i}} \int_{|k|=1} \left(|1 + R_+(k)k^{2n}|^2 + |T(k)|^2 \right) |\hat{f}(k)|^2 \frac{dk}{k},$$

which establishes $\mathbb{1} + \mathcal{F}_n^+ \geq 0$. In addition, by virtue of $|1 + R_+(k)k^{2n}|^2 + |T(k)|^2 > 0$ (a.e.), -1 is no eigenvalue and thus $\mathbb{1} + \mathcal{F}_n^+ \geq \varepsilon_n$ for some $\varepsilon_n > 0$. That $\varepsilon_n \to 1$ follows from $\|\mathcal{F}_n^+\| \to 0$.

To see that $\mathcal{F}_n^+$ is trace class we use the splitting $\mathcal{F}_n^+ = \tilde{\mathcal{F}}_n^+ + \sum_{\ell=1}^{N} \tilde{\mathcal{F}}_n^{+,\ell}$ according to (10.70). The operators $\tilde{\mathcal{F}}_n^{+,\ell}$ are positive and trace class. The operator $\tilde{\mathcal{F}}_n^+$ is given by multiplication with $k^{2n} R_+(k)$ in Fourier space and hence is trace class since $|R_+(k)|$ is integrable. □

A similar result holds for $\mathcal{F}_n^-$.

10.4. Inverse scattering theory

In this section we want to invert the process of scattering theory. Clearly, if $S_+(H)$ (and thus $\mathcal{F}_n^+$) is given, we can use the Gel'fand-Levitan-Marchenko equation (10.71) to reconstruct $a(n), b(n)$ from $\mathcal{F}_n^+$

(10.98)
$$a(n)^2 = \frac{1}{4} \frac{\langle \delta_0, (\mathbb{1} + \mathcal{F}_n^+)^{-1} \delta_0 \rangle}{\langle \delta_0, (\mathbb{1} + \mathcal{F}_{n+1}^+)^{-1} \delta_0 \rangle},$$
$$b(n) = \frac{1}{2} \left(\frac{\langle \delta_1, (\mathbb{1} + \mathcal{F}_n^+)^{-1} \delta_0 \rangle}{\langle \delta_0, (\mathbb{1} + \mathcal{F}_n^+)^{-1} \delta_0 \rangle} - \frac{\langle \delta_1, (\mathbb{1} + \mathcal{F}_{n-1}^+)^{-1} \delta_0 \rangle}{\langle \delta_0, (\mathbb{1} + \mathcal{F}_{n-1}^+)^{-1} \delta_0 \rangle} \right).$$

In other words, the scattering data $S_+(H)$ uniquely determine a, b. Since $\mathcal{F}_n^+$ is trace class we can use Kramer's rule to express the above scalar products. If we delete the first row and first column in the matrix representation of $\mathbb{1} + \mathcal{F}_n^+$ we obtain $\mathbb{1} + \mathcal{F}_{n+1}^+$. If we delete the first row and second column in the matrix representation of $\mathbb{1} + \mathcal{F}_n^+$ we obtain an operator $\mathbb{1} + \mathcal{G}_n^+$. By Kramer's rule we have

(10.99)
$$\langle \delta_0, (\mathbb{1} + \mathcal{F}_n^+)^{-1} \delta_0 \rangle = \frac{\det(\mathbb{1} + \mathcal{F}_{n+1}^+)}{\det(\mathbb{1} + \mathcal{F}_n^+)},$$
$$\langle \delta_1, (\mathbb{1} + \mathcal{F}_n^+)^{-1} \delta_0 \rangle = \frac{\det(\mathbb{1} + \mathcal{G}_n^+)}{\det(\mathbb{1} + \mathcal{F}_n^+)},$$

where the determinants have to be interpreted as Fredholm determinants.

These formulas can even be used for practical computations. Let $\mathcal{F}_n^N, \mathcal{G}_n^N$ be the finite rank operators obtained from $\mathcal{F}_n^+, \mathcal{G}_n^+$, respectively, by setting $F^+(2n+j) = 0$ for $j > N$. Then we have $\mathcal{F}_n^N \to \mathcal{F}_n^+$, $\mathcal{G}_n^N \to \mathcal{G}_n^+$ in norm and hence also

10.4. Inverse scattering theory

$(\mathbb{1} + \mathcal{F}_n^N)^{-1} \to (\mathbb{1} + \mathcal{F}_n^+)^{-1}$ as $N \to \infty$. Furthermore,

$$\langle \delta_0, (\mathbb{1} + \mathcal{F}_n^+)^{-1} \delta_0 \rangle = \lim_{N \to \infty} \langle \delta_0, (\mathbb{1} + \mathcal{F}_n^N)^{-1} \delta_0 \rangle$$
$$= \lim_{N \to \infty} \frac{\det(\mathbb{1} + \mathcal{F}_{n+1}^N)}{\det(\mathbb{1} + \mathcal{F}_n^N)},$$
$$\langle \delta_1, (\mathbb{1} + \mathcal{F}_n^+)^{-1} \delta_0 \rangle = \lim_{N \to \infty} \langle \delta_1, (\mathbb{1} + \mathcal{F}_n^N)^{-1} \delta_0 \rangle$$
(10.100)
$$= \lim_{N \to \infty} \frac{\det(\mathbb{1} + \mathcal{G}_n^N)}{\det(\mathbb{1} + \mathcal{F}_n^N)},$$

where

$$\mathcal{F}_n^N = \begin{pmatrix} F^+(2n+r+s), & r+s \leq N \\ 0, & r+s > N \end{pmatrix}_{0 \leq r,s},$$

(10.101)
$$\mathcal{G}_n^N = \begin{pmatrix} F^+(2n+s+1) - \delta_{0,s}, & r = 0 \\ F^+(2n+r+s+2), & r > 0, r+s \leq N \\ 0, & r+s > N \end{pmatrix}_{0 \leq r,s}.$$

For this procedure it is clearly interesting to know when given sets $S_\pm$ are the scattering data of some operator H. In the previous section we have seen that Hypothesis 10.9 is necessary. In this section we will show that it is also sufficient for $S_\pm$ to be the scattering data of some Jacobi operator H. We prove a preliminary lemma first.

Lemma 10.11. *Suppose a given set S_+ satisfies (H.10.9) (ii) and (iii). Then the sequences $a_+(n)$, $b_+(n)$ defined as in (10.98) satisfy $n|2a_+(n) - 1|, n|b_+(n)| \in \ell^1(\mathbb{N})$. In addition, $f_+(k,n) = A_+(n)^{-1} \sum_{m=n}^\infty K_+(n,m) k^m$, where $K_+(n,m) = A_+(n)^{-2} \langle \delta_m, (\mathbb{1} + \mathcal{F}_n^+)^{-1} \delta_0 \rangle$ and $A_+(n)^2 = \langle \delta_0, (\mathbb{1} + \mathcal{F}_n^+)^{-1} \delta_0 \rangle$, satisfies $\tau_+ f = \frac{k + k^{-1}}{2} f$.*

Proof. We abbreviate

(10.102)
$$C(n) = \sum_{j=0}^\infty |F^+(n+j) - F^+(n+j+2)|, \quad C_1(n) = \sum_{j=n}^\infty C(j)$$

and observe $C(n+1) \leq C(n)$ and

$$|F^+(n)| + |F^+(n+1)| \leq \sum_{j=0}^\infty |F^+(n+2j) - F^+(n+2j+2)|$$

(10.103)
$$+ \sum_{j=0}^\infty |F^+(n+2j+1) - F^+(n+2j+3)| = C(n).$$

Moreover, let $\tilde{K}_{+,j}(n) = \langle \delta_j, (\mathbb{1} + \mathcal{F}_n^+)^{-1} \delta_0 \rangle - \delta_{0,j} = A_+(n)^2 K_+(n, n+j) - \delta_{0,j}$. Then

(10.104)
$$\tilde{K}_{+,j}(n) + F^+(2n+j) + \sum_{\ell=0}^\infty \tilde{K}_{+,\ell}(n) F^+(2n+\ell+j) = 0$$

or equivalently $\tilde{K}_{+,j}(n) = (\mathbb{1} + \mathcal{F}_n^+)^{-1} F^+(2n+j)$. Using

$$\sum_{\ell=0}^{\infty} |\tilde{K}_{+,\ell}(n)| \leq N(n) C_1(2n), \quad N(n) = \|(\mathbb{1}+\mathcal{F}_n^+)^{-1}\|_1, \qquad (10.105)$$

we obtain

$$|\tilde{K}_{+,j}(n)| \leq C(2n+j) + C(2n+j) \sum_{\ell=0}^{\infty} |\tilde{K}_{+,\ell}(n)|$$
$$\leq C(2n+j)(1 + N(n)C_1(2n)). \qquad (10.106)$$

Taking differences in (10.104) we see

$$|\tilde{K}_{+,j}(n) - \tilde{K}_{+,j}(n+1)| \leq |F^+(2n+j) - F^+(2n+j+2)|$$
$$+ \sum_{\ell=0}^{\infty} \Big(|F^+(2n+j+\ell) - F^+(2n+j+\ell+2)| |\tilde{K}_{+,\ell}(n)|$$
$$+ |F^+(2n+j+\ell+2)| |\tilde{K}_{+,\ell}(n) - \tilde{K}_{+,\ell}(n+1)| \Big) \qquad (10.107)$$

Using

$$(\mathbb{1} + \mathcal{F}_n^+)(\tilde{K}_{+,j}(n) - \tilde{K}_{+,j}(n+1)) = F^+(2n+j+2) - F^+(2n+j)$$
$$+ \sum_{\ell=0}^{\infty} \Big(F^+(2n+j+\ell+2) - F^+(2n+j+\ell) \Big) \tilde{K}_{+,\ell}(n+1) \qquad (10.108)$$

and hence

$$\sum_{j=0}^{\infty} |\tilde{K}_{+,j}(n) - \tilde{K}_{+,j}(n+1)| \leq N(n) \sum_{j=1}^{\infty} \Big(|F^+(2n+j+2) - F^+(2n+j)|$$
$$+ \sum_{\ell=0}^{\infty} |F^+(2n+j+\ell+2) - F^+(2n+j+\ell)| |\tilde{K}_{+,\ell}(n+1)| \Big)$$
$$\leq O(1) C(2n) \qquad (10.109)$$

we finally infer

$$|\tilde{K}_{+,j}(n) - \tilde{K}_{+,j}(n+1)| \leq |F^+(2n+j) - F^+(2n+j+2)|$$
$$+ O(1) C(2n) C(2n+j). \qquad (10.110)$$

Thus $n|\tilde{K}_{+,j}(n) - \tilde{K}_{+,j}(n+1)| \in \ell^1(\mathbb{N})$ and the same is true for $4a_+(n)^2 - 1$, $b_+(n)$.

Next we turn to $f_+(k,n)$. Abbreviate

$$(\Delta K)(n,m) = 4a_+(n)^2 K(n+1,m) + K(n-1,m) + 2b_+(n) K(n,m)$$
$$- K(n,m+1) - K(n,m-1), \qquad (10.111)$$

then $f_+(k,n)$ satisfies $\tau_+ f = \frac{k+k^{-1}}{2} f$ if $\Delta K_+ = 0$ (cf. (10.8)). We have $K_+(n,n) = 1$ and $K_+(n,n+1) = -\sum_{m=n+1}^{\infty} 2b_+(m)$ by construction. Moreover, using the Gel'fand-Levitan-Marchenko equation

$$K_+(n,m) = -F^+(n+m) - \sum_{\ell=n+1}^{\infty} K_+(n,\ell) F^+(\ell+m) \qquad (10.112)$$

10.4. Inverse scattering theory

one checks

$$(10.113) \quad (\Delta K_+)(n,m) = -\sum_{\ell=n+1}^{\infty} (\Delta K_+)(n,\ell) F^+(\ell+m), \quad m > n+1.$$

By Lemma 7.8 this equation has only the trivial solution $\Delta K_+ = 0$ and hence the proof is complete. $\square$

A corresponding result holds for given data S_-. Now we can prove the main result of this section.

Theorem 10.12. *Hypothesis (H.10.9) is necessary and sufficient for two sets $S_\pm$ to be the left/right scattering data of a unique Jacobi operator H associated with sequences a, b satisfying (H.10.1).*

Proof. Necessity has been established in the previous section. Moreover, by the previous lemma we know existence of sequences $a_\pm, b_\pm$ and corresponding solutions $f_\pm(k,n)$ associated with $S_\pm$, respectively. Hence it remains to establish the relation between these two parts. The key will be the equation $T(k)f_\mp(k,n) = f_\pm(k^{-1},n) + R_\pm(k)f_\pm(k,n)$ and the consistency condition.

We start by defining

$$(10.114) \quad \phi_{+,j}(n) = \frac{1}{A_+(n)} \sum_{m=0}^{\infty} K_{+,m}(n) \tilde{F}^+(2n+m+j)$$

implying

$$(10.115) \quad \sum_{j=-\infty}^{\infty} \phi_{+,j}(n) k^{-n-j} = R_+(k) f_+(k,n).$$

Using the Gel'fand-Levitan-Marchenko equation we further infer

$$\phi_{+,j}(n) = A_+(n)\delta_0(j) - \frac{1}{A_+(n)} \left(K_{+,j}(n) + \sum_{m=0}^{\infty} K_{+,m}(n) \sum_{\ell=0}^{N} \gamma_{+,\ell} k_\ell^{2n+m+\ell} \right)$$

$$(10.116) \quad = A_+(n)\delta_0(j) - \frac{K_{+,j}(n)}{A_+(n)} - \sum_{\ell=1}^{N} \gamma_{+,\ell} f_+(k_\ell, n) k_\ell^{n+\ell}.$$

Next, computing the Fourier series we see

$$R_+(k) f_+(k,n) = \sum_{j=1} \phi_{+,-j}(n) k^{-n+j} + A_+(n) k^{-n} - f_+(k^{-1}, n)$$

$$(10.117) \quad - \sum_{\ell=1}^{N} \gamma_{+,\ell} f_+(k_\ell, n) \frac{k_\ell^n k^{-n}}{1 - k_\ell k^{-1}}$$

and hence

$$(10.118) \quad T(k) h_-(k,n) = f_+(k^{-1}, n) + R_+(k) f_+(k,n),$$

where

$$(10.119) \quad h_-(k,n) = \frac{k^{-n}}{T(k)} \left(A_+(n) + \sum_{j=1}^{\infty} \phi_{+,-j}(n) k^j - \sum_{\ell=1}^{N} \gamma_{+,\ell} f_+(k_\ell, n) \frac{k_\ell^n k}{k - k_\ell} \right).$$

Similarly, one obtains

$$(10.120) \quad h_+(k,n) = \frac{k^n}{T(k)}\left(A_-(n) + \sum_{j=1}^{\infty} \phi_{-,j}(n)k^j - \sum_{\ell=1}^{N} \gamma_{-,\ell} f_-(k_\ell, n)\frac{k_\ell^{-n}k}{k-k_\ell}\right),$$

where

$$(10.121) \quad \phi_{-,j}(n) = \frac{1}{A_-(n)} \sum_{m=0}^{\infty} K_{-,m}(n)\tilde{F}^+(2n-m-j).$$

In particular, we see that $k^{\mp n}h_\pm(k,n)$ are holomorphic inside the unit circle. In addition, by virtue of the consistency condition (H.10.9)(i) we obtain

$$(10.122) \quad h_\pm(k_\ell, n) = (\operatorname{res}_{k_j} T(k))^{-1} \gamma_{\mp,\ell} k_\ell f_\mp(k_\ell, n)$$

and (use also $|T(k)|^2 + |R_\pm(k)|^2 = 1$)

$$(10.123) \quad T(k)f_\mp(k,n) = h_\pm(k^{-1}, n) + R_\pm(k)h_\pm(k,n).$$

Eliminating (e.g.) $R_-(k)$ from the last equation and $T(k)h_+(k,n) = f_-(k^{-1}, n) + R_-(k)f_-(k,n)$ we see

$$f_-(\overline{k}, n)h_-(k,n) - f_-(k,n)h_-(\overline{k}, n) =$$
$$(10.124) \quad T(k)\Big(h_-(k,n)h_+(k,n) - f_-(k,n)f_+(k,n)\Big) = G(k,n), \quad |k|=1.$$

Equation (10.122) shows that all poles of $G(k,n) = \sum_{j\in\mathbb{Z}} G_j k^j$ inside the unit circle are removable and hence $G_j = 0$, $j < 0$. Moreover, by $G(\overline{k}, n) = -G(k,n)$ we see $G_{-j} = -G_j$ and thus $G(k,n) = 0$. If we compute $0 = \lim_{k\to 0} h_-(k,n)h_+(k,n) - f_-(k,n)f_+(k,n)$ we infer $A_-(n)A_+(n) = T(0)$ and thus $a_-(n) = a_+(n) \equiv a(n)$.

Our next aim is to show $h_\pm(k,n) = f_\pm(k,n)$. To do this we consider the function

$$(10.125) \quad H(k,n) = \sum_{j\in\mathbb{Z}} H_j k^j = \frac{h_-(k,n)}{f_-(k,n)} = \frac{f_+(k,n)}{h_+(k,n)}, \quad |k| \leq 1,$$

which is meromorphic inside the unit disk with $H_0 = 1$. However, if we choose $-n$ large enough, then $k^n f_-(k,n)$ cannot vanish and hence $H(k,n)$ is holomorphic inside the unit disc implying $H_j = 0$, $j < 0$. And since $H(\overline{k}, n) = H(k,n)$, $|k|=1$, we even have $H_j = 0$, $j > 0$, that is, $H(k,n) = 1$. Similarly, considering $H(k,n)^{-1}$, we see $H(k,n) = 1$ for $+n$ large enough.

Thus we know $h_\pm(k,n) = f_\pm(k,n)$, $|k| \leq 1$, at least for $|n|$ large enough. In particular, $h_\pm(k,n) \in \ell^2_\pm(\mathbb{Z})$, $|k| < 1$, and hence $h_+(k,n)$, $f_-(k,n)$ and $f_+(k,n)$, $h_-(k,n)$ can be used to compute the Weyl m-functions of H_- and H_+, respectively. Since the Weyl m-functions of H_- and H_+ are equal for some n, we conclude $H_- = H_+ \equiv H$ (i.e., $b_-(n) = b_+(n) \equiv b(n)$).

Up to this point we have constructed an operator H which has the correct reflection coefficients and the correct eigenvalues. To finish the proof we need to show that our operator also has the correct norming constants. This follows from (cf. (10.34))

$$(10.126) \quad \sum_{n\in\mathbb{Z}} f_-(k_\ell, n)f_+(k_\ell, n) = k_\ell (\operatorname{res}_{k_j} T(k))^{-1}$$

and (10.122). $\square$

Chapter 11

Spectral deformations – Commutation methods

11.1. Commuting first order difference expressions

The idea of this section is to factor a given Jacobi operator into the product of two simpler operators, that is, two operators associated with first order difference expressions. Since it is not clear how to do this, we will first investigate the other direction, that is, we take a first order difference expression and multiply it with its adjoint. The main reason why we take the adjoint as second operator is, since this choice ensures self-adjointness of the resulting product (i.e., we get a Jacobi operator). In addition, changing the order of multiplication produces a second operator whose spectral properties are closely related to those of the first. In fact, we have the well-known result by von Neumann (see, e.g., [**241**], Theorem 5.39 and [**54**], Theorem 1).

Theorem 11.1. *Suppose A is a bounded operator with adjoint A^*. Then AA^* and A^*A are non-negative self-adjoint operators which are unitarily equivalent when restricted to the orthogonal complements of their corresponding null-spaces. Moreover, the resolvents of AA^* and A^*A for $z \in \mathbb{C}\backslash(\sigma(A^*A) \cup \{0\})$ are related by*

$$(11.1) \quad \begin{aligned} (AA^* - z)^{-1} &= \frac{1}{z}\Big(1 - A(A^*A - z)^{-1}A^*\Big), \\ (A^*A - z)^{-1} &= \frac{1}{z}\Big(1 - A^*(AA^* - z)^{-1}A\Big). \end{aligned}$$

Let ρ_o, ρ_e be two real-valued sequences fulfilling

$$(11.2) \qquad \rho_o, \rho_e \in \ell^\infty(\mathbb{Z}), \qquad -\rho_o(n), \rho_e(n) > 0.$$

The second requirement is for convenience only (it will imply $a_j(n) < 0$ below). Now consider the (bounded) operators

$$
\begin{array}{rccc}
A & \ell^2(\mathbb{Z}) & \to & \ell^2(\mathbb{Z}) \\
 & f(n) & \mapsto & \rho_o(n)f(n+1) + \rho_e(n)f(n), \\
A^* & \ell^2(\mathbb{Z}) & \to & \ell^2(\mathbb{Z}) \\
 & f(n) & \mapsto & \rho_o(n-1)f(n-1) + \rho_e(n)f(n),
\end{array}
$$
(11.3)

where A^* is the adjoint operator of A. Using them we can construct two non-negative operators $H_1 = A^*A$ and $H_2 = AA^*$. A direct calculation shows

(11.4) $\quad H_j f(n) = \tau_j f(n) = a_j(n)f(n+1) + a_j(n-1)f(n-1) + b_j(n)f(n),$

with

(11.5) $\quad \begin{array}{ll} a_1(n) = \rho_o(n)\rho_e(n) & b_1(n) = \rho_o(n-1)^2 + \rho_e(n)^2 \\ a_2(n) = \rho_o(n)\rho_e(n+1) & b_2(n) = \rho_o(n)^2 + \rho_e(n)^2 \end{array}.$

Due to (11.2) we have $a_j, b_j \in \ell^\infty(\mathbb{Z})$ and $a_j(n) < 0$, $b_j(n) > 0$. Next, observe that the quantities

(11.6) $\quad \phi_1(n) = -\dfrac{\rho_e(n)}{\rho_o(n)}, \qquad \phi_2(n) = -\dfrac{\rho_o(n)}{\rho_e(n+1)}$

satisfy

(11.7) $\quad a_j(n)\phi_j(n) + \dfrac{a_j(n-1)}{\phi_j(n-1)} = -b_j(n).$

Hence our approach gives us positive solutions

(11.8) $\quad \tau_j u_j = 0, \qquad u_j(n) = \prod_{m=n_0}^{n-1}{}^* \phi_j(m) > 0$

for free. Moreover, it shows that a given operator H_1 can only be factorized in this manner if a positive solution u_1 of $\tau_1 u_1 = 0$ exists. On the other hand, since we have

$$\rho_o(n) = -\sqrt{-\dfrac{a_1(n)}{\phi_1(n)}} = -\sqrt{-a_2(n)\phi_2(n)},$$

(11.9) $\quad \rho_e(n) = \sqrt{-a_1(n)\phi_1(n)} = \sqrt{-\dfrac{a_2(n-1)}{\phi_2(n-1)}},$

a positive solution is the only necessary ingredient for such a factorization. Together with Section 2.3, where we have seen that positive solutions exist if $H_j \geq 0$ (see also Remark 2.10) we obtain

Corollary 11.2. *Suppose $a(n) < 0$. Then a Jacobi operator is non-negative, $H \geq 0$, if and only if there exists a positive solution of the corresponding Jacobi equation.*

11.2. The single commutation method

In this section we invert the process of the last section. This will result in a method for inserting eigenvalues below the spectrum of a given Jacobi operator H. It is usually referred to as **single commutation method** and a detailed study is the topic of this and the next section.

To assure existence of the aforementioned necessary positive solutions, we will need $a < 0$ throughout Sections 11.2 and 11.3.

Hypothesis H. 11.3. Suppose $a, b \in \ell^\infty(\mathbb{Z}, \mathbb{R})$ and $a(n) < 0$.

Let H be a given Jacobi operator. We pick $\sigma_1 \in [-1, 1]$ and $\lambda_1 < \inf \sigma(H)$. Furthermore, recall the minimal positive solutions $u_\pm(\lambda_1, n)$ constructed in Section 2.3 and set

$$(11.10) \qquad u_{\sigma_1}(\lambda_1, n) = \frac{1 + \sigma_1}{2} u_+(\lambda_1, n) + \frac{1 - \sigma_1}{2} u_-(\lambda_1, n).$$

By the analysis of Section 2.3 (cf. (2.68)), any positive solution can be written in this form up to normalization.

Motivated by (11.9) we consider the sequences

$$(11.11) \qquad \rho_{o,\sigma_1}(n) = -\sqrt{-\frac{a(n) u_{\sigma_1}(\lambda_1, n)}{u_{\sigma_1}(\lambda_1, n+1)}}, \quad \rho_{e,\sigma_1}(n) = \sqrt{-\frac{a(n) u_{\sigma_1}(\lambda_1, n+1)}{u_{\sigma_1}(\lambda_1, n)}}$$

which are bounded because of

$$(11.12) \qquad \left|\frac{a(n) u_{\sigma_1}(n+1)}{u_{\sigma_1}(n)}\right| + \left|\frac{a(n-1) u_{\sigma_1}(n-1)}{u_{\sigma_1}(n)}\right| = |\lambda_1 - b(n)|.$$

Thus we can define a corresponding operator A_{σ_1} on $\ell^2(\mathbb{Z})$ together with its adjoint $A^*_{\sigma_1}$,

$$(11.13) \qquad \begin{aligned} A_{\sigma_1} f(n) &= \rho_{o,\sigma_1}(n) f(n+1) + \rho_{e,\sigma_1}(n) f(n), \\ A^*_{\sigma_1} f(n) &= \rho_{o,\sigma_1}(n-1) f(n-1) + \rho_{e,\sigma_1}(n) f(n). \end{aligned}$$

For simplicity of notation we will not distinguish between the operator A_{σ_1} and its difference expression.

By Theorem 11.1, $A^*_{\sigma_1} A_{\sigma_1}$ is a positive self-adjoint operator. A quick calculation shows

$$(11.14) \qquad H = A^*_{\sigma_1} A_{\sigma_1} + \lambda_1.$$

Commuting $A^*_{\sigma_1}$ and A_{σ_1} yields a second positive self-adjoint operator $A_{\sigma_1} A^*_{\sigma_1}$ and further the commuted operator

$$(11.15) \qquad H_{\sigma_1} = A_{\sigma_1} A^*_{\sigma_1} + \lambda_1.$$

The next theorem characterizes H_{σ_1}.

Theorem 11.4. *Assume (H.11.3) and let $\lambda_1 < \inf \sigma(H)$, $\sigma_1 \in [-1, 1]$. Then the self-adjoint operator H_{σ_1} is associated with*

$$(11.16) \qquad (\tau_{\sigma_1} f)(n) = a_{\sigma_1}(n) f(n+1) + a_{\sigma_1}(n-1) f(n-1) + b_{\sigma_1}(n) f(n),$$

where
$$a_{\sigma_1}(n) = -\frac{\sqrt{a(n)a(n+1)u_{\sigma_1}(\lambda_1,n)u_{\sigma_1}(\lambda_1,n+2)}}{u_{\sigma_1}(\lambda_1,n+1)},$$

$$b_{\sigma_1}(n) = \lambda_1 - a(n)\Big(\frac{u_{\sigma_1}(\lambda_1,n)}{u_{\sigma_1}(\lambda_1,n+1)} + \frac{u_{\sigma_1}(\lambda_1,n+1)}{u_{\sigma_1}(\lambda_1,n)}\Big)$$

(11.17)
$$= b(n) + \partial^* \frac{a(n)u_{\sigma_1}(\lambda_1,n)}{u_{\sigma_1}(\lambda_1,n+1)}$$

satisfy (H.11.3). The equation $\tau_{\sigma_1} v = \lambda_1 v$ has the positive solution

(11.18) $$v_{\sigma_1}(\lambda_1, n) = \frac{1}{\sqrt{-a(n)u_{\sigma_1}(\lambda_1,n)u_{\sigma_1}(\lambda_1,n+1)}},$$

which is an eigenfunction of H_{σ_1} if and only if $\sigma_1 \in (-1,1)$. $H - \lambda_1$ and $H_{\sigma_1} - \lambda_1$ restricted to the orthogonal complements of their corresponding one-dimensional null-spaces are unitarily equivalent and hence

(11.19)
$$\sigma(H_{\sigma_1}) = \begin{cases} \sigma(H) \cup \{\lambda_1\}, & \sigma_1 \in (-1,1) \\ \sigma(H), & \sigma_1 \in \{-1,1\} \end{cases}, \quad \sigma_{ac}(H_{\sigma_1}) = \sigma_{ac}(H),$$

$$\sigma_{pp}(H_{\sigma_1}) = \begin{cases} \sigma_{pp}(H) \cup \{\lambda_1\}, & \sigma_1 \in (-1,1) \\ \sigma_{pp}(H), & \sigma_1 \in \{-1,1\} \end{cases}, \quad \sigma_{sc}(H_{\sigma_1}) = \sigma_{sc}(H).$$

In addition, the operator

(11.20) $$A_{\sigma_1} f(n) = \frac{W_n(u_{\sigma_1}(\lambda_1), f)}{\sqrt{-a(n)u_{\sigma_1}(\lambda_1,n)u_{\sigma_1}(\lambda_1,n+1)}}$$

acts as transformation operator

(11.21) $$\tau_{\sigma_1} A_{\sigma_1} f = A_{\sigma_1} \tau f.$$

Moreover, one obtains

(11.22) $$W_{\sigma_1}(A_{\sigma_1} u(z), A_{\sigma_1} v(z)) = (z - \lambda_1) W(u(z), v(z))$$

for solutions $u(z)$, $v(z)$ of $\tau u = zu$, where $W_{\sigma_1, n}(u,v) = a_{\sigma_1}(n)(u(n)v(n+1) - u(n+1)v(n))$. The resolvent of H_{σ_1} for $z \in \mathbb{C}\backslash(\sigma(H) \cup \{\lambda_1\})$ is given by

(11.23) $$(H_{\sigma_1} - z)^{-1} = \frac{1}{z - \lambda_1}\Big(1 - A_{\sigma_1}(H - z)^{-1} A_{\sigma_1}^*\Big),$$

or, in terms of Green's functions for $n \geq m$, $z \in \mathbb{C}\backslash(\sigma(H) \cup \{\lambda_1\})$,

(11.24) $$G_{\sigma_1}(z, n, m) = \frac{A_{\sigma_1} u_-(z, m) A_{\sigma_1} u_+(z, n)}{(z - \lambda_1) W(u_-(z), u_+(z))}.$$

Hence, $u_{\sigma_1, \pm}(z, n) = \pm A_{\sigma_1} u_\pm(z, n)$ satisfy $u_{\sigma_1, \pm}(z, n) \in \ell_\pm(\mathbb{Z})$ and are the minimal solutions of $(H_{\sigma_1} - z)u = 0$ for $z < \lambda_1$. In addition, we have

(11.25) $$\sum_{n \in \mathbb{Z}} v_{\sigma_1}(\lambda_1, n)^2 = \frac{4}{1 - \sigma_1^2} W(u_-(\lambda_1), u_+(\lambda_1))^{-1}, \quad \sigma_1 \in (-1,1),$$

and, if $\tau u(\lambda) = \lambda u(\lambda)$, $u(\lambda, .) \in \ell^2(\mathbb{Z})$,

(11.26) $$\sum_{n \in \mathbb{Z}} (A_{\sigma_1} u(\lambda, n))^2 = (\lambda - \lambda_1) \sum_{n \in \mathbb{Z}} u(\lambda, n)^2.$$

11.2. The single commutation method

Proof. The unitary equivalence together with equation (11.23) follow from Theorem 11.1. Equations (11.20) – (11.22) are straightforward calculations. Equation (11.24) follows from the obvious fact that $A_{\sigma_1} u_\pm(z, n) \in \ell^2_\pm(\mathbb{Z})$ and (11.22). Moreover,

$$(11.27) \qquad A_{\sigma_1} u_\pm(\lambda_1, n) = \mp \frac{1 \mp \sigma_1}{2} W(u_-(\lambda_1), u_+(\lambda_1)) v_{\sigma_1}(\lambda_1, n)$$

shows that $v_{\sigma_1}(\lambda_1, n)$ is in $\ell^2(\mathbb{Z})$ if and only if $\sigma_1 \in (-1, 1)$. In addition, we infer $\pm A_{\sigma_1} u_\pm(\lambda_1, n) \geq 0$ since $W(u_-(\lambda_1), u_+(\lambda_1)) < 0$ (by $G(\lambda_1, n, n) < 0$) and hence a simple continuity argument implies $\pm A_{\sigma_1} u_\pm(z, n) > 0$ for $z < \lambda_1$. Next, if $\sigma_1 \in (-1, 1)$, we can use (11.27) to compute the residue of (11.24) at $z = \lambda_1$ and compare this with (2.34) to obtain (11.25). To see (11.26) use $\langle A_{\sigma_1} u, A_{\sigma_1} u \rangle = \langle u, A^*_{\sigma_1} A_{\sigma_1} u \rangle$. The rest is simple. $\square$

Remark 11.5. (i). Multiplying u_{σ_1} with a positive constant leaves all formulas and, in particular, H_{σ_1} invariant.
(ii). We can also insert eigenvalues into the highest spectral gap, that is, above the spectrum of H, upon considering $-H$. Then $\lambda > \sup(\sigma(H))$ implies that we don't have positive but rather alternating solutions and all our previous calculations carry over with minor changes.
(iii). We can weaken (H.11.3) by requiring $a(n) \neq 0$ instead of $a(n) < 0$. Everything stays the same with the only difference that $u_\pm$ are not positive but change sign in such a way that (2.52) stays positive. Moreover, the signs of $a_{\sigma_1}(n)$ can also be prescribed arbitrarily by altering the signs of ρ_{o,σ_1} and ρ_{e,σ_1}.
(iv). The fact that $v_{\sigma_1} \in \ell^2(\mathbb{Z})$ if and only if $\sigma_1 \in (-1, 1)$ gives an alternate proof for

$$(11.28) \qquad \sum_{n=0}^{\pm\infty} \frac{1}{-a(n) u_{\sigma_1}(\lambda_1, n) u_{\sigma_1}(\lambda_1, n+1)} < \infty \text{ if and only if } \sigma_1 \in \begin{matrix} [-1, 1) \\ (-1, 1] \end{matrix}$$

(cf. Lemma 2.9 (iv)).
(v). In the case $\sigma_1 = \pm 1$ the second minimal solution for $z = \lambda_1$ is given by (cf. (1.51))

$$(11.29) \qquad u_{\sigma_1, \mp}(\lambda_1, n) = v_{\sigma_1}(n) \sum_{j=\pm\infty}^{n+1 \atop n} u_{\sigma_1}(\lambda_1, n), \quad \sigma_1 = \pm 1.$$

(vi). The formula (11.22) can be generalized to

$$W_{\sigma_1}(A_{\sigma_1} u(z), A_{\sigma_1} v(\hat{z}))$$
$$= (z - \lambda_1) W(u(z), v(\hat{z})) + \frac{(z - \hat{z}) v(\hat{z}, n+1)}{u(\lambda_1, n+1)} W_n(u(\lambda_1), u(z))$$
$$(11.30) \qquad = -\frac{a(n) a(n+1)}{u(\lambda_1, n+1)} C_n(u(\lambda_1), u(z), v(\hat{z})),$$

where $C_n(\ldots)$ denotes the N-dimensional **Casoratian**

$$(11.31) \qquad C_n(u_1, \ldots, u_N) = \det\left(u_i(n + j - 1)\right)_{1 \leq i, j \leq N}.$$

In the remainder of this section we will show some connections between the single commutation method and some other theories. We start with the Weyl m-functions.

Lemma 11.6. *Assume (H.11.3). The Weyl $\tilde{m}$-functions $\tilde{m}_{\sigma_1,\pm}(z)$ of H_{σ_1}, $\sigma_1 \in [-1,1]$ in terms of $\tilde{m}_\pm(z)$, the ones of H, read*

$$(11.32) \quad \tilde{m}_{\sigma_1,\pm}(z) = \frac{\mp u_{\sigma_1}(\lambda_1,1)}{a(1)u_{\sigma_1}(\lambda_1,2)}\left(1 \pm \frac{(z-\lambda_1)\tilde{m}_\pm(z)}{1 \mp \frac{a(0)}{\beta}\tilde{m}_\pm(z)}\right), \quad \beta = -\frac{u_{\sigma_1}(\lambda_1,1)}{u_{\sigma_1}(\lambda_1,0)}.$$

Proof. The above formulas follows after a straightforward calculation using (11.24) and (2.2). □

From Lemma 8.1 (iii) we obtain the useful corollary

Corollary 11.7. *The operator H_{σ_1} is reflectionless if and only if H is.*

Moreover, we even have the following result.

Lemma 11.8. *The operators $H_n^\infty - \lambda_1$ and $(H_{\sigma_1})_{n-1}^{\beta_{\sigma_1}} - \lambda_1$ (resp. $H_n^\beta - \lambda_1$ and $(H_{\sigma_1})_n^\infty - \lambda_1$) are unitarily equivalent when restricted to the orthogonal complements of their corresponding one-dimensional null-spaces, where*

$$(11.33) \quad \beta = -\frac{u_{\sigma_1}(\lambda_1,1)}{u_{\sigma_1}(\lambda_1,0)}, \qquad \beta_{\sigma_1} = -\frac{v_{\sigma_1}(\lambda_1,1)}{v_{\sigma_1}(\lambda_1,0)}.$$

Proof. We recall Remark 1.11 and we set $\lambda_1 = 0$ without loss of generality. Choose $\tilde{A}_{\sigma_1} = P_n A_{\sigma_1} P_n + \rho_{e,\sigma_1}(n)\langle \delta_{n+1},.\rangle\delta_n$, then one computes $\tilde{A}_{\sigma_1}^* \tilde{A}_{\sigma_1} = P_n H P_n$ and $\tilde{A}_{\sigma_1} \tilde{A}_{\sigma_1}^* = P_{n-1}^{\beta_{\sigma_1}} H_{\sigma_1} P_{n-1}^{\beta_{\sigma_1}}$. Similarly, choosing $\tilde{A}_{\sigma_1} = P_n A_{\sigma_1} P_n + \rho_{e,\sigma_1}(n-1)\langle \delta_n,.\rangle\delta_{n-1}$, one obtains $\tilde{A}_{\sigma_1}^* \tilde{A}_{\sigma_1} = P_n^\beta H P_n^\beta$ and $\tilde{A}_{\sigma_1} \tilde{A}_{\sigma_1}^* = P_n H_{\sigma_1} P_n$. □

Finally, we turn to scattering theory and assume (H.10.1) (cf. Remark 11.5). We denote the set of eigenvalues by $\sigma_p(H) = \{\lambda_j\}_{j \in J}$, where $J \subseteq \mathbb{N}$ is a finite index set, and use the notation introduced in Chapter 10.

Lemma 11.9. *Suppose H satisfies (H.10.1) and let H_{σ_1} be constructed as in Theorem 11.4 with*

$$(11.34) \quad u_{\sigma_1}(\lambda_1,n) = \frac{1+\sigma_1}{2}f_+(k_1,n) + \frac{1-\sigma_1}{2}f_-(k_1,n).$$

Then the transmission $T_{\sigma_1}(k)$ and reflection coefficients $R_{\pm,\sigma_1}(k)$ of H_{σ_1} in terms of the corresponding scattering data $T(k)$, $R_\pm(k)$ of H are given by

$$T_{\sigma_1}(k) = \frac{1-kk_1}{k-k_1}T(k), \quad R_{\pm,\sigma_1}(k) = k^{\pm 1}\frac{k-k_1}{1-kk_1}R_\pm(k), \quad \sigma_1 \in (-1,1),$$

$$(11.35) \quad T_{\sigma_1}(k) = T(k), \quad R_{\pm,\sigma_1}(k) = \frac{k_1^{\sigma_1} - k^{\mp 1}}{k_1^{\sigma_1} - k^{\pm 1}}R_\pm(k), \quad \sigma_1 \in \{-1,1\},$$

where $k_1 = \lambda_1 + \sqrt{\lambda_1^2 - 1} \in (-1,0)$. Moreover, the norming constants $\gamma_{\pm,\sigma_1,j}$ associated with $\lambda_j \in \sigma_{pp}(H_{\sigma_1})$ in terms of $\gamma_{\pm,j}$, corresponding to H, read

$$\gamma_{\pm,\sigma_1,j} = |k_j|^{\pm 1}\frac{1-k_jk_1}{(k_j-k_1)}\gamma_{\pm,j}, \quad j \in J, \, \sigma_1 \in (-1,1),$$

$$\gamma_{\pm,\sigma_1,1} = \left(\frac{1-\sigma_1}{1+\sigma_1}\right)^{\pm 1}|1-k_1^{\mp 2}|T(k_1), \quad \sigma_1 \in (-1,1),$$

$$(11.36) \quad \gamma_{\pm,\sigma_1,j} = |k_1^{\sigma_1} - k_j^{\mp 1}|\gamma_{\pm,j}, \quad j \in J, \, \sigma_1 \in \{-1,1\}.$$

Proof. The claims follow easily after observing that up to normalization the Jost solutions of H_{σ_1} are given by $A_{\sigma_1}f_\pm(k,n)$ (compare (11.24)). □

11.3. Iteration of the single commutation method

From the previous section it is clear that the single commutation method can be iterated. In fact, choosing $\lambda_2 < \lambda_1$ and $\sigma_2 \in [-1,1]$ we can define

$$(11.37) \qquad u_{\sigma_1,\sigma_2}(\lambda_2, n) = \frac{1+\sigma_2}{2} u_{+,\sigma_1}(\lambda_2, n) + \frac{1-\sigma_2}{2} u_{-,\sigma_1}(\lambda_2, n)$$

and corresponding sequences $\rho_{o,\sigma_1,\sigma_2}$, $\rho_{e,\sigma_1,\sigma_2}$. As before, the associated operators A_{σ_1,σ_2}, $A^*_{\sigma_1,\sigma_2}$ satisfy

$$(11.38) \qquad H_{\sigma_1} = A^*_{\sigma_1,\sigma_2} A_{\sigma_1,\sigma_2} - \lambda_2$$

and a further commutation then yields the operator

$$(11.39) \qquad H_{\sigma_1,\sigma_2} = A_{\sigma_1,\sigma_2} A^*_{\sigma_1,\sigma_2} - \lambda_2.$$

Clearly, if one works out the expression for a_{σ_1,σ_2} and b_{σ_1,σ_2} (by simply plugging in) we get rather clumsy formulas. Moreover, these formulas will get even worse in each step. The purpose of this section is to show that the resulting quantities after N steps can be expressed in terms of determinants.

Theorem 11.10. *Assume (H.11.3). Let H be as in Section 11.2 and choose*

$$(11.40) \qquad \lambda_N < \cdots < \lambda_2 < \lambda_1 < \inf \sigma(H), \quad \sigma_\ell \in [-1,1], \quad 1 \leq \ell \leq N \in \mathbb{N}.$$

Then we have

$$a_{\sigma_1,\ldots,\sigma_N}(n) = -\sqrt{a(n)a(n+N)} \frac{\sqrt{C_n(U_{\sigma_1,\ldots,\sigma_N}) C_{n+2}(U_{\sigma_1,\ldots,\sigma_N})}}{C_{n+1}(U_{\sigma_1,\ldots,\sigma_N})},$$

$$b_{\sigma_1,\ldots,\sigma_N}(n) = \lambda_N - a(n) \frac{C_{n+2}(U_{\sigma_1,\ldots,\sigma_{N-1}}) C_n(U_{\sigma_1,\ldots,\sigma_N})}{C_{n+1}(U_{\sigma_1,\ldots,\sigma_{N-1}}) C_{n+1}(U_{\sigma_1,\ldots,\sigma_N})}$$

$$- a(n+N-1) \frac{C_n(U_{\sigma_1,\ldots,\sigma_{N-1}}) C_{n+1}(U_{\sigma_1,\ldots,\sigma_N})}{C_{n+1}(U_{\sigma_1,\ldots,\sigma_{N-1}}) C_n(U_{\sigma_1,\ldots,\sigma_N})}$$

$$= b(n) + \partial^* a(n) \frac{D_n(U_{\sigma_1,\ldots,\sigma_N})}{C_{n+1}(U_{\sigma_1,\ldots,\sigma_N})}$$

$$(11.41) \qquad = b(n+N) - \partial^* a(n+N) \frac{\tilde{D}_{n+1}(U_{\sigma_1,\ldots,\sigma_N})}{C_{n+1}(U_{\sigma_1,\ldots,\sigma_N})},$$

where $(U_{\sigma_1,\ldots,\sigma_N}) = (u^1_{\sigma_1}, \ldots, u^N_{\sigma_N})$ and

$$(11.42) \qquad u^\ell_{\sigma_\ell}(n) = \frac{1+\sigma_\ell}{2} u_+(\lambda_\ell, n) + (-1)^{\ell+1} \frac{1-\sigma_\ell}{2} u_-(\lambda_\ell, n).$$

Here $C_n(\ldots)$ denotes the Casoratian (11.31) and

$$D_n(u_1,\ldots,u_N) = \det \begin{pmatrix} u_i(n), & j=1 \\ u_i(n+j), & j>1 \end{pmatrix}_{1 \leq i,j \leq N},$$

$$(11.43) \qquad \tilde{D}_n(u_1,\ldots,u_N) = \det \begin{pmatrix} u_i(n+j-1), & j<N \\ u_i(n+N), & j=N \end{pmatrix}_{1 \leq i,j \leq N}.$$

The corresponding Jacobi operator $H_{\sigma_1,\ldots,\sigma_N}$ is self-adjoint with spectrum

$$(11.44) \qquad \sigma(H_{\sigma_1,\ldots,\sigma_N}) = \sigma(H) \cup \{\lambda_\ell \mid \sigma_\ell \in (-1,1),\ 1 \leq \ell \leq N\}.$$

In addition, we have

$$\rho_{o,\sigma_1,\ldots,\sigma_N}(n) =$$
$$-\sqrt{-a(n)\frac{C_{n+2}(U_{\sigma_1,\ldots,\sigma_{N-1}})C_n(U_{\sigma_1,\ldots,\sigma_N})}{C_{n+1}(U_{\sigma_1,\ldots,\sigma_{N-1}})C_{n+1}(U_{\sigma_1,\ldots,\sigma_N})}},$$

$$\rho_{e,\sigma_1,\ldots,\sigma_N}(n) =$$

(11.45)
$$\sqrt{-a(n+N-1)\frac{C_n(U_{\sigma_1,\ldots,\sigma_{N-1}})C_{n+1}(U_{\sigma_1,\ldots,\sigma_N})}{C_{n+1}(U_{\sigma_1,\ldots,\sigma_{N-1}})C_n(U_{\sigma_1,\ldots,\sigma_N})}}.$$

and

(11.46) $$A_{\sigma_1,\ldots,\sigma_N}\cdots A_{\sigma_1}f(n) = \frac{\left(\prod_{j=0}^{\ell-1}\sqrt{-a(n+j)}\right)C_n(U_{\sigma_1,\ldots,\sigma_N},f)}{\sqrt{C_n(U_{\sigma_1,\ldots,\sigma_N})C_{n+1}(U_{\sigma_1,\ldots,\sigma_N})}}.$$

The principal solutions of $\tau_{\sigma_1,\ldots,\sigma_\ell}u = \lambda u$ for $1 \leq \ell \leq N$, $\lambda < \lambda_\ell$ are given by

(11.47) $$u_{\sigma_1,\ldots,\sigma_\ell,\pm}(\lambda,n) = (\pm 1)^\ell A_{\sigma_1,\ldots,\sigma_\ell}\cdots A_{\sigma_1}u_\pm(\lambda,n)$$

and

(11.48) $$u_{\sigma_1,\ldots,\sigma_\ell}(\lambda_\ell,n) = \frac{1+\sigma_\ell}{2}u_{\sigma_1,\ldots,\sigma_{\ell-1},+}(\lambda_\ell,n) + \frac{1-\sigma_\ell}{2}u_{\sigma_1,\ldots,\sigma_{\ell-1},-}(\lambda_\ell,n)$$

is used to define $H_{\sigma_1,\ldots,\sigma_\ell}$.

To prove this theorem we need a special case of Sylvester's determinant identity (cf. [**91**], Sect. II.3, [**128**]).

Lemma 11.11. *Let $\underline{M}$ be an arbitrary $n \times n$ matrix ($n \geq 2$). Denote by $\underline{M}^{(i,j)}$ the matrix with the i-th column and the j-th row deleted and by $\underline{M}^{(i,j,l,m)}$ the matrix with the i-th and l-th column and the j-th and m-th row deleted. Then we have*

(11.49)
$$\det \underline{M}^{(n,n)} \det \underline{M}^{(n-1,n-1)} - \det \underline{M}^{(n-1,n)} \det \underline{M}^{(n,n-1)}$$
$$= \det \underline{M}^{(n,n,n-1,n-1)} \det \underline{M}.$$

Proof. Obviously we can assume $\underline{M}$ to be nonsingular. First observe that interchanging the i-th row and the j-th row for $1 \leq i,j \leq n-2$ leaves the formula invariant. Furthermore, adding an arbitrary multiple of the j-th row to the i-th row with $1 \leq j \leq n-2$, $1 \leq i \leq n$, also leaves the formula unchanged. Thus we may assume $\underline{M}$ to be of the form

(11.50) $$\begin{pmatrix} M_{11} & M_{12} & \cdots & & \cdots & M_{1n} \\ & M_{22} & & & & \vdots \\ & & \ddots & & & \vdots \\ & & & M_{n-1,n-1} & M_{n-1,n} \\ & & & M_{n,n-1} & M_{n,n} \end{pmatrix},$$

11.3. Iteration of the single commutation method

if $\underline{M}^{(n,n,n-1,n-1)}$ is nonsingular, and of the form

(11.51)
$$\begin{pmatrix} 0 & M_{12} & \cdots & \cdots & M_{1n} \\ \vdots & \vdots & & & \vdots \\ 0 & M_{n-2,2} & & & \vdots \\ M_{n-1,1} & \cdots & \cdots & \cdots & M_{n-1,n} \\ M_{n,1} & \cdots & \cdots & \cdots & M_{n,n} \end{pmatrix},$$

if $\underline{M}^{(n,n,n-1,n-1)}$ is singular. But for such matrices the claim can be verified directly. $\square$

Proof. (of Theorem 11.10) It is enough to prove (11.41) and (11.46), the remaining assertions then follow from these two equations. We will use a proof by induction on N. All formulas are valid for $N = 1$ and we have

(11.52)
$$f_{\sigma_1,\ldots,\sigma_{N+1}}(n) = \frac{-\sqrt{-a_{\sigma_1,\ldots,\sigma_N}(n)}C_n(u_{\sigma_1,\ldots,\sigma_N}(\lambda_N),f)}{\sqrt{u_{\sigma_1,\ldots,\sigma_N}(\lambda_N,n)u_{\sigma_1,\ldots,\sigma_N}(\lambda_N,n+1)}},$$

$$a_{\sigma_1,\ldots,\sigma_{N+1}}(n) = -\sqrt{a_{\sigma_1,\ldots,\sigma_N}(n)a_{\sigma_1,\ldots,\sigma_N}(n+1)} \times \frac{\sqrt{u_{\sigma_1,\ldots,\sigma_N}(\lambda_N,n)u_{\sigma_1,\ldots,\sigma_N}(\lambda_N,n+2)}}{u_{\sigma_1,\ldots,\sigma_N}(\lambda_N,n+1)},$$

where $f_{\sigma_1,\ldots,\sigma_\ell} = A_{\sigma_1,\ldots,\sigma_N} \cdots A_{\sigma_1} f$. Equation (11.46) follows after a straightforward calculation using (by Lemma 11.11)

(11.53)
$$C_n(U_{\sigma_1,\ldots,\sigma_N},f)C_{n+1}(U_{\sigma_1,\ldots,\sigma_{N+1}}) - C_{n+1}(U_{\sigma_1,\ldots,\sigma_N},f)C_n(U_{\sigma_1,\ldots,\sigma_{N+1}})$$
$$= C_{n+1}(U_{\sigma_1,\ldots,\sigma_N})C_n(U_{\sigma_1,\ldots,\sigma_{N+1}},f).$$

The formula for $a_{\sigma_1,\ldots,\sigma_N}$ is then a simple calculation.

The first formula for $b_{\sigma_1,\ldots,\sigma_N}$ follows after solving the corresponding Jacobi equation $(\tau_{\sigma_1,\ldots,\sigma_N} - \lambda_N)f_{\sigma_1,\ldots,\sigma_N} = 0$ for it. The remaining two follow upon letting $\lambda \to -\infty$ in (11.47) and comparing with (6.27). The rest is straightforward. $\square$

For the sake of completeness we also extend Lemma 11.9. For brevity we assume $\sigma_\ell \in (-1,1)$.

Lemma 11.12. *Suppose H satisfies (H.10.1) and let $H_{\sigma_1,\ldots,\sigma_N}$, $\sigma_\ell \in (-1,1,)$, $1 \leq \ell \leq N$, be constructed as in Theorem 11.10 with*

(11.54)
$$u_{\sigma_\ell}^\ell(n) = \frac{1+\sigma_\ell}{2}f_+(k_\ell,n) + (-1)^{\ell+1}\frac{1-\sigma_\ell}{2}f_-(k_\ell,n).$$

Then the transmission $T_{\sigma_1,\ldots,\sigma_N}(k)$ and reflection coefficients $R_{\sigma_1,\ldots,\sigma_N,\pm}(k)$ of the operator $H_{\sigma_1,\ldots,\sigma_N}$ in terms of the corresponding scattering data $T(k)$, $R_\pm(k)$ of H are given by

$$T_{\sigma_1,\ldots,\sigma_N}(k) = \left(\prod_{\ell=1}^N \frac{1-k\,k_\ell}{k-k_\ell}\right)T(k),$$

(11.55)
$$R_{\sigma_1,\ldots,\sigma_N,\pm}(k) = k^{\pm N}\left(\prod_{\ell=1}^N \frac{k-k_\ell}{1-k\,k_\ell}\right)R_\pm(k),$$

where $k_\ell = \lambda_\ell + \sqrt{\lambda_\ell^2 - 1} \in (-1, 0)$, $1 \leq \ell \leq N$. Moreover, the norming constants $\gamma_{\pm,\sigma_1,\ldots,\sigma_N,j}$ associated with $\lambda_j \in \sigma_{pp}(H_{\sigma_1,\ldots,\sigma_N})$ in terms of $\gamma_{\pm,j}$ corresponding to H read

$$\gamma_{\pm,\sigma_1,\ldots,\sigma_N,j} = \left(\frac{1-\sigma_j}{1+\sigma_j}\right)^{\pm 1} |k_j|^{-2\mp(N-1)} \frac{\prod_{\ell=1}^N |1-k_j k_\ell|}{\prod_{\substack{\ell=1\\ \ell\neq j}}^N |k_j - k_l|} T(k_j), \quad 1 \leq j \leq N,$$

(11.56)
$$\gamma_{\pm,\sigma_1,\ldots,\sigma_N,j} = |k_j|^{\pm N} \prod_{\ell=1}^N \frac{1-k_j k_\ell}{|k_j - k_\ell|} \gamma_{\pm,j}, \quad j \in J.$$

Proof. Observe that

$$u_{\sigma_1,\sigma_2}(\lambda_2, n) = \frac{1+\sigma_2}{2} A_{\sigma_1} f_+(k_2, n) + \frac{1-\sigma_2}{2} A_{\sigma_1} f_-(k_2, n)$$

(11.57)
$$= c\left(\frac{1+\hat{\sigma}_2}{2} f_{+,\sigma_1}(k_2, n) + \frac{1-\hat{\sigma}_2}{2} f_{-,\sigma_1}(k_2, n)\right),$$

where $c > 0$ and $\sigma_2, \hat{\sigma}_2$ are related via

(11.58)
$$\frac{1+\hat{\sigma}_2}{1-\hat{\sigma}_2} = \frac{1}{k_2} \frac{1+\sigma_1}{1-\sigma_1}.$$

The claims now follow from Lemma 11.9 after extending this result by induction. $\square$

11.4. Application of the single commutation method

In this section we want to apply the single commutation method to Jacobi operators associated with quasi-periodic sequences (a, b) as in Section 9.2 (cf. (9.48)). We choose $p_1 = (\lambda_1, \sigma_1 R_{2g+2}^{1/2}(\lambda_1))$ and $\phi(n) = \phi(p_1, n)$. To ensure $\phi(n) > 0$ we require $\pi(p_1) \leq E_0$, $\sigma_1 \in \{\pm 1\}$. Using the representations of a, b, ϕ in terms of theta functions

$$a(n) = -\tilde{a}\sqrt{\frac{\theta(\underline{z}(n+1))\theta(\underline{z}(n-1))}{\theta(\underline{z}(n))^2}},$$

$$b(n) = \tilde{b} + \sum_{j=1}^g c_j(g) \frac{\partial}{\partial w_j} \ln\left(\frac{\theta(\underline{w} + \underline{z}(n))}{\theta(\underline{w} + \underline{z}(n-1))}\right)\bigg|_{\underline{w}=0},$$

(11.59) $\quad \phi(n) = \sqrt{\frac{\theta(\underline{z}_0(n-1))}{\theta(\underline{z}_0(n+1))}} \frac{\theta(\underline{z}(p_1, n+1))}{\theta(\underline{z}(p_1, n))} \exp\left(\int_{p_0}^{p_1} \tau_{\infty_+,\infty_-}\right),$

(where all roots are chosen with positive sign) yields

$$\rho_{o,\sigma_1}(n) = -\sqrt{-\frac{a(n)}{\phi(n)}}$$

$$= -\sqrt{\tilde{a}\frac{\theta(\underline{z}(n+1))}{\theta(\underline{z}(n))} \frac{\theta(\underline{z}(p_1, n))}{\theta(\underline{z}(p_1, n+1))}} \exp\left(-\int_{p_0}^{p_1} \frac{\tau_{\infty_+,\infty_-}}{2}\right),$$

$$\rho_{e,\sigma_1}(n) = \sqrt{-a(n)\phi(n)}$$

(11.60) $\quad = \sqrt{\tilde{a}\frac{\theta(\underline{z}_0(n-1))}{\theta(\underline{z}_0(n))} \frac{\theta(\underline{z}(p_1, n+1))}{\theta(\underline{z}(p_1, n))}} \exp\left(\int_{p_0}^{p_1} \frac{\tau_{\infty_+,\infty_-}}{2}\right).$

11.4. Application of the single commutation method

Next we calculate

$$\phi_{\sigma_1}(n) = \frac{\rho_{o,\sigma_1}(n)}{\rho_{e,\sigma_1}(n+1)}$$

(11.61)
$$= \sqrt{\frac{\theta(\underline{z}(p_1,n))}{\theta(\underline{z}(p_1,n+2))}} \frac{\theta(\underline{z}(n+1))}{\theta(\underline{z}(n))} \exp\left(-\int_{p_0}^{p_1} \tau_{\infty_+,\infty_-}\right).$$

This suggests to define

(11.62) $\quad \underline{z}_{\sigma_1}(p,n) = \underline{\hat{A}}_{p_0}(p) - \underline{\hat{\alpha}}_{p_0}(\mathcal{D}_{\underline{\hat{\mu}}_{\sigma_1}}) + 2n\underline{\hat{A}}_{p_0}(\infty_+) - \underline{\hat{\Xi}}_{p_0},$

associated with a new Dirichlet divisor $\mathcal{D}_{\underline{\hat{\mu}}_{\sigma_1}}$ given by

(11.63) $\quad \underline{\alpha}_{p_0}(\mathcal{D}_{\underline{\hat{\mu}}_{\sigma_1}}) = \underline{\alpha}_{p_0}(\mathcal{D}_{\underline{\hat{\mu}}}) - \underline{A}_{p_0}(p_1) - \underline{A}_{p_0}(\infty_+).$

Lemma 11.8 says that we have $p_1 = \hat{\lambda}_{g+1}^\beta(0)$ and

(11.64) $\quad \hat{\mu}_{\sigma_1,j}(0) = \hat{\lambda}_j^\beta(0), \quad 1 \le j \le g,$

where $\beta = \phi(0)$. This can also be obtained directly by comparing (11.63) and (9.11).

Upon using

(11.65) $\quad \underline{z}(p_1,n) = \underline{z}_{\sigma_1}(\infty_+, n-1), \quad \underline{z}(\infty_+,n) = \underline{z}_{\sigma_1}(p_1^*,n),$

we obtain (again all roots are chosen with positive sign)

$$a_{\sigma_1}(n) = -\tilde{a}\sqrt{\frac{\theta(\underline{z}_{\sigma_1}(n+1))\theta(\underline{z}_{\sigma_1}(n-1))}{\theta(\underline{z}_{\sigma_1}(n))^2}},$$

$$b_{\sigma_1}(n) = \tilde{b} + \sum_{j=1}^{g} c_j(g) \frac{\partial}{\partial w_j} \ln\left(\frac{\theta(\underline{w}+\underline{z}_{\sigma_1}(n))}{\theta(\underline{w}+\underline{z}_{\sigma_1}(n-1))}\right)\bigg|_{\underline{w}=0},$$

(11.66) $\quad \phi_{\sigma_1}(n) = \sqrt{\frac{\theta(\underline{z}_{\sigma_1}(n-1))}{\theta(\underline{z}_{\sigma_1}(n+1))}} \frac{\theta(\underline{z}_{\sigma_1}(p_1^*,n+1))}{\theta(\underline{z}_{\sigma_1}(p_1^*,n))} \exp\left(\int_{p_0}^{p_1^*} \tau_{\infty_+,\infty_-}\right).$

Remark 11.13. If we choose $\pi(p_1) \ge E_{2g+2}$ instead of $\pi(p_1) \le E_1$ the operator H_{σ_1} is well-defined but the sequences $\rho_{o,\sigma_1}, \rho_{e,\sigma_1}$ become purely imaginary.

Now let us choose $\pi(p_1) \in [E_{2j-1}, E_{2j}]$. Then neither $\rho_{o,\sigma_1}, \rho_{e,\sigma_1}$ nor H_{σ_1} are well-defined. But if we ignore this and make (formally) a second commutation at a point $p_2 = (\lambda_2, \sigma_2 R_{2g+2}^{1/2}(\lambda_2))$ with $\pi(p_2) \in [E_{2j-1}, E_{2j}]$, we get an operator H_2 associated with a Dirichlet divisor $\mathcal{D}_{\underline{\hat{\mu}}_{\sigma_1,\sigma_2}}$ given by

(11.67) $\quad \begin{aligned} \underline{\alpha}_{p_0}(\mathcal{D}_{\underline{\hat{\mu}}_{\sigma_1,\sigma_2}}) &= \underline{\alpha}_{p_0}(\mathcal{D}_{\underline{\hat{\mu}}_{\sigma_1}}) - \underline{A}_{p_0}(p_1) - \underline{A}_{p_0}(\infty_+) \\ &= \underline{\alpha}_{p_0}(\mathcal{D}_{\underline{\hat{\mu}}}) - \underline{A}_{p_0}(p_1) - \underline{A}_{p_0}(p_2) - 2\underline{A}_{p_0}(\infty_+). \end{aligned}$

Since $2[i\,\text{Im}(\underline{A}_{p_0}(p_1))] \in J(M)$, H_2 is a well-defined operator. Moreover, choosing $p_1 = \hat{\mu}_j$ in the first step implies that the Dirichlet eigenvalue $\hat{\mu}_j$ is formally replaced by one at ∞_-. The second step corrects this by moving the Dirichlet eigenvalue to the point $p_2^* = (\mu, \sigma R_{2g+2}^{1/2}(\mu)))$ and the factor $2\underline{A}_{p_0}(\infty_+)$ shifts the sequences by one. We will undo the shift and denote the corresponding sequences by $a_{(\mu,\sigma)}$, $b_{(\mu,\sigma)}$. From Theorem 11.10 we know that they are equivalently given by

$$a_{(\mu,\sigma)}(n+1) = -\sqrt{a(n)a(n+2)}$$
$$\times \sqrt{\frac{C_n(\psi_{\sigma_j}(\mu_j), \psi_{-\sigma}(\mu))C_{n+2}(\psi_{\sigma_j}(\mu_j), \psi_{-\sigma}(\mu))}{C_{n+1}(\psi_{\sigma_j}(\mu_j), \psi_{-\sigma}(\mu))^2}},$$
(11.68) $\quad b_{(\mu,\sigma)}(n+1) = b(n) - \partial^* \frac{\psi_{\sigma_j}(\mu_j, n)\psi_{-\sigma}(\mu, n+1)}{C_n(\psi_{\sigma_j}(\mu_j), \psi_{-\sigma}(\mu))},$

where the $n+1$ on the left-hand-side takes the aforementioned shift of reference point into account. Thus, applying this procedure g times we can replace all Dirichlet eigenvalues (μ_j, σ_j) by $(\tilde{\mu}_j, \tilde{\sigma}_j)$ proving that the sequences associated with $\tilde{\mu}_j$, $1 \le j \le g$, are given by

$$a_{(\tilde{\mu}_1, \tilde{\sigma}_1), \ldots, (\tilde{\mu}_g, \tilde{\sigma}_g)}(n) = \sqrt{a(n-g)a(n+g)}$$
$$\times \frac{\sqrt{C_{n-g}(\Psi_{(\tilde{\mu}_1, \tilde{\sigma}_1), \ldots, (\tilde{\mu}_g, \tilde{\sigma}_g)})C_{n-g+2}(\Psi_{(\tilde{\mu}_1, \tilde{\sigma}_1), \ldots, (\tilde{\mu}_g, \tilde{\sigma}_g)})}}{C_{n-g+1}(\Psi_{(\tilde{\mu}_1, \tilde{\sigma}_1), \ldots, (\tilde{\mu}_g, \tilde{\sigma}_g)})},$$
$$b_{(\tilde{\mu}_1, \tilde{\sigma}_1), \ldots, (\tilde{\mu}_g, \tilde{\sigma}_g)}(n) = b(n) - \partial^* a(n) \frac{\tilde{D}_{n-g+1}(\Psi_{(\tilde{\mu}_1, \tilde{\sigma}_1), \ldots, (\tilde{\mu}_g, \tilde{\sigma}_g)})}{C_{n-g+1}(\Psi_{(\tilde{\mu}_1, \tilde{\sigma}_1), \ldots, (\tilde{\mu}_g, \tilde{\sigma}_g)})},$
(11.69)

where $(\Psi_{(\tilde{\mu}_1, \tilde{\sigma}_1), \ldots, (\tilde{\mu}_g, \tilde{\sigma}_g)}) = (\psi_{\sigma_1}(\mu_1), \psi_{-\tilde{\sigma}_1}(\tilde{\mu}_1), \ldots, \psi_{\sigma_g}(\mu_g), \psi_{-\tilde{\sigma}_g}(\tilde{\mu}_g))$. We will generalize this method to arbitrary operators in Section 11.8.

11.5. A formal second commutation

Our next aim is to remove the condition that H is bounded from below. To do this we apply the single commutation method twice which will produce a method for inserting eigenvalues into arbitrary spectral gaps of a given Jacobi operator (not only into the lowest). This method is hence called the double commutation method.

Let H be a given operator ($a < 0$ for the moment) and let $u_\pm(\lambda_1, .)$, $\lambda_1 < \sigma(H)$, be a positive solution. Define

$$Af(n) = \rho_o(n)f(n+1) + \rho_e(n)f(n),$$
(11.70)
$$A^*f(n) = \rho_o(n-1)f(n-1) + \rho_e(n)f(n),$$

where

(11.71) $\quad \rho_o(n) = \sqrt{-\frac{a(n)u_\pm(\lambda_1, n)}{u_\pm(\lambda_1, n+1)}}, \quad \rho_e(n) = -\sqrt{-\frac{a(n)u_\pm(\lambda_1, n+1)}{u_\pm(\lambda_1, n)}}.$

Then we have $H = A^*A + \lambda_1$ and we can define a second operator

(11.72) $\qquad\qquad \tilde{H} = AA^* + \lambda_1.$

Fix $\gamma_\pm > 0$. Then by (11.18) and (11.29)

(11.73) $\qquad v(\lambda_1, n) = \frac{c_{\gamma_\pm}(\lambda_1, n)}{\sqrt{-a(n)u_\pm(\lambda_1, n)u_\pm(\lambda_1, n+1)}}$

11.5. A formal second commutation

solves $\tilde{H}v = \lambda_1 v$, where

$$(11.74) \qquad c_{\gamma_\pm}(\lambda_1, n) = \frac{1}{\gamma_\pm} \mp \sum_{j=\pm\infty}^{n}{}^* u_\pm(\lambda_1, j)^2.$$

Now define

$$(11.75) \qquad \begin{aligned} \rho_{o,\gamma_\pm}(n) &= \rho_e(n+1)\sqrt{\frac{c_{\gamma_\pm}(\lambda_1, n)}{c_{\gamma_\pm}(\lambda_1, n+1)}}, \\ \rho_{e,\gamma_\pm}(n) &= \rho_o(n)\sqrt{\frac{c_{\gamma_\pm}(\lambda_1, n+1)}{c_{\gamma_\pm}(\lambda_1, n)}} \end{aligned}$$

and introduce corresponding operators $A_{\gamma_\pm}$, $A^*_{\gamma_\pm}$ on $\ell^2(\mathbb{Z})$ by

$$(11.76) \qquad \begin{aligned} (A_{\gamma_\pm} f)(n) &= \rho_{o,\gamma_\pm}(n) f(n+1) + \rho_{e,\gamma_\pm}(n) f(n), \\ (A^*_{\gamma_\pm} f)(n) &= \rho_{o,\gamma_\pm}(n-1) f(n-1) + \rho_{e,\gamma_\pm}(n) f(n). \end{aligned}$$

A simple calculation shows that $A^*_{\gamma_\pm} A_{\gamma_\pm} = A_{\sigma_1} A^*_{\sigma_1}$ and hence

$$(11.77) \qquad H_{\sigma_1} = A^*_{\gamma_\pm} A_{\gamma_\pm} + \lambda_1.$$

Performing a second commutation yields the doubly commuted operator

$$(11.78) \qquad H_{\gamma_\pm} = A_{\gamma_\pm} A^*_{\gamma_\pm} + \lambda_1.$$

Explicitly, one verifies

$$(11.79) \qquad (H_{\gamma_\pm} f)(n) = a_{\gamma_\pm}(n) f(n+1) + a_{\gamma_\pm}(n-1) f(n-1) - b_{\gamma_\pm}(n) f(n),$$

with

$$(11.80) \qquad \begin{aligned} a_{\gamma_\pm}(n) &= a(n+1) \frac{\sqrt{c_{\gamma_\pm}(\lambda_1, n) c_{\gamma_\pm}(\lambda_1, n+2)}}{c_{\gamma_\pm}(\lambda_1, n+1)}, \\ b_{\gamma_\pm}(n) &= b(n+1) \mp \left(\frac{a(n) u_\pm(\lambda_1, n) u_\pm(\lambda_1, n+1)}{c_{\gamma_\pm}(\lambda_1, n)} \right. \\ &\qquad \left. - \frac{a(n+1) u_\pm(\lambda_1, n+1) u_\pm(\lambda_1, n+2)}{c_{\gamma_\pm}(\lambda_1, n+1)} \right). \end{aligned}$$

Note that

$$(11.81) \qquad u_{\gamma_\pm}(\lambda_1, n) = \frac{u_\pm(\lambda_1, n+1)}{\sqrt{c_{\gamma_\pm}(\lambda_1, n) c_{\gamma_\pm}(\lambda_1, n+1)}}$$

is a solution of $H_{\gamma_\pm} u = \lambda_1 u$ fulfilling $u_{\gamma_\pm}(\lambda_1) \in \ell^2(\pm\infty, n_0)$ and because of

$$(11.82) \qquad \sum_{n\in\mathbb{Z}} |u_{\gamma_\pm}(\lambda_1, n)|^2 = \gamma_\pm \pm \lim_{n\to\mp\infty} \frac{1}{c_{\gamma_\pm}(\lambda_1, n)},$$

we have $u_{\gamma_\pm}(\lambda_1) \in \ell^2(\mathbb{Z})$. Furthermore, if u is a solution of $Hu = zu$ (for arbitrary $z \in \mathbb{C}$), then

$$(11.83) \qquad \begin{aligned} u_{\gamma_\pm}(z, n) &= \frac{1}{z - \lambda_1} A A_{\gamma_\pm} u(z, n) \\ &= \frac{c_{\gamma_\pm}(\lambda_1, n) u(z, n+1) \pm \frac{1}{z-\lambda_1} u_\pm(z, n+1) W_n(u_\pm(\lambda_1), u(z))}{\sqrt{c_{\gamma_\pm}(\lambda_1, n) c_{\gamma_\pm}(\lambda_1, n+1)}} \end{aligned}$$

is a solution of $H_{\gamma_\pm} u = zu$.

Now observe that $H_{\gamma_\pm}$ remains well-defined even if $u_\pm$ is no longer positive. This applies, in particular, in the case where $u_\pm(\lambda_1)$ has zeros and hence all intermediate operators $A_{\sigma_1}, A_{\gamma_\pm}, \tilde{H}$, etc., become ill-defined. Thus, to define $H_{\gamma_\pm}$, it suffices to assume the existence of a solution $u_\pm(\lambda_1)$ which is square summable near $\pm\infty$. This condition is much less restrictive than the existence of a positive solution (e.g., the existence of a spectral gap for H around λ_1 is sufficient in this context).

One expects that formulas analogous to (11.19) will carry over to this more general setup. That this is actually the case will be shown in Theorem 11.16. Hence the double commutation method (contrary to the single commutation method) enables one to insert eigenvalues not only below the spectrum of H but into arbitrary spectral gaps of H.

11.6. The double commutation method

In this section we provide a complete characterization of the **double commutation method** for Jacobi operators. Since Theorem 11.1 only applies when $\lambda_1 < \sigma(H)$, we need a different approach. We start with a linear transformation which turns out to be unitary when restricted to proper subspaces of $\ell^2(\mathbb{Z})$. We use this transformation to construct an operator H_{γ_1} from a given background operator H. This operator H_{γ_1} will be the doubly commuted operator of H as discussed in the previous section. The following abbreviation will simplify our notation considerably

$$(11.84) \qquad \langle f, g \rangle_m^n = \sum_{j=m}^n \overline{f(j)} g(j).$$

Let $u \in \ell_-^2(\mathbb{Z})$ be a given real-valued sequence. Choose a constant $\gamma \in [-\|u\|^{-2}, \infty)$ or $\gamma = \infty$ and define

$$(11.85) \qquad c_\gamma(n) = \frac{1}{\gamma} + \langle u, u \rangle_{-\infty}^n, \quad \gamma \neq 0$$

(setting $\infty^{-1} = 0$). Consider the following linear transformation

$$(11.86) \qquad \begin{array}{rcl} U_\gamma : \ell^2(\mathbb{Z}) & \to & \ell(\mathbb{Z}) \\ f(n) & \mapsto & \sqrt{\frac{c_\gamma(n)}{c_\gamma(n-1)}} f(n) - u_\gamma(n) \langle u, f \rangle_{-\infty}^n \end{array},$$

($U_0 = \mathbb{1}$), where

$$(11.87) \qquad u_\gamma(n) = \frac{u(n)}{\sqrt{c_\gamma(n-1) c_\gamma(n)}},$$

($u_0 = 0$). We note that U_γ can be defined on $\ell_-^2(\mathbb{Z})$ and $U_\gamma u = \gamma^{-1} u_\gamma$, $\gamma \neq 0$. Furthermore,

$$(11.88) \qquad u_\gamma(n)^2 = \partial^* \frac{1}{c_\gamma(n)},$$

and hence

$$(11.89) \qquad \|u_\gamma\|^2 = \begin{cases} \gamma, & u \notin \ell^2(\mathbb{Z}) \\ \frac{\gamma^2 \|u\|^2}{1 + \gamma \|u\|^2}, & u \in \ell^2(\mathbb{Z}) \end{cases}$$

11.6. The double commutation method

implying $u_\gamma \in \ell^2(\mathbb{Z})$ if $-\|u\|^{-2} < \gamma < \infty$. If $\gamma = -\|u\|^{-2}$, $\gamma = \infty$ we only have $u_\gamma \in \ell^2_-(\mathbb{Z})$, $u_\gamma \in \ell^2_+(\mathbb{Z})$, respectively. In addition, we remark that for $f_\gamma = U_\gamma f$ we have

$$u_\gamma(n) f_\gamma(n) = -\partial^* \frac{\langle u, f \rangle^n_{-\infty}}{c_\gamma(n)},$$

(11.90)
$$|f_\gamma(n)|^2 = |f(n)|^2 + \partial^* \frac{|\langle u, f \rangle^n_{-\infty}|^2}{c_\gamma(n)}.$$

Summing over n and taking limits (if $\gamma = \infty$ use Cauchy-Schwarz) shows

$$\langle u_\gamma, f_\gamma \rangle^n_{-\infty} = \begin{cases} c_\gamma(n)^{-1} \langle u, f \rangle^n_{-\infty}, & \gamma \in \mathbb{R} \\ \frac{\langle u, f \rangle}{\|u\|^2} - c_\infty(n)^{-1} \langle u, f \rangle^\infty_{n+1}, & \gamma = \infty \end{cases},$$

(11.91)
$$\langle f_\gamma, f_\gamma \rangle^n_{-\infty} = \langle f, f \rangle^n_{-\infty} - \frac{|\langle u, f \rangle^n_{-\infty}|^2}{c_\gamma(n)}.$$

Clearly, the last equation implies $U_\gamma : \ell^2(\mathbb{Z}) \to \ell^2(\mathbb{Z})$. In addition, we remark that this also shows $U_\gamma : \ell^2_-(\mathbb{Z}) \to \ell^2_-(\mathbb{Z})$.

Denote by P, P_γ the orthogonal projections onto the one-dimensional subspaces of $\ell^2(\mathbb{Z})$ spanned by u, u_γ (set $P, P_\gamma = 0$ if $u, u_\gamma \notin \ell^2(\mathbb{Z})$), respectively. Define

(11.92)
$$U_\gamma^{-1} : \ell^2(\mathbb{Z}) \to \ell(\mathbb{Z})$$
$$g(n) \mapsto \begin{cases} \sqrt{\frac{c_\gamma(n-1)}{c_\gamma(n)}} g(n) + u(n) \langle u_\gamma, g \rangle^n_{-\infty}, & \gamma \in \mathbb{R} \\ \sqrt{\frac{c_\infty(n-1)}{c_\infty(n)}} g(n) - u(n) \langle u_\infty, g \rangle^\infty_{n+1}, & \gamma = \infty \end{cases}$$

and note

(11.93)
$$c_\gamma^{-1}(n) = \begin{cases} \gamma - \langle u_\gamma, u_\gamma \rangle^n_{-\infty}, & \gamma \in \mathbb{R} \\ \|u\|^{-2} + \langle u_\infty, u_\infty \rangle^\infty_{n+1}, & \gamma = \infty \end{cases}.$$

As before one can show $U_\gamma^{-1} : (\mathbb{1} - P_\gamma) \ell^2(\mathbb{Z}) \to \ell^2(\mathbb{Z})$ and one verifies

(11.94)
$$U_\gamma U_\gamma^{-1} = \mathbb{1}, \ U_\gamma^{-1} U_\gamma = \mathbb{1}, \quad \gamma \in \mathbb{R},$$
$$U_\infty U_\infty^{-1} = \mathbb{1}, \ U_\infty^{-1} U_\infty = \mathbb{1} - P, \quad \gamma = \infty.$$

If $P = 0$, $\gamma \in (-\|u\|^{-2}, \infty)$, then $U_\gamma U_\gamma^{-1} = \mathbb{1}$ should be understood on $(\mathbb{1} - P_\gamma) \ell^2(\mathbb{Z})$ since $U_\gamma^{-1} u_\gamma \notin \ell^2(\mathbb{Z})$ by

(11.95)
$$U_\gamma^{-1} u_\gamma = \begin{cases} \gamma u, & \gamma \in \mathbb{R} \\ \|u\|^{-2} u, & \gamma = \infty \end{cases}.$$

Summarizing,

Lemma 11.14. *The operator U_γ is unitary from $(\mathbb{1} - P)\ell^2(\mathbb{Z})$ onto $(\mathbb{1} - P_\gamma)\ell^2(\mathbb{Z})$ with inverse U_γ^{-1}. If $P, P_\gamma \neq 0$, then U_γ can be extended to a unitary transformation $\tilde{U}_\gamma$ on $\ell^2(\mathbb{Z})$ by*

(11.96)
$$\tilde{U}_\gamma = U_\gamma(\mathbb{1} - P) + \sqrt{1 + \gamma \|u\|^2}\, U_\gamma P.$$

Proof. The first identity in (11.91) shows that U_γ maps $(\mathbb{1} - P)\ell^2(\mathbb{Z})$ onto $(\mathbb{1} - P_\gamma)\ell^2(\mathbb{Z})$. Unitarity follows from (11.91) and

(11.97)
$$\lim_{n \to \infty} \frac{|\langle u, f \rangle^n_{-\infty}|^2}{\langle u, u \rangle^n_{-\infty}} = 0$$

for any $f \in \ell^2(\mathbb{Z})$ if $u \notin \ell^2(\mathbb{Z})$. In fact, suppose $\|f\| = 1$, pick m and $n > m$ so large that $\langle f, f \rangle_m^\infty \leq \varepsilon/2$ and $\langle u, u \rangle_{-\infty}^{m-1}/\langle u, u \rangle_{-\infty}^n \leq \varepsilon/2$. Splitting up the sum in the numerator and applying Cauchy's inequality then shows that the limit of (11.97) is smaller than ε. The rest follows from (11.89). □

We remark that (11.91) plus the polarization identity implies

$$(11.98) \qquad \langle f_\gamma, g_\gamma \rangle_{-\infty}^n = \langle f, g \rangle_{-\infty}^n - \frac{\langle f, u \rangle_{-\infty}^n \langle u, g \rangle_{-\infty}^n}{c_\gamma(n)},$$

where $f_\gamma = U_\gamma f$, $g_\gamma = U_\gamma g$.

Next, we take a Jacobi operator associated with bounded sequences a, b as usual. We assume (λ_1, γ_1) to be of the following kind.

Hypothesis H. 11.15. Suppose (λ, γ) satisfies the following conditions.
(i). $u_-(\lambda, n)$ exists.
(ii). $\gamma \in [-\|u_-(\lambda)\|^{-2}, \infty) \cup \{\infty\}$.

Note that by Lemma 2.2 a sufficient condition for $u_-(\lambda_1, n)$ to exist is $\lambda \in \mathbb{R} \backslash \sigma_{ess}(H_-)$.

We now use Lemma 11.14 with $u(n) = u_-(\lambda_1, n)$, $\gamma = \gamma_1$, $c_{\gamma_1}(\lambda_1, n) = c_\gamma(n)$ to prove

Theorem 11.16. Suppose (H.11.15) and let H_{γ_1} be the operator associated with

$$(11.99) \qquad (\tau_{\gamma_1} f)(n) = a_{\gamma_1}(n) f(n+1) + a_{\gamma_1}(n-1) f(n-1) + b_{\gamma_1}(n) f(n),$$

where

$$(11.100) \qquad \begin{aligned} a_{\gamma_1}(n) &= a(n) \frac{\sqrt{c_{\gamma_1}(\lambda_1, n-1) c_{\gamma_1}(\lambda_1, n+1)}}{c_{\gamma_1}(\lambda_1, n)}, \\ b_{\gamma_1}(n) &= b(n) - \partial^* \frac{a(n) u_-(\lambda_1, n) u_-(\lambda_1, n+1)}{c_{\gamma_1}(\lambda_1, n)}. \end{aligned}$$

Then

$$(11.101) \qquad H_{\gamma_1}(\mathbb{1} - P_{\gamma_1}(\lambda_1)) = U_{\gamma_1} H U_{\gamma_1}^{-1}(\mathbb{1} - P_{\gamma_1}(\lambda_1))$$

and $\tau_{\gamma_1} u_{\gamma_1, -}(\lambda_1) = \lambda_1 u_{\gamma_1, -}(\lambda_1)$, where

$$(11.102) \qquad u_{\gamma_1, -}(\lambda_1, n) = \frac{u_-(\lambda_1, n)}{\sqrt{c_{\gamma_1}(\lambda_1, n-1) c_{\gamma_1}(\lambda_1, n)}}.$$

Proof. Let f be a sequence which is square summable near $-\infty$ implying that τf is also square summable near $-\infty$. Then a straightforward calculation shows

$$(11.103) \qquad \tau_{\gamma_1}(U_{\gamma_1} f) = U_{\gamma_1}(\tau f).$$

The rest follows easily. □

Corollary 11.17. Suppose $u_-(\lambda_1) \notin \ell^2(\mathbb{Z})$.
(i). If $\gamma_1 > 0$, then H and $(\mathbb{1} - P_{\gamma_1}(\lambda_1)) H_{\gamma_1}$ are unitarily equivalent. Moreover, H_{γ_1} has the additional eigenvalue λ_1 with eigenfunction $u_{\gamma_1, -}(\lambda_1)$.
(ii). If $\gamma_1 = \infty$, then H and H_{γ_1} are unitarily equivalent.
    Suppose $u_-(\lambda_1) \in \ell^2(\mathbb{Z})$ (i.e., λ_1 is an eigenvalue of H).
(i). If $\gamma_1 \in (-\|u_-(\lambda_1)\|^{-2}, \infty)$, then H and H_{γ_1} are unitarily equivalent (using $\tilde{U}_{\gamma_1}$ from Lemma 11.14).

11.6. The double commutation method

(ii). If $\gamma_1 = -\|u_-(\lambda_1)\|^{-2}, \infty$, then $(\mathbb{1} - P(\lambda_1))H$, H_{γ_1} are unitarily equivalent, that is, the eigenvalue λ_1 is removed.

Remark 11.18. (i). By choosing $\lambda_1 \in \sigma_{ac}(H) \cup \sigma_{sc}(H)$ (provided the continuous spectrum is not empty and a solution satisfying (H.11.15) exists) we can use the double commutation method to construct operators with eigenvalues embedded in the continuous spectrum.
(ii). Multiplying γ_1 by some constant $c^2 > 0$ amounts to the same as multiplying $u_-(\lambda_1, .)$ by c^{-1}.

The previous theorem tells us only how to transfer solutions of $\tau u = zu$ into solutions of $\tau_{\gamma_1} v = zv$ if u is square summable near $-\infty$. The following lemma treats the general case.

Lemma 11.19. *The sequence*

$$(11.104) \quad u_{\gamma_1}(z,n) = \frac{c_{\gamma_1}(\lambda_1,n)u(z,n) - \frac{1}{z-\lambda_1}u_-(\lambda_1,n)W_n(u_-(\lambda_1),u(z))}{\sqrt{c_{\gamma_1}(\lambda_1,n-1)c_{\gamma_1}(\lambda_1,n)}},$$

$z \in \mathbb{C}\backslash\{\lambda_1\}$ solves $\tau_{\gamma_1} u = zu$ if $u(z)$ solves $\tau u = zu$. If $u(z)$ is square summable near $-\infty$, we have $u_{\gamma_1}(z,n) = (U_{\gamma_1} u)(z,n)$, justifying our notation. Furthermore, we note

$$(11.105) \quad |u_{\gamma_1}(z,n)|^2 = |u(z,n)|^2 + \frac{1}{|z-\lambda_1|^2}\partial^* \frac{|W_n(u_-(\lambda_1),u(z))|^2}{c_{\gamma_1}(\lambda_1,n)},$$

and

$$(11.106) \quad W_{\gamma_1,n}(u_{-,\gamma_1}(\lambda_1), u_{\gamma_1}(z)) = \frac{W_n(u_-(\lambda_1),u(z))}{c_{\gamma_1}(\lambda_1,n)},$$

where $W_{\gamma_1,n}(u,v) = a_{\gamma_1}(n)(u(n)v(n+1) - u(n+1)v(n))$. Hence u_{γ_1} is square summable near ∞ if u is. If $\hat{u}_{\gamma_1}(\hat{z})$ is constructed analogously, then

$$W_{\gamma_1,n}(u_{\gamma_1}(z), \hat{u}_{\gamma_1}(\hat{z})) = W_n(u(z), \hat{u}(\hat{z})) + \frac{1}{c_{\gamma_1}(\lambda_1,n)}\frac{z-\hat{z}}{(z-\lambda_1)(\hat{z}-\lambda_1)} \times$$
$$(11.107) \qquad W_n(u_-(\lambda_1),u(z))W_n(u_-(\lambda_1),\hat{u}(\hat{z})).$$

Proof. All facts are tedious but straightforward calculations. $\square$

As a consequence we get.

Corollary 11.20. *The Green function $G_{\gamma_1}(z,n,m)$ of H_{γ_1} is given by $(n > m)$*

$$(11.108) \quad G_{\gamma_1}(z,n,m) = \frac{u_{+,\gamma_1}(z,n)u_{-,\gamma_1}(z,m)}{W(u_-(z),u_+(z))},$$

where

$$(11.109) \quad u_{\pm,\gamma_1}(z,n) = \frac{c_{\gamma_1}(\lambda_1,n)u_\pm(z,n) - \frac{1}{z-\lambda_1}u_-(\lambda_1,n)W_n(u_-(\lambda_1),u_\pm(z))}{\sqrt{c_{\gamma_1}(\lambda_1,n-1)c_{\gamma_1}(\lambda_1,n)}}.$$

Next, we turn to Weyl m-functions. Without loss of generality we assume $u_-(\lambda_1, n) = s_\beta(\lambda_1, n, 0)$, that is,

$$(11.110) \quad u_-(\lambda_1, 0) = -\sin(\alpha), \quad u_-(\lambda_1, 0) = \cos(\alpha), \quad \alpha \in [0,\pi)$$

(change γ_1 if necessary — see Remark 11.18 (ii)).

Theorem 11.21. Let $\tilde{m}^\beta_\pm(z)$, $\tilde{m}^{\tilde\beta}_{\gamma_1,\pm}(z)$ denote the Weyl $\tilde{m}$-functions of H, H_{γ_1} respectively. Then we have

(11.111)
$$\tilde{m}^{\tilde\beta}_{\gamma_1,\pm}(z) = \frac{c_{\gamma_1}(\lambda_1,0)}{c_{\gamma_1}(\lambda_1,-1)}\frac{1+\beta^2}{1+\tilde\beta^2}\left(\tilde{m}^\beta_\pm(z) \mp \frac{c_{\gamma_1}(\lambda_1,0)^{-1}}{z-\lambda_1} \pm \frac{\beta c_{\gamma_1}(\lambda_1,1)^{-1}}{a(0)(1+\tilde\beta^2)}\right),$$

where $\beta = \cot(\alpha)$ and

(11.112)
$$\tilde\beta = \sqrt{\frac{c_{\gamma_1}(\lambda_1,-1)}{c_{\gamma_1}(\lambda_1,1)}}\beta.$$

Proof. Consider the sequences

(11.113)
$$\frac{\sqrt{c_{\gamma_1}(\lambda_1,-1)c_{\gamma_1}(\lambda_1,1)}}{c_{\gamma_1}(\lambda_1,0)}s_{\beta,\gamma_1}(z,n),$$
$$\frac{c_{\gamma_1}(\lambda_1,1)}{c_{\gamma_1}(\lambda_1,0)}\frac{1+\tilde\beta^2}{1+\beta^2}\Big(c_{\beta,\gamma_1}(z,n) - \delta s_{\beta,\gamma_1}(z,n)\Big),$$

(where δ is defined in (11.116) below) constructed from the fundamental system $c_\beta(z,n)$, $s_\beta(z,n)$ for τ (cf. (2.71)) as in Lemma 11.19. They form a fundamental system for τ_{γ_1} corresponding to the initial conditions associated with $\tilde\beta$. Now use (11.106) to evaluate (2.73). $\square$

As a straightforward consequence we obtain the following lemma.

Corollary 11.22. *(i).* The operator H_{γ_1} is a reflectionless if and only if H is (cf. Lemma 8.1).
(ii). Near $z = \lambda_1$ we have

(11.114)
$$\tilde{m}^{\tilde\beta}_{\gamma_1,+}(z) = -\frac{\gamma^\beta_+(\lambda_1) + c_{\gamma_1}(\lambda_1,0)^{-1}}{z-\lambda_1} + O(z-\lambda_1)^0,$$
$$\tilde{m}^{\tilde\beta}_{\gamma_1,-}(z) = -\frac{(c_{\gamma_1}(\lambda_1,0)(\gamma_1 c_{\gamma_1}(\lambda_1,0)-1))^{-1}}{z-\lambda_1} + O(z-\lambda_1)^0,$$

where $\gamma^\beta_+(\lambda_1) = (\langle u_-(\lambda_1), u_-(\lambda_1)\rangle_1^\infty)^{-1} \geq 0$ is the norming constant of H^β_+ at λ_1.

As a direct consequence of Theorem 11.21 we infer

Theorem 11.23. Given H and H_{γ_1} the respective Weyl M-matrices $M^\beta(z)$ and $M^{\tilde\beta}_{\gamma_1}(z)$ are related by

(11.115)
$$M^{\tilde\beta}_{\gamma_1,0,0}(z) = \frac{c_{\gamma_1}(\lambda_1,0)}{c_{\gamma_1}(\lambda_1,1)}\frac{1+\beta^2}{1+\tilde\beta^2}M^\beta_{0,0}(z),$$
$$M^{\tilde\beta}_{\gamma_1,0,1}(z) = \frac{c_{\gamma_1}(\lambda_1,0)}{\sqrt{c_{\gamma_1}(\lambda_1,-1)c_{\gamma_1}(\lambda_1,1)}}\Big(M^\beta_{0,1}(z) + \delta M^\beta_{0,0}(z)\Big),$$
$$M^{\tilde\beta}_{\gamma_1,1,1}(z) = \frac{c_{\gamma_1}(\lambda_1,0)}{c_{\gamma_1}(\lambda_1,-1)}\frac{1+\tilde\beta^2}{1+\beta^2}\Big(M^\beta_{1,1}(z) + 2\delta M^\beta_{0,1}(z)$$
$$+ \delta^2 M^\beta_{0,0}(z)\Big).$$

11.6. The double commutation method

where

(11.116) $$\delta = \frac{a(0)c_{\gamma_1}(\lambda_1, 0)^{-1}}{z - \lambda_1} - \frac{\tilde{\beta} c_{\gamma_1}(\lambda_1, 1)^{-1}}{1 + \tilde{\beta}^2}.$$

The quantities $M_{\gamma_1,0,0}^{\tilde{\beta}}(z)$ and $M_{\gamma_1,0,1}^{\tilde{\beta}}(z)$ are both holomorphic near $z = \lambda_1$. Moreover, $M_{\gamma_1,1,1}^{\tilde{\beta}}(z)$ has a simple pole with residue $-(\gamma_1 c_{\gamma_1}(\lambda_1, 0))^{-1}$ at $z = \lambda_1$.

Using the previous corollary plus weak convergence of $\pi^{-1}\mathrm{Im}(F(\lambda + i\varepsilon))d\lambda$ to the corresponding spectral measure $d\rho(\lambda)$ as $\varepsilon \downarrow 0$ (cf. Lemma B.3) implies

Corollary 11.24. *The matrix measures $d\rho^\beta$, $d\rho_{\gamma_1}^{\tilde{\beta}}$ corresponding to $M^\beta(z)$, $M_{\gamma_1}^{\tilde{\beta}}(z)$ are related by*

$$d\rho_{\gamma_1,0,0}^{\tilde{\beta}}(\lambda) = \frac{c_{\gamma_1}(\lambda_1, 0)}{c_{\gamma_1}(\lambda_1, -1)} \frac{1 + \beta^2}{1 + \tilde{\beta}^2} d\rho_{0,0}^\beta(\lambda),$$

$$d\rho_{\gamma_1,0,1}^{\tilde{\beta}}(\lambda) = \frac{c_{\gamma_1}(\lambda_1, 0)}{\sqrt{c_{\gamma_1}(\lambda_1, -1)c_{\gamma_1}(\lambda_1, 1)}}\left(d\rho_{0,1}^\beta(\lambda) + \delta d\rho_{0,0}^\beta(\lambda)\right),$$

$$d\rho_{\gamma_1,1,1}^{\tilde{\beta}}(\lambda) = \frac{c_{\gamma_1}(\lambda_1, 0)}{c_{\gamma_1}(\lambda_1, -1)} \frac{1 + \beta^2}{1 + \tilde{\beta}^2}\left(d\rho_{1,1}^\beta(\lambda) + 2\delta d\rho_{0,1}^\beta(\lambda)\right.$$

(11.117) $$\left. + \delta^2 d\rho_{0,0}^\beta(\lambda)\right) + (\gamma_1 c_{\gamma_1}(\lambda_1, 0))^{-1} d\Theta(\lambda - \lambda_1).$$

Equivalently

(11.118) $$d\rho_{\gamma_1}^{\tilde{\beta}}(\lambda) = U d\rho^\beta(\lambda) U^T + \frac{1}{\gamma_1 c_{\gamma_1}(\lambda_1, 0)}\begin{pmatrix} 0 & 0 \\ 0 & 1 \end{pmatrix},$$

where

(11.119) $$U = \sqrt{\frac{c_{\gamma_1}(\lambda_1, 0)}{c_{\gamma_1}(\lambda_1, -1)} \frac{1 + \tilde{\beta}^2}{1 + \beta^2}} \begin{pmatrix} \frac{1+\tilde{\beta}^2}{1+\beta^2}\sqrt{\frac{c_{\gamma_1}(\lambda_1, -1)}{c_{\gamma_1}(\lambda_1, 1)}} & 0 \\ \delta & 1 \end{pmatrix}.$$

Remark 11.25. *We can interchange the role of $-\infty$ and ∞ in this section by substituting $-\infty \to \infty$, $\langle .,..\rangle_{-\infty}^n \to \langle .,..\rangle_{n+1}^\infty$, $u_- \to u_+$.*

Our next aim is to show how the scattering data of the operators H, H_{γ_1} are related, where H_{γ_1} is defined as in Theorem 11.16 using the Jost solution $f_-(k, n)$ (see Theorem 10.2).

Lemma 11.26. *Let H be a given Jacobi operator satisfying (H.10.1). Then the doubly commuted operator H_{γ_1}, $\gamma_1 > 0$, defined via $u_-(\lambda_1, n) = f_-(k_1, n)$, $\lambda_1 = (k_1 + k_1^{-1})/2$ as in Theorem 11.16, has the transmission and reflection coefficients*

$$T_{\gamma_1}(k) = \mathrm{sgn}(k_1)\frac{k\, k_1 - 1}{k - k_1} T(k),$$

(11.120) $$R_{-,\gamma_1}(k) = R_-(k), \qquad R_{+,\gamma_1}(k) = \left(\frac{k - k_1}{k\, k_1 - 1}\right)^2 R_+(k),$$

where z and k are related via $z = (k + k^{-1})/2$. Furthermore, the norming constants $\gamma_{-,j}$ corresponding to $\lambda_j \in \sigma_{pp}(H)$, $j \in J$, (cf. (10.53)) remain unchanged except for an additional eigenvalue λ_1 with norming constant $\gamma_{-,1} = \gamma_1$ if $u_-(\lambda_1) \notin \ell^2(\mathbb{Z})$ respectively with norming constant $\tilde{\gamma}_{-,1} = \gamma_{-,1} + \gamma_1$ if $u_-(\lambda_1) \in \ell^2(\mathbb{Z})$ and $\gamma_{-,1}$ denotes the original norming constant of $\lambda_1 \in \sigma_p(H)$.

Proof. By Lemma 11.19 the Jost solutions $f_{\pm,\gamma_1}(k,n)$ are up to a constant given by

(11.121) $$\frac{c_{\gamma_1}(\lambda_1, n-1)f_\pm(k,n) - \frac{1}{z-\lambda_1}u_-(\lambda_1,n)W_{n-1}(u_-(\lambda_1), f_\pm(k))}{\sqrt{c_{\gamma_1}(\lambda_1, n-1)c_{\gamma_1}(\lambda_1, n)}}.$$

This constant is easily seen to be 1 for $f_{-,\gamma_1}(k,n)$. Thus we can compute $R_-(\lambda)$ using (11.107) (the second unknown constant cancels). The rest follows by a straightforward calculation. □

11.7. Iteration of the double commutation method

Finally, we demonstrate how to iterate the double commutation method. We choose a given background operator H and further (λ_1, γ_1) according to (H.11.15). Next use $u_-(\lambda_1)$ to define the transformation U_{γ_1} and the operator H_{γ_1}. In the second step, we choose (λ_2, γ_2) and another function $u_-(\lambda_2)$ to define $u_{-,\gamma_1}(\lambda_2) = U_{\gamma_1}u_-(\lambda_2)$, a corresponding transformation U_{γ_1,γ_2}, and an operator H_{γ_1,γ_2}. Applying this procedure N-times results in

Theorem 11.27. *Let H be a given background operator and let $(\lambda_\ell, \gamma_\ell)$, $1 \leq \ell \leq N$, satisfy (H.11.15). Define the following matrices ($1 \leq \ell \leq N$)*

$$C^\ell(n) = \left(\frac{\delta_{r,s}}{\gamma_r} + \langle u_-(\lambda_r), u_-(\lambda_s)\rangle_{-\infty}^n\right)_{1 \leq r,s \leq \ell},$$

$$C^\ell(f,g)(n) = \begin{pmatrix} C^{\ell-1}(n)_{r,s}, & r,s \leq \ell-1 \\ \langle f, u_-(\lambda_s)\rangle_{-\infty}^n, & s \leq \ell-1, r=\ell \\ \langle u_-(\lambda_r), g\rangle_{-\infty}^n, & r \leq \ell-1, s=\ell \\ \langle f, g\rangle_{-\infty}^n, & r=s=\ell \end{pmatrix}_{1 \leq r,s \leq \ell},$$

$$D_j^\ell(n) = \begin{pmatrix} C^\ell(n)_{r,s}, & r,s \leq \ell \\ u_-(\lambda_s, n-j), & s \leq \ell, r=\ell+1 \\ u_-(\lambda_r, n), & r \leq \ell, s=\ell+1 \\ \delta_{0,j}, & r=s=\ell+1 \end{pmatrix}_{1 \leq r,s \leq \ell+1},$$

(11.122) $$U^\ell(f(n)) = \begin{pmatrix} C^\ell(n)_{r,s}, & r,s \leq \ell \\ \langle u_-(\lambda_s), f\rangle_{-\infty}^n, & s \leq \ell, r=\ell+1 \\ u_-(\lambda_r, n), & r \leq \ell, s=\ell+1 \\ f, & r=s=\ell+1 \end{pmatrix}_{1 \leq r,s \leq \ell+1}.$$

Then we have (set $C^0(n) = 1$, $U^0(f) = f$)

(11.123) $$\langle U_{\gamma_1,\ldots,\gamma_{\ell-1}} \cdots U_{\gamma_1} f, U_{\gamma_1,\ldots,\gamma_{\ell-1}} \cdots U_{\gamma_1} g\rangle_{-\infty}^n = \frac{\det C^\ell(f,g)(n)}{\det C^{\ell-1}(n)}$$

and

(11.124) $$U_{\gamma_1,\ldots,\gamma_\ell} \cdots U_{\gamma_1} f(n) = \frac{\det U^\ell(f)(n)}{\sqrt{\det C^\ell(n-1)\det C^\ell(n)}}.$$

In particular, we obtain

$$c_{\gamma_1,\ldots,\gamma_\ell}(\lambda_\ell, n) = \frac{1}{\gamma_\ell} + \langle U_{\gamma_1,\ldots,\gamma_{\ell-1}} \cdots U_{\gamma_1}u_-(\lambda_\ell), U_{\gamma_1,\ldots,\gamma_{\ell-1}} \cdots U_{\gamma_1}u_-(\lambda_\ell)\rangle_{-\infty}^n$$

(11.125) $$= \frac{\det C^\ell(n)}{\det C^{\ell-1}(n)}$$

11.7. Iteration of the double commutation method

and hence

$$\text{(11.126)} \qquad \prod_{\ell=1}^{N} c_{\gamma_1,\dots,\gamma_\ell}(\lambda_\ell, n) = \det C^N(n).$$

The corresponding operator $H_{\gamma_1,\dots,\gamma_N}$ is associated with

$$a_{\gamma_1,\dots,\gamma_N}(n) = a(n) \frac{\sqrt{\det C^N(n-1) \det C^N(n+1)}}{\det C^N(n)},$$

$$\text{(11.127)} \qquad b_{\gamma_1,\dots,\gamma_N}(n) = b(n) - \partial^* a(n) \frac{\det D_1^N(n+1)}{\det C^N(n)},$$

and we have

$$H_{\gamma_1,\dots,\gamma_N}(\mathbb{1} - \sum_{j=1}^N P_{\gamma_1,\dots,\gamma_N}(\lambda_j))$$

$$\text{(11.128)} \qquad = (U_{\gamma_1,\dots,\gamma_N} \cdots U_{\gamma_1}) H (U_{\gamma_1}^{-1} \cdots U_{\gamma_1,\dots,\gamma_N}^{-1})(\mathbb{1} - \sum_{j=1}^N P_{\gamma_1,\dots,\gamma_N}(\lambda_j)).$$

Here $P_{\gamma_1,\dots,\gamma_N}(\lambda_j)$ denotes the projection onto the one-dimensional subspace spanned by

$$\text{(11.129)} \qquad u_{\gamma_1,\dots,\gamma_N,-}(\lambda_\ell, n) = \frac{\det U_\ell^N(n)}{\sqrt{\det C^N(n-1) \det C^N(n)}},$$

where $U_\ell^N(n)$ denotes the matrix $U^N(u_-(\lambda_\ell))(n)$ with ℓ'th column and row dropped.

Proof. We start with (11.123). Using Lemma 11.11 we obtain

$$\text{(11.130)} \quad \begin{aligned} \det C^\ell \det C^\ell(f,g) - \det C^\ell(u_-(\lambda_\ell), g) \det C^\ell(f, u_-(\lambda_\ell)) \\ = \det C^{\ell-1} \det C^{\ell+1}(f,g), \end{aligned}$$

which proves (11.123) together with a look at (11.98) by induction on N. Next, (11.125) easily follows from (11.123). Similarly,

$$\text{(11.131)} \quad \begin{aligned} \det C^\ell \det U^{\ell-1}(f) - \det U^{\ell-1}(u_-(\lambda_\ell)) \det C^\ell(u_-(\lambda_\ell), f) \\ = \det C^{\ell-1} \det U^\ell(f), \end{aligned}$$

and (11.92) prove (11.124). To show the formula for $b_{\gamma_1,\dots,\gamma_N}$ we use that (11.124) with $f = u_-(z)$ has an expansion of the type (6.27). Hence we need to expand $U^N(u_-(z))$ in powers of z and compare coefficients with (6.27). We can assume $u_-(z,0) = 1$ implying $\langle u_-(\lambda), u_-(z)\rangle_{-\infty}^n = (z-\lambda)^{-1} W_n(u_-(\lambda), u_-(z)) = O(z^n)$ as $|z| \to \infty$. Now we see that $\langle u_-(\lambda), u_-(z)\rangle_{-\infty}^n = u_-(\lambda, n) u_-(z, n) + u_-(\lambda, n-1) u_-(z, n-1) + O(z^{n-2})$ and thus

$$\text{(11.132)}$$
$$\det U^N(u_-(z))(n) = u(z,n) \det D_0^N(n) + u(z, n-1) \det D_1^N(n) + O(z^{n-2}).$$

Using (6.27) to expand $u(z,n)$ and comparing coefficients as indicated above shows $\det D_0^N(n) = \det C^N(n-1)$ plus the formula for $b_{\gamma_1,\dots,\gamma_N}$. The rest follows in a straightforward manner. □

Remark 11.28. (i). Using (11.107) instead of (11.98) one infers that

$$u_{\gamma_1,\ldots,\gamma_N}(z,n) = \frac{\det U^N(u(z,n))}{\sqrt{\det C^N(n-1)\det C^N(n)}}, \tag{11.133}$$

where

$$U^N(u(z,n)) = \begin{pmatrix} C^N(n)_{r,s} & r,s \leq N \\ \frac{1}{z-\lambda_s}W_n(u_-(\lambda_s),u(z)) & s\leq N, r=N+1 \\ u_-(\lambda_r,n) & r\leq N, s=N+1 \\ u(z,n) & r=s=N+1 \end{pmatrix}_{1\leq r,s\leq N+1} \tag{11.134}$$

satisfies $\tau_{\gamma_1,\ldots,\gamma_N}u = zu$ if $u(z,\cdot)$ satisfies $\tau u = zu$.
(ii). Equation (11.129) can be rephrased as

$$(u_{\gamma_1,\ldots,\gamma_N,-}(\lambda_1,n),\ldots,u_{\gamma_1,\ldots,\gamma_N,-}(\lambda_N,n)) =$$
$$\sqrt{\frac{\det C^N(n)}{\det C^N(n-1)}}(C^N(n))^{-1}(u_-(\lambda_1,n),\ldots,u_-(\lambda_N,n)), \tag{11.135}$$

where $(C^\ell(n))^{-1}$ is the inverse matrix of $C^\ell(n)$.
(iii). The ordering of the pairs (λ_j,γ_j), $1 \leq j \leq N$, is clearly irrelevant (interchanging row i,j and column i,j leaves all determinants unchanged). Moreover, if $\lambda_i = \lambda_j$, then $(\lambda_i,\gamma_i), (\lambda_j,\gamma_j)$ can be replaced by $(\lambda_i,\gamma_i+\gamma_j)$ (by the first assertion it suffices to verify this for $N = 2$).

Finally, we also extend Lemma 11.26. For simplicity we assume $u_-(\lambda_j,n) \notin \ell^2(\mathbb{Z})$, $\gamma_j > 0$, $1 \leq j \leq N$.

Lemma 11.29. *Let H be a given Jacobi operator satisfying (H.10.1). Then the operator $H_{\gamma_1,\ldots,\gamma_N}$, defined via $u_-(\lambda_\ell,n) = f_-(k_\ell,n)$ with $\lambda_\ell = (k_\ell + k_\ell^{-1})/2$ in $\mathbb{R}\backslash\sigma(H_{\gamma_1,\ldots,\gamma_{\ell-1}})$, $1 \leq \ell \leq N$, has the transmission and reflection coefficients*

$$T_{\gamma_1,\ldots,\gamma_N}(k) = \Big(\prod_{\ell=1}^N \operatorname{sgn}(k_\ell)\frac{k\,k_\ell - 1}{k - k_\ell}\Big)T(k),$$
$$R_{-,\gamma_1,\ldots,\gamma_N}(k) = R_-(k),$$
$$R_{+,\gamma_1,\ldots,\gamma_N}(k) = \Big(\prod_{\ell=1}^N \Big(\frac{k - k_\ell}{k\,k_\ell - 1}\Big)^2\Big)R_+(k), \tag{11.136}$$

where $z = (k + k^{-1})/2$. Furthermore, the norming constants $\gamma_{-,j}$ corresponding to $\lambda_j \in \sigma_p(H)$, $j \in J$, (cf. (10.53)) remain unchanged and the additional eigenvalues λ_ℓ have norming constants $\gamma_{-,\ell} = \gamma_\ell$.

Remark 11.30. Of special importance is the case $a(n) = 1/2$, $b(n) = 0$. Here we have $f_\pm(k,n) = k^{\pm n}$, $T(k) = 1$, and $R_\pm(k) = 0$. We know from Section 10.4 that $R_\pm(k)$, $|k| = 1$ together with the point spectrum and corresponding norming constants uniquely determine $a(n)$, $b(n)$. Hence we infer from Lemma 11.12 that $H_{\gamma_1,\ldots,\gamma_N}$ constructed from $u_-(\lambda_\ell,n) = k_\ell^{-n}$ as in Theorem 11.27 and $H_{\sigma_1,\ldots,\sigma_N}$

constructed from $u^\ell_{\sigma_\ell} = \frac{1+\sigma_\ell}{2} k_\ell^n + (-1)^{\ell+1} \frac{1-\sigma_\ell}{2} k_\ell^{-n}$ as in Theorem 11.10 coincide if

$$\gamma_j = \left(\frac{1-\sigma_j}{1+\sigma_j}\right)^{-1} |k_j|^{-1-N} \frac{\prod_{\ell=1}^{N} |1-k_j k_\ell|}{\prod_{\substack{\ell=1 \\ \ell\neq j}}^{N} |k_j - k_\ell|}, \quad 1 \leq j \leq N. \tag{11.137}$$

11.8. The Dirichlet deformation method

In this section we will see that the method of changing individual Dirichlet eigenvalues found in Section 11.4 for quasi-periodic finite-gap sequences works in a much more general setting.

Let us formulate our basic hypothesis.

Hypothesis H. 11.31. (i). Let (E_0, E_1) be a spectral gap of H, that is, $(E_0, E_1) \cap \sigma(H) = \{E_0, E_1\}$.
(ii). Suppose $\mu_0 \in \sigma_d(H_{\sigma_0}) \cap [E_0, E_1]$.
(iii). Let $(\mu, \sigma) \in [E_0, E_1] \times \{\pm\}$ and $\mu \in (E_0, E_1)$ or $\mu \in \sigma_d(H)$.

Remark 11.32. If μ_0 is an eigenvalue of two of the operators H, H_-, H_+, then it is also one of the third. Hence, if $\mu_0 \in \sigma_d(H_{\sigma_0}) \setminus \sigma_d(H_{-\sigma_0})$, then $\mu_0 \in (E_0, E_1)$ and if $\mu_0 \in \sigma_d(H_{\sigma_0}) \cap \sigma_d(H_{-\sigma_0})$, then $\mu_0 \in \{E_0, E_1\}$. (The choice of σ_0 in the latter case is irrelevant). Condition (ii) thus says that $\mu_0 = E_{0,1}$ is only allowed if $E_{0,1}$ is a discrete eigenvalue of H. Similarly in (iii) for μ.

Our next objective is to define the operator $H_{(\mu,\sigma)}$ as in Section 11.4 by performing two single commutations using $u_{\sigma_0}(\mu_0,.)$ and $u_{-\sigma}(\mu,.)$. Since $H_{(\mu_0,\sigma_0)} = H$, we will assume $(\mu,\sigma) \neq (\mu_0, \sigma_0)$ without restriction.

Due to our assumption (H.11.31), we can find solutions $u_{\sigma_0}(\mu_0,.)$, $u_{-\sigma}(\mu,.)$ (cf. Lemma 2.2) and define

$$W_{(\mu,\sigma)}(n) = \begin{cases} \frac{W_n(u_{\sigma_0}(\mu_0), u_{-\sigma}(\mu))}{\mu - \mu_0}, & \mu \neq \mu_0 \\ \sum_{m=\sigma_0\infty}^{n} u_{\sigma_0}(\mu_0, m)^2, & (\mu,\sigma) = (\mu_0, -\sigma_0) \end{cases}, \tag{11.138}$$

where $\sum_{m=+\infty}^{n} = -\sum_{m=n+1}^{\infty}$. The motivation for the case $(\mu,\sigma) = (\mu_0, -\sigma_0)$ follows from (assuming $u_{-\sigma}(\mu, m)$ holomorphic w.r.t. μ – compare (2.32))

$$\lim_{\mu \to \mu_0} \frac{W_n(u_{\sigma_0}(\mu_0), u_{\sigma_0}(\mu))}{\mu - \mu_0} = \lim_{\mu \to \mu_0} \sum_{m=\sigma_0\infty}^{n} u_{\sigma_0}(\mu_0, m) u_{\sigma_0}(\mu, m)$$

$$= \sum_{m=\sigma_0\infty}^{n} u_{\sigma_0}(\mu_0, m)^2. \tag{11.139}$$

From the proof of Theorem 4.19 we infer

Lemma 11.33. *Suppose (H.11.31), then*

$$W_{(\mu,\sigma)}(n+1) W_{(\mu,\sigma)}(n) > 0, \quad n \in \mathbb{Z}. \tag{11.140}$$

Thus the sequences

$$a_{(\mu,\sigma)}(n) = a(n) \sqrt{\frac{W_{(\mu,\sigma)}(n-1) W_{(\mu,\sigma)}(n+1)}{W_{(\mu,\sigma)}(n)^2}},$$

$$b_{(\mu,\sigma)}(n) = b(n) - \partial^* \frac{a(n) u_{\sigma_0}(\mu_0, n) u_{-\sigma}(\mu, n+1)}{W_{(\mu,\sigma)}(n)}, \tag{11.141}$$

are both well-defined and we can consider the associated difference expression

(11.142) $(\tau_{(\mu,\sigma)}u)(n) = a_{(\mu,\sigma)}(n)u(n+1) + a_{(\mu,\sigma)}(n-1)u(n-1) + b_{(\mu,\sigma)}(n)u(n).$

However, we do not know at this point whether $a_{(\mu,\sigma)}(n), b_{(\mu,\sigma)}(n)$ are bounded or not. This will be ignored for the moment. The next lemma collects some basic properties which follow either from Section 11.3 (choosing $N = 2$ in Theorem 11.10) or can be verified directly.

Lemma 11.34. *Let*

(11.143) $(A_{(\mu,\sigma)}u)(z,n) = \dfrac{W_{(\mu,\sigma)}(n)u(z,n) - \frac{1}{z-\mu_0}u_{-\sigma}(\mu,n)W_n(u_{\sigma_0}(\mu_0), u(z))}{\sqrt{W_{(\mu,\sigma)}(n-1)W_{(\mu,\sigma)}(n)}},$

where $u(z)$ solves $\tau u = zu$ for $z \in \mathbb{C}\backslash\{\mu_0\}$. Then we have

(11.144) $\tau_{(\mu,\sigma)}(A_{(\mu,\sigma)}u)(z,n) = z(A_{(\mu,\sigma)}u)(z,n)$

and

(11.145)
$|(A_{(\mu,\sigma)}u)(z,n)|^2 = |u(z,n)|^2 + \dfrac{1}{|z-\mu_0|^2}\dfrac{u_{-\sigma}(\mu,n)}{u_{\sigma_0}(\mu_0,n)}\partial^*\dfrac{|W_n(u_{\sigma_0}(\mu_0),u(z))|^2}{W_{(\mu,\sigma)}(n)}.$

Moreover, the sequences

(11.146)
$u_{\mu_0}(n) = \dfrac{u_{-\sigma}(\mu,n)}{\sqrt{W_{(\mu,\sigma)}(n-1)W_{(\mu,\sigma)}(n)}},$
$u_\mu(n) = \dfrac{u_{\sigma_0}(\mu_0,n)}{\sqrt{W_{(\mu,\sigma)}(n-1)W_{(\mu,\sigma)}(n)}},$

satisfy $\tau_{(\mu,\sigma)}u = \mu_0 u$, $\tau_{(\mu,\sigma)}u = \mu u$ respectively. Note also $u_{\sigma_0}(\mu_0,0) = u_\mu(0) = 0$ and

(11.147) $u_{\mu_0}(n)u_\mu(n) = \partial^*\dfrac{1}{W_{(\mu,\sigma)}(n)}.$

In addition, let $u(z)$, $\hat{u}(z)$ satisfy $\tau u = zu$, then

$W_{(\mu,\sigma),n}(u_{\mu_0}, A_{(\mu,\sigma)}u(z)) = \dfrac{W_n(u_{-\sigma}(\mu), u(z))}{W_{(\mu,\sigma)}(n)},$

$W_{(\mu,\sigma),n}(u_\mu, A_{(\mu,\sigma)}u(z)) = \dfrac{z-\mu}{z-\mu_0}\dfrac{W_n(u_{\sigma_0}(\mu_0), u(z))}{W_{(\mu,\sigma)}(n)},$

$W_{(\mu,\sigma),n}(A_{(\mu,\sigma)}u(z), A_{(\mu,\sigma)}\hat{u}(\hat{z})) = \dfrac{z-\mu}{z-\mu_0}W_n(u(z), \hat{u}(\hat{z}))$

(11.148) $\quad + \dfrac{z-\hat{z}}{(z-\mu_0)(\hat{z}-\mu_0)}\dfrac{W_n(u_\sigma(\mu), u(z))W_n(u_{-\sigma_0}(\mu_0), \hat{u}(\hat{z}))}{W_{(\mu,\sigma)}(n)},$

where $W_{(\mu,\sigma),n}(u,v) = a_{(\mu,\sigma)}(n)\Big(u(n)v(n+1) - u(n+1)v(n)\Big).$

Since we do not know whether $a_{(\mu,\sigma)}(n), b_{(\mu,\sigma)}(n)$ are bounded or not, we need to (temporarily) introduce suitable boundary conditions for $\tau_{(\mu,\sigma)}$ (compare Section 2.6). They will soon turn out to be superfluous.

11.8. The Dirichlet deformation method

Let $\omega \in \{\pm\}$ and

(11.149) $$BC_\omega(f) = \begin{cases} \lim_{n\to\omega\infty} W_n(u_\omega, f) = 0 & \text{if } \tau_{(\mu,\sigma)} \text{ is } l.c. \text{ at } \omega\infty \\ 0 & \text{if } \tau_{(\mu,\sigma)} \text{ is } l.p. \text{ at } \omega\infty \end{cases},$$

where u_ω is $u_\omega = u_\mu$ or $u_\omega = u_{\mu_0}$.

Using this boundary conditions we define

(11.150) $$H_{(\mu,\sigma)}: \begin{array}{c} \mathfrak{D}(H_{(\mu,\sigma)}) \to \ell^2(\mathbb{Z}) \\ f \mapsto \tau_{(\mu,\sigma)} f \end{array},$$

where the domain of $H_{(\mu,\sigma)}$ is explicitly given by

(11.151) $\mathfrak{D}(H_{(\mu,\sigma)}) = \{f \in \ell^2(\mathbb{Z}) | \tau_{(\mu,\sigma)} f \in \ell^2(\mathbb{Z}), BC_-(f) = BC_+(f) = 0\}.$

Furthermore, $H_{(\mu,\sigma),\pm}$ denote the corresponding Dirichlet half-line operators with respect to the base point $n_0 = 0$.

We first give a complete spectral characterization of the half-line operators $H_{(\mu,\sigma),\pm}$. In addition, this will provide all necessary results for the investigation of $H_{(\mu,\sigma)}$.

To begin with we compute the Weyl $\tilde{m}$-functions of $H_{(\mu,\sigma)}$.

Theorem 11.35. *Let $\tilde{m}_\pm(z)$ and $\tilde{m}_{(\mu,\sigma),\pm}(z)$ denote the Weyl $\tilde{m}$-functions of H and $H_{(\mu,\sigma)}$ respectively. Then we have*

(11.152) $$\tilde{m}_{(\mu,\sigma),\pm}(z) = \frac{1}{1+\gamma_{(\mu,\sigma)}}\left(\frac{z-\mu_0}{z-\mu}\tilde{m}_\pm(z) \mp \frac{\gamma_{(\mu,\sigma)}}{z-\mu}\right),$$

where

(11.153) $$\gamma_{(\mu,\sigma)} = \begin{cases} -\sigma(\mu-\mu_0)\tilde{m}_{-\sigma}(\mu), & \mu \neq \mu_0 \\ -\frac{\sigma_0 u_{\sigma_0}(\mu_0,\sigma_0 1)^2}{\sum_{n\in\sigma_0\mathbb{N}} u_{\sigma_0}(\mu_0,n)^2}, & (\mu,\sigma) = (\mu_0,-\sigma_0) \end{cases}.$$

In particular, $H_{(\mu,\sigma),\pm}$ and thus $a_{(\mu,\sigma)}$, $b_{(\mu,\sigma)}$ are bounded.

Proof. We first note that

(11.154) $$\begin{aligned} c_{(\mu,\sigma)}(z,n) &= \frac{z-\mu_0}{z-\mu}(A_{(\mu,\sigma)}c)(z,n) - \frac{\gamma_{(\mu,\sigma)}}{z-\mu}a(0)(A_{(\mu,\sigma)}s)(z,n), \\ s_{(\mu,\sigma)}(z,n) &= \sqrt{1+\gamma_{(\mu,\sigma)}}(A_{(\mu,\sigma)}s)(z,n) \end{aligned}$$

constructed from the fundamental system $c(z,n)$, $s(z,n)$ for τ form a fundamental system for $\tau_{(\mu,\sigma)}$ corresponding to the same initial conditions. Furthermore, note

(11.155) $$\frac{W_{(\mu,\sigma)}(1)}{W_{(\mu,\sigma)}(0)} = 1 + \gamma_{(\mu,\sigma)}, \quad \frac{W_{(\mu,\sigma)}(0)}{W_{(\mu,\sigma)}(-1)} = 1.$$

Now the result follows upon evaluating (cf. Section 2.6)

(11.156) $$\tilde{m}_{(\mu,\sigma),\pm}(z) = \frac{\pm 1}{a_{(\mu,\sigma)}(0)} \lim_{n\to\pm\infty} \frac{W_{(\mu,\sigma),n}(c_{(\mu,\sigma)}(z), u_\omega)}{W_{(\mu,\sigma),n}(s_{(\mu,\sigma)}(z), u_\omega)}.$$

Using (11.148) one obtains for $u_\omega = u_\mu(n)$, $u_{\mu_0}(n)$ and $v_\omega = u_{\sigma_0}(\mu_0,n)$, $u_{\sigma_0}(\mu_0,n)$, respectively,

(11.157) $$\tilde{m}_{(\mu,\sigma),\pm}(z) = \frac{1}{1+\gamma_{(\mu,\sigma)}}\left(\frac{z-\mu_0}{z-\mu}\frac{\pm 1}{a(0)}\lim_{n\to\pm\infty}\frac{W_n(c(z),v_\omega)}{W_n(s(z),v_\omega)} \mp \frac{\gamma_{(\mu,\sigma)}}{z-\mu}\right).$$

Hence the claim is evident. $\square$

In particular, since $H_{(\mu,\sigma)}$ is bounded (away from μ, μ_0 the spectrum of $H_{(\mu,\sigma),\pm}$ is unchanged), we can forget about the boundary conditions (11.149) and $H_{(\mu,\sigma)}$ is defined on $\ell^2(\mathbb{Z})$. As a consequence we note

Corollary 11.36. *The sequences*

(11.158) $\qquad u_{(\mu,\sigma),\pm}(z,n) = (A_{(\mu,\sigma)} u_\pm)(z,n), \quad z \in \mathbb{C}\backslash\{\mu,\mu_0\},$

are square summable near $\pm\infty$. Moreover, the same is true for

$$u_{(\mu,\sigma),-\sigma}(\mu,n) = W_{(\mu,\sigma)}(n) u_{\mu_0}(n) - u_\mu(n) \sum_{m=-\sigma\infty}^{n} u_{-\sigma}(\mu,m)^2,$$

(11.159) $\quad u_{(\mu,\sigma),\sigma_0}(\mu_0,n) = W_{(\mu,\sigma)}(n) u_\mu(n) - u_{\mu_0}(n) \sum_{m=\sigma_0\infty}^{n} u_{\sigma_0}(\mu_0,m)^2$

and

(11.160) $\qquad \begin{aligned} u_{(\mu,\sigma),\sigma}(\mu,n) &= u_\mu(n), & \mu \notin \sigma_d(H), \\ u_{(\mu,\sigma),-\sigma_0}(\mu_0,n) &= u_{\mu_0}(n), & \mu_0 \notin \sigma_d(H). \end{aligned}$

If μ or $\mu_0 \in \sigma_d(H)$ one has to replace the last formulas by

$$u_{(\mu,\sigma),\sigma}(\mu,n) = W_{(\mu,\sigma)}(n) u_{\mu_0}(n) - u_\mu(n) \sum_{m=\sigma\infty}^{n} u_{-\sigma}(\mu,m)^2,$$

(11.161) $\quad u_{(\mu,\sigma),-\sigma_0}(\mu_0,n) = W_{(\mu,\sigma)}(n) u_\mu(n) - u_{\mu_0}(n) \sum_{m=-\sigma_0\infty}^{n} u_{\sigma_0}(\mu_0,m)^2,$

respectively.

Proof. Follows immediately from

(11.162) $\quad u_{(\mu,\sigma),\pm}(z,n) = \dfrac{c_\pm(z)}{1+\gamma_{(\mu,\sigma)}} \dfrac{z-\mu_0}{z-\mu} \left(\dfrac{c_{(\mu,\sigma)}(z,n)}{a_{(\mu,\sigma)}(0)} \mp \tilde{m}_{(\mu,\sigma),\pm}(z) s_{(\mu,\sigma)}(z,n) \right)$

if

(11.163) $\qquad u_\pm(z,n) = c_\pm(z) \left(\dfrac{c(z,n)}{a(0)} \mp \tilde{m}_\pm(z) s(z,n) \right).$

If $z = \mu, \mu_0$ one can assume $u_\pm(z)$ holomorphic with respect to z near μ, μ_0 and consider limits (compare Section 2.2). $\square$

Next we are interested in the pole structure of $\tilde{m}_{(\mu,\sigma),\pm}(z)$ near $z = \mu, \mu_0$. A straightforward investigation of (11.152) using the Herglotz property of $\tilde{m}_{(\mu,\sigma),\pm}(z)$ shows

Corollary 11.37. *We have*

(11.164) $\qquad \tilde{m}_{(\mu,\sigma),\omega}(z) = \begin{cases} -\dfrac{\gamma_\mu}{z-\mu} + O(z-\mu)^0, & \omega = \sigma \\ O(z-\mu)^0, & \omega = -\sigma \end{cases}, \quad \omega \in \{\pm\},$

where

(11.165) $\qquad \gamma_\mu = \begin{cases} (\mu-\mu_0)(\tilde{m}_+(\mu) + \tilde{m}_-(\mu)), & \mu \neq \mu_0 \\ \dfrac{u_-(\mu,-1)^2}{\sum_{n=-\infty}^{-1} u_-(\mu,n)^2} + \dfrac{u_+(\mu,1)^2}{\sum_{n=1}^{+\infty} u_+(\mu,n)^2}, & (\mu,\sigma) = (\mu_0,-\sigma_0) \end{cases} \geq 0.$

11.8. The Dirichlet deformation method

In addition, $\gamma_\mu = 0$ if $\mu \in \sigma_d(H) \backslash \{\mu_0\}$ and $\gamma_\mu > 0$ otherwise. If $\mu \neq \mu_0$, then

(11.166) $$\tilde{m}_{(\mu,\sigma),\pm}(z) = O(z - \mu_0)^0,$$

and note $\gamma_\mu = a(0)^{-2}(\mu - \mu_0)g(\mu,0)^{-1}$.

Using the previous corollary plus weak convergence of $\pi^{-1}\text{Im}(\tilde{m}_\pm(\lambda + i\varepsilon))d\lambda$ to the corresponding spectral measure $d\tilde{\rho}_\pm(\lambda)$ as $\varepsilon \downarrow 0$ implies

Lemma 11.38. *Let $d\tilde{\rho}_\pm(\lambda)$ and $d\tilde{\rho}_{(\mu,\sigma),\pm}(\lambda)$ be the respective spectral measures of $\tilde{m}_\pm(z)$ and $\tilde{m}_{(\mu,\sigma),\pm}(z)$. Then we have*

(11.167)
$$d\tilde{\rho}_{(\mu,\sigma),\pm}(\lambda) = \frac{1}{1+\gamma_{(\mu,\sigma)}}\left(\frac{\lambda - \mu_0}{\lambda - \mu}d\tilde{\rho}_\pm(\lambda) + \left\{\begin{array}{ll}\gamma_\mu d\Theta(\lambda - \mu) & \sigma = \pm \\ 0 & \sigma = \mp\end{array}\right\}\right),$$

where $\gamma_{(\mu,\sigma)}, \gamma_\mu$ are defined in (11.153), (11.165) and $d\Theta$ is the unit point measure concentrated at 0.

Let $P_\pm(\mu_0)$, $P_{(\mu,\sigma),\pm}(\mu)$ denote the orthogonal projections onto the subspaces spanned by $u_{\sigma_0}(\mu_0,.)$, $u_\mu(.)$ in $\ell^2(\pm\mathbb{N})$, respectively. Then the above results clearly imply

Theorem 11.39. *The operators $(\mathbb{1} - P_\pm(\mu_0))H_\pm$ and $(\mathbb{1} - P_{(\mu,\sigma),\pm}(\mu))H_{(\mu,\sigma),\pm}$ are unitarily equivalent. Moreover, $\mu \notin \sigma(H_{(\mu,\sigma),-\sigma})$ and $\mu_0 \notin \sigma(H_{(\mu,\sigma),\pm})\backslash\{\mu\}$.*

If $\mu \notin \sigma_d(H)$ or $(\mu,\sigma) = (\mu_0, -\sigma_0)$, then $\mu \in \sigma(H_{(\mu,\sigma),\sigma})$ and thus

(11.168) $$\sigma(H_{(\mu,\sigma),\pm}) = \Big(\sigma(H_\pm)\backslash\{\mu_0\}\Big) \cup \left\{\begin{array}{ll}\{\mu\}, & \sigma = \pm \\ \emptyset, & \sigma = \mp\end{array}\right..$$

Otherwise, that is, if $\mu \in \sigma_d(H)\backslash\{\mu_0\}$, then $\mu \notin \sigma(H_{(\mu,\sigma),\sigma})$ and thus

(11.169) $$\sigma(H_{(\mu,\sigma),\pm}) = \sigma(H_\pm)\backslash\{\mu_0\}.$$

In essence, Theorem 11.39 says that, as long as $\mu \notin \sigma_d(H)\backslash\{\mu_0\}$, the Dirichlet datum (μ_0, σ_0) is rendered into (μ, σ), whereas everything else remains unchanged. If $\mu \in \sigma_d(H)\backslash\{\mu_0\}$, that is, if we are trying to move μ_0 to an eigenvalue, then μ_0 is removed. This latter case reflects the fact that we cannot move μ_0 to an eigenvalue E without moving the Dirichlet eigenvalue on the other side of E to E at the same time (compare Section 8.2).

Remark 11.40. (i). For $f \in \ell(\mathbb{N})$ set

$$(A_{(\mu,\sigma),+}f)(n) = \sqrt{\frac{W_{(\mu,\sigma)}(n)}{W_{(\mu,\sigma)}(n-1)}}f(n)$$
$$- u_{-\sigma_0,(\mu,\sigma)}(\mu_0,n)\sum_{j=1}^n u_{\sigma_0}(\mu_0,j)f(j),$$

$$(A^{-1}_{(\mu,\sigma),+}f)(n) = \sqrt{\frac{W_{(\mu,\sigma)}(n-1)}{W_{(\mu,\sigma)}(n)}}f(n)$$

(11.170)
$$- u_{-\sigma}(\mu,n)\sum_{j=1}^n u_{\sigma,(\mu,\sigma)}(\mu,j)f(j).$$

Then we have $A_{(\mu,\sigma),+} A^{-1}_{(\mu,\sigma),+} = A^{-1}_{(\mu,\sigma),+} A_{(\mu,\sigma),+} = \mathbb{1}_{\ell(\mathbb{N})}$ and $A_{(\mu,\sigma),+}$ acts as transformation operator

(11.171) $$\tau_{(\mu,\sigma),+} = A_{(\mu,\sigma),+} \tau_+ A^{-1}_{(\mu,\sigma),+}.$$

Similarly, for $f \in \ell(-\mathbb{N})$ set

$$(A_{(\mu,\sigma),-} f)(n) = \sqrt{\frac{W_{(\mu,\sigma)}(n)}{W_{(\mu,\sigma)}(n-1)}} f(n)$$
$$- u_{-\sigma_0,(\mu,\sigma)}(\mu_0, n) \sum_{j=n+1}^{1} u_{\sigma_0}(\mu_0, j) f(j),$$

$$(A^{-1}_{(\mu,\sigma),-} f)(n) = \sqrt{\frac{W_{(\mu,\sigma)}(n-1)}{W_{(\mu,\sigma)}(n)}} f(n)$$

(11.172) $$- u_{-\sigma}(\mu, n) \sum_{j=n+1}^{1} u_{\sigma,(\mu,\sigma)}(\mu, j) f(j).$$

Then we have $A_{(\mu,\sigma),-} A^{-1}_{(\mu,\sigma),-} = A^{-1}_{(\mu,\sigma),-} A_{(\mu,\sigma),-} = \mathbb{1}_{\ell(-\mathbb{N})}$ and

(11.173) $$\tau_{(\mu,\sigma),-} = A_{(\mu,\sigma),-} \tau_- A^{-1}_{(\mu,\sigma),-}.$$

(ii). Note that the case $(\mu, \sigma) = (\mu_0, -\sigma_0)$ corresponds to the double commutation method with $\gamma_1 = \infty$ (cf. Section 11.6). Furthermore, the operators $A_{(\mu,\sigma),\pm}$ are unitary when restricted to proper subspaces of $\ell^2(\pm\mathbb{N})$ in this case.
(iii). Due to the factor $\frac{z-\mu_0}{z-\mu}$ in front of $\tilde{m}_{(\mu,\sigma),\pm}(z)$, all norming constants (i.e., the negative residues at each pole of $\tilde{m}_{(\mu,\sigma),\pm}(z)$) are altered.
(iv). Clearly, by Theorem 11.35, the Dirichlet deformation method preserves reflectionless properties (cf. Lemma 8.1).

Having these results at our disposal, we can now easily deduce all spectral properties of the operator $H_{(\mu,\sigma)}$. First of all, Theorem 11.35 yields for the Weyl matrix of $H_{(\mu,\sigma)}$.

Theorem 11.41. *The respective Weyl M-matrices $M(z)$, $M_{(\mu,\sigma)}(z)$ of H, $H_{(\mu,\sigma)}$ are related by*

$$M_{(\mu,\sigma),0,0}(z) = \frac{1}{(1+\gamma_{(\mu,\sigma)})^2} \frac{z-\mu}{z-\mu_0} M_{0,0}(z),$$

$$M_{(\mu,\sigma),0,1}(z) = \frac{1}{1+\gamma_{(\mu,\sigma)}} \left(M_{0,1}(z) + \frac{\gamma_{(\mu,\sigma)}}{z-\mu_0} a(0) M_{0,0}(z) \right),$$

$$M_{(\mu,\sigma),1,1}(z) = \frac{z-\mu_0}{z-\mu} M_{1,1}(z) - 2 \frac{\gamma_{(\mu,\sigma)}}{z-\mu_0} a(0) M_{0,1}(z)$$

(11.174) $$+ \frac{\gamma^2_{(\mu,\sigma)}}{(z-\mu_0)(z-\mu)} a(0)^2 M_{0,0}(z).$$

Moreover, $M_{(\mu,\sigma),j,k}(z, m, n)$, $j, k \in \{0, 1\}$, are holomorphic near $z = \mu, \mu_0$.

Given the connection between $M(z)$ and $M_{(\mu,\sigma)}(z)$, we can compute the corresponding Herglotz matrix measure of $M_{(\mu,\sigma)}(z)$ as in Lemma 11.38.

11.8. The Dirichlet deformation method

Lemma 11.42. *The matrix measures $d\rho$ and $d\rho_{(\mu,\sigma)}$ corresponding to $M(z)$ and $M_{(\mu,\sigma)}(z)$ are related by*

$$d\rho_{(\mu,\sigma),0,0}(\lambda) = \frac{1}{(1+\gamma_{(\mu,\sigma)})^2} \frac{\lambda-\mu}{\lambda-\mu_0} d\rho_{0,0}(\lambda),$$

$$d\rho_{(\mu,\sigma),0,1}(\lambda) = \frac{1}{1+\gamma_{(\mu,\sigma)}} \left(d\rho_{0,1}(\lambda) + \frac{\gamma_{(\mu,\sigma)}}{\lambda-\mu_0} a(0) d\rho_{0,0}(\lambda) \right),$$

$$d\rho_{(\mu,\sigma),1,1}(\lambda) = \frac{\lambda-\mu_0}{\lambda-\mu} d\rho_{1,1}(\lambda) - 2\frac{\gamma_{(\mu,\sigma)}}{\lambda-\mu_0} a(0) d\rho_{0,1}(\lambda)$$

$$(11.175) \qquad + \frac{\gamma_{(\mu,\sigma)}^2}{(\lambda-\mu_0)(\lambda-\mu)} a(0)^2 d\rho_{0,0}(\lambda).$$

Equivalently

$$(11.176) \qquad d\rho_{(\mu,\sigma)}(\lambda) = \frac{1}{(z-\mu)(z-\mu_0)} \times \begin{pmatrix} \frac{z-\mu}{1+\gamma_{(\mu,\sigma)}} & 0 \\ a(0)\gamma_{(\mu,\sigma)} & z-\mu_0 \end{pmatrix} d\rho(\lambda) \begin{pmatrix} \frac{z-\mu}{1+\gamma_{(\mu,\sigma)}} & a(0)\gamma_{(\mu,\sigma)} \\ 0 & z-\mu_0 \end{pmatrix}.$$

This finally leads to our main theorem in this section

Theorem 11.43. *Let H, $H_{(\mu,\sigma)}$ be defined as before. Denote by $P(\mu_0)$ and $P(\mu)$ the orthogonal projections corresponding to the spaces spanned by $u_{\sigma_0}(\mu_0,.)$ and $u_{-\sigma}(\mu,0)$ in $\ell^2(\mathbb{Z})$ respectively. Then $(\mathbb{1} - P(\mu_0) - P(\mu))H$ and $H_{(\mu,\sigma)}$ are unitarily equivalent. In particular, H and $H_{(\mu,\sigma)}$ are unitarily equivalent if $\mu, \mu_0 \notin \sigma_d(H)$.*

Remark 11.44. By inspection, Dirichlet deformations produce the commuting diagram

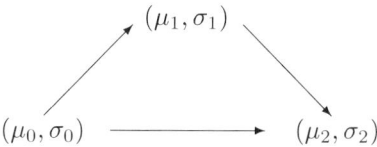

for (μ_j, σ_j), $0 \leq j \leq 2$, according to (H.11.31). In fact, this can be easily verified using (the first two) formulas for the spectral measures (11.175) which uniquely determine the corresponding operator.

Remark 11.45. We have seen in Theorem 11.39 that the Dirichlet deformation method cannot create a situation where a discrete eigenvalue E of H is rendered into a Dirichlet eigenvalue (i.e., moving μ_0 to the eigenvalue E). However, one can use the following three-step procedure to generate a prescribed degeneracy at an eigenvalue E of H:

(i). Use the Dirichlet deformation method to move μ to a discrete eigenvalue E of H. (This removes both the discrete eigenvalue E of H and the (Dirichlet) eigenvalue μ of $H_- \oplus H_+$).

(ii) As a consequence of step (i), there is now another eigenvalue $\tilde{\mu}$ of $H_- \oplus H_+$ in the resulting larger spectral gap of H. Move $\tilde{\mu}$ to E using the Dirichlet deformation method.

(iii) Use the double commutation method to insert an eigenvalue of H at E and to change σ_0 into any allowed value.

Theorem 11.16 then shows that the resulting operator is unitarily equivalent to the original operator H, and Theorem 11.21 proves that the remaining Dirichlet eigenvalues remain invariant.

Finally, we briefly comment on how to iterate the Dirichlet deformation method. All ingredients have already been derived in Section 11.3. Suppose

$$(11.177) \qquad (E_{0,j}, E_{1,j}), \quad (\mu_{0,j}, \sigma_{0,j}), (\mu_j, \sigma_j) \in [E_{0,j}, E_{1,j}] \times \{\pm\}$$

satisfy (H.11.31) for each $j = 1, \ldots, N$, $N \in \mathbb{N}$. Then the Dirichlet deformation result after N steps, denoted by $H_{(\mu_1, \sigma_1), \ldots, (\mu_N, \sigma_N)}$, is associated with the sequences

$$a_{(\mu_1, \sigma_1), \ldots, (\mu_N, \sigma_N)}(n) = \sqrt{a(n-N)a(n+N)}$$
$$\times \frac{\sqrt{C_{n-N}(U_{1,\ldots,N}) C_{n-N+2}(U_{1,\ldots,N})}}{C_{n-N+1}(U_{1,\ldots,N})},$$

$$(11.178) \qquad b_{(\mu_1, \sigma_1), \ldots, (\mu_N, \sigma_N)}(n) = b(n) - \partial^* a(n) \frac{\tilde{D}_{n-N+1}(U_{1,\ldots,N})}{C_{n-N+1}(U_{1,\ldots,N})},$$

where $(U_{1,\ldots,N}) = (u_{\sigma_{0,1}}(\mu_{0,1}), u_{\sigma_1}(\mu_1), \ldots, u_{\sigma_{0,N}}(\mu_{0,N}), u_{\sigma_N}(\mu_N))$.

Notes on literature

Chapter 1. The literature on second order difference equations is enormous and tracing all results would require a historian. Hence I can only give some hints which serve as starting points. Classical references are the books by Atkinson [**19**], Berezanskiĭ [**27**] (Chapter 7), and Fort [**87**]. A more recent reference would be the book by Carmona and Lacroix [**40**] which focuses on random operators. Clearly, there are far more books on this topic available, but these are the ones I found most useful for my purpose. However, note that not everything in the first chapter can be found in these references. I have added some results, needed in later chapters, which I could not find in the literature (e.g., I don't know a reference for Lemma 1.2).

In addition, recent and easy introductions to difference equations can be found in, for instance, Jordan [**141**], Kelley and Peterson [**147**], and Mickens [**171**]. A book which contains numerous examples and special results is the one by Agarwal [**4**].

Chapter 2. The theory presented in this chapter is the discrete analog of what is known as Weyl-Titchmarsh-Kodaira theory for Sturm-Liouville operators. The continuous theory originated in the celebrated work of Weyl [**244**] with subsequent extensions by Titchmarsh ([**227**], [**228**]), Kodaira and numerous others (see any book on Sturm-Liouville operators for further details). The origins of the discrete theory go back to Hellinger [**131**] and Nevanlinna [**182**]. Many results have been rediscovered several times and the literature on this subject is far to extensive to be quoted in full.

Again classical references are Atkinson [**19**], Berezanskiĭ [**27**] (Chapter 7), Fort [**87**], and the more recent monograph by Carmona and Lacroix [**40**]. My personal favorite though is Berezanskiĭ [**27**]. For connections with spectral theory on the half line and the moment problem I suggest the classical book by Akhiezer [**9**]. There is also a very good and recent survey article on the moment problem by Simon [**209**]. For the view point of orthogonal polynomials, the standard reference is the book of Szegö [**217**]. You might also have a look at [**211**], which has emphasis on applications, or [**57**], which uses a Riemann-Hilbert approach.

Especially for relations between Weyl m-functions and the theory of continued fractions see Atkinson [**9**] and also [**24**], [**240**]. A paper which closely follows the original ideas of Titchmarsh is [**18**]. There is a large bulk of papers on the question of essential self-adjointness of Jacobi operators (i.e., limit point, limit circle criteria). See for instance [**9**] (addenda and problems to Chapter 1), [**19**], [**27**], [**110**] (Appendix E), [**132**], [**135**],

[188]. The characterization of boundary conditions in the limit circle case can be found in [242], [243] (however, a much more cumbersome approach is used there). A paper which focuses on the contributions by Krein is [186].

The investigation of positive operators respectively positive solutions has attracted quite some interest and many results have been discovered more than once. Our section compiles results from [26], [111], and [188]. I have streamlined most arguments and simplified the proofs. Further references are [6], [133], and [191].

The section on inverse problems is taken from [222]. Additional results can be found in [88], [89], [92] and, in the case of finite Jacobi matrices, [104].

There are extensions of many results to block Jacobi matrices (i.e., Jacobi matrices whose coefficients are again matrices). The interested reader might look at (e.g.) [5], [7], or [245].

Chapter 3. The first section summarizes some well-known results, which however are often hard to find. Some of them don't seem to be written up at all and I can't give a good reference for this section. For the continuous case you should look at the beautiful paper of Simon [204] and, again, at the book of Berezanskiĭ [27].

The second section extends results of [215] to the case $a(n) \neq 1$ and provides a few basic facts for easy reference.

The absolutely (resp. singularly) continuous spectrum is one of the most investigated object in spectral theory. The contents of this section clearly has its origin in the beautiful theory of Gilbert and Pearson [120] which is known as the principle of subordinacy. This theory has since then been simplified and extended by various people (see e.g., [119], [48], [145]). I follow ideas of [68], [139], [163], and [207]. For some general results on singular continuous spectra see [208].

Bounds on the number of eigenvalues are considered in [95] and [96] (see also [195], Theorem XIII.10).

Additional topics like embedded eigenvalues, slowly decreasing potentials, or some explicitly solvable problems can be found in [179], [180], [58], [149], [215], or [77]. A different approach to find classes of operators with absolutely continuous spectral measures is presented in [71] (see also [72]).

Chapter 4. Oscillation theory originates in the work of Sturm [216]. Numerous papers have been written on this subject in the case of differential equations and several monographs are devoted solely to this topic. In the discrete case there is a large amount of papers investigating whether a given Jacobi equation is oscillatory or not (e.g., [128], [132], [133], [134], [188], and [189]). The case of finite matrices is treated (e.g.) in the textbook by Fort [87]. For the case of infinite matrices only the connection between (non)oscillation and the infimum of the essential spectrum was known ([121], Theorem 32). The precise connection between the number of oscillations and the spectrum was only clarified recently in [223]. This article is also the main reference for this chapter. Additional references are [19], [26], and [147].

Chapter 5. The first main theorem, that the spectrum is a.s. the same for all operators is due to Pastur [187]. The results concerning the Lyapunov exponent are mainly taken from [49], the ones for the integrated density of states respectively the rotation number are mainly collected from [69], [70], and [22]. It is sometimes claimed that the process (5.48) is subadditve (which *is* true for differential operators) so I tried to clarify this point. The results on the absolutely continuous spectrum originate from work of Kotani and are taken from [205]. The section on almost periodic operators summarizes some results from [21] and [122]. There have recently been some efforts to prove uniform spectral results (i.e, for all operators in the hull), see, for example, [51], [52] and the references therein. In this respect, it should be mentioned that the absolutely continuous spectrum is constant for all operators in the hull ([**163**], Theorem 1.5). It is also known that the other parts need not be constant (see [**140**]). If you want to find out almost everything about the almost Mathieu operator, I suggest [**138**] and [**162**].

The reader who wants to learn more about this subject should consult the monograph by Carmona and Lacroix [**40**] which is devoted entirely to this subject. Another brief introduction can be found in Cycon et al. [**50**] (Chapters 9 and 10). For some recent generalizations to orthogonal polynomials on the unit circle see [**100**].

Chapter 6. The first trace formulas for Schrödinger operators on a finite interval were given by Gel'fand and Levitan [**93**]. Since then numerous extensions have been made. In particular, recently Gesztesy and Simon showed [**105**] that all main results can be viewed as special cases of Krein's spectral shift theory [**155**]. Most parts of this section are taken from [**222**].

Chapter 7. Floquet (resp. Bloch) theory is one of the classical results found in most textbooks on differential equations. In the case of periodic Jacobi operators the first spectral characterization was given by [**173**]. The first two sections contain the analogs of several results from periodic Sturm-Liouville operators. The third section is somewhat inspired by [**169**] (see also Chapter 8). Further results on perturbed periodic operators can be found in [**98**], [**121**] (Section 67), [**150**], and [**197**]. For extensions to complex or operator valued coefficients see [**176**] or [**165**], respectively.

Chapter 8. Reflectionless operators have attracted several mathematicians since they allow an explicit treatment. In the first two sections I follow closely ideas of [**113**] and [**222**]. However, I have added some additional material. Complementary results to this sections, namely for the case of Cantor spectra, can be found in [**12**] – [**14**], and [**212**], [**213**].

The remaining sections are of importance for the second part. The approach is inspired by ideas of McKean [**169**]. Some parts are taken from [**35**] (see also [**173**]), but again the material has been extended to a larger class of operators and augmented.

Chapter 9. The use of Riemann surfaces usually appears in connection with the Toda lattice. The first results seem due to Krichever (see, e.g., [**156**] – [**160**]). The main reference for this chapter is again [**35**]. Additional references are [**74**], [**169**], [**173**], [**230**], and [**175**]. This technique can also be extended to surfaces with infinitely many branch points [**12**] – [**14**], [**212**] and [**213**].

For the explicit calculations in the genus one case, I have used the tables for elliptic integrals in [**38**]. In addition, you might also find [**10**], [**123**], and [**127**] useful.

Chapter 10. Discrete scattering theory has been first developed on an informal level by Case in a series of papers [**42**] – [**47**]. The first rigorous results were given by Guseinov [**124**], but unfortunately without proof. Additional results are given in [**125**], [**126**], and [**224**]. This chapter extends the results of [**124**] and gives (to the best of my knowledge) the first proof in print. Most of the other results are also new and have not previously appeared elsewhere. Further references are [**80**] (Chap. 2.III), [**94**], [**95**], [**97**], [**102**], and [**230**]. For generalizations to polynomials on the unit circle see [**99**].

Chapter 11. The single commutation method for Jacobi operators first appeared in Gesztesy et al. [**117**]. A special case of the double commutation method can be found in [**235**]. The general case of the double commutation method seems due to [**110**] which is one of the the main sources for this chapter. It also contains the analogue of the FIT-formula (Section 11.4) derived in [**83**] for Schrödinger operators. The second main source is [**221**] where the discrete Dirichlet deformation method is introduced.

Part 2

Completely Integrable Nonlinear Lattices

Chapter 12

The Toda system

This chapter is devoted to the Toda hierarchy. The first section gives an informal introduction and is mainly for background and motivation. The following sections contain a rigorous treatment based on the Lax pair formalism. The basic existence and uniqueness theorem for solutions of the initial value problem is proven and the connection with hyperelliptic Riemann surfaces is established.

12.1. The Toda lattice

The Toda lattice is a simple model for a nonlinear one-dimensional crystal. It describes the motion of a chain of particles with nearest neighbor interaction. The equation of motion for such a system is given by

$$(12.1) \quad m\frac{d^2}{dt^2}x(n,t) = V'(x(n+1,t) - x(n,t)) - V'(x(n,t) - x(n-1,t)),$$

where m denotes the mass of each particle, $x(n,t)$ is the displacement of the n-th particle from its equilibrium position, and $V(r)$ ($V'(r) = \frac{dV}{dr}(r)$) is the interaction potential. As discovered by M. Toda, this system gets particularly interesting if one chooses an exponential interaction,

$$(12.2) \quad V(r) = \frac{m\rho^2}{\tau^2}\left(e^{-r/\rho} + \frac{r}{\rho} - 1\right) = \frac{m\rho^2}{\tau^2}\left(\left(\frac{r}{\rho}\right)^2 + O\left(\frac{r}{\rho}\right)^3\right), \quad \tau, \rho \in \mathbb{R}.$$

This model is of course only valid as long as the relative displacement is not too large (i.e., at least smaller than the distance of the particles in the equilibrium position). For small displacements it is equal to a harmonic crystal with force constant $\frac{m}{\tau^2}$ (cf. Section 1.5).

After a scaling transformation, $t \mapsto t/\tau$, $x \mapsto x/\rho$, we can assume $m = \tau = \rho = 1$. If we suppose $x(n,t) - x(n-1,t) \to 0$, $\dot{x}(n,t) \to 0$ sufficiently fast as $|n| \to \infty$, we can introduce the Hamiltonian ($q = x$, $p = \dot{x}$)

$$(12.3) \quad \mathcal{H}(p,q) = \sum_{n \in \mathbb{Z}} \left(\frac{p(n,t)^2}{2} + (e^{-(q(n+1,t) - q(n,t))} - 1)\right)$$

221

and rewrite the equations of motion in the Hamiltonian form

$$\frac{d}{dt}p(n,t) = -\frac{\partial \mathcal{H}(p,q)}{\partial q(n,t)}$$
$$= e^{-(q(n,t) - q(n-1,t))} - e^{-(q(n+1,t) - q(n,t))},$$
(12.4) $$\frac{d}{dt}q(n,t) = \frac{\partial \mathcal{H}(p,q)}{\partial p(n,t)} = p(n,t).$$

We remark that these equations are invariant under the transformation

(12.5) $\quad p(n,t) \to p(n,t) + p_0 \quad$ and $\quad q(n,t) \to q(n,t) + q_0 + p_0 t, \quad (p_0, q_0) \in \mathbb{R}^2,$

which reflects the fact that the dynamics remains unchanged by a uniform motion of the entire crystal.

The fact which makes the Toda lattice particularly interesting is the existence of soliton solutions. These are pulslike waves traveling through the crystal without changing their shape. Such solutions are rather special since from a generic linear equation one would expect spreading of wave packets (see Section 1.5) and from a generic nonlinear wave equation one would expect that solutions only exist for a finite time (breaking of waves).

The simplest example of such a solitary wave is the one-soliton solution

(12.6) $\quad q_1(n,t) = q_0 - \ln \dfrac{1 + \gamma \exp(-2\kappa n \pm 2\sinh(\kappa)t)}{1 + \gamma \exp(-2\kappa(n-1) \pm 2\sinh(\kappa)t)}, \quad \kappa, \gamma > 0.$

It describes a single bump traveling trough the crystal with speed $\pm \sinh(\kappa)/\kappa$ and width proportional to $1/\kappa$. That is, the smaller the soliton the faster it propagates. It results in a total displacement

(12.7) $$\lim_{n \to \infty} \big(q_1(n,t) - q_1(-n,t)\big) = 2\kappa$$

of the crystal, which can equivalently be interpreted as the total compression of the crystal around the bump. The total moment and energy are given by

(12.8) $$\sum_{n \in \mathbb{Z}} p_1(n,t) = 2\sinh(\kappa),$$
$$\mathcal{H}(p_1, q_1) = 2(\sinh(\kappa)\cosh(\kappa) - \kappa).$$

Existence of such solutions is usually connected to complete integrability of the system which is indeed the case here. To see this, we introduce Flaschka's variables

(12.9) $\quad a(n,t) = \dfrac{1}{2} e^{-(q(n+1,t) - q(n,t))/2}, \quad b(n,t) = -\dfrac{1}{2} p(n,t)$

and obtain the form most convenient for us

(12.10) $$\dot{a}(n,t) = a(n,t)\big(b(n+1,t) - b(n,t)\big),$$
$$\dot{b}(n,t) = 2\big(a(n,t)^2 - a(n-1,t)^2\big).$$

12.1. The Toda lattice

The inversion is given by

$$p(n,t) = -2b(n,t),$$

$$q(n,t) = q(n,0) - 2\int_0^t b(n,s)ds$$

$$= q(0,0) - 2\int_0^t b(0,s)ds - 2\sum_{j=0}^{n-1}{}^* \ln(2a(j,t))$$

(12.11)
$$= q(0,0) - 2\int_0^t b(n,s)ds - 2\sum_{j=0}^{n-1}{}^* \ln(2a(j,0)).$$

To show complete integrability it suffices to find a so-called Lax pair, that is, two operators $H(t)$, $P(t)$ such that the Lax equation

(12.12)
$$\frac{d}{dt}H(t) = P(t)H(t) - H(t)P(t)$$

is equivalent to (12.10). One can easily convince oneself that the choice

(12.13)
$$H(t): \ell^2(\mathbb{Z}) \to \ell^2(\mathbb{Z})$$
$$f(n) \mapsto a(n,t)f(n+1) + a(n-1,t)f(n-1) + b(n,t)f(n),$$
$$P(t): \ell^2(\mathbb{Z}) \to \ell^2(\mathbb{Z})$$
$$f(n) \mapsto a(n,t)f(n+1) - a(n-1,t)f(n-1)$$

does the trick. Now the Lax equation implies that the operators $H(t)$ for different $t \in \mathbb{R}$ are all unitarily equivalent and that

(12.14)
$$\mathrm{tr}\big(H(t)^j - H_0^j\big), \quad j \in \mathbb{N},$$

are conserved quantities, where H_0 is the operator corresponding to the constant solution $a_0(n,t) = \frac{1}{2}$, $b_0(n,t) = 0$ (it is needed to make the trace converge). For example,

$$\mathrm{tr}\big(H(t) - H_0\big) = \sum_{n \in \mathbb{Z}} b(n,t) = -\frac{1}{2}\sum_{n \in \mathbb{Z}} p(n,t) \text{ and}$$

(12.15)
$$\mathrm{tr}\big(H(t)^2 - H_0^2\big) = \sum_{n \in \mathbb{Z}} b(n,t)^2 + 2(a(n,t)^2 - \frac{1}{4}) = \frac{1}{2}\mathcal{H}(p,q)$$

correspond to conservation of the total momentum and the total energy, respectively.

This reformulation of the Toda equations as a Lax pair is the key to methods of solving the Toda equations based on spectral and inverse spectral theory for the Jacobi operator H.

Using these methods one can find the general N-soliton solution

(12.16)
$$q_N(n,t) = q_0 - \ln \frac{\det(\mathbb{1} + C_N(n,t))}{\det(\mathbb{1} + C_N(n-1,t))},$$

where
(12.17)
$$C_N(n,t) = \left(\frac{\sqrt{\gamma_i \gamma_j}}{1 - e^{-(\kappa_i + \kappa_j)}} e^{-(\kappa_i + \kappa_j)n - (\sigma_i \sinh(\kappa_i) + \sigma_j \sinh(\kappa_j))t} \right)_{1 \leq i,j \leq N}$$

with $\kappa_j, \gamma_j > 0$ and $\sigma_j \in \{\pm 1\}$. One can also find (quasi-)periodic solutions using techniques from Riemann surfaces (respectively algebraic curves). Each such solution is associated with an hyperelliptic curve of the type

(12.18)
$$w^2 = \prod_{j=0}^{2g+2} (z - E_j), \quad E_j \in \mathbb{R},$$

where E_j, $0 \leq j \leq 2g+1$, are the band edges of the spectrum of H (which is independent of t and hence determined by the initial conditions). One obtains

(12.19) $$q(n,t) = q_0 - 2(t\tilde{b} + n\ln(2\tilde{a})) - \ln \frac{\theta(\underline{z}_0 - 2n\underline{A}_{p_0}(\infty_+) - 2t\underline{c}(g))}{\theta(\underline{z}_0 - 2(n-1)\underline{A}_{p_0}(\infty_+) - 2t\underline{c}(g))},$$

where $\underline{z}_0 \in \mathbb{R}^g$, $\theta : \mathbb{R}^g \to \mathbb{R}$ is the Riemann theta function associated with the hyperelliptic curve (12.18), and $\tilde{a}, \tilde{b} \in \mathbb{R}$, $\underline{A}_{p_0}(\infty_+), \underline{c}(g) \in \mathbb{R}^g$ are constants depending only on the curve (i.e., on E_j, $0 \leq j \leq 2g+1$). If $q(n,0), p(n,0)$ are (quasi-) periodic with average 0, then $\tilde{a} = \frac{1}{2}$, $\tilde{b} = 0$.

The rest of this monograph is devoted to a rigorous mathematical investigation of these methods.

12.2. Lax pairs, the Toda hierarchy, and hyperelliptic curves

In this section we introduce the Toda hierarchy using a recursive approach for the standard Lax formalism and derive the Burchnall-Chaundy polynomials in connection with the stationary Toda hierarchy.

We let the sequences a, b depend on an additional parameter $t \in \mathbb{R}$ and require

Hypothesis H. 12.1. Suppose $a(t), b(t)$ satisfy

(12.20) $$a(t), b(t) \in \ell^\infty(\mathbb{Z}, \mathbb{R}), \quad a(n,t) \neq 0, \quad (n,t) \in \mathbb{Z} \times \mathbb{R},$$

and let $t \in \mathbb{R} \mapsto (a(t), b(t)) \in \ell^\infty(\mathbb{Z}) \oplus \ell^\infty(\mathbb{Z})$ be differentiable.

We introduce the corresponding operator $H(t)$ as usual, that is,

(12.21) $$\begin{array}{rcl} H(t) : & \ell^2(\mathbb{Z}) & \to \quad \ell^2(\mathbb{Z}) \\ & f(n) & \mapsto \quad a(n,t)f(n+1) + a(n-1,t)f(n-1) + b(n,t)f(n) \end{array}.$$

The idea of the Lax formalism is to find a finite, skew-symmetric operator $P_{2r+2}(t)$ such that the **Lax equation**

(12.22) $$\frac{d}{dt}H(t) - [P_{2r+2}(t), H(t)] = 0, \quad t \in \mathbb{R},$$

(here $[.,..]$ denotes the commutator, i.e., $[P, H] = PH - HP$) holds. More precisely, we seek an operator $P_{2r+2}(t)$ such that the commutator with $H(t)$ is a symmetric difference operator of order at most two. Equation (12.22) will then give an evolution equation for $a(t)$ and $b(t)$. Our first theorem tells us what to choose for $P_{2r+2}(t)$.

12.2. The Toda hierarchy

Theorem 12.2. *Suppose $P(t)$ is of order at most $2r+2$ and the commutator with $H(t)$ is of order at most 2. Then $P(t)$ is of the form*

$$(12.23) \quad P(t) = \sum_{j=0}^{r} \left(c_{r-j} \tilde{P}_{2j+2}(t) + d_{r-j} H(t)^{j+1} \right) + d_{r+1} \mathbb{1},$$

where $c_j, d_j \in \mathbb{C}$, $0 \le j \le r$, $d_{r+1} \in \ell(\mathbb{Z})$, and

$$(12.24) \quad \tilde{P}_{2j+2}(t) = [H(t)^{j+1}]_+ - [H(t)^{j+1}]_-$$

(cf. the notation in (1.11)) is called homogeneous Lax operator. Moreover, denote by $P_{2r+2}(t)$ the operator $P(t)$ with $c_0 = 1$ and $d_j = 0$, $0 \le j \le r+1$. Then we have

$$(12.25) \quad P_{2r+2}(t) = -H(t)^{r+1} + \sum_{j=0}^{r}(2a(t)g_j(t)S^+ - h_j(t))H(t)^{r-j} + g_{r+1}(t),$$

where $(g_j(n,t))_{0 \le j \le r+1}$ and $(h_j(n,t))_{0 \le j \le r+1}$ are given by

$$g_j(n,t) = \sum_{\ell=0}^{j} c_{j-\ell} \langle \delta_n, H(t)^\ell \delta_n \rangle,$$

$$(12.26) \quad h_j(n,t) = 2a(n,t) \sum_{\ell=0}^{j} c_{j-\ell} \langle \delta_{n+1}, H(t)^\ell \delta_n \rangle + c_{j+1}$$

and satisfy the recursion relations

$$g_0 = 1, \quad h_0 = c_1,$$
$$2g_{j+1} - h_j - h_j^- - 2bg_j = 0, \quad 0 \le j \le r,$$
$$(12.27) \quad h_{j+1} - h_{j+1}^- - 2(a^2 g_j^+ - (a^-)^2 g_j^-) - b(h_j - h_j^-) = 0, \quad 0 \le j < r.$$

For the commutator we obtain

$$[P_{2r+2}(t), H(t)] = a(t)(g_{r+1}^+(t) - g_{r+1}(t))S^+ + a^-(t)(g_{r+1}(t) - g_{r+1}^-(t))S^-$$
$$(12.28) \quad + (h_{r+1}(t) - h_{r+1}^-(t)).$$

Proof. By Lemma 1.2 we can write

$$(12.29) \quad P(t) = -h_{-1}(t) H(t)^{r+1} + \sum_{j=0}^{r}(2a(t)g_j(t)S^+ - h_j(t))H(t)^{r-j} + g_{r+1}(t),$$

where $g_{r+1}(t)$ is only added for convenience and hence can be chosen arbitrarily. Now we insert this ansatz into $[P, H]$. Considering the term $(S^-)^{r+2}$ we see that $h_{-1}(t)$ must be independent of n, say $h_{-1}(t) = c_0 - d_0$. Next, we obtain after a long but straightforward calculation

$$[P,H] = 2a(g_0^+ - g_0)S^+ H^{r+1} - (h_0 - h_0^-)H^{r+1}$$
$$- \sum_{j=0}^{r-1} a \Big(\partial(2g_{j+1} - h_j - h_j^- - 2bg_j) \Big) S^+ H^{r-j}$$
$$- \sum_{j=0}^{r-1} \Big(h_{j+1} - h_{j+1}^- - 2(a^2 g_j^+ - (a^-)^2 g_j^-) - b(h_j - h_j^-) \Big) H^{r-j}$$
$$(12.30) \quad + a(g_{r+1}^+ - g_{r+1})S^+ + S^- a(g_{r+1}^+ - g_{r+1}) + (h_{r+1} - h_{r+1}^-),$$

where g_{r+1}, h_{r+1} have been chosen according to

$$\partial(2g_{r+1} - h_r - h_r^- - 2bg_r) = 0,$$

(12.31) $\quad h_{r+1} - h_{r+1}^- - 2(a^2 g_r^+ - (a^-)^2 g_r^-) - b(h_r - h_r^-) = 0.$

(Recall $\partial f = f^+ - f$.) But (12.30) is of order 2 if and only if (compare Lemma 1.2)

$$g_0 = c_0, \quad h_0 = c_1 - d_1,$$
$$2g_{j+1} - h_j - h_j^- - 2bg_j = 2d_{j+1}, \quad 0 \leq j \leq r-1,$$

(12.32) $\quad h_{j+1} - h_{j+1}^- - 2(a^2 g_j^+ - (a^-)^2 g_j^-) - b(h_j - h_j^-) = 0, \quad 0 \leq j < r-2,$

where c_0, c_1, d_j, $1 \leq j \leq r$, are constants. By Lemma 6.4

(12.33) $\quad \tilde{g}_j(n,t) = \langle \delta_n, H(t)^j \delta_n \rangle, \quad \tilde{h}_j(n,t) = 2a(n,t)\langle \delta_{n+1}, H(t)^j \delta_n \rangle$

is a solution of this system for $c_0 = 1$, $c_j = d_j = 0$, $1 \leq j \leq r$. It is called the homogeneous solution for if we assign the weight one to a and b, then $\tilde{g}_j(n,t)$, $\tilde{h}_j(n,t)$ are homogeneous of degree j, $j+1$, respectively. The general solution of the above system (12.32) is hence given by $g_j = \sum_{\ell=0}^{j} c_\ell \tilde{g}_{j-\ell}$, $1 \leq j \leq r$, and $h_j = \sum_{\ell=0}^{j} c_\ell \tilde{h}_{j-\ell} + c_{j+1} - d_{j+1}$, $1 \leq j \leq r-1$. Introducing another arbitrary sequence d_{r+1} it is no restriction to assume that the formula for h_j also holds for $j = r$.

It remains to verify (12.24). We use induction on r. The case $r = 0$ is easy. By (12.25) we need to show

(12.34) $\quad \tilde{P}_{2r+2} = \tilde{P}_{2r} H + \left(2a\tilde{g}_r S^+ - \tilde{h}_r\right) - \tilde{g}_r H + \tilde{g}_{r+1}.$

This can be done upon considering $\langle \delta_m, \tilde{P}_{2r+2} \delta_n \rangle$ and making case distinctions $m < n-1, m = n-1, m = n, m = n+1, m > n+1$. Explicitly, one verifies, for instance, in the case $m = n$,

$\langle \delta_n, \tilde{P}_{2r+2} \delta_n \rangle$
$= \langle \delta_n, \tilde{P}_{2r}(a\delta_{n-1} + a^- \delta_{n+1} + b\delta_n) \rangle - \tilde{h}_r(n) - b(n)\tilde{g}_r(n) + \tilde{g}_{r+1}(n)$
$= \langle \delta_n, ([H^r]_+ - [H^r]_-)(a\delta_{n-1} + a^- \delta_{n+1} + b\delta_n) \rangle - \tilde{h}_r(n) - b(n)\tilde{g}_r(n) + \tilde{g}_{r+1}(n)$
$= \langle \delta_n, [H^r]_+ a^- \delta_{n+1} \rangle - \langle \delta_n, [H^r]_- a\delta_{n-1} \rangle - \tilde{h}_r(n) - b(n)\tilde{g}_r(n) + \tilde{g}_{r+1}(n)$
$= a(n)\langle \delta_n, H^r \delta_{n+1} \rangle - a(n-1)\langle \delta_n, H^r \delta_{n-1} \rangle - \tilde{h}_r(n) - b(n)\tilde{g}_r(n) + \tilde{g}_{r+1}(n)$
$= -\dfrac{\tilde{h}_r(n) + \tilde{h}_r(n-1)}{2} - b(n)\tilde{g}_r(n) + \tilde{g}_{r+1}(n) = 0$

(12.35)
using (12.27),

(12.36) $\quad \langle \delta_m, [H^j]_\pm \delta_n \rangle = \begin{cases} \langle \delta_m, H^j \delta_n \rangle, & \pm(m-n) > 0 \\ 0, & \pm(m-n) \leq 0 \end{cases},$

and $H\delta_m = a(m)\delta_{m+1} + a(m-1)\delta_{m-1} + b(m)\delta_m$. This settles the case $m = n$. The remaining cases are settled one by one in a similar fashion. $\square$

Remark 12.3. It is also easy to obtain (12.28) from (12.23) and (12.24). In fact, simply evaluate

$$\langle \delta_m, [\tilde{P}_{2j+2}, H]\delta_n \rangle = \langle H\delta_m, [H^{j+1}]_+ \delta_n \rangle + \langle H\delta_n, [H^{j+1}]_+ \delta_m \rangle$$
(12.37) $\quad\quad\quad\quad - \langle H\delta_m, [H^{j+1}]_- \delta_n \rangle - \langle H\delta_n, [H^{j+1}]_- \delta_m \rangle$

12.2. The Toda hierarchy

as in the proof of Theorem 12.2.

Since the self-adjoint part of $P(t)$ does not produce anything interesting when inserted into the Lax equation, we will set $d_j = 0$, $0 \le j \le r+1$, and take

$$(12.38) \qquad P_{2r+2}(t) = \sum_{j=0}^{r} c_{r-j} \tilde{P}_{2j+2}(t)$$

as our **Lax operator** in (12.22). Explicitly we have

$$P_2(t) = a(t)S^+ - a^-(t)S^-$$
$$P_4(t) = a(t)a^+(t)S^{++} + a(t)(b^+(t) + b(t))S^+ - a^-(t)(b(t) + b^-(t))S^-$$
$$\qquad - a^-(t)a^{--}(t)S^{--} + c_1(a(t)S^+ - a^-(t)S^-)$$

(12.39) etc. .

Clearly, $H(t)$ and $iP_{2r+2}(t)$ are bounded, self-adjoint operators.

Even though the expression (12.25) for $P_{2r+2}(t)$ looks much more complicated and clumsy in comparison to (12.24), we will see that this ruse of expanding $P_{2r+2}(t)$ in powers of $H(t)$ will turn out most favorable for our endeavor. But before we can see this, we need to make sure that the Lax equation is well-defined, that is, that $H(t)$ is differentiable.

First of all, please recall the following facts. Denote by $\mathfrak{B}(\ell^2(\mathbb{Z}))$ the C^*-algebra of bounded linear operators. Suppose $A, B : \mathbb{R} \to \mathfrak{B}(\ell^2(\mathbb{Z}))$ are differentiable with derivative $\dot{A}$, $\dot{B}$, respectively, then we have

- $A + B$ is differentiable with derivative $\dot{A} + \dot{B}$,
- AB is differentiable with derivative $\dot{A}B + A\dot{B}$,
- A^* is differentiable with derivative $\dot{A}^*$,
- A^{-1} (provided A is invertible) is differentiable with derivative $-A^{-1}\dot{A}A^{-1}$.

In addition, $f : \mathbb{R} \to \ell^\infty(\mathbb{Z})$ is differentiable if and only if the associated multiplication operator $f : \mathbb{R} \to \mathfrak{B}(\ell^2(\mathbb{Z}))$ is (since the embedding $\ell^\infty(\mathbb{Z}) \hookrightarrow \mathfrak{B}(\ell^2(\mathbb{Z}))$ is isometric).

For our original problem this implies that $H(t)$ and $P_{2r+2}(t)$ are differentiable since they are composed of differentiable operators. Hence the Lax equation (12.22) is well-defined and by (12.30) it is equivalent to

$$\mathrm{TL}_r(a(t), b(t))_1 = \dot{a}(t) - a(t)\big(g_{r+1}^+(t) - g_{r+1}(t)\big) = 0,$$
(12.40)
$$\mathrm{TL}_r(a(t), b(t))_2 = \dot{b}(t) - \big(h_{r+1}(t) - h_{r+1}^-(t)\big) = 0,$$

where the dot denotes a derivative with respect to t. Or, in integral form we have

$$a(t) = a(0) \exp\Big(\int_0^t \big(g_{r+1}^+(t) - g_{r+1}(t)\big) ds \Big),$$
(12.41)
$$b(t) = b(0) + \int_0^t \big(h_{r+1}(t) - h_{r+1}^-(t)\big) ds.$$

Varying $r \in \mathbb{N}_0$ yields the **Toda hierarchy** (TL hierarchy)

$$(12.42) \qquad \mathrm{TL}_r(a,b) = (\mathrm{TL}_r(a,b)_1, \mathrm{TL}_r(a,b)_2) = 0, \quad r \in \mathbb{N}_0.$$

Notice that multiplying $P_{2r+2}(t)$ with $c_0 \ne 0$ gives only a rescaled version of the Toda hierarchy which can be reduced to the original one by substituting $t \to t/c_0$. Hence our choice $c_0 = 1$.

Explicitly, one obtains from (12.27),

$$g_1 = b + c_1,$$
$$h_1 = 2a^2 + c_2,$$
$$g_2 = a^2 + (a^-)^2 + b^2 + c_1 b + c_2,$$
$$h_2 = 2a^2(b^+ + b) + c_1 2a^2 + c_3,$$
$$g_3 = a^2(b^+ + 2b) + (a^-)^2(2b + b^-) + b^3$$
$$\quad + c_1(a^2 + (a^-)^2 + b^2) + c_2 b + c_3,$$
$$h_3 = 2a^2((a^+)^2 + a^2 + (a^-)^2 + b^2 + b^+ b + (b^+)^2)$$
$$\quad + c_1 2a^2(b^+ + b) + c_2 2a^2 + c_4,$$

(12.43) etc.

and hence

$$\mathrm{TL}_0(a,b) = \begin{pmatrix} \dot{a} - a(b^+ - b) \\ \dot{b} - 2(a^2 - (a^-)^2) \end{pmatrix},$$

$$\mathrm{TL}_1(a,b) = \begin{pmatrix} \dot{a} - a((a^+)^2 - (a^-)^2 + (b^+)^2 - b^2) \\ \dot{b} - 2a^2(b^+ + b) + 2(a^-)^2(b + b^-) \end{pmatrix}$$
$$\quad - c_1 \begin{pmatrix} a(b^+ - b) \\ 2(a^2 - (a^-)^2) \end{pmatrix},$$

$$\mathrm{TL}_2(a,b) = \begin{pmatrix} \dot{a} - a((a^+)^2(b^{++} + 2b^+) + a^2(2b^+ + b) + (b^+)^3 - a^2(b^+ + 2b) - (a^-)^2(2b + b^-) - b^3) \\ \dot{b} + 2((a^-)^4 + (a^{--}a^-)^2 - a^2(a^2 + (a^+)^2 + (b^+)^2 + bb^+ + b^2) + (a^-)^2(b^2 + b^- b + (b^-)^2)) \end{pmatrix}$$
$$\quad - c_1 \begin{pmatrix} a((a^+)^2 - (a^-)^2 + (b^+)^2 - b^2) \\ 2a^2(b^+ + b) - 2(a^-)^2(b + b^-) \end{pmatrix} - c_2 \begin{pmatrix} a(b^+ - b) \\ 2(a^2 - (a^-)^2) \end{pmatrix},$$

(12.44) etc.

represent the first few equations of the Toda hierarchy. We will require $c_j \in \mathbb{R}$ even though c_j could depend on t. The corresponding **homogeneous Toda equations** obtained by taking all summation constants equal to zero, $c_\ell \equiv 0$, $1 \leq \ell \leq r$, are then denoted by

$$(12.45) \qquad \widetilde{\mathrm{TL}}_r(a,b) = \mathrm{TL}_r(a,b)\Big|_{c_\ell \equiv 0,\ 1 \leq \ell \leq r}.$$

We are interested in investigating the **initial value problem** associated with the Toda equations, that is,

$$(12.46) \qquad \mathrm{TL}_r(a,b) = 0, \quad \big(a(t_0), b(t_0)\big) = (a_0, b_0),$$

where (a_0, b_0) are two given (bounded) sequences. Since the Toda equations are autonomous, we will choose $t_0 = 0$ without restriction.

In order to draw a number of fundamental consequences from the Lax equation (12.22), we need some preparations.

Let $P(t)$, $t \in \mathbb{R}$, be a family of bounded skew-adjoint operators in $\ell^2(\mathbb{Z})$. A two parameter family of operators $U(t,s)$, $(t,s) \in \mathbb{R}^2$, is called a **unitary propagator** for $P(t)$, if

1. $U(t,s)$, $(s,t) \in \mathbb{R}^2$, is unitary.
2. $U(t,t) = \mathbb{1}$ for all $t \in \mathbb{R}$.
3. $U(t,s)U(s,r) = U(t,r)$ for all $(r,s,t) \in \mathbb{R}^3$.

12.2. The Toda hierarchy

4. The map $t \mapsto U(t,s)$ is differentiable in the Banach space $\mathfrak{B}(\ell^2(\mathbb{Z}))$ of bounded linear operators and

(12.47) $$\frac{d}{dt}U(t,s) = P(t)U(t,s), \quad (t,s) \in \mathbb{R}^2.$$

Note $U(s,t) = U(t,s)^{-1} = U(t,s)^*$ and $d/dt\, U(s,t) = -U(s,t)P(t)$.

With this notation the following well-known theorem from functional analysis holds:

Theorem 12.4. *Let $P(t)$, $t \in \mathbb{R}$, be a family of bounded skew-adjoint operators such that $t \mapsto P(t)$ is differentiable. Then there exists a unique unitary propagator $U(t,s)$ for $P(t)$.*

Proof. Consider the equation $\dot{U}(t) = P(t)U(t)$. By standard theory of differential equations, solutions for the initial value problem exist locally and are unique (cf., e.g., Theorem 4.1.5 of [1]). Moreover, since $\|P(t)\|$ is uniformly bounded on compact sets, all solutions are global. Hence we have a unique solution $U(t,s)$, $(t,s) \in \mathbb{R}^2$ such that $U(s,s) = \mathbb{1}$. It remains to verify that this propagator $U(t,s)$ is unitary. Comparing the adjoint equation

(12.48) $$\frac{d}{dt}U(t,s)^* = \Big(\frac{d}{dt}U(t,s)\Big)^* = (P(t)U(t,s))^* = -U(t,s)^*P(t)$$

and

(12.49) $$\frac{d}{dt}U(t,s)^{-1} = -U(t,s)^{-1}\Big(\frac{d}{dt}U(t,s)\Big)U(t,s)^{-1} = -U(t,s)^{-1}P(t)$$

we infer $U(t,s)^* = U(t,s)^{-1}$ by unique solubility of the initial value problem and $U(s,s)^* = U(s,s)^{-1} = \mathbb{1}$. $\square$

If $P(t) = P$ is actually time-independent (stationary solutions), then the unitary propagator is given by Stone's theorem, that is, $U(t,s) = \exp((t-s)P)$.

The situation for unbounded operators is somewhat more difficult and requires the operators $P(t)$, $t \in \mathbb{R}$, to have a common dense domain (cf. [**218**], Corollary on page 102, [**193**], Theorem X.69).

Now we can apply this fact to our situation.

Theorem 12.5. *Let $a(t), b(t)$ satisfy $\mathrm{TL}_r(a,b) = 0$ and (H.12.1). Then the Lax equation (12.22) implies existence of a unitary propagator $U_r(t,s)$ for $P_{2r+2}(t)$ such that*

(12.50) $$H(t) = U_r(t,s)H(s)U_r(t,s)^{-1}, \quad (t,s) \in \mathbb{R}^2.$$

Thus all operators $H(t)$, $t \in \mathbb{R}$, are unitarily equivalent and we might set

(12.51) $$\sigma(H) \equiv \sigma(H(t)) = \sigma(H(0)), \quad \rho(H) \equiv \rho(H(t)) = \rho(H(0)).$$

(Here $\sigma(.)$ and $\rho(.) = \mathbb{C}\backslash\sigma(.)$ denote the spectrum and resolvent set of an operator, respectively.)

In addition, if $\psi(s) \in \ell^2(\mathbb{Z})$ solves $H(s)\psi(s) = z\psi(s)$, then the function

(12.52) $$\psi(t) = U_r(t,s)\psi(s)$$

fulfills

(12.53) $$H(t)\psi(t) = z\psi(t), \quad \frac{d}{dt}\psi(t) = P_{2r+2}(t)\psi(t).$$

Proof. Let $U_r(t,s)$ be the unitary propagator for $P_{2r+2}(t)$. We need to show that $\tilde{H}(t) = U_r(t,s)^{-1} H(t) U_r(t,s)$ is equal to $H(s)$. Since $\tilde{H}(s) = H(s)$ it suffices to show that $\tilde{H}(t)$ is independent of t, which follows from

$$\frac{d}{dt}\tilde{H}(t) = U_r(t,s)^{-1}\Big(\frac{d}{dt}H(t) - [P_{2r+2}(t), H(t)]\Big)U_r(t,s) = 0. \tag{12.54}$$

The rest is immediate from the properties of the unitary propagator. □

To proceed with our investigation of the Toda equations, we ensure existence and uniqueness of global solutions next. To do this, we consider the Toda equations as a flow on the Banach space $M = \ell^\infty(\mathbb{Z}) \oplus \ell^\infty(\mathbb{Z})$.

Theorem 12.6. *Suppose $(a_0, b_0) \in M$. Then there exists a unique integral curve $t \mapsto (a(t), b(t))$ in $C^\infty(\mathbb{R}, M)$ of the Toda equations, that is, $\mathrm{TL}_r(a(t), b(t)) = 0$, such that $(a(0), b(0)) = (a_0, b_0)$.*

Proof. The r-th Toda equation gives rise to a vector field X_r on M, that is,

$$\frac{d}{dt}(a(t), b(t)) = X_r(a(t), b(t)) \quad \Leftrightarrow \quad \mathrm{TL}_r(a(t), b(t)) = 0. \tag{12.55}$$

Since this vector field has a simple polynomial dependence in a and b it is differentiable and hence (cf. again [1], Theorem 4.1.5) solutions of the initial value problem exist locally and are unique. In addition, by equation (12.50) we have $\|a(t)\|_\infty + \|b(t)\|_\infty \leq 2\|H(t)\| = 2\|H(0)\|$ (at least locally). Thus any integral curve $(a(t), b(t))$ is bounded on finite t-intervals implying global existence (see, e.g., [1], Proposition 4.1.22). □

Let $\tau(t)$ denote the differential expression associated with $H(t)$. If $\mathrm{Ker}(\tau(t)-z)$, $z \in \mathbb{C}$, denotes the two-dimensional nullspace of $\tau(t) - z$ (in $\ell(\mathbb{Z})$), we have the following representation of $P_{2r+2}(t)$ restricted to $\mathrm{Ker}(\tau(t)-z)$,

$$P_{2r+2}(t)\Big|_{\mathrm{Ker}(\tau(t)-z)} = 2a(t)G_r(z,t)S^+ - H_{r+1}(z,t), \tag{12.56}$$

where $G_r(z,n,t)$ and $H_{r+1}(z,n,t)$ are monic (i.e. the highest coefficient is one) polynomials given by

$$G_r(z,n,t) = \sum_{j=0}^{r} g_{r-j}(n,t) z^j,$$

$$H_{r+1}(z,n,t) = z^{r+1} + \sum_{j=0}^{r} h_{r-j}(n,t) z^j - g_{r+1}(n,t). \tag{12.57}$$

One easily obtains

$$\dot{a} = a\big(H_{r+1}^+ + H_{r+1} - 2(z - b^+)G_r^+\big),$$
$$\dot{b} = 2\big(a^2 G_r^+ - (a^-)^2 G_r^-\big) + (z-b)^2 G_r - (z-b) H_{r+1}. \tag{12.58}$$

As an illustration we record a few of the polynomials G_r and H_{r+1},

$$G_0 = 1 = \tilde{G}_0,$$
$$H_1 = z - b = \tilde{H}_1,$$
$$G_1 = z + b + c_1 = \tilde{G}_1 + c_1\tilde{G}_0,$$
$$H_2 = z^2 + a^2 - (a^-)^2 - b^2 + c_1(z-b) = \tilde{H}_2 + c_1\tilde{H}_1,$$
$$G_2 = z^2 + bz + a^2 + (a^-)^2 + b^2 + c_1(z+b) + c_2 = \tilde{G}_2 + c_1\tilde{G}_1 + c_2\tilde{G}_0,$$
$$H_3 = z^3 + 2a^2 z - 2(a^-)^2 b - b^3 + a^2 b^+ - (a^-)^2 b^-$$
$$+ c_1(z^2 + a^2 - (a^-)^2 - b^2) + c_2(z-b) = \tilde{H}_3 + c_1\tilde{H}_2 + c_2\tilde{H}_1,$$

(12.59)   etc. .

Here $\tilde{G}_r(z,n)$ and $\tilde{H}_{r+1}(z,n)$ are the homogeneous quantities corresponding to $G_r(z,n)$ and $H_{r+1}(z,n)$, respectively. By (12.38) we have

$$(12.60) \qquad G_r(z,n) = \sum_{\ell=0}^{r} c_{r-\ell}\tilde{G}_\ell(z,n), \quad H_{r+1}(z,n) = \sum_{\ell=0}^{r} c_{r-\ell}\tilde{H}_{\ell+1}(z,n).$$

Remark 12.7. (i). Since, by (12.27), $a(t)$ enters quadratically in $g_j(t), h_j(t)$, respectively $G_r(z,.,t), H_{r+1}(z,.,t)$, the Toda hierarchy (12.42) is invariant under the substitution

$$(12.61) \qquad a(n,t) \to \varepsilon(n)a(n,t),$$

where $\varepsilon(n) \in \{+1, -1\}$. This result should be compared with Lemma 1.6.

(ii). If $a(n_0, 0) = 0$ we have $a(n_0, t) = 0$ for all $t \in \mathbb{R}$ (by (12.41)). This implies $H = H_{-,n_0+1} \oplus H_{+,n_0}$ with respect to the decomposition $\ell^2(\mathbb{Z}) = \ell^2(-\infty, n_0+1) \oplus \ell^2(n_0, \infty)$. Hence $P_{2r+2} = P_{-,n_0+1,2r+2} \oplus P_{+,n_0,2r+2}$ decomposes as well and we see that the Toda lattice also splits up into two independent parts $\text{TL}_{\pm,n_0,r}$. In this way, the half line Toda lattice follows from our considerations as a special case. Similarly, we can obtain Toda lattices on finite intervals.

12.3. Stationary solutions

In this section we specialize to the stationary Toda hierarchy characterized by $\dot{a} = \dot{b} = 0$ in (12.42) or, equivalently, by commuting difference expressions

$$(12.62) \qquad [P_{2r+2}, H] = 0$$

of order $2r+2$ and 2, respectively. Equations (12.58) then yield the equivalent conditions

$$(z-b)(H_{r+1} - H_{r+1}^-) = 2a^2 G_r^+ - 2(a^-)^2 G_r,$$
$$(12.63) \qquad H_{r+1}^+ + H_{r+1} = 2(z - b^+)G_r^+.$$

Comparison with Section 8.3 suggests to define

$$(12.64) \qquad R_{2r+2} = (H_{r+1}^2 - 4a^2 G_r G_r^+).$$

A simple calculation using (12.63)

$$(z-b)(R_{2r+2} - R_{2r+2}^-)$$
$$= (z-b)\big((H_{r+1} + H_{r+1}^-)(H_{r+1} - H_{r+1}^-) - 4G_r(a^2 G_r^+ - (a^-)^2 G_r^-)\big)$$
$$(12.65) \quad = 2\big(H_{r+1} + H_{r+1}^- - 2(z-b)G_r\big)\big(a^2 G_r^+ - (a^-)^2 G_r^-\big) = 0$$

then proves that R_{2r+2} is independent of n. Thus one infers

$$(12.66) \qquad R_{2r+2}(z) = \prod_{j=0}^{2r+1}(z - E_j), \quad \{E_j\}_{0 \le j \le 2r+1} \subset \mathbb{C}.$$

The resulting hyperelliptic curve of (arithmetic) genus r obtained upon compactification of the curve

$$(12.67) \qquad w^2 = R_{2r+2}(z) = \prod_{j=0}^{2r+1}(z - E_j)$$

will be the basic ingredient in our algebro-geometric treatment of the Toda hierarchy in Section 13.1.

Equations (12.63), (12.64) plus Theorem 2.31 imply

$$(12.68) \qquad g(z,n) = \frac{G_r(z,n)}{R_{2r+2}^{1/2}(z)}, \quad h(z,n) = \frac{H_{r+1}(z,n)}{R_{2r+2}^{1/2}(z)},$$

where $g(z,n) = \langle \delta_n, (H-z)^{-1}\delta_n \rangle$, $h(z,n) = 2a(n)\langle \delta_{n+1}, (H-z)^{-1}\delta_n \rangle - 1$ as usual.

Despite these similarities we need to emphasize that the numbers E_m, $0 \le m \le 2r+1$, do not necessarily satisfy (8.54). This is no contradiction but merely implies that there must be common factors in the denominators and numerators of (12.68) which cancel. That is, if the number of spectral gaps of H is $s+2$, then there is a monic polynomial $Q_{r-s}(z)$ (independent of n) such that $G_r(z,n) = Q_{r-s}(z)G_s(z,n)$, $H_{r+1}(z,n) = Q_{r-s}(z)H_{s+1}(z,n)$, and $R_{2r+2}^{1/2}(z) = Q_{r-s}(z)^2 R_{2s+2}^{1/2}(z)$.

We have avoided these factors in Section 8.3 (which essentially correspond to closed gaps (cf. Remark 7.6)). For example, in case of the constant solution $a(n,t) = 1/2$, $b(n,t) = 0$ of $\mathrm{TL}_r(a,b) = 0$ we have for $r = 0, 1, \ldots$

$$G_0(z) = 1, \quad H_1(z) = z, \quad R_2(z) = z^2 - 1,$$
$$G_1(z) = z + c_1, \quad H_2(z) = (z + c_1)z, \quad R_4(z) = (z + c_1)^2(z^2 - 1),$$

(12.69) etc. .

Conversely, any given reflectionless finite gap sequences (a,b) satisfy (12.63) and hence give rise to a stationary solution of some equation in the Toda hierarchy and we obtain

Theorem 12.8. *The stationary solutions of the Toda hierarchy are precisely the reflectionless finite-gap sequences investigated in Section 8.3.*

In addition, with a little more work, we can even determine which equation (i.e., the constants c_j). This will be done by relating the polynomials $G_r(z,n)$, $H_{r+1}(z,n)$ to the homogeneous quantities $\tilde{G}_r(z,n)$, $\tilde{H}_{r+1}(z,n)$. We introduce the constants $c_j(\underline{E})$, $\underline{E} = (E_0, \ldots, E_{2r+1})$, by

$$(12.70) \qquad R_{2r+2}^{1/2}(z) = -z^{r+1}\sum_{j=0}^{\infty} c_j(\underline{E})z^{-j}, \quad |z| > \|H\|,$$

12.3. Stationary solutions

implying

$$c_0(\underline{E}) = 1, \quad c_1(\underline{E}) = -\frac{1}{2}\sum_{j=0}^{2r+1} E_j, \quad \text{etc.}. \tag{12.71}$$

Lemma 12.9. *Let $a(n)$, $b(n)$ be given reflectionless finite-gap sequences (see Section 8.3) and let $G_r(z,n)$, $H_{r+1}(z,n)$ be the associated polynomials (see (8.66) and (8.70)). Then we have (compare (12.60))*

$$G_r(z,n) = \sum_{\ell=0}^{r} c_{r-\ell}(\underline{E})\tilde{G}_\ell(z,n), \quad H_{r+1}(z,n) = \sum_{\ell=0}^{r} c_{r-\ell}(\underline{E})\tilde{H}_{\ell+1}(z,n). \tag{12.72}$$

In addition, $\tilde{g}_\ell(n)$ can be expressed in terms of E_j and $\mu_j(n)$ by

$$\tilde{g}_1(n) = b^{(1)}(n), \quad \tilde{g}_\ell(n) = \frac{1}{\ell}\Big(b^{(\ell)}(n) + \sum_{j=1}^{\ell-1}\frac{j}{\ell}\tilde{g}_{\ell-j}(n)b^{(j)}(n)\Big), \quad \ell > 1, \tag{12.73}$$

where

$$b^{(\ell)}(n) = \frac{1}{2}\sum_{j=0}^{2r+1} E_j^\ell - \sum_{j=1}^{r} \mu_j(n)^\ell. \tag{12.74}$$

Proof. From (8.96) we infer for $|z| > \|H\|$, using Neumann's expansion for the resolvent of H and the explicit form of $\tilde{g}, \tilde{h}$ given in Theorem 12.2, that

$$G_r(z,n) = -\frac{R_{2r+2}^{1/2}(z)}{z}\sum_{\ell=0}^{\infty} \tilde{g}_\ell(n)z^{-\ell},$$

$$H_{r+1}(z,n) = R_{2r+2}^{1/2}(z)\Big(1 - \frac{1}{z}\sum_{\ell=0}^{\infty} \tilde{h}_\ell(n)z^{-\ell}\Big). \tag{12.75}$$

This, together with (12.70), completes the first part. The rest follows from (6.59) and Theorem 6.10. $\square$

Corollary 12.10. *Let $a(n)$, $b(n)$ be given reflectionless finite-gap sequences with corresponding polynomial $R_{2s+2}^{1/2}(z)$. Then (a,b) is a stationary solution of TL_r if and only if there is a constant polynomial $Q_{r-s}(z)$ of degree $r-s$ such that $c_j = c_j(\underline{E})$, where $\underline{E}$ is the vector of zeros of $R_{2r+2}^{1/2}(z) = Q_{r-s}(z)^2 R_{2s+2}^{1/2}(z)$.*

It remains to show how all stationary solutions for a given equation of the Toda hierarchy can be found. In fact, up to this point we don't even know whether the necessary conditions (12.63) can be satisfied for arbitrary choice of the constants $(c_j)_{1 \leq j \leq r}$.

Suppose $(c_j)_{1 \leq j \leq r}$ is given and define $d_0 = c_0 = 1$,

$$d_j = 2c_j + \sum_{\ell=1}^{j-1} c_\ell c_{j-\ell}, \quad 1 \leq j \leq r. \tag{12.76}$$

Choose $d_j \in \mathbb{R}$, $r+1 \leq j \leq 2r+2$, and define $(E_j)_{0 \leq j \leq 2r+1}$ by

$$R_{2r+2}(z) = \sum_{j=0}^{2r+2} d_{2r+2-j}z^j = \prod_{j=0}^{2r+1}(z - E_j). \tag{12.77}$$

Note that our choice of d_j implies $c_j(\underline{E}) = c_j$, $1 \leq j \leq r$, which is a necessary condition by Lemma 12.9. Since $g(z,n)$ cannot be meromorphic, $R_{2r+2}(z)$ must not be a complete square. Hence those choices of $d_j \in \mathbb{R}$, $r+1 \leq j \leq 2r+2$, have to be discarded. For any other choice we obtain a list of band edges satisfying (8.54) after throwing out all closed gaps. In particular, setting

$$\Sigma(\underline{d}) = \bigcup_{j=0}^{r} [E_{2j}, E_{2j+1}] \tag{12.78}$$

any operator in $\mathrm{Iso}_R(\Sigma(\underline{d}))$ (see Section 8.3) produces a stationary solution. Thus, the procedure of Section 8.3 can be used to compute all corresponding stationary gap solutions.

Theorem 12.11. *Fix* TL_r, *that is, fix* $(c_j)_{1 \leq j \leq r}$ *and* r. *Let* $(d_j)_{1 \leq j \leq r}$ *be given by (12.76). Then all stationary solutions of* TL_r *can be obtained by choosing* $(d_j)_{r+1 \leq j \leq 2r+2}$ *such that* $R_{2r+2}(z)$ *defined as in (12.77) is not a complete square and then choosing any operator in the corresponding isospectral class* $\mathrm{Iso}_R(\Sigma(\underline{d}))$.

Remark 12.12. The case where $R_{2r+2}(z)$ is a complete square corresponds to stationary solutions with $a(n) = 0$ for some n. For instance, $G_0(n) = 1$, $H_1(n) = z - b(n)$, $R_{2r+2}(z) = (z - b(n))^2$ corresponds to $a(n) = 0$, $n \in \mathbb{Z}$.

Finally, let us give a further interpretation of the polynomial $R_{2r+2}(z)$.

Theorem 12.13. *The polynomial* $R_{2r+2}(z)$ *is the* **Burchnall-Chaundy polynomial** *relating* P_{2r+2} *and* H, *that is,*

$$P_{2r+2}^2 = R_{2r+2}(H) = \prod_{j=0}^{2r+1} (H - E_j). \tag{12.79}$$

Proof. Because of (12.62) one computes

$$\left(P_{2r+2}\Big|_{\mathrm{Ker}(\tau-z)}\right)^2 = \left((2aG_rS^+ - H_{r+1})\Big|_{\mathrm{Ker}(\tau-z)}\right)^2$$
$$= \left(2aG_r(2(z-b^+)G_r^+ - H_{r+1}^+ - H_{r+1})S^+ + H_{r+1}^2 - 4a^2G_rG_r^+\right)\Big|_{\mathrm{Ker}(\tau-z)}$$
$$= R_{2r+2}(z)\Big|_{\mathrm{Ker}(\tau-z)}$$

(12.80)

and since $z \in \mathbb{C}$ is arbitrary, the rest follows from Corollary 1.3. □

This result clearly shows again the close connection between the Toda hierarchy and hyperelliptic curves of the type $M = \{(z,w)|w^2 = \prod_{j=0}^{2r+2}(z - E_j)\}$.

12.4. Time evolution of associated quantities

For our further investigations in the next two chapters, it will be important to know how several quantities associated with the Jacobi operator $H(t)$ vary with respect to t.

First we will try to calculate the time evolution of the fundamental solutions $c(z,.,t)$, $s(z,.,t)$ if $a(t)$, $b(t)$ satisfy $\mathrm{TL}_r(a,b) = 0$. For simplicity of notation we will not distinguish between $H(t)$ and its differential expression $\tau(t)$. To emphasize

12.4. Time evolution of associated quantities

that a solution of $H(t)u = zu$ is not necessarily in $\ell^2(\mathbb{Z})$ we will call such solutions **weak solutions**. Similarly for $P_{2r+2}(t)$.

First, observe that (12.22) implies

$$(12.81) \qquad (H(t) - z)(\frac{d}{dt} - P_{2r+2}(t))\Phi(z, ., t) = 0,$$

where $\Phi(z, n, t)$ is the transfer matrix from (1.30). But this means

$$(12.82) \qquad (\frac{d}{dt} - P_{2r+2}(t))\Phi(z, ., t) = \Phi(z, ., t)C_r(z, t)$$

for a certain matrix $C_r(z, t)$. If we evaluate the above expression at $n = 0$, using $\Phi(z, 0, t) = \mathbb{1}$, we obtain

$$(12.83) \begin{aligned} C_r(z, t) &= \bigl(P_{2r+2}(t)\Phi(z, ., t)\bigr)(0) \\ &= \begin{pmatrix} -H_{r+1}(z, 0, t) & 2a(0, t)G_r(z, 0, t) \\ -2a(0, t)G_r(z, 1, t) & 2(z - b(1, t))G_r(z, 1, t) - H_{r+1}(z, 1, t) \end{pmatrix}. \end{aligned}$$

The time evolutions of $c(z, n, t)$ and $s(z, n, t)$ now follow from

$$(12.84) \qquad \dot\Phi(z, ., t) = P_{2r+2}\Phi(z, ., t) + \Phi(z, ., t)C_r(z, t)$$

or more explicitly

$$\begin{aligned} \dot c(z, n, t) &= 2a(n, t)G_r(z, n, t)c(z, n+1, t) - \bigl(H_{r+1}(z, n, t) \\ &\quad + H_{r+1}(z, 0, t)\bigr)c(z, n, t) - 2a(0, t)G_r(z, 1, t)s(z, n, t), \\ \dot s(z, n, t) &= 2a(n, t)G_r(z, n, t)s(z, n+1, t) - (H_{r+1}(z, n, t) + H_{r+1}(z, 1, t) \end{aligned}$$
$$(12.85) \quad - 2(z - b(1, t))G_r(z, 1, t))s(z, n, t) + 2a(0, t)G_r(z, 0, t)c(z, n, t).$$

Remark 12.14. In case of periodic coefficients, this implies for the time evolution of the monodromy matrix $M(z, t)$

$$(12.86) \qquad \frac{d}{dt}M(z, t) = [M(z, t), C_r(z, n, t)],$$

where

$$(12.87) \qquad C_r(z, n, t) = \begin{pmatrix} -H_{r+1} & 2aG_r \\ -2aG_r^+ & 2(z - b^+)G_r^+ - H_{r+1} \end{pmatrix}.$$

This shows (take the trace) that the discriminant is time independent and that $\{E_j\}_{j=1}^{2N}$, A, and B (cf. Section 7.1) are time independent.

Evaluating (12.86) explicitly and expressing everything in terms of our polynomials yields (omitting some dependencies)

$$(12.88) \quad \begin{aligned} \frac{d}{dt}\begin{pmatrix} -H/2 & aG \\ aG^+ & H/2 \end{pmatrix} &= \\ \begin{pmatrix} 4a^2(G_rG^+ - G_r^+G) & 2aG_rH + (z - 2b^+)G_r^+G \\ -2aG_r^+H - a(z - 2b^+)G_r^+G^+) & -4a^2(G_rG^+ - G_r^+G) \end{pmatrix}. \end{aligned}$$

Equation (12.84) enables us to prove

Lemma 12.15. *Assume (H.12.1) and suppose* $\mathrm{TL}_r(a, b) = 0$. *Let* $u_0(z, n)$ *be a weak solution of* $H(0)u_0 = zu_0$. *Then the system*

$$(12.89) \qquad H(t)u(z, n, t) = zu(z, n, t), \quad \frac{d}{dt}u(z, n, t) = P_{2r+2}(t)u(z, n, t)$$

has a unique weak solution fulfilling the initial condition

(12.90) $$u(z, n, 0) = u_0(z, n).$$

If $u_0(z, n)$ is continuous (resp. holomorphic) with respect to z, then so is $u(z, n, t)$.
Furthermore, if $u_{1,2}(z, n, t)$ both solve (12.89), then

(12.91) $$W(u_1(z), u_2(z)) = W_n(u_1(z, t), u_2(z, t)) = a(n, t)\Big(u_1(z, n, t)u_2(z, n+1, t) - u_1(z, n+1, t)u_2(z, n, t)\Big)$$

depends neither on n nor on t.

Proof. Any solution $u(z, n, t)$ of the system (12.89) can be written as

(12.92) $$u(z, n, t) = u(z, 0, t)c(z, n, t) + u(z, 1, t)s(z, n, t)$$

and from (12.84) we infer that (12.89) is equivalent to the ordinary differential equation

(12.93) $$\begin{pmatrix} \dot{u}(z, 0, t) \\ \dot{u}(z, 1, t) \end{pmatrix} = -C_r(z, t) \begin{pmatrix} u(z, 0, t) \\ u(z, 1, t) \end{pmatrix}, \quad \begin{pmatrix} u(z, 0, 0) \\ u(z, 1, 0) \end{pmatrix} = \begin{pmatrix} u_0(z, 0) \\ u_0(z, 1) \end{pmatrix},$$

which proves the first assertion. The second is a straightforward calculation using (12.56) and (12.58). □

The next lemma shows that solutions which are square summable near $\pm\infty$ for one $t \in \mathbb{R}$ remain square summable near $\pm\infty$ for all $t \in \mathbb{R}$, respectively.

Lemma 12.16. *Let $u_{\pm,0}(z, n)$ be a solution of $H(0)u = zu$ which is square summable near $\pm\infty$. Then the solution $u_\pm(z, n, t)$ of the system (12.89) with initial data $u_{\pm,0}(z, n) \in \ell^2_\pm(\mathbb{Z})$ is square summable near $\pm\infty$ for all $t \in \mathbb{R}$, respectively.*

Denote by $G(z, n, m, t)$ the Green function of $H(t)$. Then we have $(z \in \rho(H))$

(12.94) $$G(z, m, n, t) = \frac{1}{W(u_-(z), u_+(z))} \begin{cases} u_+(z, n, t)u_-(z, m, t) & \text{for } m \leq n \\ u_+(z, m, t)u_-(z, n, t) & \text{for } n \leq m \end{cases}.$$

Especially, if $z < \sigma(H)$ and $a(n, t) < 0$ we can choose $u_{\pm,0}(z, n) > 0$, implying $u_\pm(z, n, t) > 0$.

Proof. We only prove the u_- case (the u_+ case follows from reflection) and drop z for notational simplicity. By Lemma 12.15 we have a solution $u(n, t)$ of (12.89) with initial condition $u(n, 0) = u_{+,0}(n)$ and hence

(12.95) $$S(n, t) = S(n, 0) + 2\int_0^t \text{Re} \sum_{j=-n}^{0} \overline{u(j, s)} P_{2r+2}(s) u(j, s) ds,$$

where $S(n, t) = \sum_{j=-n}^{0} |u(j, t)|^2$. Next, by boundedness of $a(t)$, $b(t)$, we can find a constant $C > 0$ such that $4|H_{r+1}(n, t)| \leq C$ and $8|a(n, t)G_r(n, t)| \leq C$. Using (12.56) and the Cauchy-Schwarz inequality yields

(12.96) $$\Big|\sum_{j=-n}^{0} \overline{u(j, s)} P_{2r+2}(s) u(j, s)\Big| \leq \frac{C}{2}\Big(|u(1, s)|^2 + S(n, s)\Big).$$

Invoking Gronwall's inequality shows

(12.97) $$S(n, t) \leq \Big(S(n, 0) + C\int_0^t |u(1, s)|^2 e^{-Cs} ds\Big) e^{Ct}$$

12.4. Time evolution of associated quantities

and letting $n \to \infty$ implies that $u(z,.,t) \in \ell_-^2(\mathbb{Z})$.

Since $u(z,n,t) = 0$, $z < \sigma(H)$, is not possible by Lemma 2.6, positivity follows as well and we are done. □

Using the Lax equation (12.22) one infers ($z \in \rho(H)$)

$$\frac{d}{dt}(H(t)-z)^{-1} = [P_{2r+2}(t),(H(t)-z)^{-1}]. \tag{12.98}$$

Furthermore, using (12.94) we obtain for $m \le n$

$$\frac{d}{dt}G(z,n,m,t) =$$
$$= \frac{u_-(z,m)(P_{2r+2}u_+(z,.))(n) + u_+(z,n)(P_{2r+2}u_-(z,.))(m)}{W(u_-(z),u_+(z))}. \tag{12.99}$$

As a consequence (use (12.56) and (12.58)) we also have

$$\frac{d}{dt}g(z,n,t) = 2\big(G_r(z,n,t)h(z,n,t) - g(z,n,t)H_{r+1}(z,n,t)\big),$$
$$\frac{d}{dt}h(z,n,t) = 4a(n,t)^2\big(G_r(z,n,t)g(z,n+1,t)$$
$$- g(z,n,t)G_r(z,n+1,t)\big) \tag{12.100}$$

and

$$\frac{d}{dt}\tilde{g}_j(t) = -2\tilde{g}_{r+j+1}(t) + 2\sum_{\ell=0}^r \big(g_{r-\ell}(t)\tilde{h}_{\ell+j}(t) - \tilde{g}_{\ell+j}(t)h_{r-\ell}(t)\big)$$
$$+ 2g_{r+1}(t)\tilde{g}_j(t),$$
$$\frac{d}{dt}\tilde{h}_j(t) = 4a(t)^2\sum_{\ell=0}^r \big(g_{r-\ell}(t)\tilde{g}_{\ell+j}^+(t) - \tilde{g}_{\ell+j}(t)g_{r-\ell}^+(t)\big). \tag{12.101}$$

Chapter 13

The initial value problem for the Toda system

In the previous chapter we have seen that the initial value problem associated with the Toda equations has a unique (global) solution. In this section we will consider certain classes of initial conditions and derive explicit formulas for solutions in these special cases.

13.1. Finite-gap solutions of the Toda hierarchy

In this section we want to construct reflectionless finite-gap solutions (see Section 8.3) for the Toda hierarchy. Our starting point will be an r-gap stationary solution (a_0, b_0) of the type (8.83). This r-gap stationary solution (a_0, b_0) represents the initial condition for our Toda flow (cf. (12.42)),

$$(13.1) \qquad \widehat{\mathrm{TL}}_s(a(t), b(t)) = 0, \quad (a(0), b(0)) = (a_0, b_0)$$

for some $s \in \mathbb{N}_0$, whose explicit solution we seek. That is, we take a stationary solution of the Toda hierarchy and consider the time evolution with respect to a (in general) different equation of the Toda hierarchy. To stress this fact, we use a hat for the Toda equation (and all associated quantities) which gives rise to the time evolution.

From our treatment in Section 8.3 we know that (a_0, b_0) is determined by the band edges $(E_j)_{0 \leq j \leq 2r+1}$ and the Dirichlet eigenvalues $(\hat{\mu}_{0,j}(n_0))_{1 \leq j \leq r}$ at a fixed point $n_0 \in \mathbb{Z}$. Since we do not know whether the reflectionless property of the initial condition is preserved by the Toda flow, $(a(t), b(t))$ might not be reflectionless. However, if we suppose that $(a(t), b(t))$ is reflectionless for all t, we have

$$(13.2) \qquad g(z, n_0, t) = \frac{\prod_{j=1}^{r} z - \mu_j(n_0, t)}{R_{2r+2}^{1/2}(z)}.$$

Plugging (13.2) into (12.100) and evaluating at $\mu_j(n_0, t)$ we arrive at the following time evolution for the Dirichlet eigenvalues $\hat{\mu}_j(n_0, t)$

(13.3)
$$\frac{d}{dt}\mu_j(n_0, t) = -2\hat{G}_s(\mu_j(n_0, t), n_0, t)\sigma_j(n_0, t)R_j(n_0, t),$$
$$\hat{\mu}_j(n_0, 0) = \hat{\mu}_{0,j}(n_0), \quad 1 \leq j \leq r, \, t \in \mathbb{R},$$

(recall (8.67)). Here $\hat{G}_s(z, n_0, t)$ has to be expressed in terms of $\hat{\mu}_j(n_0, t)$, $1 \leq j \leq r$ (cf. Lemma 12.9).

Now our strategy consists of three steps.

1. Show solvability of the system (13.3).
2. Construct $a(n, t)$, $b(n, t)$ (and associated quantities) from $(\hat{\mu}_j(n_0, t))_{1 \leq j \leq r}$ as in Section 8.3.
3. Show that $a(n, t)$, $b(n, t)$ solve $\hat{\mathrm{TL}}_r(a, b) = 0$.

In order to show our first step, that is, solubility of the system (13.3) we will first introduce a suitable manifold M^D for the Dirichlet eigenvalues $\hat{\mu}_j$. For the sake of simplicity we will only consider the case $r = 2$ with $E_0 = E_1 < E_2 = E_3 < E_4 < E_5$, the general case being similar. We define M^D as the set of all points $(\hat{\mu}_1, \hat{\mu}_2)$ satisfying (H.8.12) together with the following charts (compare Section A.7).

(i). Set $U_{\tilde{\sigma}_1, \tilde{\sigma}_2} = \{(\hat{\mu}_1, \hat{\mu}_2) \in M^D | \mu_{1,2} \neq E_2, \mu_2 \neq E_4, \sigma_{1,2} = \tilde{\sigma}_{1,2}\}$, $U'_{\tilde{\sigma}_1, \tilde{\sigma}_2} = (E_0, E_2) \times (E_2, E_4)$.

(13.4)
$$\begin{array}{cccc}
U_{\tilde{\sigma}_1, \tilde{\sigma}_2} & \to & U'_{\tilde{\sigma}_1, \tilde{\sigma}_2} & \quad U'_{\tilde{\sigma}_1, \tilde{\sigma}_2} \to U_{\tilde{\sigma}_1, \tilde{\sigma}_2} \\
(\hat{\mu}_1, \hat{\mu}_2) & \mapsto & (\mu_1, \mu_2) & \quad (\mu_1, \mu_2) \mapsto ((\mu_1, \tilde{\sigma}_1), (\mu_2, \tilde{\sigma}_2))
\end{array}.$$

(ii). Set $U_{E_2} = \{(\hat{\mu}_1, \hat{\mu}_2) \in M^D | \mu_2 \neq E_4, \sigma_1 = -\sigma_2\}$, $U'_{E_2} = (E_0 - E_4, E_4 - E_0) \times (0, \infty)$.

(13.5)
$$\begin{array}{ll}
U_{E_2} & \to U'_{E_2} \\
(\hat{\mu}_1, \hat{\mu}_2) & \mapsto \left(\sigma_1(\mu_2 - \mu_1), -\left(\frac{\mu_2 - E_2}{\mu_1 - E_2}\right)^{\sigma_1}\right), \\
((E_2, \sigma), (E_2, \sigma)) & \mapsto (0, \frac{1-\sigma}{1+\sigma}) \\
U'_{E_2} & \to U_{E_2} \\
(\nu_1 > 0, \nu_2) & \mapsto \left((E_2 - \frac{\nu_1}{1+\nu_2}, +), (E_2 + \frac{\nu_1\nu_2}{1+\nu_2}, -)\right) \\
(0, \nu_2) & \mapsto \left((E_2, \frac{1-\nu_2}{1+\nu_2}), (E_2, \frac{1-\nu_2}{1+\nu_2})\right) \\
(\nu_1 < 0, \nu_2) & \mapsto \left((E_2 + \frac{\nu_1\nu_2}{1+\nu_2}, -), (E_2 - \frac{\nu_1}{1+\nu_2}, +)\right)
\end{array}.$$

(iii). Set $U_{E_4, \sigma} = \{(\hat{\mu}_1, \hat{\mu}_2) \in M^D | \mu_{1,2} \neq E_2, \sigma_1 = \sigma\}$, $U'_{E_4, \sigma} = (E_0, E_2) \times (-\sqrt{E_4 - E_2}, \sqrt{E_4 - E_2})$.

(13.6)
$$\begin{array}{l}
U_{E_4, \sigma} \to U'_{E_4, \sigma} \\
(\hat{\mu}_1, \hat{\mu}_2) \mapsto (\mu_1, \sigma_2\sqrt{E_4 - \mu_2}) \\
U'_{E_4, \sigma} \to U_{E_4, \sigma} \\
(\zeta_1, \zeta_2) \mapsto ((\zeta_1, \sigma), (E_4 - \zeta_2^2, \mathrm{sgn}(\zeta_2)))
\end{array}.$$

13.1. Finite-gap solutions of the Toda hierarchy

Next, we look at our system (13.3). In the chart (i) it reads (omitting the dependence on t and n_0)

$$
\begin{aligned}
\dot{\mu}_1 &= -2\hat{G}_s(\mu_1)\frac{\sigma_1 R_6^{1/2}(\mu_1)}{\mu_1 - \mu_2}, \\
\dot{\mu}_2 &= -2\hat{G}_s(\mu_2)\frac{\sigma_2 R_6^{1/2}(\mu_2)}{\mu_2 - \mu_1}.
\end{aligned}
\tag{13.7}
$$

In the chart (ii) it reads ($\nu_1 \neq 0$)

$$
\begin{aligned}
\dot{\nu}_1 &= \frac{1}{1+\nu_2}\left(\nu_2 F_s(\frac{\nu_1 \nu_2}{1+\nu_2}) - F_s(\frac{\nu_1}{1+\nu_2})\right), \\
\dot{\nu}_2 &= \nu_2 \frac{F_s(\frac{\nu_1 \nu_2}{1+\nu_2}) - F_s(\frac{\nu_1}{1+\nu_2})}{\nu_1},
\end{aligned}
\tag{13.8}
$$

where

$$
F_s(z) = \frac{\hat{G}_s(z)R_6^{1/2}(z)}{z - E_2} = -2\hat{G}_s(z)(z-E_0)\sqrt{z-E_4}\sqrt{z-E_5}.
\tag{13.9}
$$

Note that $F_s(z)$ remains invariant if μ_1 and μ_2 are exchanged. Moreover, the singularity at $\nu_1 = 0$ in the second equation is removable and hence this vector field extends to the whole of U'_{E_2}.

Finally, in the chart (iii) it reads

$$
\begin{aligned}
\dot{\zeta}_1 &= -2\hat{G}_s(\xi_1)\frac{\sigma_1 R_6^{1/2}(\xi_1)}{\xi_1 - E_4 + \xi_2^2}, \\
\dot{\zeta}_2 &= -2\hat{G}_s(E_4 - \xi_2^2)\frac{(E_4 - E_0 + \xi_2^2)(E_4 - E_2 + \xi_2^2)\sqrt{E_5 - E_4 - \xi_2^2}}{\xi_1 - E_4 + \xi_2^2}.
\end{aligned}
\tag{13.10}
$$

Hence our system (13.3) is a C^∞ vector field on the manifold M^D. Since M^D is not compact, this implies unique solubility only locally. Thus we need to make sure that any solution with given initial conditions in M^D stays within M^D. Due to the factor $(\mu_1 - E_0)$ (hidden in $R_6^{1/2}(\mu_1)$) in (13.7), μ_1 cannot reach E_0. If μ_2 hits E_4 it simply changes its sign σ_2 and moves back in the other direction. Because of the factor ν_2 in (13.8), ν_2 remains positive. Finally, we have to consider the case where $\mu_{1,2}$ is close to E_2 and $\sigma_1 = \sigma_2$. In this case we can make the coordinate change $\nu_1 = \mu_1 + \mu_2$, $\nu_2 = (E_2 - \mu_1)(\mu_2 - E_2)$. Again we get $\dot{\nu}_2 = \nu_2(\dots)$ and hence ν_2 remains positive or equivalently neither μ_1 nor μ_2 can reach E_2 if $\sigma_1 = \sigma_2$.

Concluding, for given initial conditions $(\hat{\mu}_1(0), \hat{\mu}_2(0)) = (\hat{\mu}_{0,1}, \hat{\mu}_{0,2}) \in M^D$ we get a unique solution $(\hat{\mu}_1(t), \hat{\mu}_2(t)) \in C^\infty(\mathbb{R}, M^D)$. Due to the requirement $\mu_1 \in (E_0, E_2]$ and $\mu_2 \in [E_2, E_4]$, the functions $\mu_{1,2}(t)$ are not $C^\infty(\mathbb{R}, \mathbb{R})$! However, note that the transformation (13.5) amounts to exchanging μ_1 and μ_2 whenever μ_1 and μ_2 cross E_2. Hence if $F(\mu_1, \mu_2) \in C^\infty(\mathbb{R}^2, \mathbb{R})$ is symmetric with respect to μ_1 and μ_2, then we have $F(\mu_1(.), \mu_2(.)) \in C^\infty(\mathbb{R}, \mathbb{R})$.

In summary,

Theorem 13.1. *The system (13.3) has a unique global solution*

$$
\left(\hat{\mu}_j(n_0, .)\right)_{1 \leq j \leq r} \in C^\infty(\mathbb{R}, M^D)
\tag{13.11}
$$

for each initial condition satisfying (H.8.12).

If $s = r$ and $\hat{c}_\ell = c_\ell$, $1 \leq \ell \leq s$, we obtain, as expected from the stationary r-gap outset, $\hat{\mu}_j(n_0, t) = \hat{\mu}_{0,j}(n_0)$ from (13.17) since $G_r(\hat{\mu}_{0,j}(n_0), n_0, 0) = 0$.

Now we come to the second step. Using (13.11) we can define $a(n,t)$, $b(n,t)$ and associated polynomials $G_r(z, n_0, t)$, $H_{r+1}(z, n_0, t)$ as in Section 8.3 (cf. (8.83) and (8.66), (8.70)).

Rather than trying to compute the time derivative of $a(n,t)$ and $b(n,t)$ directly, we consider $G_r(z, n, t)$ and $H_{r+1}(z, n, t)$ first. We start with the calculation of the time derivative of $G_r(z, n_0, t)$. With the help of (13.3) we obtain

$$(13.12) \quad \frac{d}{dt} G_r(z, n_0, t)\Big|_{z=\mu_j(n_0,t)} = 2\hat{G}_s(\mu_j(n_0, t), n_0, t) H_{r+1}(\mu_j(n_0, t), n_0, t),$$

$1 \leq j \leq r$. Since two polynomials of (at most) degree $r - 1$ coinciding at r points are equal, we infer

$$(13.13) \quad \frac{d}{dt} G_r(z, n_0, t) = 2\big(\hat{G}_s(z, n_0, t) H_{r+1}(z, n_0, t) - G_r(z, n_0, t) \hat{H}_{s+1}(z, n_0, t)\big),$$

provided we can show that the right-hand side of (13.13) is a polynomial of degree less or equal to $r - 1$. By (12.60) it suffices to consider the homogeneous case. We divide (13.13) by $R_{2r+2}^{1/2}(z)$ and use (12.75) to express $G_r(z, n_0, t)$ and $H_{r+1}(z, n_0, t)$ respectively (12.57) to express $\tilde{G}_s(z, n_0, t)$ and $\tilde{H}_{s+1}(z, n_0, t)$. Collecting powers of z shows that the coefficient of z^j, $0 \leq j \leq r$, is

$$(13.14) \quad \tilde{g}_{s-j} - \sum_{\ell=j+1}^{s} \tilde{g}_{\ell-j-1} \tilde{h}_{s-\ell} - \tilde{g}_{s-j} + \sum_{\ell=j+1}^{s} \tilde{h}_{s-\ell} \tilde{g}_{\ell-j-1} = 0,$$

which establishes the claim.

To obtain the time derivative of $H_{r+1}(z, n_0, t)$ we use

$$(13.15) \quad H_{r+1}(z, n_0, t)^2 - 4a(n_0, t)^2 G_r(z, n_0, t) G_r(z, n_0 + 1, t) = R_{2r+2}^{1/2}(z)$$

as in (8.73). Again, evaluating the time derivative first at $\mu_j(n_0, t)$, one obtains

$$(13.16) \quad \frac{d}{dt} H_{r+1}(z, n_0, t) = 4a(n_0, t)^2 \big(\hat{G}_s(z, n_0, t) G_r(z, n_0 + 1, t) \\ - G_r(z, n_0, t) \hat{G}_s(z, n_0 + 1, t)\big)$$

for those $t \in \mathbb{R}$ such that $\mu_j(n_0, t) \notin \{E_{2j-1}, E_{2j}\}$, provided the right-hand side of (13.16) is of degree at most $r - 1$. This can be shown as before. Since the exceptional set is discrete, the identity follows for all $t \in \mathbb{R}$ by continuity.

Similarly, differentiating (13.15) and evaluating $(d/dt)G_r(z, n_0 + 1, t)$ at $z = \mu_j(n_0 + 1, t)$ (the zeros of $G_r(z, n_0 + 1, t)$), we see that (13.13) also holds with n_0 replaced by $n_0 + 1$ and finally that $(\hat{\mu}_j(n_0 + 1, t))_{1 \leq j \leq r}$ satisfies (13.3) with initial condition $\hat{\mu}_j(n_0 + 1, 0) = \hat{\mu}_{0,j}(n_0 + 1)$. Proceeding inductively, we obtain this result for all $n \geq n_0$ and with a similar calculation (cf. Section 8.3) for all $n \leq n_0$.

Summarizing, we have constructed the points $(\hat{\mu}_j(n, t))_{1 \leq j \leq r}$ for all $(n, t) \in \mathbb{Z} \times \mathbb{R}$ such that

$$\frac{d}{dt} \mu_j(n, t) = -2\hat{G}_s(\mu_j(n, t), n, t) \sigma_j(n, t) R_j(n, t),$$

$$(13.17) \quad \hat{\mu}_j(n, 0) = \hat{\mu}_{0,j}(n), \quad 1 \leq j \leq r, \ (n, t) \in \mathbb{Z} \times \mathbb{R},$$

with $(\hat{\mu}_j(n, .))_{1 \leq j \leq r} \in C^\infty(\mathbb{R}, M^D)$.

13.1. Finite-gap solutions of the Toda hierarchy

Furthermore, we have corresponding polynomials $G_r(z,n,t)$ and $H_{r+1}(z,n,t)$ satisfying

$$
\begin{aligned}
\frac{d}{dt}G_r(z,n,t) &= 2\Big(\hat{G}_s(z,n,t)H_{r+1}(z,n,t) \\
&\quad - G_r(z,n,t)\hat{H}_{s+1}(z,n,t)\Big)
\end{aligned}
\tag{13.18}
$$

and

$$
\begin{aligned}
\frac{d}{dt}H_{r+1}(z,n,t) &= 4a(n,t)^2\Big(\hat{G}_s(z,n,t)G_r(z,n+1,t) \\
&\quad - G_r(z,n,t)\hat{G}_s(z,n+1,t)\Big).
\end{aligned}
\tag{13.19}
$$

It remains to verify that the sequences $a(t)$, $b(t)$ solve the Toda equation $\widehat{\mathrm{TL}}_s(a,b) = 0$.

Theorem 13.2. *The solution $(a(t), b(t))$ of the $\widehat{\mathrm{TL}}_s$ equations (13.1) with r-gap initial conditions (a_0, b_0) of the type (8.83) is given by*

$$
\begin{aligned}
a(n,t)^2 &= \frac{1}{2}\sum_{j=1}^{r}\hat{R}_j(n,t) + \frac{1}{8}\sum_{j=0}^{2r+1}E_j^2 - \frac{1}{4}\sum_{j=1}^{r}\mu_j(n,t)^2 - \frac{1}{4}b(n,t)^2, \\
b(n,t) &= \frac{1}{2}\sum_{j=0}^{2r+1}E_j - \sum_{j=1}^{r}\mu_j(n,t),
\end{aligned}
\tag{13.20}
$$

where $(\hat{\mu}_j(n,t))_{1\le j \le r}$ is the unique solution of (13.17).

Proof. Differentiating (13.15) involving (13.18), (13.19) and (8.82) yields the first equation of (12.58). Differentiating (8.82) using (13.18), (13.19) and (8.101) yields the second equation of (12.58). $\square$

Observe that this algorithm is constructive and can be implemented numerically (it only involves finding roots of polynomials and integrating ordinary differential equations).

Next, let us deduce some additional information from the results obtained thus far.

Clearly (13.18) and (13.19) imply

$$
\begin{aligned}
\frac{d}{dt}K^\beta_{r+1}(z,n,t) &= 2a(n,t)\Big(\big(\beta^{-1}G_r(z,n+1,t) - \beta G_r(z,n,t)\big)\hat{K}^\beta_{s+1}(z,n,t) \\
&\quad - \big(\beta^{-1}\hat{G}_s(z,n+1,t) - \beta\hat{G}_s(z,n,t)\big)K^\beta_{r+1}(z,n,t)\Big) \\
&\quad + \frac{\dot{a}(n,t)}{a(n,t)}\Big(K^\beta_{r+1}(z,n,t) + \beta\big(\beta^{-1}G_r(z,n+1,t) - \beta G_r(z,n,t)\big)\Big)
\end{aligned}
\tag{13.21}
$$

and as a consequence we note (use (8.113))

Lemma 13.3. *The time evolution of the zeros* $\lambda_j^\beta(n,t)$, $\beta \in \mathbb{R}\backslash\{0\}$, *of the polynomial* $K_{r+1}^\beta(.,n,t)$ *is given by*

$$\frac{d}{dt}\lambda_j^\beta(n,t) = \Big(\frac{a(n,t)}{\beta}\hat{K}_{s+1}^\beta(\lambda_j^\beta(n,t),n,t) + \frac{\dot{a}(n,t)}{2a(n,t)}\Big)$$
(13.22)
$$\times \frac{R_{2r+2}^{1/2}(\lambda_j^\beta(n,t))}{\prod_{\ell \neq j}^{r+1} \lambda_j^\beta(n,t) - \lambda_\ell^\beta(n,t)}.$$

If $\lambda_j^\beta(n,t) = \lambda_{j+1}^\beta(n,t)$ *the right-hand side has to be replaced by a limit as in (8.67).*

Finally, let us compute the time dependent Baker-Akhiezer function. Using

(13.23) $\quad \phi(p,n,t) = \dfrac{H_{r+1}(p,n,t) + R_{2r+2}^{1/2}(p)}{2a(n,t)G_r(p,n,t)} = \dfrac{2a(n,t)G_r(p,n+1,t)}{H_{r+1}(p,n,t) - R_{2r+2}^{1/2}(p)}$

we define for $\psi(p,n,n_0,t)$

$$\psi(p,n,n_0,t) = \exp\Big(\int_0^t \big(2a(n_0,x)\hat{G}_s(p,n_0,x)\phi(p,n_0,x)$$
(13.24)
$$- \hat{H}_{s+1}(p,n_0,x)\big)dx\Big)\prod_{m=n_0}^{n-1}{}^* \phi(p,m,t).$$

Straightforward calculations then imply

(13.25) $\quad a(n,t)\phi(p,n,t) + a(n-1,t)\phi(p,n-1,t)^{-1} = \pi(p) - b(n,t),$

$$\frac{d}{dt}\ln\phi(p,n,t) = -2a(n,t)\big(\hat{G}_s(p,n,t)\phi(p,n,t) + \hat{G}_s(p,n+1,t)\phi(p,n,t)^{-1}\big)$$
$$+ 2(\pi(p) - b(n+1,t))\hat{G}_s(p,n+1,t)$$
$$- \hat{H}_{s+1}(p,n+1,t) + \hat{H}_{s+1}(p,n,t)$$
$$= 2a(n+1,t)\hat{G}_s(p,n+1,t)\phi(p,n+1,t)$$
$$- 2a(n,t)\hat{G}_s(p,n,t)\phi(p,n,t)$$
(13.26)
$$- \hat{H}_{s+1}(p,n+1,t) + \hat{H}_{s+1}(p,n,t).$$

Similarly, (compare Lemma 12.15)

(13.27) $\quad\begin{aligned}a(n,t)\psi(p,n+1,n_0,t) + a(n-1,t)\psi(p,n-1,n_0,t)\\= (\pi(p) - b(n,t))\psi(p,n,n_0,t),\end{aligned}$

and

$$\frac{d}{dt}\psi(p,n,n_0,t) = 2a(n,t)\hat{G}_s(p,n,t)\psi(p,n+1,n_0,t)$$
$$- \hat{H}_{s+1}(p,n,t)\psi(p,n,n_0,t)$$
(13.28)
$$= \hat{P}_{2s+2}(t)\psi(p,.,n_0,t)(n).$$

The analogs of all relations found in Section 8.3 clearly extend to the present time dependent situation.

13.2. Quasi-periodic finite-gap solutions and the time-dependent Baker-Akhiezer function

In this section we again consider the case with no eigenvalues, that is $r = g$, since we do not want to deal with singular curves. The manifold for our Dirichlet eigenvalues is in this case simply the submanifold $\otimes_{j=1}^{g} \pi^{-1}([E_{2j-1}, E_{2j}]) \subset M_g$ introduced in Lemma 9.1.

For simplicity we set $n_0 = 0$ and omit it in the sequel. In order to express $\phi(p, n, t)$ and $\psi(p, n, t)$ in terms of theta functions, we need a bit more of notation. Let $\omega_{\infty_\pm, j}$ be the normalized Abelian differential of the second kind with a single pole at $\infty_\pm$ of the form (using the chart (A.102))

$$(13.29) \qquad \omega_{\infty_\pm, j} = \bigl(w^{-2-j} + O(1)\bigr) dw \quad \text{near } \infty_\pm, \quad j \in \mathbb{N}_0.$$

For $\omega_{\infty_+, j} - \omega_{\infty_-, j}$ we can make the following ansatz

$$(13.30) \qquad \omega_{\infty_+, j} - \omega_{\infty_-, j} = \sum_{k=0}^{g+j+1} \tilde{c}_{g+j-k+1} \frac{\pi^k d\pi}{R_{2g+2}^{1/2}}.$$

From the asymptotic requirement (13.29) we obtain $\tilde{c}_k = c_k(\underline{E})$, $0 \le k \le j+1$, where $c_k(\underline{E})$ is defined in (12.70). The remaining g constants $\tilde{c}_{k+j+1}$, $1 \le k \le g$, have to be determined from the fact that the a-periods must vanish.

Given the summation constants $\hat{c}_1, \ldots, \hat{c}_s$ in $\hat{G}_s$, see (12.38), we then define

$$(13.31) \qquad \Omega_s = \sum_{j=0}^{s} (j+1)\hat{c}_{s-j}(\omega_{\infty_+, j} - \omega_{\infty_-, j}), \quad \hat{c}_0 = 1.$$

Since the differentials $\omega_{\infty_\pm, j}$ were supposed to be normalized, we have

$$(13.32) \qquad \int_{a_j} \Omega_s = 0, \quad 1 \le j \le g.$$

Next, let us write down the expansion for the abelian differentials of the first kind ζ_j,

$$(13.33) \qquad \zeta_j = \left(\sum_{\ell=0}^{\infty} \eta_{j,\ell}(\infty_\pm) w^\ell\right) dw = \pm \left(\sum_{\ell=0}^{\infty} \eta_{j,\ell}(\infty_+) w^\ell\right) dw \quad \text{near } \infty_\pm.$$

Using (A.116) we infer

$$(13.34) \qquad \underline{\eta}_\ell = \sum_{k=\max\{1, g-\ell\}}^{g} \underline{c}(k) d_{\ell-g+k}(\underline{E}),$$

where $d_k(\underline{E})$ is defined by

$$(13.35) \qquad \frac{1}{R_{2g+2}^{1/2}(z)} = \frac{-1}{z^{g+1}} \sum_{k=0}^{\infty} \frac{d_k(\underline{E})}{z^k}, \quad d_0(\underline{E}) = 1, \; d_1(\underline{E}) = \frac{1}{2} \sum_{k=0}^{2g+1} E_k, \text{ etc. }.$$

In addition, relation (A.19) yields

$$(13.36) \qquad U_{s,j} = \frac{1}{2\pi i} \int_{b_j} \Omega_s = 2 \sum_{\ell=0}^{s} \hat{c}_{s-\ell} \eta_{j,\ell}(\infty_+), \quad 1 \le j \le g.$$

As in Lemma 9.7 we will show uniqueness of the Baker-Akhiezer function.

Lemma 13.4. *Let $\psi(.,n,t)$, $(n,t) \in \mathbb{Z} \times \mathbb{R}$, be meromorphic on $M\backslash\{\infty_+,\infty_-\}$ with essential singularities at $\infty_\pm$ such that*

(13.37) $$\tilde{\psi}(p,n,t) = \psi(p,n,t) \exp\left(-t \int_{p_0}^p \Omega_s\right)$$

is multivalued meromorphic on M and its divisor satisfies

(13.38) $$(\tilde{\psi}(.,n,t)) \geq -\mathcal{D}_{\hat{\underline{\mu}}(0,0)} + n(\mathcal{D}_{\infty_+} - \mathcal{D}_{\infty_-}).$$

Define a divisor $\mathcal{D}_{\mathrm{zer}}(n,t)$ by

(13.39) $$(\tilde{\psi}(.,n,t)) = \mathcal{D}_{\mathrm{zer}}(n,t) - \mathcal{D}_{\hat{\underline{\mu}}(0,0)} + n(\mathcal{D}_{\infty_+} - \mathcal{D}_{\infty_-}).$$

If $\mathcal{D}_{\hat{\underline{\mu}}(0,0)}$ is nonspecial, then so is $\mathcal{D}_{\mathrm{zer}}(n,t)$ for all $(n,t) \in \mathbb{Z} \times \mathbb{R}$, that is, if

(13.40) $$i(\mathcal{D}_{\mathrm{zer}}(n,t)) = 0, \quad (n,t) \in \mathbb{Z} \times \mathbb{R},$$

and $\psi(.,n,t)$ is unique up to a constant multiple (which may depend on n and t).

Proof. By the Riemann-Roch theorem for multivalued functions (cf. [81], III.9.12) there exists at least one Baker-Akhiezer function. The rest follows as in Lemma 9.7. $\square$

Given these preparations we obtain the following characterization of $\phi(p,n,t)$ and $\psi(p,n,t)$ in (13.23) and (13.24).

Theorem 13.5. *Introduce*

$$\underline{z}(p,n,t) = \underline{\hat{A}}_{p_0}(p) - \underline{\hat{\alpha}}_{p_0}(\mathcal{D}_{\hat{\underline{\mu}}(0,0)}) + 2n\underline{\hat{A}}_{p_0}(\infty_+) + t\underline{U}_s - \underline{\hat{\Xi}}_{p_0},$$
(13.41) $$\underline{z}(n,t) = \underline{z}(\infty_+,n,t).$$

Then we have

$$\phi(p,n,t) = C(n,t) \frac{\theta(\underline{z}(p,n+1,t))}{\theta(\underline{z}(p,n,t))} \exp\left(\int_{p_0}^p \omega_{\infty_+,\infty_-}\right),$$

(13.42) $\psi(p,n,t) = C(n,0,t) \dfrac{\theta(\underline{z}(p,n,t))}{\theta(\underline{z}(p,0,0))} \exp\left(n \int_{p_0}^p \omega_{\infty_+,\infty_-} + t \int_{p_0}^p \Omega_s\right),$

where $C(n,t)$, $C(n,0,t)$ are real-valued,

(13.43) $$C(n,t)^2 = \frac{\theta(\underline{z}(n-1,t))}{\theta(\underline{z}(n+1,t))}, \quad C(n,0,t)^2 = \frac{\theta(\underline{z}(0,0))\theta(\underline{z}(-1,0))}{\theta(\underline{z}(n,t))\theta(\underline{z}(n-1,t))},$$

and the sign of $C(n,t)$ is opposite to that of $a(n,t)$. Moreover,

(13.44) $$\underline{\alpha}_{p_0}(\mathcal{D}_{\hat{\underline{\mu}}(n,t)}) = \underline{\alpha}_{p_0}(\mathcal{D}_{\hat{\underline{\mu}}(0,0)}) - 2n\underline{A}_{p_0}(\infty_+) - [t\underline{U}_s],$$

where $[.]$ denotes the equivalence class in $J(M)$ (cf. (A.53)).

Hence the flows (13.11) are linearized by the Abel map

(13.45) $$\frac{d}{dt}\underline{\hat{\alpha}}_{p_0}(\mathcal{D}_{\hat{\underline{\mu}}(n,t)}) = -\underline{U}_s.$$

Proof. First of all note that both functions in (13.42) are well-defined due to (9.34), (13.32), (13.36), and (A.69).

Denoting the right-hand side of (13.42) by $\Psi(p,n,t)$, our goal is to prove $\psi = \Psi$. By Theorem 9.2 it suffices to identify

(13.46) $$\psi(p,0,t) = \Psi(p,0,t).$$

13.2. Quasi-periodic finite-gap solutions

We start by noting that (13.23), (13.24), and (13.18) imply

$$\psi(p,0,t) = \exp\left(\int_0^t \Big(2a(0,x)\hat{G}_s(z,0,x)\phi(p,0,x) - \hat{H}_{s+1}(z,0,x)\Big)dx\right)$$

(13.47) $$= \exp\left(\int_0^t \Big(\frac{\hat{G}_s(z,0,x)R_{2g+2}^{1/2}(p)H_{g+1}(z,0,x)}{G_g(z,0,x)} - \hat{H}_{s+1}(z,0,x)\Big)dx\right).$$

In order to spot the zeros and poles of ψ on $M\backslash\{\infty_+,\infty_-\}$ we need to expand the integrand in (13.47) near its singularities (the zeros $\mu_j(0,x)$ of $G_g(z,0,x)$). Using (13.17) one obtains

$$\psi(p,0,t) = \exp\left(\int_0^t \Big(\frac{\frac{d}{dx}\mu_j(0,x)}{\mu_j(0,x)-\pi(p)} + O(1)\Big)dx\right)$$

(13.48) $$= \begin{cases} (\mu_j(0,t)-\pi(p))O(1) & \text{for } p \text{ near } \hat{\mu}_j(0,t) \neq \hat{\mu}_j(0,0) \\ O(1) & \text{for } p \text{ near } \hat{\mu}_j(0,t) = \hat{\mu}_j(0,0) \\ (\mu_j(0,0)-\pi(p))^{-1}O(1) & \text{for } p \text{ near } \hat{\mu}_j(0,0) \neq \hat{\mu}_j(0,t) \end{cases},$$

with $O(1) \neq 0$. Hence all zeros and all poles of $\psi(p,0,t)$ on $M\backslash\{\infty_+,\infty_-\}$ are simple and the poles coincide with those of $\Psi(p,0,t)$. Next, we need to identify the essential singularities of $\psi(p,0,t)$ at $\infty_\pm$. For this purpose we use (13.25) and rewrite (13.47) in the form

$$\psi(p,0,t) = \exp\left(\int_0^t \Big(\frac{1}{2}\frac{\frac{d}{dx}G_g(z,0,x)}{G_g(z,0,x)} + R_{2g+2}^{1/2}(p)\frac{\hat{G}_s(z,0,x)}{G_g(z,0,x)}\Big)dx\right)$$

(13.49) $$= \left(\frac{G_g(z,0,t)}{G_g(z,0,0)}\right)^{1/2} \exp\left(R_{2g+2}^{1/2}(p)\int_0^t \frac{\hat{G}_s(z,0,x)}{G_g(z,0,x)}dx\right).$$

We claim that

(13.50) $$R_{2g+2}^{1/2}(p)\hat{G}_s(z,n,t)/G_g(z,n,t) = \mp\sum_{\ell=0}^s \hat{c}_{s-\ell}z^{1+\ell} + O(1) \text{ for } p \text{ near } \infty_\pm.$$

By (12.38), in order to prove (13.50), it suffices to prove the homogeneous case $\hat{c}_0 = 1$, $\hat{c}_\ell = 0$, $1 \leq \ell \leq s$, that is $\hat{G}_s = \tilde{G}_s$. Using (8.96), we may rewrite (13.50) in the form

$$\tilde{G}_s(z,n,t)z^{-s-1} = z^{-1}\sum_{\ell=0}^s \tilde{f}_{s-\ell}(n,t)z^{\ell-s}$$

(13.51) $$= -g(z,n,t) + O(z^{-s-1}) \quad \text{as } z \to \infty.$$

But this follows from (6.7) and hence we conclude $\psi = const(t)\Psi$. The remaining constant can be determined as in Theorem 9.2. □

It is instructive to give a direct proof that (13.17) is linearized by Abel's map. It suffices to assume that all $\mu_j(n,t)$ are away from the band edges E_m (the general case follows from continuity) and that $\hat{c}_j = 0$, $1 \leq j \leq s$. Therefore we calculate (in the cart induced by the projection) using (9.17), (13.17)

$$
\begin{aligned}
\frac{d}{dt}\alpha_{p_0}(\mathcal{D}_{\hat{\underline{\mu}}(n,t)}) &= \sum_{j=1}^{g} \dot{\mu}_j(n,t) \sum_{k=1}^{g} \underline{c}(k) \frac{\mu_j(n,t)^{k-1}}{R_{2g+2}^{1/2}(\hat{\mu}_j(n,t))} \\
&= -2 \sum_{j,k=1}^{g} \underline{c}(k) \frac{\tilde{G}_s(\mu_j(n,t),n,t)}{\prod_{\ell \neq j}(\mu_j(n,t) - \mu_\ell(n,t))} \mu_j(n,t)^{k-1}
\end{aligned}
$$

$$
(13.52) \qquad = \frac{-2}{2\pi i} \sum_{k=1}^{g} \underline{c}(k) \int_\Gamma \frac{\tilde{G}_s(z,n,t)}{\prod_{\ell=1}^{g}(z - \mu_\ell(n,t))} z^{k-1} dz,
$$

where Γ is a closed path in $\mathbb{C}$ encircling all Dirichlet eigenvalues $\mu_j(n,t)$. The last integral can be evaluated by calculating the residue at ∞. Using (see (13.51))

$$
(13.53) \qquad \frac{\tilde{G}_s(z,n,t)}{\prod_{j=1}^{g}(z - \mu_k(n,t))} = \frac{\tilde{G}_s(z,n,t)}{g(z,n,t)} \frac{1}{R_{2g+2}^{1/2}(z)} = \frac{z^{s+1}(1+O(z^{-s}))}{-R_{2g+2}^{1/2}(z)}
$$

we infer (cf. (13.34))

$$
(13.54) \qquad \frac{d}{dt}\alpha_{p_0}(\mathcal{D}_{\hat{\underline{\mu}}(n,t)}) = -2 \sum_{k=\max\{1,g-s\}}^{g} \underline{c}(k) d_{s-g+k}(\underline{E}) = -2\underline{\eta}_s = -\underline{U}_s.
$$

Equation (9.13) shows that the flows (13.22) are linearized by the Abel map as well

$$
(13.55) \qquad \frac{d}{dt}\hat{\alpha}_{p_0}(\mathcal{D}_{\hat{\underline{\lambda}}^\beta(n,t)}) = -\underline{U}_s.
$$

Finally, the θ-function representation for the time dependent quasi-periodic g-gap solutions of the Toda hierarchy follows as in Theorem 9.4.

Theorem 13.6. *The solution $(a(t), b(t))$ of the $\widehat{\mathrm{TL}}_s$ equations (13.1) with g-gap initial conditions (a_0, b_0) is given by*

$$
a(n,t)^2 = \tilde{a}^2 \frac{\theta(\underline{z}(n-1,t))\theta(\underline{z}(n+1,t))}{\theta(\underline{z}(n,t))^2},
$$

$$
b(n,t) = \tilde{b} + \sum_{j=1}^{g} c_j(g) \frac{\partial}{\partial w_j} \ln\left(\frac{\theta(\underline{w} + \underline{z}(n,t))}{\theta(\underline{w} + \underline{z}(n-1,t))}\right)\bigg|_{\underline{w}=0}
$$

$$
= E_0 + \tilde{a} \frac{\theta(\underline{z}(n-1,t))\theta(\underline{z}(p_0, n+1, t))}{\theta(\underline{z}(n,t))\theta(\underline{z}(p_0, n, t))}
$$

$$
(13.56) \qquad + \tilde{a} \frac{\theta(\underline{z}(n,t))\theta(\underline{z}(p_0, n-1, t))}{\theta(\underline{z}(n-1,t))\theta(\underline{z}(p_0, n, t))},
$$

with $\tilde{a}$, $\tilde{b}$ introduced in (9.42).

13.3. A simple example – continued

We now continue our simple example of Section 1.3. The sequences

$$
(13.57) \qquad a(n,t) = a_0 = \frac{1}{2}, \qquad b(n,t) = b_0 = 0,
$$

solve the Toda equation $\mathrm{TL}_r(a_0, b_0) = 0$ since we have

$$G_{0,r}(z, n, t) = G_{0,r}(z) = \prod_{j=1}^{r}(z - E_{2j}),$$

$$H_{0,r+1}(z, n, t) = H_{0,r+1}(z) = z\prod_{j=1}^{r}(z - E_{2j}),$$

(13.58) $$R_{2r+2}(z) = \prod_{j=0}^{2r-1}(z - E_j),$$

where $E_0 = -1$, $E_{2r+1} = 1$, $E_{2j-1} = E_{2j}$, $1 \leq j \leq r$. In addition, we have

(13.59) $$\tilde{g}_{0,2j} = \binom{2j}{j}, \quad \tilde{g}_{0,2j+1} = 0, \quad \tilde{h}_{0,j} = \tilde{g}_{0,j-1}.$$

Furthermore, the sequences

(13.60) $$\psi_\pm(z, n, t) = k^{\pm n} \exp\left(\frac{\pm \alpha_r(k)t}{2}\right), \quad z = \frac{k + k^{-1}}{2},$$

where

(13.61) $$\alpha_r(k) = 2\big(kG_{0,r}(z) - H_{0,r+1}(z)\big) = (k - k^{-1})G_{0,r}(z)$$

satisfy (compare Section 1.3)

$$H_0(t)\psi_\pm(z, n, t) = z\psi_\pm(z, n, t),$$
$$\frac{d}{dt}\psi_\pm(z, n, t) = P_{0,2r+2}(t)\psi_\pm(z, n, t)$$
(13.62) $$= 2a_0 G_{0,r}(z)\psi_\pm(z, n+1, t) - H_{0,r+1}(z)\psi_\pm(z, n, t).$$

Note $\alpha_r(k) = -\alpha_r(k^{-1})$. Explicitly we have

(13.63)
$$\tilde{G}_{0,2r}(z) = \sum_{j=0}^{r} \frac{(2j)!}{4^j (j!)^2} z^{2(r-j)}, \quad \tilde{G}_{0,2r+1}(z) = z\tilde{G}_{0,2r}(z), \quad \tilde{H}_{0,r+1}(z) = z\tilde{G}_{0,r}(z)$$

and hence

$$\alpha_0(k) = k - k^{-1},$$
$$\alpha_1(k) = k^2 - k^{-2} + c_1(k - k^{-1}),$$
(13.64) $$\text{etc. .}$$

13.4. Inverse scattering transform

This section is an application of Chapter 10 to the Toda case. Our strategy is similar to that of the previous section. Again we will look for the time evolution of suitable spectral data, namely the scattering data associated with $H(t)$. We first show that it suffices to check (10.1) for one $t_0 \in \mathbb{R}$.

Lemma 13.7. *Suppose $a(n, t), b(n, t)$ is a solution of the Toda system satisfying (10.1) for one $t_0 \in \mathbb{R}$, then (10.1) holds for all $t \in \mathbb{R}$, that is,*

(13.65) $$\sum_{n \in \mathbb{Z}} |n|\big(|1 - 2a(n, t)| + |b(n, t)|\big) < \infty.$$

Proof. Without loss of generality we choose $t_0 = 0$. Shifting $a \to a - \frac{1}{2}$ we can consider the norm

$$\|(a,b)\|_* = \sum_{n \in \mathbb{Z}}(1+|n|)(|1-2a(n)|+|b(n)|). \tag{13.66}$$

Proceeding as in Theorem 12.6 we conclude local existence of solutions with respect to this norm. Next, we note that we have the estimate

$$\sum_{n \in \mathbb{Z}}(1+|n|)|g_r(n,t) - g_{0,r}| \leq C_r(\|H(0)\|)\|(a(t),b(t))\|_*,$$

$$\sum_{n \in \mathbb{Z}}(1+|n|)|h_r(n,t) - h_{0,r}| \leq C_r(\|H(0)\|)\|(a(t),b(t))\|_*, \tag{13.67}$$

where $C_r(\|H(0)\|)$ is some positive constant. It suffices to consider the homogeneous case $c_j = 0$, $1 \leq j \leq r$. In this case the claim follows by induction from equations (6.9) and (6.13) (note that we have $g_i(n,t)g_j(m,t) - g_{0,i}g_{0,j} = (g_i(n,t) - g_{0,i})g_j(m,t) - g_{0,i}(g_j(m,t) - g_{0,j}))$. Hence we infer from (12.40)

$$|a(n,t) - \frac{1}{2}| \leq |a(n,0) - \frac{1}{2}| + \|H(0)\| \int_0^t |g_{r+1}(n,s) - g_{0,r+1}|$$
$$+ |g_{r+1}(n+1,s) - g_{0,r+1}|ds,$$

$$|b(n,t)| \leq |b(n,0)| + \int_0^t |h_{r+1}(n,s) - h_{0,r+1}|$$
$$+ |h_{r+1}(n-1,s) - h_{0,r+1}|ds \tag{13.68}$$

and thus

$$\|(a(t),b(t))\|_* \leq \|(a(0),b(0))\|_* + \tilde{C}\int_0^t \|(a(s),b(s))\|_* ds, \tag{13.69}$$

where $\tilde{C} = 2(1+\|H(0)\|)C_{r+1}(\|H(0)\|)$. The rest follows from Gronwall's inequality. $\square$

Thus we can define Jost solutions, transmission and reflection coefficients as in Chapter 10 with the only difference that they now depend on an additional parameter $t \in \mathbb{R}$. For example, we set

$$S_\pm(H(t)) = \{R_\pm(k,t), |k|=1;\ k_\ell, \gamma_{\pm,\ell}(t),\ 1 \leq \ell \leq N\}. \tag{13.70}$$

Clearly we are interested how the scattering data vary with respect to t.

Theorem 13.8. *Suppose $a(n,t)$, $b(n,t)$ is a solution of the Toda system satisfying (10.1) for one $t_0 \in \mathbb{R}$. Then the functions*

$$\exp(\pm\alpha_r(k)t)f_\pm(k,n,t) \tag{13.71}$$

fulfill (12.89) with $z = (k+k^{-1})/2$, where $f_\pm(k,n,t)$ are the Jost solutions and $\alpha_r(k)$ is defined in (13.61). In addition, we have

$$T(k,t) = T(k,0),$$
$$R_\pm(k,t) = R_\pm(k,0)\exp(\pm\alpha_r(k)t),$$
$$\gamma_{\pm,\ell}(t) = \gamma_{\pm,\ell}(0)\exp(\mp 2\alpha_r(k_\ell)t),\quad 1\leq \ell \leq N. \tag{13.72}$$

13.4. Inverse scattering transform

Proof. As in Lemma 7.10 one shows that $f_\pm(k,n,t)$ is continuously differentiable with respect to t and that $\lim_{n\to\pm\infty} k^{\mp n} \hat{f}_\pm(k,n,t) \to 0$ (use the estimates (13.67)). Now let $(k+k^{-1})/2 \in \rho(H)$, then Lemma 12.16 implies that the solution of (12.89) with initial condition $f_\pm(k,n,0)$, $|k| < 1$, is of the form

$$(13.73) \qquad C_\pm(t) f_\pm(k,n,t).$$

Inserting this into (12.89), multiplying with $k^{\mp n}$, and evaluating as $n \to \pm\infty$ yields

$$(13.74) \qquad C_\pm(t) = \exp(\pm \alpha_r(k) t).$$

The general result for all $|k| \leq 1$ now follows from continuity. This immediately implies the formulas for $T(k,t), R_\pm(k,t)$. Finally, let $k = k_\ell$. Then we have

$$(13.75) \qquad \exp(\pm \alpha_r(k_\ell) t) f_\pm(k_\ell, n, t) = U_r(t,0) f_\pm(k_\ell, n, 0)$$

(compare (12.52)), which implies

$$(13.76) \qquad \frac{d}{dt} \frac{\exp(\pm 2\alpha_r(k_\ell) t)}{\gamma_{\pm,\ell}(t)} = \frac{d}{dt} \|U_r(t,0) f_\pm(k_\ell,.,0)\| = 0$$

and concludes the proof. $\square$

Thus $S_\pm(H(t))$ can be expressed in terms of $S_\pm(H(0))$ as follows

$$(13.77) \qquad S_\pm(H(t)) = \{\; R_\pm(k,0)\exp(\pm\alpha_r(k)t), |k|=1;\\ k_\ell, \gamma_{\pm,\ell}(0)\exp(\mp 2\alpha_r(k_\ell)t),\; 1\leq \ell \leq N\}$$

and $a(n,t), b(n,t)$ can be computed from $a(n,0), b(n,0)$ using the Gel'fand-Levitan-Marchenko equations. This procedure is known as inverse scattering transform and can be summarized as follows:

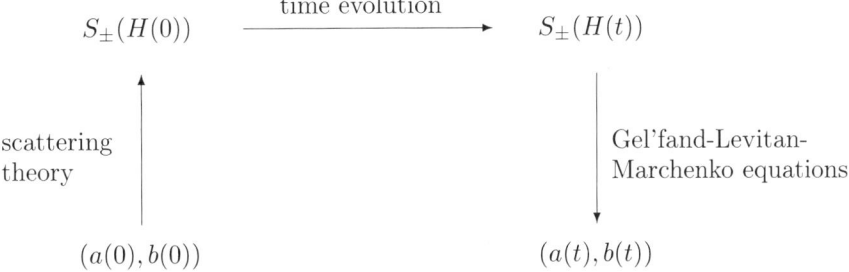

Hence, in the short range situation, the initial value problem for the Toda lattice equations can be reduced to solving three linear problems. The mathematical pillars of this method are formed by Theorem 12.6, Lemma 13.7, Theorem 13.8, and Theorem 10.10.

Since the transmission coefficient is time independent, so is the perturbation determinant (10.36)

$$(13.78) \qquad \alpha(k(z),t) = \alpha(k(z),0) = \exp\left(\int_\mathbb{R} \frac{\xi_\alpha(\lambda) d\lambda}{\lambda - z}\right).$$

Moreover, the traces

(13.79) $$\text{tr}\Big(H(t)^\ell - (H_0)^\ell\Big) = \text{tr}\Big(H(0)^\ell - (H_0)^\ell\Big) = \ell \int_{\mathbb{R}} \lambda^{\ell-1} \xi_\alpha(\lambda) d\lambda.$$

produce an infinite series of **conservation laws** for the Toda lattice. Or, one could also take the expansion coefficients K_m (see (10.17)) as conserved quantities.

13.5. Some additions in case of the Toda lattice

The rest of this chapter is devoted to the Toda lattice, that is, to the case $\text{TL}_0(a, b) = 0$.

Combining

(13.80) $$\frac{d}{dt}\underline{\alpha}_{p_0}(\mathcal{D}_{\underline{\hat{\mu}}(n,t)}) = -2\underline{c}(g).$$

with (9.16) we obtain

(13.81) $$\underline{\alpha}_{p_0}(\mathcal{D}_{\underline{\hat{\mu}}(n,t)}) = \underline{\alpha}_{p_0}(\mathcal{D}_{\underline{\hat{\mu}}}) - 2n\underline{A}_{p_0}(\infty_+) - 2[t\underline{c}(g)].$$

Hence we can rewrite (13.56) as

(13.82) $$\begin{aligned} a(n,t)^2 &= \tilde{a}^2 \frac{\theta(\underline{z}(n+1,t))\theta(\underline{z}(n-1,t))}{\theta(\underline{z}(n,t))^2}, \\ b(n,t) &= \tilde{b} + \frac{1}{2}\frac{d}{dt}\ln\left(\frac{\theta(\underline{z}(n,t))}{\theta(\underline{z}(n-1,t))}\right). \end{aligned}$$

Now let us verify $\text{TL}_0(a, b) = 0$ directly. We start with the first equation

$$\frac{d}{dt}a(n,t)^2 = \tilde{a}^2 \frac{d}{dt}\frac{\theta(\underline{z}(n+1,t))\theta(\underline{z}(n-1,t))}{\theta(\underline{z}(n,t))^2}$$

$$= \tilde{a}^2 \frac{\theta(\underline{z}(n+1,t))\theta(\underline{z}(n-1,t))}{\theta(\underline{z}(n,t))^2} \frac{d}{dt}\ln\left(\frac{\theta(\underline{z}(n+1,t))\theta(\underline{z}(n-1,t))}{\theta(\underline{z}(n,t))^2}\right)$$

(13.83) $$= 2a(n,t)^2\Big(b(n+1,t) - b(n,t)\Big).$$

The second equation is more complicated. We compare the expansions (7.30) and (9.45), (9.42) to obtain

$$-a^2(n-1,t) = \tilde{b}^2 - \tilde{c} + \tilde{b}\sum_{j=1}^g B_j(n,t) + \frac{1}{2}\bigg(\Big(\sum_{j=1}^g B_j(n,t)\Big)^2$$

$$- \sum_{j=1}^g \Big(\sum_{k=1}^{2g+2} \frac{E_k}{2} + \frac{c_j(g-1)}{c_j(g)}\Big) B_j(n,t) + \sum_{j,k=1}^g BB_{j,k}(n,t)\bigg),$$

$$a^2(n,t) + b^2(n,t) = \tilde{c} + \tilde{b}\sum_{j=1}^g B_j(n,t) + \frac{1}{2}\bigg(\Big(\sum_{j=1}^g B_j(n,t)\Big)^2$$

(13.84) $$+ \sum_{j=1}^g \Big(\sum_{k=1}^{2g+2} \frac{E_k}{2} + \frac{c_j(g-1)}{c_j(g)}\Big) B_j(n,t) + \sum_{j,k=1}^g BB_{j,k}(n,t)\bigg).$$

13.5. Some additions in case of the Toda lattice

Adding these equations and eliminating $b(n,t)^2$ yields

$$a^2(n,t) - a^2(n-1,t) = \sum_{j,k=1}^{g} BB_{j,k}(n,t) = \frac{1}{4}\frac{d^2}{dt^2}\ln\left(\frac{\theta(\underline{z}(n,t))}{\theta(\underline{z}(n-1,t))}\right)$$

$$(13.85) \qquad = \frac{1}{2}\frac{d}{dt}b(n,t).$$

The Baker-Akhiezer function for the Toda lattice ψ is given by

$$(13.86) \quad \psi(p,n,t) = C(n,0,t)\frac{\theta(\underline{z}(p,n,t))}{\theta(\underline{z}(p,0,0))}\exp\left(n\int_{p_0}^{p}\omega_{\infty_+,\infty_-} + t\int_{p_0}^{p}\Omega_0\right),$$

where

$$(13.87) \qquad C(n,0,t)^2 = \frac{\theta(\underline{z}(0,0))\theta(\underline{z}(-1,0))}{\theta(\underline{z}(n,t))\theta(\underline{z}(n-1,t))}$$

and $\Omega_0 = \omega_{\infty_+,0} - \omega_{\infty_-,0}$. Since we have the following expansion near $\infty_\pm$

$$(13.88) \qquad \int_{p_0}^{p(z)}\Omega_0 = \mp(z + \tilde{d} + O(\frac{1}{z})), \quad \tilde{d} \in \mathbb{R},$$

we can make the following ansatz for Ω_0

$$(13.89) \qquad \Omega_0 = \frac{\prod_{j=0}^{g}(\pi - \tilde{\lambda}_j)}{R_{2g+2}^{1/2}}d\pi.$$

The constants $\tilde{\lambda}_j$ have to be determined from the normalization

$$(13.90) \qquad \int_{a_j}\Omega_0 = 2\int_{E_{2j-1}}^{E_{2j}}\frac{\prod_{\ell=0}^{g}(z - \tilde{\lambda}_\ell)}{R_{2g+2}^{1/2}(z)}dz = 0,$$

and the requirement

$$(13.91) \qquad \sum_{j=0}^{g}\tilde{\lambda}_j = \frac{1}{2}\sum_{j=0}^{2g+1}E_j.$$

It follows that the b-periods are given by

$$(13.92) \qquad \int_{b_j}\Omega_0 = -2\int_{E_{2j-2}}^{E_{2j-1}}\frac{\prod_{\ell=0}^{g}(z - \tilde{\lambda}_\ell)}{R_{2g+2}^{1/2}(z)}dz = 4\pi\mathrm{i}\, c_j(g).$$

Remark 13.9. Expanding equation (13.86) around $\infty_\pm$ shows that

$$(13.93) \qquad \int_{p_0}^{p}\Omega_0 = \mp\left(z + \tilde{b} + O(z^{-1})\right) \quad \text{near } \infty_\pm,$$

where $\tilde{b}$ is defined in (9.42). Conversely, proving (13.28) as in (9.61) (by expanding both sides in (13.28) around $\infty_\pm$ and using Lemma 13.4) turns out to be equivalent to proving (13.93). Comparison with the expansion (13.88) further yields $\tilde{b} = \tilde{d}$.

13.6. The elliptic case – continued

In this section we extend our calculations for the special case $g = 1$ started in Section 9.3.

We first express the differential Ω_0. We need to determine the constants $\tilde{\lambda}_0, \tilde{\lambda}_1$ from

(13.94) $$\tilde{\lambda}_0 + \tilde{\lambda}_1 = \frac{E_0 + E_1 + E_2 + E_3}{2}$$

and ([**38**] 254.00, 254.10, 340.01, 340.02, 336.02)

$$\int_{a_1} \Omega_0 = 2 \int_{E_1}^{E_2} \frac{x^2 - (\tilde{\lambda}_0 + \tilde{\lambda}_1)x + \tilde{\lambda}_0\tilde{\lambda}_1}{R_4^{1/2}(x)} dx$$

$$= \frac{C}{2}\Big((E_0 E_3 + E_1 E_2 + 2\tilde{\lambda}_0\tilde{\lambda}_1)K(k)$$

(13.95) $$+ (E_2 - E_0)(E_3 - E_1)E(k)\Big) = 0,$$

which implies

(13.96) $$2\tilde{\lambda}_0\tilde{\lambda}_1 = (E_2 - E_0)(E_1 - E_3)\frac{E(k)}{K(k)} + (E_0 E_3 + E_1 E_2).$$

The b-period is given by (cf. (13.92))

(13.97) $$\int_{b_1} \Omega_0 = -2 \int_{E_0}^{E_1} \frac{x^2 - (\tilde{\lambda}_0 + \tilde{\lambda}_1)x + \tilde{\lambda}_0\tilde{\lambda}_1}{R_4^{1/2}(x)} dx = 4\pi \mathrm{i}\, c(1) = \frac{2\pi \mathrm{i}}{CK(k)}.$$

And for $p = (z, \pm R_4^{1/2}(z))$ near p_0 we have ([**38**] 251.00, 251.03, 340.01, 340.02, 336.02)

$$\int_{p_0}^{p} \Omega_0 = \pm \int_{E_0}^{z} \frac{x^2 - (\tilde{\lambda}_0 + \tilde{\lambda}_1)x + \tilde{\lambda}_0\tilde{\lambda}_1}{R_4^{1/2}(x)} dx$$

$$= \frac{C(E_2 - E_0)}{2}\Bigg((E_3 - E_1)\frac{E(k)F(u(z),k) - K(k)E(u(z),k)}{K(k)}$$

(13.98) $$+ (E_3 - E_0)\frac{u(z)\sqrt{1-u(z)^2}\sqrt{1-k^2u(z)^2}}{1 - \frac{E_3-E_0}{E_3-E_1}u(z)^2}\Bigg).$$

Finally we remark that (9.85) now reads

(13.99) $$\mu(n,t) = E_1 \frac{1 - \frac{E_2-E_1}{E_2-E_0}\frac{E_0}{E_1}\mathrm{sn}(2K(k)\delta + 2nF(\sqrt{\frac{E_3-E_1}{E_3-E_0}},k) + \frac{2t}{C})^2}{1 - \frac{E_2-E_1}{E_2-E_0}\mathrm{sn}(2K(k)\delta + 2nF(\sqrt{\frac{E_3-E_1}{E_3-E_0}},k) + \frac{2t}{C})^2},$$

since we have

(13.100) $$A_{(E_0,0)}(\hat{\mu}(n,t)) = A_{(E_0,0)}(\hat{\mu}(0,0)) - 2nA_{(E_0,0)}(\infty_+) - 2[tc(1)].$$

Chapter 14

The Kac-van Moerbeke system

In this chapter we will introduce the Kac-van Moerbeke (KM) hierarchy. Using a commutation (sometimes also called a supersymmetric) approach we will show that the Kac-van Moerbeke hierarchy is a modified Toda hierarchy precisely in the way that the modified Korteweg-de Vries (mKdV) hierarchy is related to the Korteweg-de Vries (KdV) hierarchy. The connection between the TL hierarchy and its modified counterpart, the KM hierarchy, is given by a Bäcklund transformation (the Miura transformation in case of the KdV and mKdV equation) based on the factorization of difference expressions found in Section 11.1. This will allow us to compute new solutions from known ones. In particular, we will be able to compute the N-soliton solution starting from the trivial solution (see Section 13.3) of the TL equation.

14.1. The Kac-van Moerbeke hierarchy and its relation to the Toda hierarchy

This section is devoted to the Kac-van Moerbeke hierarchy and its connection with the Toda hierarchy.

We will suppose that $\rho(t)$ satisfies the following hypothesis throughout this chapter.

Hypothesis H. 14.1. Let

(14.1) $$\rho(t) \in \ell^\infty(\mathbb{Z}, \mathbb{R}), \quad \rho(n,t) \neq 0, \ (n,t) \in \mathbb{Z} \times \mathbb{R},$$

and let $t \in \mathbb{R} \mapsto \rho(t) \in \ell^\infty(\mathbb{Z})$ be differentiable.

Define the "even" and "odd" parts of $\rho(t)$ by

(14.2) $$\rho_e(n,t) = \rho(2n,t), \ \rho_o(n,t) = \rho(2n+1,t), \quad (n,t) \in \mathbb{Z} \times \mathbb{R},$$

and consider the bounded operators (in $\ell^2(\mathbb{Z})$)

(14.3) $$A(t) = \rho_o(t)S^+ + \rho_e(t), \quad A(t)^* = \rho_o^-(t)S^- + \rho_e(t).$$

As in Section 11.1 we have the factorization

(14.4) $$H_1(t) = A(t)^* A(t), \qquad H_2(t) = A(t) A(t)^*,$$

with

(14.5) $$H_k(t) = a_k(t) S^+ + a_k^-(t) S^- + b_k(t), \qquad k = 1, 2,$$

and

(14.6) $$\begin{aligned} a_1(t) &= \rho_e(t) \rho_o(t), & b_1(t) &= \rho_e(t)^2 + \rho_o^-(t)^2, \\ a_2(t) &= \rho_e^+(t) \rho_o(t), & b_2(t) &= \rho_e(t)^2 + \rho_o(t)^2. \end{aligned}$$

In the sequel it will be important to observe that the transformation $\hat{\rho}(n,t) = \rho(n+1,t)$ implies

(14.7) $$\hat{\rho}_e(n,t) = \rho_o(n,t), \qquad \hat{\rho}_o(n,t) = \rho_e(n+1,t)$$

and hence $\hat{a}_1(t) = a_2(t)$, $\hat{b}_1(t) = b_2(t)$, $\hat{a}_2(t) = a_1^+(t)$, $\hat{b}_2(t) = b_1^+(t)$. In addition, we have $\hat{g}_{1,j}(t) = g_{2,j}(t)$, $\hat{h}_{1,j}(t) = h_{2,j}(t)$ and $\hat{g}_{2,j}(t) = g_{1,j}^+(t)$, $\hat{h}_{2,j}(t) = h_{1,j}^+(t)$.

Using $\langle \delta_m, H_2^{j+1} \delta_n \rangle = \langle \delta_m, A H_1^j A^* \delta_n \rangle$ we can express $g_{2,j}, h_{2,j}$, $j \in \mathbb{N}$, in terms of $g_{1,j}, h_{1,j}$, $j \in \mathbb{N}$. Explicit evaluation with $m = n$ gives (provided we pick the same summation constant $c_{1,\ell} = c_{2,\ell} \equiv c_\ell$, $1 \le \ell \le r$)

(14.8) $$g_{2,j+1}(t) = \rho_o(t)^2 g_{1,j}^+(t) + \rho_e(t)^2 g_{1,j}(t) + h_{1,j}(t).$$

Similarly, for $m = n+1$ we obtain

(14.9) $$\begin{aligned} h_{2,j+1}(t) - h_{1,j+1} &= \rho_o(t)^2 \Big(2(\rho_e^+(t)^2 g_{1,j}^+(t) - \rho_e(t)^2 g_{1,j}(t)) \\ &\quad + (h_{1,j}^+(t) - h_{1,j}(t)) \Big). \end{aligned}$$

By virtue of (14.7) we can even get two more equations

$$g_{1,j+1}(t) = \rho_e(t)^2 g_{2,j}(t) + \rho_o^-(t)^2 g_{2,j}^-(t) + h_{2,j}^-(t),$$

(14.10) $$\begin{aligned} h_{2,j+1}^-(t) - h_{1,j+1} &= \rho_e(t)^2 \Big(-2(\rho_o(t)^2 g_{2,j}(t) - \rho_o^-(t)^2 g_{2,j}^-(t)) \\ &\quad - (h_{2,j}(t) - h_{2,j}^-(t)) \Big). \end{aligned}$$

These relations will turn out useful below.

Now we define matrix-valued operators $D(t)$, $Q_{2r+2}(t)$ in $\ell^2(\mathbb{Z}, \mathbb{C}^2)$ as follows,

(14.11) $$D(t) = \begin{pmatrix} 0 & A(t)^* \\ A(t) & 0 \end{pmatrix}, \qquad Q_{2r+2}(t) = \begin{pmatrix} P_{1,2r+2}(t) & 0 \\ 0 & P_{2,2r+2}(t) \end{pmatrix},$$

$r \in \mathbb{N}_0$. Here $P_{k,2r+2}(t)$, $k = 1, 2$, are defined as in (12.38), that is,

$$P_{k,2r+2}(t) = -H_k(t)^{r+1} + \sum_{j=0}^r \big(2 a_k(t) g_{k,j}(t) S^+ - h_{k,j}(t) \big) H_k(t)^j + g_{k,r+1},$$

(14.12) $$P_{k,2r+2}(t) \Big|_{\mathrm{Ker}(\tau_k(t) - z)} = 2 a_k(t) G_{k,r}(z,t) S^+ - H_{k,r+1}(z,t).$$

The sequences $(g_{k,j}(n,t))_{0 \le j \le r}$, $(h_{k,j}(n,t))_{0 \le j \le r+1}$ are defined as in (12.27), and the polynomials $G_{k,r}(z,n,t)$, $H_{k,r+1}(z,n,t)$ are defined as in (12.57). Moreover,

14.1. The Kac-van Moerbeke hierarchy

we choose the same summation constants in $P_{1,2r+2}(t)$ and $P_{2,2r+2}(t)$ (i.e., $c_{1,\ell} = c_{2,\ell} \equiv c_\ell$, $1 \le \ell \le r$). Explicitly we have

$$Q_2(t) = \begin{pmatrix} \rho_e(t)\rho_o(t)S^+ - \rho_e^-(t)\rho_o^-(t)S^- & 0 \\ 0 & \rho_e^+(t)\rho_o(t)S^+ - \rho_e(t)\rho_o^-(t)S^- \end{pmatrix}$$
(14.13)   etc. .

Analogous to equation (12.22) we now look at

(14.14)
$$\frac{d}{dt}D = [Q_{2r+2}, D] = \begin{pmatrix} 0 & P_{1,2r+2}A^* - A^*P_{2,2r+2} \\ P_{2,2r+2}A - AP_{1,2r+2} & 0 \end{pmatrix}.$$

Using $H_1^j A^* = A^* H_2^j$ and $S^- = \frac{1}{a_2^-}(H_2 - a_2 S^+ - b_2)$ we obtain after a little calculation

$$A^* P_{2,2r+2} - P_{1,2r+2} A^* =$$
$$\sum_{j=1}^{r} \frac{\rho_o \rho_e^+}{\rho_e}\Big(2\rho_e^2(g_{2,j} - g_{1,j}) + (h_{2,j}^- - h_{1,j})\Big) S^+ H_2^{r-j}$$
$$+ \sum_{j=1}^{r} \frac{1}{\rho_e}\Big((h_{2,j+1}^- - h_{1,j+1}) + 2((a_1)^2 g_{1,j} - (a_2^-)^2 g_{2,j}^-)$$
$$- \rho_o^2(h_{2,j}^- - h_{1,j}) + \rho_e^2(h_{2,j} - h_{2,j}^-)\Big) H_2^{r-j} + (h_{2,r+1}^- - h_{1,r+1})$$
(14.15) $$- \rho_o^-(g_{2,r+1}^- - g_{1,r+1})S^- + \rho_e(g_{2,r+1} - g_{1,r+1}).$$

Evaluating $\langle \delta_n, (H_1^\ell A^* - A^* H_2^\ell)\delta_n \rangle = 0$ explicitly gives

(14.16) $$h_{2,j}^- - h_{1,j} + 2\rho_e^2(g_{2,j} - g_{1,j}) = 0$$

and the second identity in (14.10) together with (14.16) implies

(14.17)
$$(h_{2,j+1}^- - h_{1,j+1}) + 2((a_1)^2 g_{1,j} - (a_2^-)^2 g_{2,j}^-) - \rho_o^2(h_{2,j}^- - h_{1,j}) + \rho_e^2(h_{2,j} - h_{2,j}^-) = 0.$$

Hence we obtain

(14.18) $P_{1,2r+2}A^* - A^*P_{2,2r+2} = -\rho_o^-(g_{2,r+1}^- - g_{1,r+1})S^- + \rho_e(g_{2,r+1} - g_{1,r+1}).$

Or, in other words,

(14.19) $$\frac{d}{dt}D(t) - [Q_{2r+2}(t), D(t)] = 0$$

is equivalent to

(14.20)
$$\mathrm{KM}_r(\rho) = (\mathrm{KM}_r(\rho)_e, \mathrm{KM}_r(\rho)_o)$$
$$= \begin{pmatrix} \dot\rho_e - \rho_e(g_{2,r+1} - g_{1,r+1}) \\ \dot\rho_o + \rho_o(g_{2,r+1} - g_{1,r+1}^+) \end{pmatrix} = 0.$$

Here the dot denotes a derivative with respect to t. One look at the transformation (14.7) verifies that the equations for ρ_o, ρ_e are in fact one equation for ρ. More explicitly, combining $g_{k,j}$, respectively $h_{k,j}$, into one sequence

(14.21) $\begin{array}{ll} G_j(2n) = g_{1,j}(n) \\ G_j(2n+1) = g_{2,j}(n) \end{array}$, respectively $\begin{array}{ll} H_j(2n) = h_{1,j}(n) \\ H_j(2n+1) = h_{2,j}(n) \end{array}$, we can rewrite (14.20) as

(14.22) $$\mathrm{KM}_r(\rho) = \dot{\rho} - \rho(G^+_{r+1} - G_{r+1}).$$

From (12.27) we see that G_j, H_j satisfy the recursions

$$G_0 = 1, \quad H_0 = c_1,$$
$$2G_{j+1} - H_j - H_j^{--} - 2(\rho^2 + (\rho^-)^2)G_j = 0, \quad 0 \le j \le r,$$
$$H_{j+1} - H_{j+1}^{--} - 2((\rho\rho^+)^2 G_j^+ - (\rho^-\rho)^2 G_j^{--})$$
(14.23) $$-(\rho^2 + (\rho^-)^2)(H_j - H_j^{--}) = 0, \quad 0 \le j < r.$$

As in the Toda context (12.42), varying $r \in \mathbb{N}_0$ yields the **Kac-van Moerbeke hierarchy** (KM hierarchy) which we denote by

(14.24) $$\mathrm{KM}_r(\rho) = 0, \quad r \in \mathbb{N}_0.$$

Using (14.23) we compute

$$G_1 = \rho^2 + (\rho^-)^2 + c_1,$$
$$H_1 = 2(\rho\rho^+)^2 + c_2,$$
$$G_2 = (\rho\rho^+)^2 + (\rho^2 + (\rho^-)^2)^2 + (\rho^-\rho^{--})^2 + c_1(\rho^2 + (\rho^-)^2) + c_2,$$
$$H_2 = 2(\rho\rho^+)^2((\rho^{++})^2 + (\rho^+)^2 + \rho^2(\rho^-)^2) + c_1 2(\rho\rho^+)^2 + c_3,$$
(14.25) $$\text{etc. .}$$

and hence

$$\mathrm{KM}_0(\rho) = \dot{\rho} - \rho\big((\rho^+)^2 - (\rho^-)^2\big) = 0,$$
$$\mathrm{KM}_1(\rho) = \dot{\rho} - \rho\big((\rho^+)^4 - (\rho^-)^4 + (\rho^{++})^2(\rho^+)^2 + (\rho^+)^2\rho^2 - \rho^2(\rho^-)^2$$
$$- (\rho^-)^2(\rho^{--})^2\big) + c_1(-\rho)\big((\rho^+)^2 - (\rho^-)^2\big) = 0,$$
(14.26) $$\text{etc. .}$$

Again the Lax equation (14.19) implies

Theorem 14.2. *Let ρ satisfy (H.14.1) and $\mathrm{KM}(\rho) = 0$. Then the Lax equation (14.19) implies the existence of a unitary propagator $V_r(t, s)$ such that we have*

(14.27) $$D(t) = V_r(t,s) D(s) V_r(t,s)^{-1}, \qquad (t,s) \in \mathbb{R}^2,$$

and thus all operators $D(t)$, $t \in \mathbb{R}$, are unitarily equivalent. Clearly, we have

(14.28) $$V_r(t,s) = \begin{pmatrix} U_{1,r}(t,s) & 0 \\ 0 & U_{2,r}(t,s) \end{pmatrix}.$$

And as in Theorem 12.6 we infer

Theorem 14.3. *Suppose $\rho_0 \in \ell^\infty(\mathbb{Z})$. Then there exists a unique integral curve $t \mapsto \rho(t)$ in $C^\infty(\mathbb{R}, \ell^\infty(\mathbb{Z}))$ of the Kac-van Moerbeke equations, $\mathrm{KM}_r(\rho) = 0$, such that $\rho(0) = \rho_0$.*

Remark 14.4. In analogy to Remark 12.7 one infers that ρ_e and ρ_o enter a_r, b_r quadratically so that the KM hierarchy (14.24) is invariant under the substitution

(14.29) $$\rho(t) \to \rho_\varepsilon(t) = \{\varepsilon(n)\rho(n,t)\}_{n \in \mathbb{Z}}, \quad \varepsilon(n) \in \{+1,-1\}, \; n \in \mathbb{Z}.$$

This result should again be compared with Lemma 1.6 and Lemma 14.9 below.

As in Section 12.2 (cf. (12.45)) we use a tilde to distinguish between inhomogeneous and **homogeneous Kac-van Moerbeke equations**, that is,

$$\widetilde{\mathrm{KM}}_r(\rho) = \mathrm{KM}_r(\rho)\big|_{c_\ell \equiv 0,\ 1\leq \ell \leq r}, \tag{14.30}$$

with c_ℓ the summation constants in $P_{k,2r+2}$.

Clearly, (14.20) can also be rewritten in terms of polynomials

$$\begin{aligned}\frac{d}{dt}\rho_e &= 2\rho_e \rho_o^2 (G_{1,r}^+(z) - G_{2,r}(z)) - \rho_e(H_{1,r+1}(z) - H_{2,r+1}(z)), \\ \frac{d}{dt}\rho_o &= -2\rho_o(\rho_e^+)^2 (G_{1,r}^+(z) - G_{2,r}^+(z)) + \rho_o(H_{1,r+1}^+(z) - H_{2,r+1}(z)),\end{aligned} \tag{14.31}$$

$r \in \mathbb{N}_0$. One only has to use (14.16) and

$$h_{2,j} - h_{1,j} + 2\rho_e^2(g_{2,j} - g_{1,j}^+) = 0, \tag{14.32}$$

which follows from (14.16) invoking the transformation (14.7).

The connection between $P_{k,2r+2}(t)$, $k = 1, 2$, and $Q_{2r+2}(t)$ is clear from (14.11), the corresponding connection between $H_k(t)$, $k = 1, 2$, and $D(t)$ is provided by the elementary observation

$$D(t)^2 = \begin{pmatrix} H_1(t) & 0 \\ 0 & H_2(t) \end{pmatrix}. \tag{14.33}$$

Moreover, since we have

$$\frac{d}{dt}D(t)^2 = [Q_{2r+2}(t), D(t)^2] \tag{14.34}$$

we obtain the implication

$$\mathrm{KM}_r(\rho) = 0 \Rightarrow \mathrm{TL}_r(a_k, b_k) = 0, \quad k = 1, 2. \tag{14.35}$$

Using (14.16) and (14.32) this can also be written as

$$\mathrm{TL}_r(a_k, b_k) = W_k \underline{\mathrm{KM}}_r(\rho), \quad k = 1, 2, \tag{14.36}$$

where $W_k(t)$ denote the matrix-valued difference expressions

$$W_1(t) = \begin{pmatrix} \rho_o(t) & \rho_e(t) \\ 2\rho_e(t) & 2\rho_o^-(t)S^- \end{pmatrix}, \quad W_2(t) = \begin{pmatrix} \rho_o(t)S^+ & \rho_e^+(t) \\ 2\rho_e(t) & 2\rho_o(t) \end{pmatrix}. \tag{14.37}$$

That is, given a solution ρ of the KM_r equation (14.24) (respectively (14.31)), one obtains two solutions, (a_1, b_1) and (a_2, b_2), of the TL_r equations (12.42) related to each other by the Miura-type transformations (14.6). Hence we have found a **Bäcklund transformation** relating the Toda and Kac-van Moerbeke systems. Note that due to (H.14.1), (a_1, b_1) and (a_2, b_2) both fulfill (H.12.1).

In addition, we can define

$$\phi_1(n, t) = -\frac{\rho_e(n, t)}{\rho_o(n, t)}, \quad \phi_2(n, t) = -\frac{\rho_o(n, t)}{\rho_e(n+1, t)}. \tag{14.38}$$

This implies

$$a_k(n, t)\phi_k(n, t) + \frac{a_k(n-1, t)}{\phi_k(n-1, t)} + b_k(n, t) = 0 \tag{14.39}$$

and using (14.8), the first identity in (14.10), and (12.27) one verifies (cf. (13.26))

$$\frac{d}{dt}\ln\phi_k(n,t) = -2a_k(n,t)\big(g_{k,r}(n,t)\phi_k(n,t) + g_{k,r}(n+1,t)\phi_k(n,t)^{-1}\big)$$
$$- 2b_k(n+1,t)g_{k,r}(n+1,t) + (g_{k,r+1}(n+1,t) - g_{k,r+1}(n,t))$$
(14.40) $\quad - (h_{k,r}(n+1,t) - h_{k,r}(n,t)).$

Hence we infer

(14.41) $\qquad H_k(t)u_k(n,t) = 0, \qquad \frac{d}{dt}u_k(n,t) = P_{k,2r+2}(t)u_k(n,t)$

(in the weak sense, i.e., u_k is not necessarily square summable), where

$$u_k(n,t) = \exp\Big(\int_0^t \big(2a_k(n_0,x)g_{k,r}(n_0,x)\phi_k(n_0,x)$$

(14.42) $\qquad\qquad\qquad - h_{k,r}(n_0,x) + g_{k,r+1}(n_0,x)\big)dx\Big)\prod_{m=n_0}^{n-1}{}^*\phi_k(m,t).$

Furthermore, explicitly writing out (14.19) shows that if

(14.43) $\quad H_k(t)u_k(z,n,t) = zu_k(z,n,t), \quad \frac{d}{dt}u_k(z,n,t) = P_{k,2r+2}(t)u_k(z,n,t)$

holds for the solution $u_1(z,n,t)$ (resp. $u_2(z,n,t)$), then it also holds for the solution $u_2(z,n,t) = A(t)u_1(z,n,t)$ (resp. $u_1(z,n,t) = A(t)^*u_2(z,n,t)$).

Summarizing,

Theorem 14.5. *Suppose ρ satisfies (H.14.1) and $\mathrm{KM}_r(\rho) = 0$. Then (a_k,b_k), $k=1,2$, satisfies (H.12.1) and $\mathrm{TL}_r(a_k,b_k) = 0$, $k=1,2$. In addition, if (14.43) holds for the solution $u_1(z,n,t)$ (resp. $u_2(z,n,t)$), then it also holds for the solution $u_2(z,n,t) = A(t)u_1(z,n,t)$ (resp. $u_1(z,n,t) = A(t)^*u_2(z,n,t)$).*

Finally, let us extend Lemma 12.16 to the case $\lambda \leq \sigma(H)$.

Lemma 14.6. *Suppose $\lambda \leq \sigma(H)$ and $a(n,t) < 0$. Then $u_0(\lambda,n) > 0$ implies that the solution $u(\lambda,n,t)$ of (12.89) with initial condition $u_0(\lambda,n)$ is positive.*

Proof. Shifting $H(t) \to H(t) - \lambda$ we can assume $\lambda = 0$. Now use $u_0(0,n) > 0$ to define $\rho_0(n)$ by

$$\rho_{0,o}(n) = -\sqrt{\frac{-a(n,0)u_0(0,n)}{u_0(0,n+1)}},$$

(14.44) $\qquad\qquad \rho_{0,e}(n) = \sqrt{\frac{-a(n,0)u_0(0,n+1)}{u_0(0,n)}}.$

By Theorem 14.3 we have a corresponding solution $\rho(n,t)$ of the KM hierarchy and hence (by (14.35)) two solutions $a_k(n,t), b_k(n,t)$ of the TL hierarchy. Since $a_1(n,0) = a(n,0)$ and $b_1(n,0) = b(n,0)$ we infer $a_1(n,t) = a(n,t)$ and $b_1(n,t) = b(n,t)$ by uniqueness (Theorem 12.6). Finally, we conclude $u(0,n,t) = u_0(0,n_0)u_1(n,t) > 0$ (with $u_1(n,t)$ as in (14.42)) again by uniqueness (Theorem 12.15). □

Another important consequence is

Lemma 14.7. *Let $\lambda \leq \sigma(H(0))$ and $a(n,t) < 0$. Suppose $u(\lambda, n, t)$ solves (12.89) and is a minimal positive solution near $\pm\infty$ for one $t = t_0$, then this holds for all $t \in \mathbb{R}$. In particular, $H(t) - \lambda$ is critical (resp. subcritical) for all $t \in \mathbb{R}$ if and only if it is critical (resp. subcritical) for one $t = t_0$.*

Proof. Since linear independence and positivity is preserved by the system (12.89) (by (12.91) and Lemma 14.6) $H(t) - \lambda$ is critical (resp. subcritical) for all $t \in \mathbb{R}$ if and only if it is critical (resp. subcritical) for one $t = t_0$. If $H(t) - \lambda$ is subcritical, we note that the characterization (2.70) of minimal solutions is independent of t. Hence it could only happen that $u_+(\lambda, n, t)$ and $u_-(\lambda, n, t)$ change place during time evolution. But this would imply $u_+(\lambda, n, t)$ and $u_-(\lambda, n, t)$ are linearly dependent at some intermediate time t contradicting $H(t) - \lambda$ subcritical. $\square$

In the special case of the stationary KM hierarchy characterized by $\dot{\rho} = 0$ in (14.24) (resp. (14.31)), or equivalently, by commuting matrix difference expressions of the type

$$[Q_{2r+2}, D] = 0, \tag{14.45}$$

the analogs of (12.79) and (12.67) then read

$$Q_{2r+2}^2 = \prod_{j=0}^{2r+1}(D - \sqrt{E_j})(D + \sqrt{E_j}) = \prod_{j=0}^{2r+1}(D^2 - E_j) \tag{14.46}$$

and

$$y^2 = \prod_{j=0}^{2r+1}(w - \sqrt{E_j})(w + \sqrt{E_j}) = \prod_{j=0}^{2r+1}(w^2 - E_j). \tag{14.47}$$

Here $\pm\sqrt{E_j}$ are the band edges of the spectrum of D, which appear in pairs since D and $-D$ are unitarily equivalent (see Lemma 14.9 below).

We note that the curve (14.47) becomes singular if and only if $E_j = 0$ for some $0 \leq j \leq 2r+1$. Since we have $0 \leq E_0 < E_1 < \cdots < E_{2r+1}$ this happens if and only if $E_0 = 0$.

We omit further details at this point since we will show in the following sections how solutions of the KM hierarchy can be computed from the corresponding ones of the TL hierarchy.

14.2. Kac and van Moerbeke's original equations

The Kac-van Moerbeke equation is originally given by

$$\frac{d}{dt}R(n,t) = \frac{1}{2}\left(e^{-R(n-1,t)} - e^{-R(n+1,t)}\right). \tag{14.48}$$

The transformation

$$\rho(n,t) = \frac{1}{2}e^{-R(n,t)/2}, \tag{14.49}$$

yields the form most convenient for our purpose

$$\dot{\rho}(n,t) = \rho(n,t)\left(\rho(n+1,t)^2 - \rho(n-1,t)^2\right), \tag{14.50}$$

which is precisely $KM_0(\rho) = 0$. The form obtained via the transformation $c(n,t) = 2\rho(n,t)^2$ is called Langmuir lattice,

$$\dot{c}(n,t) = c(n,t)\Big(c(n+1,t) - c(n-1,t)\Big), \tag{14.51}$$

and is used for modeling Langmuir oscillations in plasmas.

14.3. Spectral theory for supersymmetric Dirac-type difference operators

In this section we briefly study spectral properties of self-adjoint $\ell^2(\mathbb{Z}, \mathbb{C}^2)$ realizations associated with supersymmetric Dirac-type difference expressions.

Given $\rho \in \ell^\infty(\mathbb{Z}, \mathbb{R})$ we start by introducing

$$A = \rho_o S^+ + \rho_e, \qquad A^* = \rho_o^- S^- + \rho_e, \tag{14.52}$$

with $\rho_e(n) = \rho(2n)$, $\rho_o(n) = \rho(2n+1)$. We denote by D the bounded self-adjoint supersymmetric **Dirac operator**

$$\begin{array}{rcl} D: \ell^2(\mathbb{Z}, \mathbb{C}^2) & \to & \ell^2(\mathbb{Z}, \mathbb{C}^2) \\ f & \mapsto & \begin{pmatrix} 0 & A^* \\ A & 0 \end{pmatrix} f \end{array} \tag{14.53}$$

and by U the corresponding unitary involution

$$\begin{array}{rcl} U: \ell^2(\mathbb{Z}, \mathbb{C}^2) & \to & \ell^2(\mathbb{Z}, \mathbb{C}^2) \\ f & \mapsto & \begin{pmatrix} 0 & \mathbb{1} \\ \mathbb{1} & 0 \end{pmatrix} f \end{array}. \tag{14.54}$$

Theorem 14.8. *Suppose $\rho \in \ell^\infty(\mathbb{Z}, \mathbb{R})$, then D is a bounded self-adjoint operator with spectrum*

$$\sigma(D) = \{w \in \mathbb{R} | w^2 \in \sigma(H_1) \cup \sigma(H_2)\} \tag{14.55}$$

and resolvent

$$(D-w)^{-1} = \begin{pmatrix} w(H_1 - w^2)^{-1} & A^*(H_2 - w^2)^{-1} \\ A(H_1 - w^2)^{-1} & w(H_2 - w^2)^{-1} \end{pmatrix}, \quad w^2 \in \mathbb{C}\backslash\sigma(D), \tag{14.56}$$

where $H_1 = A^*A$, $H_2 = AA^*$.

Proof. The spectrum of D follows from the spectral mapping theorem since $D^2 = H_1 \oplus H_2$ and since D and $-D$ are unitarily equivalent (see the lemma below). The formula for the resolvent is easily vindicated using $A(H_1 - w^2)^{-1} = (H_2 - w^2)^{-1}A$ and $(H_1 - w^2)^{-1}A^* = A^*(H_2 - w^2)^{-1}$ (since $AH_1 = H_2A$ and $H_1A^* = A^*H_2$) (see also Theorem 11.1). □

Hence one can reduce the spectral analysis of supersymmetric Dirac operators D to that of H_1 and H_2. Moreover, since $H_1|_{\text{Ker}(H_1)^\perp}$ and $H_2|_{\text{Ker}(H_2)^\perp}$ are unitarily equivalent by Theorem 11.1, a complete spectral analysis of D in terms of that of H_1 and $\text{Ker}(H_2)$ can be given.

The analog of Lemma 1.6 reads

Lemma 14.9. *Suppose ρ satisfies (H.14.1) and introduce $\rho_\varepsilon \in \ell^\infty_\mathbb{R}(\mathbb{Z})$ by*

$$\rho_\varepsilon(n) = \varepsilon(n)\rho(n), \quad \varepsilon(n) \in \{+1, -1\}, \quad n \in \mathbb{Z}. \tag{14.57}$$

Define D_ε in $\ell^2(\mathbb{Z}, \mathbb{C}^2)$ as in (14.53) with ρ replaced by ρ_ε. Then D and D_ε are unitarily equivalent, that is, there exists a unitary operator U_ε in $\ell^2(\mathbb{Z}, \mathbb{C}^2)$ such that

$$\text{(14.58)} \qquad D = U_\varepsilon D_\varepsilon U_\varepsilon^{-1}.$$

Especially taking $\varepsilon(n) = -1$ shows that D and $-D$ are unitarily equivalent.

Proof. U_ε is explicitly represented by

$$\text{(14.59)} \qquad U_\varepsilon = \begin{pmatrix} U_{1,\varepsilon} & 0 \\ 0 & U_{2,\varepsilon} \end{pmatrix}, \quad U_{k,\varepsilon} = \{\tilde{\varepsilon}_k(n)\delta_{m,n}\}_{m,n \in \mathbb{Z}}, \quad k = 1, 2,$$

$\tilde{\varepsilon}_1(n+1)\tilde{\varepsilon}_2(n) = \varepsilon(2n+1)$, $\tilde{\varepsilon}_1(n)\tilde{\varepsilon}_2(n) = \varepsilon(2n)$, $n \in \mathbb{Z}$. □

14.4. Associated solutions

In Theorem 14.5 we saw, that from one solution ρ of $\mathrm{KM}_r(\rho) = 0$ we can get two solutions (a_1, b_1), (a_2, b_2) of $\mathrm{TL}_r(a, b)$. In this section we want to invert this process (cf. Section 11.2).

Suppose (a_1, b_1) satisfies (H.12.1), $a_1(n, t) < 0$ and $\mathrm{TL}_r(a_1, b_1) = 0$. Suppose $H_1 > 0$ and let $u_1(n, t) > 0$ solve

$$\text{(14.60)} \qquad H_1(t)u_1(n, t) = 0, \qquad \frac{d}{dt} u_1(n, t) = P_{1, 2r+2}(t) u_1(n, t)$$

(cf. Lemma 12.16). This implies that $\phi_1(n, t) = u_1(n+1, t)/u_1(n, t)$ fulfills

$$\text{(14.61)} \qquad a_1(n, t)\phi_1(n, t) + \frac{a_1(n-1, t)}{\phi_1(n-1, t)} = -b_1(n, t),$$

and

$$\frac{d}{dt} \ln \phi_1(n, t) = -2a_1(n, t)\big(g_{1,r}(n, t)\phi_1(n, t) + g_{1,r}(n+1, t)\phi_1(n, t)^{-1}\big)$$
$$+ 2b_1(n+1, t)g_{1,r}(n+1, t) + (g_{1,r+1}(n+1, t) - g_{1,r+1}(n, t))$$
$$\text{(14.62)} \qquad - (h_{1,r}(n+1, t) - h_{1,r}(n, t)).$$

Now define

$$\text{(14.63)} \qquad \rho_o(n) = -\sqrt{-\frac{a_1(n, t)}{\phi_1(n, t)}}, \qquad \rho_e(n, t) = \sqrt{-a_1(n, t)\phi_1(n, t)}.$$

Then, using (12.40), (12.27), and (14.8) a straightforward calculation shows that the sequence

$$\text{(14.64)} \qquad \rho(n, t) = \begin{cases} \rho_e(m, t) & \text{for } n = 2m \\ \rho_o(m, t) & \text{for } n = 2m+1 \end{cases} = (-1)^{n+1} \sqrt{-a_1(n, t)\phi_1(n, t)^{(-1)^n}},$$

fulfills (H.14.1) and $\mathrm{KM}_r(\rho) = 0$. Hence by (14.35)

$$\text{(14.65)} \qquad a_2(n, t) = \rho_e(n+1, t)\rho_o(n, t), \qquad b_2(n, t) = \rho_e(n, t)^2 + \rho_o(n, t)^2$$

satisfy $\mathrm{TL}_r(a_2, b_2) = 0$.

Since we already know some solutions of $\mathrm{TL}_r(a_1, b_1) = 0$, we can illustrate this process. Taking a finite-gap solution

$$a_1(n,t) = -\tilde{a}\sqrt{\frac{\theta(\underline{z}_1(n+1,t))\theta(\underline{z}_1(n-1,t))}{\theta(\underline{z}_1(n,t))^2}},$$

$$b_1(n,t) = \tilde{b} + \sum_{j=1}^{g} c_j(g) \frac{\partial}{\partial w_j} \ln\left(\frac{\theta(\underline{w} + \underline{z}_1(n,t))}{\theta(\underline{w} + \underline{z}_1(n-1,t))}\right)\bigg|_{\underline{w}=0},$$

(14.66) $\quad \phi_1(n,t) = \sqrt{\frac{\theta(\underline{z}_1(n-1,t))}{\theta(\underline{z}_1(n+1,t))}} \frac{\theta(\underline{z}_1(p_1,n+1,t))}{\theta(\underline{z}_1(p_1,n,t))} \exp\left(\int_{p_0}^{p_1} \tau_{+\infty,-\infty}\right),$

with $\pi(p_1) \leq E_0$ yields as in Section 11.4

$$\rho_o(n,t) = -\sqrt{-\frac{a_1(n,t)}{\phi_1(n,t)}}$$

$$= -\sqrt{\tilde{a}\frac{\theta(\underline{z}_1(n+1,t))}{\theta(\underline{z}_1(n,t))} \frac{\theta(\underline{z}_1(p_1,n,t))}{\theta(\underline{z}_1(p_1,n+1,t))}} \exp\left(\frac{-1}{2}\int_{p_0}^{p_1} \tau_{+\infty,-\infty}\right),$$

$$\rho_e(n,t) = \sqrt{-a_1(n,t)\phi_1(n,t)}$$

(14.67) $\quad = \sqrt{\tilde{a}\frac{\theta(\underline{z}_1(n-1,t))}{\theta(\underline{z}_1(n,t))} \frac{\theta(\underline{z}_1(p_1,n+1,t))}{\theta(\underline{z}_1(p_1,n,t))}} \exp\left(\frac{1}{2}\int_{p_0}^{p_1} \tau_{+\infty,-\infty}\right),$

and

$$a_2(n,t) = -\tilde{a}\sqrt{\frac{\theta(\underline{z}_2(n+1,t))\theta(\underline{z}_2(n-1,t))}{\theta(\underline{z}_2(n,t))^2}},$$

$$b_2(n,t) = \tilde{b} + \sum_{j=1}^{g} c_j(g) \frac{\partial}{\partial w_j} \ln\left(\frac{\theta(\underline{w} + \underline{z}_2(n,t))}{\theta(\underline{w} + \underline{z}_2(n-1,t))}\right)\bigg|_{\underline{w}=0},$$

(14.68) $\quad \phi_2(n,t) = \sqrt{\frac{\theta(\underline{z}_2(n-1,t))}{\theta(\underline{z}_2(n+1,t))}} \frac{\theta(\underline{z}_2(p_1^*,n+1,t))}{\theta(\underline{z}_2(p_1^*,n,t))} \exp\left(\int_{p_0}^{p_1^*} \tau_{+\infty,-\infty}\right),$

where $\underline{z}_2(n) = \underline{z}_1(p_1, n+1)$ respectively $\underline{z}_2(p_1^*, n) = \underline{z}_1(n)$. We remark that if we interchange the role of a_1 and a_2 and use p^* instead of p we get a sequence $\hat{\rho}(n,t)$ which is related to $\rho(n,t)$ via $\hat{\rho}(n,t) = \rho(n+1,t)$ (cf. (14.7)). This implies that we can assume p to lie on the upper sheet.

In the simplest case, where we start with the constant solution $a_1(n,t) = a$ and $b_1(n,t) = b$, we obtain for $p = (\lambda, \pm R_2^{1/2}(\lambda))$

(14.69) $\quad \rho_e(n,t) = \rho_o(n,t) = \frac{-1}{\sqrt{2}}\sqrt{|\lambda - b| \pm \sqrt{|\lambda - b|^2 - 4a^2}}$

and hence

(14.70) $\quad \rho(n,t) = \frac{(-1)^{n+1}}{\sqrt{2}}\sqrt{|\lambda - b| \pm \sqrt{|\lambda - b|^2 - 4a^2}}.$

14.5. N-soliton solutions on arbitrary background

In this section we want to extend the calculations of Section 14.4 and of Chapter 11.

Suppose (a,b) satisfies (H.12.1), $a(n,t) < 0$, and $\mathrm{TL}_r(a,b) = 0$. Suppose $\lambda_1 \leq \sigma(H)$, let $u_\pm(\lambda_1, n, t) > 0$ be the minimal positive solutions of (12.89) found in Lemma 14.7, and set

$$(14.71) \qquad u_{\sigma_1}(\lambda_1, n, t) = \frac{1+\sigma_1}{2} u_+(\lambda_1, n, t) + \frac{1-\sigma_1}{2} u_-(\lambda_1, n, t).$$

Note that the dependence on σ_1 will drop out in what follows if $u_+(\lambda_1, n, t)$ and $u_-(\lambda_1, n, t)$ are linearly dependent (for one and hence for all t). Now define

$$(14.72) \qquad \rho_{\sigma_1, o}(n, t) = -\sqrt{-\frac{a(n,t)}{\phi_{\sigma_1}(\lambda_1, n, t)}}, \qquad \rho_{\sigma_1, e}(n, t) = \sqrt{-a(n,t)\phi_{\sigma_1}(\lambda_1, n, t)},$$

where $\phi_{\sigma_1}(\lambda_1, n, t) = u_{\sigma_1}(\lambda_1, n+1, t)/u_{\sigma_1}(\lambda_1, n, t)$.

Then, proceeding as in the proof of Lemma 14.6 shows that the sequence

$$(14.73) \qquad \rho_{\sigma_1}(n, t) = \begin{cases} \rho_{\sigma_1, e}(m, t) & \text{for } n = 2m \\ \rho_{\sigma_1, o}(m, t) & \text{for } n = 2m+1 \end{cases},$$

fulfills (H.14.1) and $\mathrm{KM}_r(\rho) = 0$. Hence by (14.35)

$$(14.74)$$
$$a_{\sigma_1}(n, t) = \rho_{\sigma_1, e}(n+1, t) \rho_{\sigma_1, o}(n, t), \qquad b_{\sigma_1}(n, t) = \rho_{\sigma_1, e}(n, t)^2 + \rho_{\sigma_1, o}(n, t)^2$$

satisfy $\mathrm{TL}_r(a_{\sigma_1}, b_{\sigma_1}) = 0$.

We summarize this result in our first main theorem.

Theorem 14.10. *Suppose (a,b) satisfies (H.12.1) and $\mathrm{TL}_r(a,b) = 0$. Pick $\lambda_1 \leq \sigma(H)$, $\sigma_1 \in [-1, 1]$ and let $u_\pm(\lambda_1, n, t)$ be the minimal positive solutions of (12.89). Then the sequences*

$$a_{\sigma_1}(n, t) = \sqrt{\frac{a(n,t) a(n+1, t) u_{\sigma_1}(\lambda_1, n, t) u_{\sigma_1}(\lambda_1, n+2, t)}{u_{\sigma_1}(\lambda_1, n+1, t)^2}},$$

$$(14.75) \qquad b_{\sigma_1}(n, t) = b(n, t) + \partial^* \frac{a(n, t) u_{\sigma_1}(\lambda_1, n, t)}{u_{\sigma_1}(\lambda_1, n+1, t)}$$

with

$$(14.76) \qquad u_{\sigma_1}(\lambda_1, n+1, t) = \frac{1+\sigma_1}{2} u_+(\lambda_1, n, t) + \frac{1-\sigma_1}{2} u_-(\lambda_1, n, t),$$

satisfy $\mathrm{TL}_r(a_{\sigma_1}, b_{\sigma_1}) = 0$. Here $\partial^* f(n) = f(n-1) - f(n)$. In addition,

$$(14.77) \qquad \frac{a(n,t)(u_{\sigma_1}(\lambda_1, n, t) u(z, n+1, t) - u_{\sigma_1}(\lambda_1, n+1, t) u(z, n, t))}{\sqrt{-a(n,t) u_{\sigma_1}(\lambda_1, n, t) u_{\sigma_1}(\lambda_1, n+1, t)}}$$

satisfies $H_{\sigma_1} u = zu$ and $d/dt\, u = P_{\sigma_1, 2r+2} u$ (weakly) (in obvious notation).

Remark 14.11. (i). Alternatively, one could give a direct algebraic proof of the above theorem using $H_{\sigma_1} = A_{\sigma_1} H A_{\sigma_1}^*$ to express the quantities $g_{\sigma_1, j}, h_{\sigma_1, j}$ in terms of g_j, h_j.
(ii). We have omitted the requirement $a(n,t) < 0$ since the formulas for $a_{\sigma_1}, b_{\sigma_1}$ are actually independent of the sign of $a(n,t)$. In addition, we could even allow $\lambda_1 \geq \sigma(H)$. However, $\rho_{\sigma_1, e}(n,t)$ and $\rho_{\sigma_1, o}(n,t)$ would be purely imaginary in this case.

Iterating this procedure (cf. Theorem 11.10) gives

Theorem 14.12. *Let $a(t), b(t)$ satisfy (H.12.1) and $\mathrm{TL}_r(a,b) = 0$. Let $H(t)$ be the corresponding Jacobi operators and choose*

(14.78) $\qquad \lambda_N < \cdots < \lambda_2 < \lambda_1 \leq \sigma(H), \quad \sigma_\ell \in [-1,1], \quad 1 \leq \ell \leq N, \ N \in \mathbb{N}.$

Suppose $u_\pm(\lambda, n, t)$, are the principal solutions of (12.89). Then

$$a_{\sigma_1,\ldots,\sigma_N}(n,t) = -\sqrt{a(n,t)a(n+N,t)}$$
$$\times \frac{\sqrt{C_n(U_{\sigma_1,\ldots,\sigma_N}(t))C_{n+2}(U_{\sigma_1,\ldots,\sigma_N}(t))}}{C_{n+1}(U_{\sigma_1,\ldots,\sigma_N}(t))},$$

(14.79) $\qquad b_{\sigma_1,\ldots,\sigma_N}(n,t) = b(n,t) + \partial^* a(n,t) \frac{D_n(U_{\sigma_1,\ldots,\sigma_N}(t))}{C_{n+1}(U_{\sigma_1,\ldots,\sigma_N}(t))}$

satisfies $\mathrm{TL}_r(a_{\sigma_1,\ldots,\sigma_N}, b_{\sigma_1,\ldots,\sigma_N}) = 0$. Here C_n denotes the n-dimensional Casoratian

$$C_n(u_1,\ldots,u_N) = \det(u_i(n+j-1))_{1 \leq i,j \leq N},$$

(14.80) $\qquad D_n(u_1,\ldots,u_N) = \det \begin{pmatrix} u_i(n), & j=1 \\ u_i(n+j), & j>1 \end{pmatrix}_{1 \leq i,j \leq N},$

and $(U_{\sigma_1,\ldots,\sigma_N}(t)) = (u^1_{\sigma_1}(t), \ldots, u^N_{\sigma_N}(t))$ with

(14.81) $\qquad u^\ell_{\sigma_\ell}(n,t) = \frac{1+\sigma_\ell}{2} u_+(\lambda_\ell, n, t) + (-1)^{\ell+1} \frac{1-\sigma_\ell}{2} u_-(\lambda_\ell, n, t).$

Moreover, for $1 \leq \ell \leq N$, $\lambda < \lambda_\ell$,

(14.82) $\qquad u_{\pm,\sigma_1,\ldots,\sigma_\ell}(\lambda, n, t) = \frac{\pm \prod_{j=0}^{\ell-1} \sqrt{-a(n+j,t)} C_n(U_{\sigma_1,\ldots,\sigma_\ell}(t), u_\pm(\lambda, t))}{\sqrt{C_n(U_{\sigma_1,\ldots,\sigma_\ell}(t))C_{n+1}(U_{\sigma_1,\ldots,\sigma_\ell}(t))}},$

are the minimal solutions of $\tau_{\sigma_1,\ldots,\sigma_\ell}(t)u = \lambda u$ and satisfy

(14.83) $\qquad \frac{d}{dt} u_{\pm,\sigma_1,\ldots,\sigma_\ell}(\lambda, n, t) = P_{\sigma_1,\ldots,\sigma_\ell, 2r+2}(t) u_{\pm,\sigma_1,\ldots,\sigma_\ell}(\lambda, n, t).$

Defining

$$\rho_{\sigma_1,\ldots,\sigma_N,o}(n,t) =$$
$$-\sqrt{-a(n,t) \frac{C_{n+2}(U_{\sigma_1,\ldots,\sigma_{N-1}}(t)) C_n(U_{\sigma_1,\ldots,\sigma_N}(t))}{C_{n+1}(U_{\sigma_1,\ldots,\sigma_{N-1}}(t)) C_{n+1}(U_{\sigma_1,\ldots,\sigma_N}(t))}},$$

$$\rho_{\sigma_1,\ldots,\sigma_N,e}(n,t) =$$

(14.84) $\qquad \sqrt{-a(n+N-1,t) \frac{C_n(U_{\sigma_1,\ldots,\sigma_{N-1}}(t)) C_{n+1}(U_{\sigma_1,\ldots,\sigma_N}(t))}{C_{n+1}(U_{\sigma_1,\ldots,\sigma_{N-1}}(t)) C_n(U_{\sigma_1,\ldots,\sigma_N}(t))}},$

the corresponding sequence $\rho_{\sigma_1,\ldots,\sigma_N}(n)$ solves $\mathrm{KM}_r(\rho_{\sigma_1,\ldots,\sigma_N}) = 0$.

Clearly, if we drop the requirement $\lambda \leq \sigma(H)$ the solution $u_{\sigma_1}(\lambda_1, n, t)$ used to perform the factorization will no longer be positive. Hence the sequences $a_{\sigma_1}(n,t)$, $b_{\sigma_1}(n,t)$ can be complex valued and singular. Nevertheless there are two situations where a second factorization step produces again real-valued non-singular solutions.

Firstly we perform two steps with $\lambda_{1,2}$ in the same spectral gap of $H(0)$ (compare Section 11.8).

14.5. N-soliton solutions on arbitrary background

Theorem 14.13. *Suppose (a,b) satisfies (H.12.1) and $\mathrm{TL}_r(a,b) = 0$. Pick $\lambda_{1,2}$, $\sigma_{1,2} \in \{\pm 1\}$ and let $\lambda_{1,2}$ lie in the same spectral gap of $H(0)$ ($(\lambda_1, \sigma_1) \neq (\lambda_2, -\sigma_2)$ to make sure we get something new). Then the sequences*

$$a_{\sigma_1,\sigma_2}(n,t) = a(n,t)\sqrt{\frac{W_{\sigma_1,\sigma_2}(n-1,t)W_{\sigma_1,\sigma_2}(n+1,t)}{W_{\sigma_1,\sigma_2}(n,t)^2}},$$

(14.85) $\quad b_{\sigma_1,\sigma_2}(n,t) = b(n,t) - \partial^* \dfrac{a(n,t)u_{\sigma_1}(\lambda_1,n,t)u_{\sigma_2}(\lambda_2,n+1,t)}{W_{\sigma_1,\sigma_2}(n,t)},$

where

(14.86) $\quad W_{\sigma_1,\sigma_2}(n,t) = \begin{cases} \dfrac{W_n(u_{\sigma_1}(\lambda_1,t),u_{\sigma_2}(\lambda_2,t))}{\lambda_2 - \lambda_1}, & \lambda_1 \neq \lambda_2 \\ \sum_{m=\sigma_1\infty}^{n} u_{\sigma_1}(\lambda_1,m,t)^2, & (\lambda_1,\sigma_1) = (\lambda_2,\sigma_2) \end{cases},$

are real-valued non-singular solutions of $\mathrm{TL}(a_{\sigma_1,\sigma_2}, b_{\sigma_1,\sigma_2}) = 0$. In addition, the sequence

(14.87) $\quad \dfrac{W_{\sigma_1,\sigma_2}(n,t)u(z,n,t) - \frac{1}{z-\lambda_1}u_{\sigma_2}(\lambda_2,n,t)W_n(u_{\sigma_1}(\lambda_1,t),u(z,t))}{\sqrt{W_{\sigma_1,\sigma_2}(n-1,t)W_{\sigma_1,\sigma_2}(n,t)}},$

satisfies $H_{\sigma_1,\sigma_2}(t)u = zu$, $d/dt\, u = P_{\sigma_1,\sigma_2,2r+2}(t)u$ (weakly).

Proof. Theorem 4.19 implies $W_{\sigma_1,\sigma_2}(n,t)W_{\sigma_1,\sigma_2}(n+1,t) > 0$ and hence the sequences $a_{\sigma_1,\sigma_2}(t), b_{\sigma_1,\sigma_2}(t)$ satisfy (H.12.1). The rest follows from the previous theorem (with $N=2$) as follows. Replace λ_1 by $z \in (\lambda_1 - \varepsilon, \lambda_1 + \varepsilon)$ and observe that $a_{\sigma_1,\sigma_2}(n,t)$, $b_{\sigma_1,\sigma_2}(n,t)$ and $\dot{a}_{\sigma_1,\sigma_2}(n,t)$, $\dot{b}_{\sigma_1,\sigma_2}(n,t)$ are meromorphic with respect to z. From the algebraic structure we have simply performed two single commutation steps. Hence, provided Theorem 14.10 applies to this more general setting of meromorphic solutions, we can conclude that our claims hold except for a discrete set with respect to z where the intermediate operators are ill-defined due to singularities of the coefficients. However, the proof of Theorem 14.10 uses these intermediate operators and in order to see that Theorem 14.10 still holds, one has to resort to the direct algebraic proof outlined in Remark 14.11(i). Continuity with respect to z takes care of the remaining points. $\square$

Secondly, we consider again two commutation steps but now with $\lambda_1 = \lambda_2$ (compare Section 11.6).

Theorem 14.14. *Suppose (a,b) satisfies (H.12.1) and $\mathrm{TL}_r(a,b) = 0$. Pick λ_1 in a spectral gap of $H(0)$ and $\gamma_1 \in [-\|u_-(\lambda_1)\|^{-2}, \infty) \cup \{\infty\}$. Then the sequences*

$$a_{\gamma_1}(n,t) = a(n,t)\dfrac{\sqrt{c_{\gamma_1}(\lambda_1,n-1,t)c_{\gamma_1}(\lambda_1,n+1,t)}}{c_{\gamma_1}(\lambda_1,n,t)},$$

(14.88) $\quad b_{\gamma_1}(n,t) = b(n,t) - \partial^* \dfrac{a(n,t)u_-(\lambda_1,n,t)u_-(\lambda_1,n+1,t)}{c_{\gamma_1}(\lambda_1,n,t)}.$

satisfy $\mathrm{TL}(a_{\gamma_1}, b_{\gamma_1}) = 0$, where

(14.89) $\quad c_{\gamma_1}(\lambda_1,n,t) = \dfrac{1}{\gamma_1} + \sum_{m=-\infty}^{n} u_-(\lambda_1,m,t)^2.$

In addition, the sequence

$$\text{(14.90)} \quad \frac{c_{\gamma_1}(\lambda_1,n,t)u(z,n,t) - \frac{1}{z-\lambda_1}u_-(\lambda_1,n,t)W_n(u_-(\lambda_1,t),u(z,t))}{\sqrt{c_{\gamma_1}(\lambda_1,n-1,t)c_{\gamma_1}(\lambda_1,n,t)}},$$

satisfies $H_{\gamma_1}(t)u = zu$, $d/dt\, u = P_{\gamma_1, 2r+2}(t)u$ *(weakly).*

Proof. Following Section 11.5 we can obtain the double commutation method from two single commutation steps. We pick $\sigma_1 = -1$ for the first factorization. Considering $A_{\sigma_1} u_-(z, n+1, t)/(z - \lambda_1)$ and performing the limit $z \to \lambda_1$ shows that

$$\text{(14.91)} \quad v(\lambda_1, n, t) = \frac{c_{\gamma_1}(\lambda_1, n, t)}{\sqrt{-a(n,t)u_-(\lambda_1,n,t)u_-(\lambda_1,n+1,t)}}$$

is a solution of the new (singular) operator which can be used to perform a second factorization. The resulting operator is associated with $a_{\gamma_1}, b_{\gamma_1}$. Now argue as before. □

Again we point out that one can also prove this theorem directly as follows. Without restriction we choose $\lambda_1 = 0$. Then one computes

$$\text{(14.92)} \quad \frac{d}{dt}c_{\gamma_1}(0,n,t) = 2a(n,t)^2\Big(g_{r-1}(n+1,t)u_-(0,n,t)^2 + g_{r-1}(n,t)u_-(0,n+1,t)^2\Big) + 2h_{r-1}(n,t)a(n,t)u_-(0,n,t)u_-(0,n+1,t)$$

and it remains to relate $g_{\gamma_1,j}, h_{\gamma_1,j}$ and g_j, h_j. Since these quantities arise as coefficients of the Neumann expansion of the respective Green functions it suffices to relate the Green functions of H_{γ_1} and H. This can be done using (11.108).

Iterating this procedure (cf. Theorem 11.27) gives

Theorem 14.15. *Let* $a(n,t), b(n,t)$ *satisfy (H.12.1). Suppose* $\mathrm{TL}_r(a,b) = 0$ *and let* $H(t)$ *be the corresponding Jacobi operators. Choose* $\lambda_j \in \rho(H)$ *and* $\gamma_j \in [-\|u_-(\lambda_j)\|^{-2}, \infty) \cup \{\infty\}$, $1 \le j \le N$, *and assume*

$$\text{(14.93)} \quad \frac{d}{dt}u_-(\lambda_j, n, t) = P_{2r+2}(t)u_-(\lambda_j, n, t).$$

We define the following matrices

$$C^N(n,t) = \left(\frac{\delta_{ij}}{\gamma_i} + \sum_{m=-\infty}^n u_-(\lambda_i, m, t)u_-(\lambda_j, m, t)\right)_{1 \le i,j \le N},$$

$$\text{(14.94)} \quad D^N(n,t) = \begin{pmatrix} C^N(n,t)_{i,j}, & i,j \le N \\ u_-(\lambda_j, n-1, t), & j \le N, i = N+1 \\ u_-(\lambda_i, n, t), & i \le N, j = N+1 \\ 0, & i = j = N+1 \end{pmatrix}_{1 \le i,j \le N+1}.$$

Then the sequences

$$a_{\gamma_1,\ldots,\gamma_N}(n,t) = a(n,t)\frac{\sqrt{\det C^N(n-1,t)\det C^N(n+1,t)}}{\det C^N(n,t)},$$

$$\text{(14.95)} \quad b_{\gamma_1,\ldots,\gamma_N}(n,t) = b(n,t) - \partial^* a(n,t)\frac{\det D^N(n+1,t)}{\det C^N(n,t)}$$

14.5. N-soliton solutions on arbitrary background   269

satisfy $\mathrm{TL}_r(a_{\gamma_1,\dots,\gamma_N}, b_{\gamma_1,\dots,\gamma_N}) = 0$. Moreover, $(1 \leq \ell \leq N)$

$$
(14.96) \qquad u_{\gamma_1,\dots,\gamma_\ell}(z,n,t) = \frac{\det U^N(u(z,n,t))}{\sqrt{\det C^\ell(n-1,t)\det C^\ell(n,t)}}
$$

satisfies

$$
(14.97) \qquad H_{\gamma_1,\dots,\gamma_N} u = \lambda_j u, \qquad \frac{d}{dt} u = P_{\gamma_1,\dots,\gamma_N, 2r+2}(t) u
$$

(in the weak sense) if $u(z,n,t)$ satisfies (12.89).

Remark 14.16. In the case $r = 0$ we even obtain

$$
(14.98) \qquad \frac{d}{dt} c_{\gamma_1}(\lambda_1, n, t) = 2a(n,t) u_-(\lambda_1, n, t) u_-(\lambda_1, n+1, t)
$$

and hence

$$
(14.99) \qquad b_{\gamma_1}(n,t) = b(n,t) + \frac{1}{2}\frac{d}{dt}\ln\frac{c_{\gamma_1}(\lambda_1, n, t)}{c_{\gamma_1}(\lambda_1, n-1, t)},
$$

where $c_{\gamma_1}(\lambda_1, n, t) = \gamma_1^{-1} + \sum_{m=-\infty}^n u_-(\lambda_1, m, t)$. Or, by induction,

$$
(14.100) \qquad b_{\gamma_1,\dots,\gamma_N}(n,t) = b(n,t) + \frac{1}{2}\frac{d}{dt}\ln\frac{\det C^N(n,t)}{\det C^N(n-1,t)}.
$$

We conclude this section with an example; the N-soliton solution of the TL and KM hierarchies. We take the constant solution of the Toda hierarchy

$$
(14.101) \qquad a_0(n,t) = \frac{1}{2}, \qquad b_0(n,t) = 0,
$$

as our background. Let H_0, $P_{0,2r+2}$ denote the associated Lax pair and recall

$$
(14.102) \qquad H_0(t) u_{0,\pm}(z,n,t) = z u_{0,\pm}(z,n,t), \qquad \frac{d}{dt} u_{0,\pm}(z,n,t) = P_{0,2r+2}(t) u_{0,\pm}(z,n,t),
$$

where

$$
(14.103) \qquad u_{0,\pm}(z,n,t) = k^{\pm n} \exp\left(\pm \frac{\alpha_r(k) t}{2}\right), \quad k = z - \sqrt{z^2 - 1}, \; |k| \leq 1,
$$

and

$$
(14.104) \qquad \alpha_r(k) = 2(k G_{0,r}(z) - H_{r+1,0}(z)) = (k - k^{-1}) G_{0,r}(z).
$$

Then the N-soliton solution of the Toda hierarchy is given by

$$
(14.105) \qquad \begin{aligned} a_{0,\sigma_1,\dots,\sigma_N}(n,t) &= \frac{\sqrt{C_n(U_{\sigma_1,\dots,\sigma_N}(t)) C_{n+2}(U_{\sigma_1,\dots,\sigma_N}(t))}}{2 C_{n+1}(U_{\sigma_1,\dots,\sigma_N}(t))}, \\ b_{0,\sigma_1,\dots,\sigma_N}(n,t) &= \partial^* \frac{D_n(U_{\sigma_1,\dots,\sigma_N}(t))}{2 C_{n+1}(U_{\sigma_1,\dots,\sigma_N}(t))}, \end{aligned}
$$

where $(U_{0,\sigma_1,\dots,\sigma_N}(t)) = (u_{\sigma_1}^1(t), \dots, u_{\sigma_N}^N(t))$ with

$$
(14.106) \qquad u_{0,\sigma_j}^j(n,t) = k_j^n + (-1)^{j+1} \frac{1-\sigma_j}{1+\sigma_j} \exp(\alpha_r(k_j) t) k_j^{-n}, \quad k_j = \lambda_j - \sqrt{\lambda_j^2 - 1}.
$$

The corresponding N-soliton solution $\rho_{\sigma_1,\ldots,\sigma_N}(n)$ of the Kac-van Moerbeke hierarchy reads

$$\rho_{0,\sigma_1,\ldots,\sigma_N,o}(n,t) =$$
$$-\sqrt{-\frac{C_{n+2}(U_{0,\sigma_1,\ldots,\sigma_{N-1}}(t))C_n(U_{0,\sigma_1,\ldots,\sigma_N}(t))}{2C_{n+1}(U_{0,\sigma_1,\ldots,\sigma_{N-1}}(t))C_{n+1}(U_{0,\sigma_1,\ldots,\sigma_N}(t))}},$$
$$\rho_{0,\sigma_1,\ldots,\sigma_N,e}(n,t) =$$
(14.107)
$$\sqrt{-\frac{C_n(U_{0,\sigma_1,\ldots,\sigma_{N-1}}(t))C_{n+1}(U_{0,\sigma_1,\ldots,\sigma_N}(t))}{2C_{n+1}(U_{0,\sigma_1,\ldots,\sigma_{N-1}}(t))C_n(U_{0,\sigma_1,\ldots,\sigma_N}(t))}}.$$

Introducing the time dependent norming constants

(14.108)
$$\gamma_j(t) = \gamma_j \exp(\alpha_r(k_j)t)$$

we obtain the following alternate expression for the N-soliton solution of the Toda hierarchy

$$a_{0,\gamma_1,\ldots,\gamma_N}(n,t) = \frac{\sqrt{\det C_0^N(n-1,t)\det C_0^N(n+1,t)}}{2\det C_0^N(n,t)},$$
(14.109)
$$b_{0,\gamma_1,\ldots,\gamma_N}(n,t) = -\partial^* \frac{\det D_0^N(n+1,t)}{2\det C_0^N(n,t)},$$

where

$$C_0^N(n,t) = \left(\frac{\delta_{rs}}{\gamma_r(t)} + \frac{(k_r k_s)^{-n}}{1-k_r k_s}\right)_{1\leq r,s\leq N},$$

(14.110)
$$D_0^N(n,t) = \begin{pmatrix} C_0^N(n,t)_{r,s}, & r,s\leq N \\ k_s^{1-n}, & s\leq N, r=N+1 \\ k_r^{-n}, & r\leq N, s=N+1 \\ 0, & r=s=N+1 \end{pmatrix}_{1\leq r,s\leq N+1}.$$

The sequences $a_{0,\gamma_1,\ldots,\gamma_N}$, $b_{0,\gamma_1,\ldots,\gamma_N}$ coincide with $a_{0,\sigma_1,\ldots,\sigma_N}$, $b_{0,\sigma_1,\ldots,\sigma_N}$ provided (cf. Remark 11.30)

(14.111)
$$\gamma_j = \left(\frac{1-\sigma_j}{1+\sigma_j}\right)^{-1} |k_j|^{-1-N} \frac{\prod_{\ell=1}^N |1-k_j k_\ell|}{\prod_{\substack{\ell=1 \\ \ell\neq j}}^N |k_j - k_\ell|}, \quad 1\leq j\leq N.$$

Notes on literature

Chapter 12. The Toda lattice was first introduced by Toda in [**233**]. For further material see his selected papers [**232**] or his monograph [**230**].

The Toda hierarchy was first introduced in [**237**] and [**238**] (see also [**236**]). Our introduction uses a recursive approach for the standard Lax formalism ([**164**]) which was first advocated by Al'ber [**11**], Jacobi [**137**], McKean [**169**], and Mumford [**178**] (Sect. I-II a). It follows essentially the monograph [**35**] (with some extensions). An existence and uniqueness theorem for the semi infinite case seems to appear first in [**65**], the form stated here is taken from [**220**]. The Burchnall-Chaundy polynomial in the case of commuting differential expressions was first obtained in [**36**], [**37**] (cf. also [**236**]).

Further references I refer to a few standard monographs such as [**78**], [**79**] [**184**], and [**190**].

Chapter 13. Section 1 and 2 closely follow [**35**] with several additions made. The Baker-Akhiezer function, the fundamental object of our approach, goes back to the work of Baker [**23**], Burchnall and Chaundy [**36**], [**37**], and Akhiezer [**8**]. The modern approach was initiated by Its and Matveev [**136**] in connection with the Korteweg-de Vries equation and further developed into a powerful machinery by Krichever (see, e.g., [**156**]–[**158**]) and others. Gesztesy and Holden are also working on a monograph on hierarchies of soliton equations and their algebro-geometric solutions [**103**]. We refer, in particular, to the extensive treatments in [**25**], [**74**], [**75**], [**76**], [**168**], [**177**], and [**184**]. In the special context of the Toda equations we refer to [**3**], [**75**], [**76**], [**156**], [**159**], [**168**], [**170**], and [**175**].

The periodic case was first investigated by Kac and van Moerbeke [**143**], [**144**]. Further references for the periodic case are [**3**], [**33**], [**53**], [**56**], [**146**], [**168**], and [**173**]–[**175**].

In the case of the Toda lattice ($r=0$) the formulas for the quasi periodic finite gap case in terms of Riemann theta functions seem due to [**210**]. However, some norming constants and the fact $\tilde{b}=\tilde{d}$ (see Remark 13.9) are not established there.

In case of the Toda equation on the half line a different approach using the spectral measure of H_+ is introduced and extended in [**28**] – [**32**] (see also [**15**]). It turns out that the time evolution of the spectral measure can be computed explicitly as the Freud transform of the initial measure.

The inverse scattering transform for the case $r=0$ has been formally developed by Flaschka [**85**] (see also [**230**] and [**80**] for the case of rapidly decaying sequences) for the

Toda lattice. In addition, Flaschka also worked out the inverse procedure in the reflectionless case (i.e., $R_\pm(k,t) = 0$). His formulas clearly apply to the entire Toda hierarchy upon using the t dependence of the norming constants given in (13.72) and coincide with those obtained using the double commutation method (i.e., the last example in Section 14.5). All results from the fourth section are taken from [**219**] and [**224**].

For additional results where the Toda equation is driven from one end by a particle see [**63**], [**64**] and also [**67**]. For results on a continuum limit of the Toda lattice see [**59**]. For connections of the Riemann-Hilbert problem and the inverse scattering method and applications to the Toda lattice see [**61**] and [**62**]. In addition, there are also very nice lecture notes by Deift [**57**] which give an introduction to the Riemann-Hilbert problem and its applications.

Chapter 14. The Kac-van Moerbeke equation has been first introduced in [**142**]. A connection between the Kac-van Moerbeke and Toda systems was already known to Hénon in 1973. The Bäcklund transformation connecting the Toda and the Kac-van Moerbeke equations has first been considered in [**234**] and [**239**]. (see also [**56**], [**151**], [**152**]). The results presented here are mainly taken from [**35**], [**117**] and [**220**]. An alternative approach to the modified Toda hierarchy, using the discrete analog of the formal pseudo differential calculus, can be found in [**161**], Ch. 4.

For applications of the Kac-van Moerbeke equation I refer to(e.g.) [**25**] (Ch. 8), [**55**], [**66**], [**130**], [**159**], [**183**], or [**246**]

Appendix A

Compact Riemann surfaces
− a review

The facts presented in this section can be found in almost any book on Riemann surfaces and their theta functions (cf., e.g. [**74**], [**81**], [**82**], [**86**], [**154**], [**178**], [**181**]). However, no book contains all results needed. Hence, in order to save you the tedious work of looking everything up from different sources and then translate results from one notation to another, I have included this appendix. In addition, I have worked out some facts on hyperelliptic surfaces in more detail than you will find them in the references. Most of the time we will closely follow [**81**] in the first sections.

A.1. Basic notation

Let M be a **compact Riemann surface** (i.e., a compact, second countable, connected Hausdorff space together with a holomorphic structure) of **genus** $g \in \mathbb{N}$. We restrict ourselves to $g \geq 1$ since most of our definitions only make sense in this case. We will think of our Riemann surface M often as it's **fundamental polygon** (e.g., for $g = 2$):

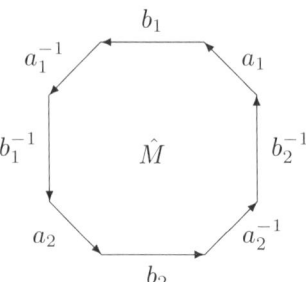

We denote the closure of the polygon by $\hat{M}$ and its boundary by $\partial \hat{M} = \prod_{\ell=1}^{g}(a_\ell b_\ell a_\ell^{-1} b_\ell^{-1})$, where $\{a_\ell, b_\ell\}_{1 \leq \ell \leq g}$ are (fixed) representatives of a **canonical homology basis** for M. We get M from $\hat{M}$ by identifying corresponding points, that is, $M = \hat{M}/\simeq$.

We denote the set of **meromorphic functions** on M by $\mathcal{M}(M)$, the set of **meromorphic differentials** by $\mathcal{M}^1(M)$, and the set of **holomorphic differentials** by $\mathcal{H}^1(M)$. If we expand $f \in \mathcal{M}(M)$ respectively $\omega \in \mathcal{M}^1(M)$ in a local chart (U, z) centered at $p \in M$ we get $(z(p) = 0)$

$$(A.1) \qquad f = \sum_{\ell=m}^{\infty} c_\ell z^\ell, \quad \text{respectively} \quad \omega = \Big(\sum_{\ell=m}^{\infty} c_\ell z^\ell\Big) dz, \qquad c_m \neq 0, \, m \in \mathbb{Z}.$$

The number m is independent of the chart chosen and so we can define $\operatorname{ord}_p f = m$ respectively $\operatorname{ord}_p \omega = m$. The number m is called the **order** of f, respectively ω, at p. We will also need the **residue** $\operatorname{res}_p \omega = c_{-1}$, which is well-defined and satisfies

$$(A.2) \qquad \sum_{p \in M} \operatorname{res}_p \omega = 0.$$

The above sum is finite since ω can only have finitely many poles (M is compact!). Next, we define the **branch number** $b_f(p) = \operatorname{ord}_p(f - f(p)) - 1$ if $f(p) \neq \infty$ and $b_f(p) = -\operatorname{ord}_p(f) - 1$ if $f(p) = \infty$. There is a connection between the **total branching number** $B = \sum_{p \in M} b_f(p)$ (the sum is again finite), the **degree** n of f

$$(A.3) \qquad n = \operatorname{degree}(f) = \sum_{p \in f^{-1}(c)} (1 + b_f(p)), \qquad c \in \mathbb{C} \cup \{\infty\}$$

(we will show that n is independent of c), and the genus g, the **Riemann-Hurwitz relation**

$$(A.4) \qquad 2g - 2 = B - 2n.$$

We omit the proof which is not difficult but involves a lot of definitions which are of no further use for our purpose. (Triangulate $\mathbb{C} \cup \{\infty\}$ such that each pole and each zero of f is a vertex. Lift the triangulation to M via f and calculate the Euler characteristic ...)

A.2. Abelian differentials

Now consider two closed differentials θ, φ on M and let us try to compute

$$\iint_M \theta \wedge \varphi. \tag{A.5}$$

We define

$$\hat{A}(p) = \int_{p_0}^{p} \hat{\varphi} \qquad p_0, p \in \hat{M}, \tag{A.6}$$

where the hat indicates that we regard φ as a differential on to $\hat{M}$. $\hat{A}$ is well-defined since $\hat{M}$ is simply connected and $d\hat{\varphi} = 0$ (of course we require the path of integration to lie in $\hat{M}$). Using Stokes theorem we obtain

$$\begin{aligned}\iint_M \theta \wedge \varphi &= \iint_{\hat{M}} \theta \wedge d\hat{A} = -\iint_{\hat{M}} d(\theta \hat{A}) = -\int_{\partial \hat{M}} \theta \hat{A} \\ &= -\sum_{\ell=1}^{g} \left(\int_{a_\ell} \theta \hat{A} + \int_{b_\ell} \theta \hat{A} + \int_{a_\ell^{-1}} \theta \hat{A} + \int_{b_\ell^{-1}} \theta \hat{A} \right).\end{aligned} \tag{A.7}$$

Now, what can we say about $\hat{A}(p)$ and $\hat{A}(\tilde{p})$ if $p, \tilde{p}$ are corresponding points on a_ℓ, a_ℓ^{-1}, respectively? If $\gamma, \tilde{\gamma}$ are paths (inside $\hat{M}$) from p_0 to $p, \tilde{p}$, respectively, we have the following situation:

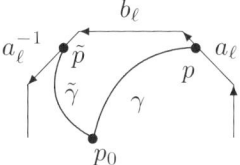

Thus we have $\hat{A}(\tilde{p}) = \hat{A}(p) + \int_{b_\ell} \varphi$ (the parts on a_ℓ and a_ℓ^{-1} cancel). A similar argument for corresponding points $p, \tilde{p}$ on b_ℓ, b_ℓ^{-1} yields $\hat{A}(\tilde{p}) = \hat{A}(p) - \int_{a_\ell} \varphi$ and we finally get

$$\iint_M \theta \wedge \varphi = \sum_{\ell=1}^{g} \left(\int_{a_\ell} \theta \int_{b_\ell} \varphi - \int_{a_\ell} \varphi \int_{b_\ell} \theta \right). \tag{A.8}$$

We will mainly be interested in meromorphic differentials $\omega \in \mathcal{M}^1(M)$, also called **abelian differentials**. We call ω of the **first kind** if ω is holomorphic, of the **second kind** if all residues vanish, and of the **third kind** else.

Unfortunately, formula (A.8) makes only sense in the special case of holomorphic differentials. Nevertheless there is one thing we can do. Let $\zeta \in \mathcal{H}^1(M)$, $\omega \in \mathcal{M}^1(M)$ and assume that ω has no poles on the boundary – which can always be achieved by slightly (continuously) deforming the (representatives of the) homology basis. The residue theorem now yields

$$2\pi\mathrm{i} \sum_{p \in M} \mathrm{res}_p \hat{A}\omega = \int_{\partial \hat{M}} \hat{A}\omega = \sum_{\ell=1}^{g} \left(\int_{a_\ell} \zeta \int_{b_\ell} \omega - \int_{a_\ell} \omega \int_{b_\ell} \zeta \right). \tag{A.9}$$

For $\zeta \in \mathcal{H}^1(M)$ we define

$$\text{(A.10)} \qquad \|\zeta\|^2 = \mathrm{i} \iint_M \zeta \wedge \overline{\zeta} = \mathrm{i} \sum_{\ell=1}^g \left(\int_{a_\ell} \zeta \int_{b_\ell} \overline{\zeta} - \int_{a_\ell} \overline{\zeta} \int_{b_\ell} \zeta \right),$$

(the bar denotes complex conjugation) which is indeed a norm (in local coordinates we get: $\mathrm{i}\zeta \wedge \overline{\zeta} = 2|\eta|^2 dx \wedge dy$ with $\zeta = \eta\, dz$, $z = x + \mathrm{i}y$).

Therefore we conclude that $\zeta \equiv 0$ if and only if (e.g.) all the a-periods are zero and further that the map from the space of holomorphic differentials $\mathcal{H}^1(M)$ (which is g-dimensional) to the a-periods ($\cong \mathbb{C}^g$) is an isomorphism. Thus we may choose a basis $\{\zeta_j\}_{j=1}^g$ for $\mathcal{H}^1(M)$ fulfilling

$$\text{(A.11)} \qquad \int_{a_k} \zeta_j = \delta_{jk}, \qquad j,k = 1,\ldots,g.$$

For the b-periods we get

$$\text{(A.12)} \qquad \int_{b_k} \zeta_j = \tau_{j,k}, \qquad j,k = 1,\ldots,g,$$

with $\tau_{j,k} \in \mathbb{C}$. Now (A.9) implies

$$\text{(A.13)} \qquad \tau_{j,k} = \tau_{k,j}$$

and (for $\zeta = \sum_{\ell=1}^g c_\ell \zeta_\ell \in \mathcal{H}^1(M)\setminus\{0\}$ with $c_\ell \in \mathbb{C}$)

$$0 < \|\zeta\|^2 = \mathrm{i} \sum_{j,k} c_j \overline{c_k} \sum_\ell \left(\int_{a_\ell} \zeta_j \int_{b_\ell} \overline{\zeta_k} - \int_{b_\ell} \zeta_j \int_{a_\ell} \overline{\zeta_k} \right)$$

$$\text{(A.14)} \qquad = 2 \sum_{j,k} c_j \overline{c_k} \operatorname{Im}(\tau_{j,k}).$$

Especially for $c_\ell = \delta_{\ell j}$ we have $\operatorname{Im}(\tau_{j,j}) > 0$. Thus $\underline{\tau}$ is symmetric with positive definite imaginary part.

We will now ensure the existence of meromorphic differentials (without proof, cf. [81], II.5.3):

Theorem A.1. *For every set $\{(p_j, (c_{j,\ell})_{1\leq \ell \leq n_j})\}_{1\leq j\leq m}$ with distinct $p_j \in M$, $c_{j,\ell} \in \mathbb{C}$, and $m, n_j \in \mathbb{N}$ we can find a meromorphic differential ω with poles of the form*

$$\text{(A.15)} \qquad \omega = \left(\sum_{\ell=1}^{n_j} \frac{c_{j,\ell}}{z_j^\ell} + O(1) \right) dz_j,$$

near each p_j in a given chart (U_j, z_j) centered at p_j, $1 \leq j \leq m$, and is holomorphic elsewhere, if and only if

$$\text{(A.16)} \qquad \sum_{p \in M} \operatorname{res}_p \omega = \sum_{j=1}^m c_{j,1} = 0.$$

Hence, for a meromorphic differential we can prescribe the poles and the Laurent principal parts at each pole with the only restriction, that the sum of the residues has to be zero. Two differentials constructed in that way may only differ by a holomorphic differential. From now on we will assume that no poles lie on the boundary – otherwise just deform the homology basis – so that we can make it unique by the following normalization

$$\text{(A.17)} \qquad \int_{a_\ell} \omega = 0, \qquad \ell = 1, \ldots, g.$$

We are interested in some special cases. First let $\omega_{p,n}$ be holomorphic in $M\backslash\{p\}$ and let (U, z) be a chart centered at p

$$\text{(A.18)} \qquad \omega_{p,n} = (\frac{1}{z^{2+n}} + O(1))dz, \qquad \int_{a_\ell} \omega_{p,n} = 0, \quad \ell = 1, \ldots, g, \quad n \in \mathbb{N}_0.$$

This definition is obviously depending on the chart around p. The b-periods follow easily from (A.9) with $\zeta = \zeta_\ell$ and $\omega = \omega_{p,n}$

$$\text{(A.19)} \qquad \int_{b_\ell} \omega_{p,n} = 2\pi\mathrm{i} \sum_{\tilde{p}\in M} \mathrm{res}_{\tilde{p}}\hat{A}\omega_{p,n} = \frac{2\pi\mathrm{i}}{(n+1)!} \left(\frac{d}{dz}\right)^n \eta_\ell(z)\Big|_{z=0},$$

where $\zeta_\ell = \eta_\ell(z)dz$. Second suppose ω_{pq} is holomorphic in $M\backslash\{p,q\}$ (with $p \neq q$) and let

$$\text{(A.20)} \qquad \begin{array}{ll} \mathrm{ord}_p\omega_{pq} = -1 & \mathrm{ord}_q\omega_{pq} = -1 \\ \mathrm{res}_p\omega_{pq} = +1 & \mathrm{res}_q\omega_{pq} = -1 \end{array}, \quad \int_{a_\ell} \omega_{pq} = 0, \quad \ell = 1, \ldots, g.$$

This definition depends on the specific homology basis chosen. Because if we deform, for example, a_ℓ we may cross a pole, thus changing the value of $\int_{a_\ell} \omega_{pq}$. The b-periods again follow from (A.9) with $\zeta = \zeta_\ell$ and $\omega = \omega_{pq}$

$$\text{(A.21)} \qquad \int_{b_\ell} \omega_{pq} = 2\pi\mathrm{i} \sum_{\tilde{p}\in M} \mathrm{res}_{\tilde{p}}\hat{A}\omega_{pq} = 2\pi\mathrm{i}(\hat{A}(p) - \hat{A}(q)) = 2\pi\mathrm{i}\int_q^p \hat{\zeta}_\ell.$$

Recall that the path of integration in the last integral lies in $\hat{M}$.

A.3. Divisors and the Riemann-Roch theorem

This section is also valid for $g = 0$. A **divisor** $\mathcal{D}$ is a map

$$\text{(A.22)} \qquad \begin{array}{rcl} \mathcal{D}: M & \to & \mathbb{Z} \\ p & \mapsto & \mathcal{D}(p) \end{array}$$

being nonzero only for a finite number of points in M. The set of all divisors $\mathrm{Div}(M)$ becomes a module over the ring $\mathbb{Z}$ by defining each operation in the natural way (i.e., pointwise). The point divisors $\mathcal{D}_p$, $p \in M$,

$$\text{(A.23)} \qquad \mathcal{D}_p(p) = 1, \qquad \mathcal{D}_p(q) = 0, \qquad q \in M\backslash\{p\}$$

form a basis for $\mathrm{Div}(M)$. We can also establish a group homomorphism by assigning each divisor a **degree**

$$\text{(A.24)} \qquad \begin{array}{rcl} \deg: \mathrm{Div}(M) & \to & \mathbb{Z} \\ \mathcal{D} & \mapsto & \deg(\mathcal{D}) = \sum_{p\in M} \mathcal{D}(p) \end{array}$$

(the sum is finite and thus well-defined). We associate a divisor with every nonzero meromorphic function and with every nonzero meromorphic differential on M

(A.25)
$$(\ ): \mathcal{M}(M)\backslash\{0\} \to \mathrm{Div}(M), \quad (\ ): \mathcal{M}^1(M)\backslash\{0\} \to \mathrm{Div}(M),$$
$$f \mapsto (f) \qquad \omega \mapsto (\omega)$$

where $(f)(p) = \mathrm{ord}_p f$ and $(\omega)(p) = \mathrm{ord}_p \omega$.

It is easy to verify the following formulas

(A.26) $\quad (f_1 f_2) = (f_1) + (f_2), \quad (\frac{1}{f_1}) = -(f_1), \quad f_1, f_2 \in \mathcal{M}(M)\backslash\{0\},$

implying $(f_1) = (f_2) \Leftrightarrow (f_1/f_2) = 0 \Leftrightarrow f_1 = c\, f_2,\, c \in \mathbb{C}$ since the only holomorphic functions on a compact Riemann surface are constants. Similarly,

(A.27) $\quad (f\, \omega_1) = (f) + (\omega_1), \quad (\frac{\omega_1}{\omega_2}) = (\omega_1) - (\omega_2), \quad \omega_1, \omega_2 \in \mathcal{M}^1(M)\backslash\{0\}$

implying $(\omega_1) = (\omega_2) \Leftrightarrow (\omega_1/\omega_2) = 0 \Leftrightarrow \omega_1 = c\, \omega_2,\, c \in \mathbb{C}$.

We claim

(A.28)
$$\deg((f)) = 0, \qquad f \in \mathcal{M}(M)\backslash\{0\},$$
$$\deg((\omega)) = 2g-2, \qquad \omega \in \mathcal{M}^1(M)\backslash\{0\}.$$

To obtain the first equation observe that $\deg((f)) = \sum_{p \in M} \mathrm{res}_p df/f = 0$. This implies that f attains every value exactly n times for a certain integer n called the degree of f (counting multiplicities). In fact, if 0 is attained n times the same is true for ∞ (otherwise $\deg((f))$ would not be zero). Now consider the function $f - c$ with $c \in \mathbb{C}$. It attains ∞ as often as f and thus zero as often as f.

For the second equation observe $\deg((\omega_1)) = \deg((\omega_2))$ which follows from $\omega_1/\omega_2 \in \mathcal{M}(M)\backslash\{0\}$. Now let f be a meromorphic function of degree n and let us calculate (df/f). At points p with $f(p) \notin \{0, \infty\}$, $\mathrm{ord}_p df/f$ is just the branch number $b_f(p)$ and at each other point it is -1. The desired result now follows from the Riemann-Hurwitz relation (A.4) and

$$\deg((\frac{df}{f})) = \sum_{p \in M} b_f(p) - \sum_{p \in f^{-1}(0)} (1+b_f(p)) - \sum_{p \in f^{-1}(\infty)} (1+b_f(p))$$
(A.29)
$$= B - 2n.$$

Divisors corresponding to meromorphic functions $\mathcal{D} = (f)$ are called **principal**, divisors corresponding to meromorphic differentials $\mathcal{D} = (\omega)$ **canonical**. It follows from the above considerations that the principal divisors form a subgroup. It is clearly normal (since all groups in sight are abelian) and we may consider its factor group induced by the following equivalence relation

(A.30) $\qquad \mathcal{D}_1 \simeq \mathcal{D}_2 \quad \Leftrightarrow \quad \mathcal{D}_1 - \mathcal{D}_2 = (f).$

The factor group is called **divisor class group** and the equivalence classes $[\mathcal{D}]$ divisor classes. Observe, that all canonical divisors lie in the same divisor class (by (A.27)). The subset formed by the divisors of degree zero $\mathrm{Div}_0(M)$ is a subgroup of $\mathrm{Div}(M)$ and it's factor group is called **Picard group** $\mathrm{Pic}(M)$.

The set $\mathrm{Div}(M)$ can be partially ordered by defining

(A.31) $\qquad \mathcal{D}_1 \geq \mathcal{D}_2 \quad \Leftrightarrow \quad \mathcal{D}_1(p) \geq \mathcal{D}_2(p), \quad p \in M.$

A.3. Divisors and the Riemann-Roch theorem

We write $\mathcal{D}_1 > \mathcal{D}_2$ if in addition $\mathcal{D}_1 \neq \mathcal{D}_2$. Note: $\mathcal{D}_1 \geq \mathcal{D}_2 \Rightarrow \deg(\mathcal{D}_1) \geq \deg(\mathcal{D}_2)$ and $\mathcal{D}_1 > \mathcal{D}_2 \Rightarrow \deg(\mathcal{D}_1) > \deg(\mathcal{D}_2)$.

A divisor is called **positive** if $\mathcal{D} \geq 0$ and **strictly positive** if $\mathcal{D} > 0$. Two divisors $\mathcal{D}_1, \mathcal{D}_2$ are called **relatively prime** if $\mathcal{D}_1(p)\mathcal{D}_2(p) = 0$, $p \in M$. A divisor can always be uniquely written as the difference of two relatively prime positive divisors

(A.32) $$\mathcal{D} = \mathcal{D}^+ - \mathcal{D}^-, \qquad \mathcal{D}^+ \geq 0,\ \mathcal{D}^- \geq 0.$$

For a meromorphic function $(f)^+$ and $(f)^-$ are called the **divisor of zeros** and the **divisor of poles** of f, respectively.

For any divisor $\mathcal{D}$ we now define

(A.33) $$\begin{aligned}\mathcal{L}(\mathcal{D}) &= \{f \in \mathcal{M}(M) | f = 0 \text{ or } (f) \geq \mathcal{D}\}, \\ \mathcal{L}^1(\mathcal{D}) &= \{\omega \in \mathcal{M}^1(M) | \omega = 0 \text{ or } (\omega) \geq \mathcal{D}\}.\end{aligned}$$

Hence $\mathcal{L}(\mathcal{D})$ is the set of all meromorphic functions, which are holomorphic at points $p \in M$ with $\mathcal{D}(p) \geq 0$, have zeros of order at least $\mathcal{D}(p)$ at points p with $\mathcal{D}(p) > 0$ and poles of order at most $-\mathcal{D}(p)$ at points p with $\mathcal{D}(p) < 0$. Similarly for $\mathcal{L}^1(\mathcal{D})$. Obviously we have

(A.34) $$\mathcal{D}_1 \geq \mathcal{D}_2 \quad \Rightarrow \quad \begin{cases} \mathcal{L}(\mathcal{D}_1) \subset \mathcal{L}(\mathcal{D}_2) \\ \mathcal{L}^1(\mathcal{D}_1) \subset \mathcal{L}^1(\mathcal{D}_2) \end{cases}$$

and $\mathcal{L}(\mathcal{D}), \mathcal{L}^1(\mathcal{D})$ inherit the vector space structure from $\mathcal{M}(M), \mathcal{M}^1(M)$, respectively. What can we say about their dimensions

(A.35) $$r(\mathcal{D}) = \dim \mathcal{L}(\mathcal{D}), \qquad i(\mathcal{D}) = \dim \mathcal{L}^1(\mathcal{D}).$$

$i(\mathcal{D})$ is called **index of specialty** of $\mathcal{D}$ and $\mathcal{D}$ is called **nonspecial** if $i(\mathcal{D}) = 0$.

First of all we have

(A.36) $$r(0) = 1, \qquad i(0) = g,$$

since a holomorphic function on a compact Riemann surface must be constant and since $\mathcal{H}^1(M)$ is g-dimensional.

If we take two divisors in the same divisor class, we can easily establish an isomorphism

(A.37) $$\begin{aligned}g \in \mathcal{L}(\mathcal{D}) &\Rightarrow fg \in \mathcal{L}(\mathcal{D} + (f)), \\ g \in \mathcal{L}(\mathcal{D} + (f)) &\Rightarrow f^{-1}g \in \mathcal{L}(\mathcal{D}),\end{aligned}$$

saying

(A.38) $$\mathcal{L}(\mathcal{D}) \cong \mathcal{L}(\mathcal{D} + (f)), \qquad r(\mathcal{D}) = r(\mathcal{D} + (f)).$$

Proceeding in a similar manner, we also obtain

(A.39) $$\begin{aligned}\mathcal{L}^1(\mathcal{D}) &\cong \mathcal{L}^1(\mathcal{D} + (f)), & i(\mathcal{D}) &= i(\mathcal{D} + (f)), \\ \mathcal{L}^1(\mathcal{D}) &\cong \mathcal{L}(\mathcal{D} - (\omega)), & i(\mathcal{D}) &= r(\mathcal{D} - (\omega)).\end{aligned}$$

If $\deg(\mathcal{D}) > 0$ it follows from $(f) \geq \mathcal{D} \Rightarrow \deg((f)) \geq \deg(\mathcal{D}) > 0$, that $r(\mathcal{D})$ has to be zero. And using (A.39) and (A.28) we get

(A.40) $$\begin{aligned}\deg(\mathcal{D}) > 0 &\Rightarrow r(\mathcal{D}) = 0, \\ \deg(\mathcal{D}) > 2g - 2 &\Rightarrow i(\mathcal{D}) = 0.\end{aligned}$$

If $\deg(\mathcal{D}) = 0$ and $f \in \mathcal{L}(\mathcal{D})$ it follows $(f) = \mathcal{D}$ (since $(f) > \mathcal{D}$ is not possible by the same argument as above) and thus

$$\deg(\mathcal{D}) = 0 \Leftrightarrow \begin{cases} \mathcal{L}(\mathcal{D}) = \mathrm{span}\{f\} & \text{if } \mathcal{D} = (f) \text{ is principal} \\ \mathcal{L}(\mathcal{D}) = \{0\} & \text{if } \mathcal{D} \text{ is not principal} \end{cases},$$

(A.41) $\deg(\mathcal{D}) = 2g - 2 \Leftrightarrow \begin{cases} \mathcal{L}^1(\mathcal{D}) = \mathrm{span}\{\omega\} & \text{if } \mathcal{D} = (\omega) \text{ is canonical} \\ \mathcal{L}^1(\mathcal{D}) = \{0\} & \text{if } \mathcal{D} \text{ is not canonical} \end{cases}.$

Next, let us assume $r(-\mathcal{D}) \leq 1 + \deg(\mathcal{D})$ for all $\mathcal{D}$ with $\deg(\mathcal{D}) \geq 0$ (we already know it for $\deg(\mathcal{D}) = 0$). If we subtract a point divisor we get

(A.42) $$r(-\mathcal{D}) \leq r(-\mathcal{D} - \mathcal{D}_p) \leq 1 + r(-\mathcal{D})$$

by (A.34) and because if f is new in our space, there is for every $g \in \mathcal{L}(-\mathcal{D} - \mathcal{D}_p)$ a $c \in \mathbb{C}$ with $g - cf \in \mathcal{L}(\mathcal{D})$. Hence using induction we infer

(A.43) $$\begin{aligned} r(-\mathcal{D}) &\leq 1 + \deg(\mathcal{D}), & \deg(\mathcal{D}) &\geq 0, \\ i(-\mathcal{D}) &\leq 2g - 1 + \deg(\mathcal{D}), & \deg(\mathcal{D}) &\geq 2 - 2g. \end{aligned}$$

By virtue of Theorem A.1 and (A.36) we have

(A.44) $$i(\mathcal{D}) = g - 1 - \deg(\mathcal{D}), \qquad \mathcal{D} < 0.$$

If we add a point divisor $\mathcal{D}_p$ to $\mathcal{D}$, we obtain from (A.39) and (A.42)

(A.45) $$i(\mathcal{D}) - 1 \leq i(\mathcal{D} + \mathcal{D}_p) \leq i(\mathcal{D}).$$

Thus, using induction we have

(A.46) $$i(\mathcal{D}) \geq g - 1 - \deg(\mathcal{D}),$$

which is by (A.39) and (A.28) equivalent to (**Riemann inequality**)

(A.47) $$r(-\mathcal{D}) \geq 1 - g + \deg(\mathcal{D}).$$

Now we will prove the famous **Riemann-Roch Theorem**:

Theorem A.2. (Riemann-Roch) *Let $\mathcal{D} \in \mathrm{Div}(M)$, then we have*

(A.48) $$r(-\mathcal{D}) = \deg(\mathcal{D}) + 1 - g + i(\mathcal{D}).$$

Proof. By (A.44) we know that it is true for $\mathcal{D} \leq 0$. So what happens if we add a point divisor to $\mathcal{D}$? We already know (A.42) and (A.45) but can we have $i(\mathcal{D} + \mathcal{D}_p) = i(\mathcal{D}) - 1$ and $r(-\mathcal{D} - \mathcal{D}_p) = 1 + r(-\mathcal{D})$? If this were the case, we would have a function $f \in \mathcal{L}(\mathcal{D} + \mathcal{D}_p)$ with a pole of order $(\mathcal{D}(p) + 1)$ at p and a differential $\omega \in \mathcal{L}^1(\mathcal{D})$ with a zero of order $\mathcal{D}(p)$ at p. Now consider the differential $f\omega$. It would have a pole of order 1 at p and would be holomorphic elsewhere, $(f\omega) = (f) + (\omega) \geq -\mathcal{D} - \mathcal{D}_p + \mathcal{D} = -\mathcal{D}_p$, contradicting our assumption. Thus we have by induction

(A.49) $$r(-\mathcal{D}) \leq \deg(\mathcal{D}) + 1 - g + i(\mathcal{D}).$$

Using (A.39) and (A.28):

(A.50) $$\begin{aligned} i(\mathcal{D}) = r(\mathcal{D} - (\omega)) &\leq \deg((\omega) - \mathcal{D}) + 1 - g + i((\omega) - \mathcal{D}) \\ &= -\deg(\mathcal{D}) + g - 1 + r(-\mathcal{D}) \end{aligned}$$

completes the proof of the Riemann-Roch theorem. □

Corollary A.3. *A Riemann surface of genus 0 is isomorphic to the Riemann sphere ($\cong \mathbb{C} \cup \{\infty\}$).*

Proof. Let $p \in M$. Since $r(\mathcal{D}_p) = 2$, there exists a (non constant) meromorphic function with a single pole. It is injective (its degree is 1) and surjective (since every non constant holomorphic map between compact Riemann surfaces is). Hence it provides the desired isomorphism. □

Corollary A.4. *($g \geq 1$) There are no meromorphic functions with a single pole of order 1 on M or equivalently, there is no point in M where all $\zeta \in \mathcal{H}^1(M)$ vanish.*

Proof. If f were such a function with pole at p we would have $r(-n\mathcal{D}_p) = n+1$ for all $n \in \mathbb{N}_0$ (consider f^n) and from Riemann-Roch $i(n\mathcal{D}_p) = g$. Hence all holomorphic differentials would vanish of arbitrary order at p which is impossible. Thus $r(-\mathcal{D}_p) = 1$ for all $p \in M$ or equivalently $i(\mathcal{D}_p) = g-1$, $p \in M$. □

If we have a sequence of points $\{p_\ell\}_{\ell \in \mathbb{N}} \subset M$ and define $\mathcal{D}_0 = 0$ and $\mathcal{D}_n = \mathcal{D}_{n-1} + \mathcal{D}_{p_n}$ for $n \in \mathbb{N}$, then there are exactly g ($g \geq 1$) integers $\{n_\ell\}_{\ell=1}^g$ such that $i(\mathcal{D}_{n_\ell}) = i(\mathcal{D}_{n_\ell - 1}) - 1$ (have a look at the proof of Riemann-Roch and note $i(\mathcal{D}_0) = g$) or equivalently, there exists no meromorphic function f with $(f)^- = \mathcal{D}_{n_\ell}$. Moreover,

$$(A.51) \qquad 1 = n_1 < \cdots < n_g < 2g.$$

(The first equality follows from our last corollary and the second from (A.41).) The numbers $\{n_\ell\}_{\ell=1}^g$ are called **Noether gaps** and in the special case where $p_1 = p_2 = \ldots$ **Weierstrass gaps**.

A surface admitting a function of lowest possible degree (i.e., 2 for $g \geq 1$) is called **hyperelliptic**. Note that if $g \leq 2$, M is always hyperelliptic ($g = 0, 1$ is obvious, for $g = 2$ choose $0 \neq \omega \in \mathcal{L}^1(0)$ and calculate $r((\omega)) = 2 + 1 - 2 + 1 = 2$).

A point $p \in M$ with $i(g\mathcal{D}_p) > 0$ is called a **Weierstrass point** (there are no Weierstrass points for $g \leq 1$). We mention (without proof) that the number of Weierstrass points W is finite and (for $g \geq 2$) satisfies $2g + 2 \leq W \leq g^3 - g$. The first inequality is attained if and only if at each Weierstrass point the gap sequence is $1, 3, \ldots, 2g - 1$ (which is the case if and only if M is hyperelliptic).

A.4. Jacobian variety and Abel's map

We start with the following (discrete) subset of $\mathbb{C}^g$

$$(A.52) \qquad L(M) = \{\underline{m} + \underline{\tau}\,\underline{n} \mid \underline{m}, \underline{n} \in \mathbb{Z}^g\} \subset \mathbb{C}^g$$

and define the **Jacobian variety** of M

$$(A.53) \qquad J(M) = \mathbb{C}^g / L(M).$$

It is a compact, commutative, g-dimensional, complex Lie group. Further we define **Abel's map** (with base point p_0):

$$(A.54) \qquad \begin{array}{rcl} \underline{A}_{p_0} : M & \to & J(M) \\ p & \mapsto & [\int_{p_0}^p \underline{\zeta}] = [(\int_{p_0}^p \zeta_1, \ldots, \int_{p_0}^p \zeta_g)] \end{array},$$

where $[\underline{z}] \in J(M)$ denotes the equivalence class of $\underline{z} \in \mathbb{C}^g$. It is clearly holomorphic and well-defined as long as we choose the same path of integration for all ζ_ℓ ($[\gamma] - [\gamma'] = \underline{m}[\underline{a}] + \underline{n}[\underline{b}]$) since

(A.55) $$\int_\gamma \underline{\zeta} - \int_{\gamma'} \underline{\zeta} = \underline{m} + \underline{\tau}\,\underline{n} \in L(M), \qquad \underline{m}, \underline{n} \in \mathbb{Z}^g.$$

Observe that $d\underline{A}_{p_0}$ is of maximal rank (i.e., 1) by Corollary A.4 (compare also (A.86)) and that a change of base point results in a global shift of the image. Note: $\underline{A}_{p_0}(p) = -\underline{A}_p(p_0)$.

We will also need a slight modification of $\underline{A}_{p_0}$

(A.56) $$\begin{aligned} \underline{\hat{A}}_{p_0} : \hat{M} &\to \mathbb{C}^g \\ p &\mapsto \int_{p_0}^p \underline{\hat{\zeta}} = \left(\int_{p_0}^p \hat{\zeta}_1, \ldots, \int_{p_0}^p \hat{\zeta}_g \right), \end{aligned}$$

with $p_0 \in \hat{M}$ and the path of integration lying in $\hat{M}$. We can easily extend Abel's map to the set of divisors $\mathrm{Div}(M)$

(A.57) $$\begin{aligned} \underline{\alpha}_{p_0} : \mathrm{Div}(M) &\to J(M) \\ \mathcal{D} &\mapsto \sum_{p \in M} \mathcal{D}(p) \underline{A}_{p_0}(p) \end{aligned}.$$

The sum is to be understood in $J(M)$ and multiplication with integers is also well-defined in $J(M)$. A similar extension $\underline{\hat{\alpha}}_{p_0}$ is defined for $\underline{\hat{A}}_{p_0}$. The natural domain for $\underline{\hat{\alpha}}_{p_0}$ would be $\mathrm{Div}(\hat{M})$, if we want to define it on $\mathrm{Div}(M)$ we have to make it unique on the boundary. But we will avoid this problem by requiring that divisors in $\mathrm{Div}(\hat{M})$ vanish on the boundary. We can now state **Abel's theorem**.

Theorem A.5. (Abel) $\mathcal{D} \in \mathrm{Div}(M)$ *is principal if and only if*

(A.58) $$\underline{\alpha}_{p_0}(\mathcal{D}) = [0] \qquad \text{and} \qquad \deg(\mathcal{D}) = 0.$$

Note that a change of the base point p_0 to p_1 amounts to adding a constant $\underline{A}_{p_1}(p) = \underline{A}_{p_0}(p) - \underline{A}_{p_0}(p_1)$, which yields $\underline{\alpha}_{p_1}(\mathcal{D}) = \underline{\alpha}_{p_0}(\mathcal{D}) - \deg(\mathcal{D})\underline{A}_{p_0}(p_1)$. Thus $\underline{\alpha}_{p_0}$ is independent of p_0 for divisors with $\deg(\mathcal{D}) = 0$.

Proof. Throughout this whole proof we choose $p_0 \notin \partial\hat{M}$ such that $\mathcal{D}(p_0) = 0$. Moreover, we assume that $\mathcal{D}(p) = 0$, $p \in \partial\hat{M}$ (otherwise deform $\{a_\ell, b_\ell\}_{1 \leq \ell \leq g}$ slightly).

Let $\mathcal{D} = (f)$ be the divisor of the meromorphic function f. Since df/f has only simple poles with residues $\mathrm{res}_p df/f = \mathrm{ord}_p f \in \mathbb{Z}$, we can use (A.20) to write

(A.59) $$\frac{df}{f} = \sum_{p \in M} \mathcal{D}(p) \omega_{pp_0} + \sum_{\ell=1}^g c_\ell \zeta_\ell.$$

A.4. Jacobian variety and Abel's map

Note that the poles at p_0 cancel. Using that the a- and b-periods must be integer multiples of $2\pi i$ by the residue theorem and (A.21) we infer

$$\int_{a_j} \frac{df}{f} = c_j = 2\pi i m_j,$$

$$\int_{b_j} \frac{df}{f} = 2\pi i \sum_{p \in M} \mathcal{D}(p) \int_{p_0}^{p} \zeta_j + \sum_{\ell=1}^{g} c_\ell \tau_{\ell,j} = 2\pi i \hat{\alpha}_{p_0,j}(\mathcal{D}) + \sum_{\ell=1}^{g} c_\ell \tau_{\ell,j}$$

(A.60) $\qquad = 2\pi i n_j,$

with $m_j, n_j \in \mathbb{Z}$. So we finally end up with

(A.61) $\qquad \underline{\alpha}_{p_0}(\mathcal{D}) = [\underline{\hat{\alpha}}_{p_0}(\mathcal{D})] = [\underline{n} - \underline{\tau}\,\underline{m}] = [\underline{0}].$

To prove the converse let $\mathcal{D}$ be given and consider (motivated by (A.59))

(A.62) $\quad f(q) = \exp\left(\sum_{p \in M} \mathcal{D}(p) \int_{q_0}^{q} \omega_{pp_0} + \sum_{\ell=1}^{g} \tilde{c}_\ell \int_{q_0}^{q} \zeta_\ell\right), \quad \tilde{c}_\ell \in \mathbb{C}, \quad q \in \hat{M}.$

where $p_0 \neq q_0 \notin \partial \hat{M}$ and $\mathcal{D}(q_0) = 0$. The constants $\tilde{c}_\ell$ are to be determined. The function f is a well-defined meromorphic function on $\hat{M}$ since a change of path (within $\hat{M}$) in the first integral amounts to a factor $2\pi i\, n$ ($n \in \mathbb{Z}$) which is swallowed by the exponential function. Clearly we have $(f) = \mathcal{D}$. For f to be a meromorphic function on M its value has to be independent of the path (in M) chosen

$$\sum_{p \in M} \mathcal{D}(p) \int_{a_j} \omega_{pp_0} + \sum_{\ell=1}^{g} \tilde{c}_\ell \int_{a_j} \zeta_\ell = \tilde{c}_j,$$

(A.63) $\quad \sum_{p \in M} \mathcal{D}(p) \int_{b_j} \omega_{pp_0} + \sum_{\ell=1}^{g} \tilde{c}_\ell \int_{b_j} \zeta_\ell = 2\pi i \hat{\alpha}_{p_0,j}(\mathcal{D}) + \sum_{\ell=1}^{g} \tilde{c}_\ell \omega_{\ell,j}.$

Thus if $\underline{\hat{\alpha}}_{p_0}(\mathcal{D}) = \underline{m} + \underline{\tau}\,\underline{n}$, we have to choose $\tilde{c}_\ell = -2\pi i n_\ell \in \mathbb{Z}$. $\square$

Corollary A.6. *The Picard group is isomorphic to the Jacobian variety*

(A.64) $\qquad\qquad\qquad \mathrm{Pic}(M) \cong J(M).$

Proof. The map $\underline{\alpha}_{p_0}$ provides an isomorphism since it is injective by Abel's theorem and surjective by Corollary A.15 below. $\square$

Corollary A.7. $\underline{A}_{p_0}$ *is injective.*

Proof. Let $p, q \in M$ with $\underline{A}_{p_0}(p) = \underline{A}_{p_0}(q)$. By Abel's theorem $\mathcal{D}_p - \mathcal{D}_q$ would be principal which is impossible because of Corollary A.4. $\square$

Thus $\underline{A}_{p_0}$ is an embedding of M into $J(M)$. For $g = 1$ it is even an isomorphism (it is surjective since every non constant map between compact Riemann surfaces is). For $g \geq 1$ we mention (without proof) **Torelli's theorem**:

Theorem A.8. (Torelli) *Two Riemann surfaces M and M' are isomorphic if and only if $J(M)$ and $J(M')$ are,*

(A.65) $\qquad\qquad M \cong M' \quad \Leftrightarrow \quad J(M) \cong J(M').$

A.5. Riemann's theta function

We first define the **Riemann theta function** associated with M

$$\theta : \mathbb{C}^g \to \mathbb{C}$$
(A.66)
$$\underline{z} \mapsto \sum_{\underline{m} \in \mathbb{Z}^g} \exp 2\pi \mathrm{i} \left(\langle \underline{m}, \underline{z} \rangle + \frac{\langle \underline{m}, \underline{\tau} \, \underline{m} \rangle}{2} \right),$$

where $\langle \underline{z}, \underline{z}' \rangle = \sum_{\ell=1}^g \overline{z_\ell} z'_\ell$. It is holomorphic since the sum converges nicely due to (A.14). We will now show some simple properties of $\theta(\underline{z})$. First observe that

(A.67)
$$\theta(-\underline{z}) = \theta(\underline{z})$$

and second let $\underline{n}, \underline{n}' \in \mathbb{Z}^g$, then

$$\theta(\underline{z} + \underline{n} + \underline{\tau} \, \underline{n}') = \sum_{\underline{m} \in \mathbb{Z}^g} \exp 2\pi \mathrm{i} \left(\langle \underline{m}, \underline{z} + \underline{n} + \underline{\tau} \, \underline{n}' \rangle + \frac{\langle \underline{m}, \underline{\tau} \, \underline{m} \rangle}{2} \right)$$

$$= \sum_{\underline{m} \in \mathbb{Z}^g} \exp 2\pi \mathrm{i} \Bigg(\langle \underline{m} + \underline{n}', \underline{z} \rangle - \langle \underline{n}', \underline{z} \rangle +$$

$$+ \frac{\langle \underline{m} + \underline{n}', \underline{\tau}(\underline{m} + \underline{n}') \rangle - \langle \underline{n}', \underline{\tau} \, \underline{n}' \rangle}{2} \Bigg)$$

(A.68)
$$= \exp 2\pi \mathrm{i} \left(-\langle \underline{n}', \underline{z} \rangle - \frac{\langle \underline{n}', \underline{\tau} \, \underline{n}' \rangle}{2} \right) \theta(\underline{z}).$$

(One has to use $\langle \underline{m}, \underline{\tau} \, \underline{n}' \rangle = \langle \underline{n}', \underline{\tau} \, \underline{m} \rangle$ which follows from symmetry of $\underline{\tau}$.) As special cases we mention

$$\theta(z + \delta_j) = \theta(z),$$
(A.69)
$$\theta(z + \tau_j) = e^{2\pi \mathrm{i}(-z_j - \tau_{j,j}/2)} \theta(z).$$

Here τ_j denotes the j'th row of $\underline{\tau}$. Notice that if we choose $\underline{z} = -(\underline{n} + \underline{\tau} \, \underline{n}')/2$, we get $\theta(-\underline{z}) = \exp(\mathrm{i}\pi \langle \underline{n}, \underline{n}' \rangle)\theta(\underline{z})$, implying that $\theta(-(\underline{n} + \underline{\tau} \, \underline{n}')/2) = 0$ if $\langle \underline{n}, \underline{n}' \rangle$ is odd.

Let $\underline{z} \in \mathbb{C}^g$. We will study the (holomorphic) function $\hat{F}$

$$\hat{F} : \hat{M} \to \mathbb{C}$$
(A.70)
$$p \mapsto \theta(\underline{\hat{A}}_{p_0}(p) - \underline{z}).$$

If $\hat{F}$ is not identically zero, we can compute the number of zeros in the following manner

$$\frac{1}{2\pi \mathrm{i}} \int_{\partial \hat{M}} \frac{d\hat{F}}{\hat{F}} = \frac{1}{2\pi \mathrm{i}} \sum_{\ell=1}^g \Bigg(\int_{a_\ell} \frac{d\hat{F}}{\hat{F}} + \int_{b_\ell} \frac{d\hat{F}}{\hat{F}}$$

(A.71)
$$+ \int_{a_\ell^{-1}} \frac{d\hat{F}}{\hat{F}} + \int_{b_\ell^{-1}} \frac{d\hat{F}}{\hat{F}} \Bigg).$$

Here and in what follows we will always assume that no zero of $\hat{F} \not\equiv 0$ lies on the boundary – otherwise deform the representatives of the homology basis. From (A.69) together with a similar argument as in (A.8) we get

(A.72) $\quad \hat{A}_{p_0,j}(\tilde{p}) = \hat{A}_{p_0,j}(p) + \tau_{\ell,j} \;\Rightarrow\; \hat{F}(\tilde{p}) = e^{2\pi \mathrm{i}(z_\ell - \hat{A}_{p_0,\ell}(p) - \tau_{\ell,\ell}/2)} \hat{F}(p)$

A.5. Riemann's theta function

if $p, \tilde{p}$ are corresponding points on a_ℓ, a_ℓ^{-1} and

(A.73) $\qquad \hat{A}_{p_0,j}(\tilde{p}) = \hat{A}_{p_0,j}(p) - \delta_{\ell j} \quad \Rightarrow \quad \hat{F}(\tilde{p}) = \hat{F}(p)$

if $p, \tilde{p}$ are corresponding points on b_ℓ, b_ℓ^{-1}. Using this we see

(A.74) $\qquad \dfrac{1}{2\pi i} \displaystyle\int_{\partial \hat{M}} \dfrac{d\hat{F}}{\hat{F}} = \sum_{\ell=1}^g \int_{a_\ell} d\hat{A}_{p_0,\ell} = \sum_{\ell=1}^g \int_{a_\ell} \zeta_\ell = g.$

Next, let us compute $\hat{\underline{\alpha}}_{p_0}((\hat{F}))$ (Since we have assumed that no zero lies on the boundary we may write $(\hat{F}) \in \mathrm{Div}(M)$.)

$$\hat{\underline{\alpha}}_{p_0}((\hat{F})) = \dfrac{1}{2\pi i} \int_{\partial \hat{M}} \hat{\underline{A}}_{p_0} \dfrac{d\hat{F}}{\hat{F}}$$

$$= \dfrac{1}{2\pi i} \sum_{\ell=1}^g \left(\int_{a_\ell} \hat{\underline{A}}_{p_0} \dfrac{d\hat{F}}{\hat{F}} + \int_{b_\ell} \hat{\underline{A}}_{p_0} \dfrac{d\hat{F}}{\hat{F}} \right.$$

(A.75) $\qquad \left. + \displaystyle\int_{a_\ell^{-1}} \hat{\underline{A}}_{p_0} \dfrac{d\hat{F}}{\hat{F}} + \int_{b_\ell^{-1}} \hat{\underline{A}}_{p_0} \dfrac{d\hat{F}}{\hat{F}} \right).$

Proceeding as in the last integral (using (A.72), (A.73)) yields

(A.76) $\quad \hat{\alpha}_{p_0,j}((\hat{F})) = \dfrac{1}{2\pi i} \displaystyle\sum_{\ell=1}^g \left(\int_{a_\ell} \left(2\pi i (\hat{A}_{p_0,j} + \tau_{\ell,j})\zeta_\ell - \tau_{\ell,j} \dfrac{d\hat{F}}{\hat{F}} \right) + \delta_{\ell j} \int_{b_\ell} \dfrac{d\hat{F}}{\hat{F}} \right).$

Hence we have to perform integrals of the type $\int_p^{\tilde{p}} d\ln(\hat{F})$, which is just $\ln(\hat{F}(\tilde{p})) - \ln(\hat{F}(p))$. Of course we do not know which branch we have to choose, so we have to add a proper multiple of $2\pi i$. Computing $\hat{F}(p), \hat{F}(\tilde{p})$ as in (A.8) we get

$$\dfrac{1}{2\pi i} \sum_{\ell=1}^g \left(\tau_{\ell,j} \int_{a_\ell} \dfrac{d\hat{F}}{\hat{F}} + \delta_{\ell j} \int_{b_\ell} \dfrac{d\hat{F}}{\hat{F}} \right) =$$

(A.77) $\qquad -\hat{A}_{p_0,j}(p) + z_j - \dfrac{\tau_{j,j}}{2} + m_j + \displaystyle\sum_{\ell=1}^g \tau_{\ell,j} n_\ell,$

with $\underline{m}, \underline{n} \in \mathbb{Z}^g$ and p is the point in $\hat{M}$ that a_ℓ and b_ℓ have in common – we have omitted the dependence of p on ℓ since $\hat{A}_{p_0,j}(p)$ is independent of ℓ. As the last (unknown) factor is in $L(M)$, we can get rid of it by changing to $J(M)$

(A.78) $\qquad \alpha_{p_0,j}((\hat{F})) = [\hat{\alpha}_{p_0,j}((\hat{F}))] = [-\hat{A}_{p_0,j}(p) + z_j - \dfrac{\tau_{j,j}}{2} + \displaystyle\sum_{\ell=1}^g \int_{a_\ell} \hat{A}_{p_0,j} \zeta_\ell].$

There is one more thing we can do,

$$\int_{a_\ell} \hat{A}_{p_0,\ell} \zeta_\ell = \dfrac{1}{2}(\hat{A}_{p_0,\ell}^2(p) - \hat{A}_{p_0,\ell}^2(\tilde{p}))$$

(A.79) $\qquad = \dfrac{1}{2}(\hat{A}_{p_0,\ell}^2(p) - (\hat{A}_{p_0,\ell}(p) - 1)^2) = \hat{A}_{p_0,\ell}(p) - \dfrac{1}{2}.$

(Compare the picture on page 2.) Thus our final result is

(A.80) $\qquad \underline{\alpha}_{p_0}((\hat{F})) = [\underline{z}] - \Xi_{p_0},$

with

$$\Xi_{p_0,j} = [\frac{1+\tau_{j,j}}{2} - \sum_{\ell \neq j} \int_{a_\ell} \hat{A}_{p_0,j} \zeta_\ell] \tag{A.81}$$

the **vector of Riemann constants**. Finally, we mention that evaluating the integral

$$\frac{1}{2\pi i} \int_{\partial \hat{M}} f \frac{d\hat{F}}{\hat{F}}, \tag{A.82}$$

where $f \in \mathcal{M}(M)$ is an arbitrary meromorphic function yields the useful formula

$$\sum_{\ell=1}^{g} \int_{a_\ell} f \zeta_\ell = \sum_{\ell=1}^{g} f(p_j) + \sum_{p \in f^{-1}(\infty)} \mathrm{res}_p\left(f \, d\ln \hat{F}\right), \tag{A.83}$$

where $\{p_j\}_{j=1}^{g}$ are the zeros of the function $\hat{F}$ which (for simplicity) are away from the poles of f.

A.6. The zeros of the Riemann theta function

We begin with some preparations. Let M_n be the set of positive divisors of degree $n \in \mathbb{N}_0$. (The case $n = 0$ is of course trivial, but it is practical to include it.) We can make M_n a complex (n-dimensional) manifold if we identify M_n with $\sigma^n M$ the n-th **symmetric power** of M. If $\mathcal{D}_n \in M_n$ and if the points $\{\tilde{p}_j\}_{1 \leq j \leq n}$ with $\mathcal{D}_n(p) \neq 0$ are distinct (i.e., $\mathcal{D}_n(p) \leq 1$, $p \in M$), a local chart is given by

$$U = \{[(p_1,\ldots,p_n)] | p_j \in U_j\}, \quad \text{with } U_j \cap U_k = \emptyset \text{ for } j \neq k$$
$$[(p_1,\ldots,p_n)] \mapsto (z_1(p_1),\ldots,z_n(p_n)), \quad \text{with } p_j \in U_j, \tag{A.84}$$

where (U_j, z_j) are charts in M centered at $\tilde{p}_j$. If $\mathcal{D}(p) \geq 2$ for some points, one has to choose more sophisticated charts (elementary symmetric functions). Note, that $\underline{\alpha}_{p_0} : M_n \to J(M)$ is a holomorphic mapping. If $\mathcal{D}_n \in M_n$ as above, one can easily compute

$$\{d\underline{\alpha}_{p_0}(\mathcal{D}_n)\}_{jk} = \zeta_j(\tilde{p}_k), \quad 1 \leq j \leq g, \; 1 \leq k \leq n \tag{A.85}$$

(with some abuse of notation: $\zeta_j(\tilde{p}_k)$ means the value of $\eta_j(z(\tilde{p}_k))$ if $\zeta_j = \eta_j(z)dz$). What can we say about $\dim \mathrm{Ker}(d\underline{\alpha}_{p_0})$? If $\zeta = \sum_{j=1}^{g} c_j \zeta_j \in \mathcal{L}^1(\mathcal{D}_n)$ we have $\zeta(\tilde{p}_k) = \sum_{j=1}^{g} c_j \zeta_j(\tilde{p}_k) = 0$. But we can even reverse this argument, ending up with

$$i(\mathcal{D}_n) = \dim \mathrm{Ker}(d\underline{\alpha}_{p_0}|_{\mathcal{D}_n}) = g - \mathrm{rank}(d\underline{\alpha}_{p_0}|_{\mathcal{D}_n}). \tag{A.86}$$

(We remark that this formula remains true for arbitrary $\mathcal{D}_n \in M_n$.)

Now we will prove a small (but useful) lemma:

Lemma A.9. *Let $\mathcal{D}_1, \mathcal{D}_2 \in M_g$ with $\underline{\alpha}_{p_0}(\mathcal{D}_1) = \underline{\alpha}_{p_0}(\mathcal{D}_2)$ and $i(\mathcal{D}_2) = 0$, then $\mathcal{D}_1 = \mathcal{D}_2$.*

Proof. By Abel's theorem $\mathcal{D}_1 - \mathcal{D}_2$ is principal and thus $r(\mathcal{D}_1 - \mathcal{D}_2) = 1$. From $r(-\mathcal{D}_2) = 1 + i(\mathcal{D}_2) = 1$ (Riemann-Roch) we conclude $\mathcal{L}(\mathcal{D}_1 - \mathcal{D}_2) = \mathcal{L}(-\mathcal{D}_2) \stackrel{(\sim)}{=} \mathbb{C}$ and therefrom $\mathcal{D}_1 - \mathcal{D}_2 = 0$. □

Next, we will show that $i(\mathcal{D}) = 0$ is no serious restriction for $\mathcal{D} \in M_g$.

A.6. The zeros of the Riemann theta function

Lemma A.10. *Let $\mathcal{D}_n \in M_n$ ($0 \leq n \leq g$), then in every neighborhood of $\mathcal{D}_n$ there is a $\mathcal{D}'_n \in M_n$ with $i(\mathcal{D}'_n) = g - n$. We may even require $\mathcal{D}'_n(p) \leq 1$, $p \in M$.*

Proof. Our lemma is true for $n = 1$, so let us assume it is also true for some $n - 1 \geq 1$. Let $\mathcal{D}_n = \mathcal{D}_{n-1} + \mathcal{D}_p$ and let $\{\omega_\ell\}_{1 \leq \ell \leq g-n+1}$ be a basis for $\mathcal{L}^1(\mathcal{D}'_{n-1})$. We can always find a neighborhood U_p of p such that none of the basis elements vanishes on $U_p \setminus \{p\}$ - so just take $\mathcal{D}'_n = \mathcal{D}'_{n-1} + \mathcal{D}_{\tilde{p}}$ with a suitable $\tilde{p} \in U_p \setminus \{p\}$. $\square$

We have shown, that the divisors $\mathcal{D}'_n$ with $i(\mathcal{D}'_n) = g - n$ and $\mathcal{D}'_n(p) \leq 1$, $p \in M$, are dense in M_n. Hence $\underline{\alpha}_{p_0} : M_g \to J(M)$ has maximal rank for such divisors and $\theta(\underline{\hat{\alpha}}_{p_0}(.) - \underline{z})$ does not vanish identically on any open subset of M_g. We set $W_n^{p_0} = \underline{\alpha}_{p_0}(M_n) \subset J(M)$ and $\hat{W}_n^{p_0} = \underline{\hat{\alpha}}_{p_0}(M_n) \subset \mathbb{C}^g$. (Our last statement now reads $\theta(\hat{W}_g^{p_0} - \underline{z}) \not\equiv 0$, where $\hat{W}_g^{p_0} - \underline{z}$ stands for $w - \underline{z}$ with $w \in \hat{W}_g^{p_0}$.) We observe $W_n^{p_0} \subset W_{n+1}^{p_0}$ (just add $\mathcal{D}_{p_0}$ to each divisor in $W_n^{p_0}$) and $\hat{W}_n^{p_0} \subset \hat{W}_{n+1}^{p_0}$.

We will now characterize the set of zeros of $\theta(\underline{z})$:

Theorem A.11. *Let $\underline{z} \in \mathbb{C}^g$. Then $\theta(\underline{z}) = 0$ if and only if $[\underline{z}] \in W_{g-1}^{p_0} + \Xi_{p_0}$.*

Note: Though θ is not well-defined on $J(M)$ the set of zeros is a well-defined subset of $J(M)$. Observe also that $[\underline{z}] \in W_{g-1}^{p_0} + \Xi_{p_0}$ is not dependent on the base point p_0 chosen!

Proof. Choose $\mathcal{D} \in M_{g-1}$ and $\tilde{p} \in M$ such that $i(\mathcal{D} + \mathcal{D}_{\tilde{p}}) = 0$. Choose a $\underline{z} \in \mathbb{C}^g$ with $[\underline{z}] = \underline{\alpha}_{p_0}(\mathcal{D} + \mathcal{D}_{\tilde{p}}) + \Xi_{p_0}$ and consider (still assuming that $\hat{F}$ has no zeros on the boundary)

$$\hat{F}(p) = \theta(\underline{\hat{A}}_{p_0}(p) - \underline{z}). \tag{A.87}$$

If $\hat{F}$ is not identically zero, we have $\underline{\alpha}_{p_0}((\hat{F})) = [\underline{z}] - \Xi_{p_0} = \underline{\alpha}_{p_0}(\mathcal{D} + \mathcal{D}_{\tilde{p}})$. Using Lemma A.9 we get $(\hat{F}) = \mathcal{D} + \mathcal{D}_{\tilde{p}}$ and thus $\hat{F}(\tilde{p}) = 0$. As this is trivially true if $\hat{F}$ vanishes identically, we are ready with the first part because the divisors $\mathcal{D}$ under consideration are dense and because $\underline{\hat{\alpha}}_{p_0} : M_{g-1} \to \mathbb{C}^g$ is continuous.

Conversely, suppose $\theta(\underline{z}) = 0$ and let s be the integer with $\theta(\hat{W}_{s-1}^{p_0} - \hat{W}_{s-1}^{p_0} - \underline{z}) \equiv 0$ and $\theta(\hat{W}_s^{p_0} - \hat{W}_s^{p_0} - \underline{z}) \not\equiv 0$. (Here $\hat{W}_g^{p_0} - \hat{W}_s^{p_0} - \underline{z}$ stands for $w_1 - w_2 - \underline{z}$ with $w_1, w_2 \in \hat{W}_g^{p_0}$.) We have $1 \leq s \leq g$. Thus we may choose two divisors $\mathcal{D}_s^1, \mathcal{D}_s^2 \in M_s$ such that $\theta(\underline{\hat{\alpha}}_{p_0}(\mathcal{D}_s^1) - \underline{\hat{\alpha}}_{p_0}(\mathcal{D}_s^2) - \underline{z}) \neq 0$ and $\mathcal{D}_s^2(p) \leq 1$, $p \in M$. Denote $\mathcal{D}_{s-1}^1 = \mathcal{D}_s^1 - \mathcal{D}_{\tilde{p}}$ with $\tilde{p}$ such that $\mathcal{D}_s^1(\tilde{p}) \neq 0$ and consider the function

$$\hat{F}(p) = \theta(\underline{\hat{A}}_{p_0}(p) + \underline{\hat{\alpha}}_{p_0}(\mathcal{D}_{s-1}^1) - \underline{\hat{\alpha}}_{p_0}(\mathcal{D}_s^2) - \underline{z}). \tag{A.88}$$

Observe that $\hat{F}(\tilde{p}) \neq 0$ and that $\hat{F}(p) = 0$ if $\mathcal{D}_s^2(p) = 1$. Thus $(\hat{F})$ may be written as $\mathcal{D}_s^2 + \mathcal{D}_{g-s}$ with $\mathcal{D}_{g-s} \in M_{g-s}$. And calculating $\underline{\alpha}_{p_0}((\hat{F}))$ yields

$$\underline{\alpha}_{p_0}((\hat{F})) = -\underline{\alpha}_{p_0}(\mathcal{D}_{s-1}^1) + \underline{\alpha}_{p_0}(\mathcal{D}_s^2) + [\underline{z}] - \Xi_{p_0} = \underline{\alpha}_{p_0}(\mathcal{D}_s^2) - \underline{\alpha}_{p_0}(\mathcal{D}_{g-s}). \tag{A.89}$$

Thus $[\underline{z}] = \underline{\alpha}_{p_0}(\mathcal{D}_{s-1}^1 + \mathcal{D}_{g-s}) + \Xi_{p_0}$. $\square$

Before we can proceed we need:

Lemma A.12. *Let $\mathcal{D}$ be an positive divisor on M. $r(-\mathcal{D}) \geq s \geq 1$ if and only if given any positive divisor $\mathcal{D}'$ of degree $< s$, there is an positive divisor $\mathcal{D}''$ such that $\mathcal{D}' + \mathcal{D}'' - \mathcal{D}$ is principal. For the if part it suffices to restrict $\mathcal{D}'$ to an open subset $U \subset M_{s-1}$.*

Proof. Since $-\mathcal{D} + \mathcal{D}' + \mathcal{D}'' = (f)$ we have $f \in \mathcal{L}(-\mathcal{D} + \mathcal{D}')$. But if $r(\mathcal{D}) < s$ we can always find a $\mathcal{D}' \in U$ with $r(-\mathcal{D} + \mathcal{D}') = 0$ (Let $\{f_\ell\}_{1 \leq \ell \leq d}$ be a basis for $\mathcal{L}(-\mathcal{D})$. Start with a suitable point p_1 where none of the functions f_ℓ vanishes, construct a new basis $\tilde{f}_\ell$ such that $\tilde{f}_\ell(p_1) = 0$ for $\ell \geq 2$ – repeat $r(\mathcal{D})$ times and set $\mathcal{D}' = \sum_{\ell=1}^{d} \mathcal{D}_{p_\ell}$.).

Now assume $r(-\mathcal{D}) = s$ and let $\mathcal{D}' \in M_{s-1}$. It follows that $r(\mathcal{D}' - \mathcal{D}) \geq s - (s-1) = 1$ (by a simple induction argument – compare (A.42)) and we can choose $\mathcal{D}'' = (f) + \mathcal{D} - \mathcal{D}'$ for a nonzero $f \in \mathcal{L}(\mathcal{D} - \mathcal{D}')$. $\square$

Note: $\mathcal{D}$ and $\mathcal{D}' + \mathcal{D}''$ have the same image under Abel's map and $r(-\mathcal{D}) = i(\mathcal{D}) + 1$ for $\mathcal{D} \in M_g$ (by Riemann-Roch).

Now let $[\underline{z}] = \underline{\alpha}_{p_0}(\mathcal{D}) + \underline{\Xi}_{p_0}$ with $\mathcal{D} \in M_{g-1}$ and $i(\mathcal{D}) = s(\geq 1)$. Any point $\underline{\xi} \in (\hat{W}_{s-1}^{p_0} - \hat{W}_{s-1}^{p_0} - \underline{z})$ can be written as $\hat{\underline{\alpha}}_{p_0}(\mathcal{D}_{s-1}^1) - \hat{\underline{\alpha}}_{p_0}(\mathcal{D}_{s-1}^2) - \underline{z}$ with $\mathcal{D}_{s-1}^1, \mathcal{D}_{s-1}^2 \in M_{s-1}$. Due to our lemma we can also write $[\underline{z}]$ as $\underline{\alpha}_{p_0}(\mathcal{D}_{s-1}^1 + \mathcal{D}_{g-s}'') + \underline{\Xi}_{p_0}$ and we get $[\underline{\xi}] = -\underline{\alpha}_{p_0}(\mathcal{D}_{s-1}^2 + \mathcal{D}_{g-s}'') - \underline{\Xi}_{p_0} \in -(W_{g-1}^{p_0} + \underline{\Xi}_{p_0})$ or equivalently $\theta(\underline{\xi}) = 0$.

On the other hand we may also vary the divisor $\mathcal{D}_s^1$ of the last theorem in a sufficiently small neighborhood getting $\mathcal{D}_{s-1}^{1'}$ and a corresponding $\mathcal{D}_{g-s}'$ also fulfilling $[\underline{z}] = \underline{\alpha}_{p_0}(\mathcal{D}_{s-1}^{1'} + \mathcal{D}_{g-s}') + \underline{\Xi}_{p_0}$. From Abel's theorem we conclude that $(\mathcal{D}_{s-1}^1 + \mathcal{D}_{g-s}) - (\mathcal{D}_{s-1}^{1'} + \mathcal{D}_{g-s}')$ is principal and from our last lemma: $i(\mathcal{D}_{s-1}^1 + \mathcal{D}_{g-s}) \geq s$. Thus we have proved most of **Riemann's vanishing theorem**:

Theorem A.13. (Riemann) *Let $\underline{z} \in \mathbb{C}^g$ and $s \geq 1$. $\theta(\hat{W}_{s-1}^{p_0} - \hat{W}_{s-1}^{p_0} - \underline{z}) \equiv 0$ but $\theta(\hat{W}_s^{p_0} - \hat{W}_s^{p_0} - \underline{z}) \not\equiv 0$ is equivalent to $[\underline{z}] = \underline{\alpha}_{p_0}(\mathcal{D}) + \underline{\Xi}_{p_0}$ with $\mathcal{D} \in M_{g-1}$ and $i(\mathcal{D}) = s$ which again is equivalent to the vanishing of all partial derivatives of θ of order less than s at $\underline{z}$ and to the non-vanishing of at least one derivative of order s at $\underline{z}$.*

Note: $s \geq 1$ implies $\theta(\underline{z}) = 0$, we always have $s \leq g$ and $\hat{W}_{s-1}^{p_0} - \hat{W}_{s-1}^{p_0}$ is independent of the base point p_0.

Proof. We omit the proof of the part concerning the derivatives (cf. [81]), but we will prove the other part: In the one direction we already know the first part and $i(\mathcal{D}) \geq s$. If $i(\mathcal{D}) > s$ we would get $\theta(\hat{W}_s^{p_0} - \hat{W}_s^{p_0} - \underline{z}) \equiv 0$ from the first statement above. In the other direction we would get from $\theta(\hat{W}_s^{p_0} - \hat{W}_s^{p_0} - \underline{z}) \equiv 0$ that $i(\mathcal{D}) > s$ using the first direction. $\square$

With some simple changes in the proof of the last theorem, we can also get another one:

Theorem A.14. *Let $\underline{z} \in \mathbb{C}^g$. s is the least integer such that $\theta(\hat{W}_{s+1}^{p_0} - \hat{W}_s^{p_0} - \underline{z}) \not\equiv 0$ is equivalent to $[\underline{z}] = \underline{\alpha}_{p_0}(\mathcal{D}) + \underline{\Xi}_{p_0}$ with $\mathcal{D} \in M_g$ and $i(\mathcal{D}) = s$.*

Note: We now have $0 \leq s \leq g - 1$. Since $\underline{z}$ was arbitrary we have solved **Jacobi's inversion problem**:

Corollary A.15. (Jacobi) $\underline{\alpha}_{p_0} : M_g \to J(M)$ *is surjective.*

Using these last theorems we can say a lot about the function

(A.90) $$\hat{F}(p) = \theta(\hat{\underline{A}}_{p_0}(p) - \underline{z}), \qquad \underline{z} \in \mathbb{C}^g, \quad p \in \hat{M}.$$

Theorem A.16. *Let $\underline{z} \in \mathbb{C}^g$ and let s be the least integer such that $\theta(\hat{W}_{s+1}^{p_0} - \hat{W}_s^{p_0} - \underline{z}) \not\equiv 0$. Then $\hat{F} \not\equiv 0$ is equivalent to $s = 0$. We can write $[\underline{z}] = \underline{\alpha}_{p_0}(\mathcal{D}) + \underline{\Xi}_{p_0}$ with $\mathcal{D} \in M_g$ and $i(\mathcal{D}) = s$. Moreover, if $\hat{F} \not\equiv 0$ we have $(\hat{F}) = \mathcal{D}$ (and hence $i((\hat{F})) = 0$).*

Proof. Observe, that both sides are equivalent to $\hat{F}(p) = \theta(\underline{\hat{A}}_{p_0}(p) - \underline{z}) \not\equiv 0$ and $\mathcal{D}(p_0) = 0$. Uniqueness follows from Lemma A.9. $\square$

Corollary A.17. *$\theta(\underline{z}) \neq 0$ is equivalent to $[\underline{z}] = \underline{\alpha}_{p_0}(\mathcal{D}) + \underline{\Xi}_{p_0}$ with a unique $\mathcal{D} \in M_g$, $i(\mathcal{D}) = 0$ and $\mathcal{D}(p_0) = 0$.*

Finally we prove Abel's theorem for differentials:

Theorem A.18. *$\mathcal{D} \in \mathrm{Div}(M)$ is canonical if and only if*

(A.91) $$\underline{\alpha}_{p_0}(\mathcal{D}) = -2\underline{\Xi}_{p_0} \quad \text{and} \quad \deg(\mathcal{D}) = 2g - 2.$$

Proof. Since all canonical divisors lie in the same divisor class their image under Abel's map is a constant (by Abel's theorem). To determine this constant let $\underline{z}$ be a zero of θ. Hence $-\underline{z}$ is a zero too, and we conclude that $[\underline{z}] = \underline{\alpha}_{p_0}(\mathcal{D}_1) + \underline{\Xi}_{p_0}$ and $[-\underline{z}] = \underline{\alpha}_{p_0}(\mathcal{D}_2) + \underline{\Xi}_{p_0}$ with $\mathcal{D}_1, \mathcal{D}_2 \in M_{g-1}$. Combining these equations we get $\underline{\alpha}_{p_0}(\mathcal{D}_1 + \mathcal{D}_2) = -2\underline{\Xi}_{p_0}$. Since $\mathcal{D}_1$ is arbitrary (because $\underline{z}$ is) our last lemma yields $r(-\mathcal{D}_1 - \mathcal{D}_2) \geq g$ and from Riemann Roch $i(\mathcal{D}_1 + \mathcal{D}_2) \geq 1$. Thus $\mathcal{D}_1 + \mathcal{D}_2$ is the divisor of a holomorphic differential (since $\deg(\mathcal{D}_1 + \mathcal{D}_2) = 2g - 2$).

Conversely, suppose $\mathcal{D}$ is of degree $2g - 2$ and $\underline{\alpha}_{p_0}(\mathcal{D}) = -2\underline{\Xi}_{p_0}$. Let ω be a meromorphic differential. From Abel's theorem we get that $\mathcal{D} - (\omega)$ is the divisor of a meromorphic function f. Thus $\mathcal{D}$ is the divisor of $f\omega$. $\square$

Corollary A.19. *$2\underline{\Xi}_{p_0} = [0]$ if and only if $(2g - 2)\mathcal{D}_{p_0}$ is canonical.*

A.7. Hyperelliptic Riemann surfaces

Finally, we want to give a constructive approach to hyperelliptic Riemann surfaces. Let

(A.92) $$\{E_n\}_{0 \leq n \leq 2g+1} \subset \mathbb{R}, \qquad E_0 < E_1 < \cdots < E_{2g+1}, \qquad g \in \mathbb{N}_0,$$

be some fixed points. Using them we may define

(A.93) $$\Pi = \mathbb{C} \setminus \left(\bigcup_{j=0}^{g} [E_{2j}, E_{2j+1}] \right),$$

which is obviously a domain (connected open subset of $\mathbb{C}$) and a holomorphic function

(A.94) $$\begin{aligned} R_{2g+2}^{1/2}(.) : \Pi &\to \mathbb{C} \\ z &\mapsto -\prod_{n=0}^{2g+1} \sqrt{z - E_n}, \end{aligned}$$

where the square root is defined as follows,

(A.95) $$\sqrt{z} = |\sqrt{z}| e^{\mathrm{i} \arg(z)/2}, \qquad \arg(z) \in (-\pi, \pi].$$

We first extend $R_{2g+2}^{1/2}(z)$ to the whole of $\mathbb{C}$ by

(A.96) $$R_{2g+2}^{1/2}(z) = \lim_{\varepsilon \downarrow 0} R_{2g+2}^{1/2}(x + i\varepsilon)$$

for $z \in \mathbb{C}\backslash\Pi$. This implies $R_{2g+2}^{1/2}(z) = i^{2g-n}|R_{2g+2}^{1/2}(z)|$ for $z \in (E_{n-1}, E_n)$, $0 \leq n \leq 2g+2$, if we set $E_{-1} = -\infty$ and $E_{2g+2} = \infty$.

We may now define the following set

(A.97) $$M = \{(z, \sigma R_{2g+2}^{1/2}(z))|z \in \mathbb{C}, \sigma \in \{-1, +1\}\} \cup \{\infty_+, \infty_-\}$$

and call $B = \{(E_n, 0)\}_{0 \leq n \leq 2g+1}$ the set of branch points.

Defining the following charts (ζ_{p_0}, U_{p_0}), we can make the set M into a Riemann surface. Abbreviate

(A.98) $$p_0 = (z_0, \sigma_0 R_{2g+2}^{1/2}(z_0)), \ p = (z, \sigma R_{2g+2}^{1/2}(z)) \in U_{p_0} \subset M,$$

$U'_{p_0} = \zeta_{p_0}(U_{p_0}) \subset \mathbb{C}$.

If $p_0 \notin B$ we set:

$U_{p_0} = \{p \in M||z - z_0| < C$ and $\sigma R_{2g+2}^{1/2}(z) \xrightarrow{\gamma} \sigma_0 R_{2g+2}^{1/2}(z_0)\}$, $U'_{p_0} = \{\zeta \in \mathbb{C}||\zeta| < C\}$, $C = \min_n |z_0 - E_n| > 0$,

(A.99) $$\begin{array}{cccc} \zeta_{p_0}: & U_{p_0} & \to & U'_{p_0} \\ & p & \mapsto & z - z_0 \end{array} \quad \begin{array}{cccc} \zeta_{p_0}^{-1}: & U'_{p_0} & \to & U_{p_0} \\ & \zeta & \mapsto & (\zeta + z_0, \sigma R_{2g+2}^{1/2}(\zeta + z_0)) \end{array},$$

where $\sigma R_{2g+2}^{1/2}(z) \xrightarrow{\gamma} \sigma_0 R_{2g+2}^{1/2}(z_0)$ means that $\sigma R_{2g+2}^{1/2}(z)$ is the branch reached by analytic continuation along γ, the straight line from z to z_0.

If $p_0 = (E_m, 0)$ we set:

$U_{p_0} = \{p \in M||z - E_m| < C\}$, $U'_{p_0} = \{\zeta \in \mathbb{C}||\zeta| < \sqrt{C}\}$, $C = \min_{n \neq m} |E_m - E_n| > 0$,

$$\begin{array}{cccc} \zeta_{p_0}: & U_{p_0} & \to & U'_{p_0} \\ & p & \mapsto & \sigma(z - E_m)^{1/2} \end{array},$$

(A.100) $$\begin{array}{cccc} \zeta_{p_0}^{-1}: & U'_{p_0} & \to & U_{p_0} \\ & \zeta & \mapsto & (\zeta^2 + E_m, \zeta \prod_{n \neq m}(\zeta^2 + E_m - E_n)^{1/2}) \end{array},$$

where the left root is defined as $(z - E_m)^{1/2} = \sqrt{|z - E_m|} \exp(i \arg(z - E_m)/2)$, with $0 \leq \arg(z - E_m) < 2\pi$ if m is even and $-\pi < \arg(z - E_m) \leq \pi$ if m is odd. The right root is holomorphic on U_{p_0} with the sign fixed by

$$\prod_{n \neq m}(\zeta^2 + E_m - E_n)^{1/2} = i^{2g-m-1}\left|\prod_{n \neq m}\sqrt{E_m - E_n}\right| \times$$

(A.101) $$\times \left(1 - \frac{1}{2}\sum_{n \neq m}\frac{1}{E_m - E_n}\zeta^2 + O(\zeta^4)\right).$$

If $p_0 = \infty_\pm$ we set:

A.7. Hyperelliptic Riemann surfaces

$U_{p_0} = \{p \in M | |z| > C\}$ $U'_{p_0} = \{\zeta \in \mathbb{C} | |\zeta| < \frac{1}{C}\}$ $C = \max_n |E_n| < \infty$

$$
\text{(A.102)} \quad \begin{aligned} \zeta_{p_0} : U_{p_0} &\to U'_{p_0} \\ p &\mapsto \frac{1}{z} \\ \infty_\pm &\mapsto 0 \end{aligned} \qquad \begin{aligned} \zeta_{p_0}^{-1} : U'_{p_0} &\to U_{p_0} \\ \zeta &\mapsto \left(\frac{1}{\zeta}, \frac{\pm\prod_n (1-\zeta E_n)^{1/2}}{\zeta^{g+1}}\right) \\ 0 &\mapsto \infty_\pm \end{aligned}
$$

where the right root is holomorphic on U_{p_0} with the sign fixed by

$$
\text{(A.103)} \qquad \prod_n (1 - \zeta E_n)^{1/2} = -1 + \frac{1}{2}\sum_n E_n \zeta + O(\zeta^2).
$$

Let us take two subsets

$$
\text{(A.104)} \qquad \Pi_\pm = \{(z, \pm R_{2g+2}^{1/2}(z)) | z \in \Pi\} \subset M,
$$

and define two more quite useful charts

$$
\text{(A.105)} \qquad \begin{aligned} \zeta_\pm : \Pi_\pm &\to \Pi \\ p &\mapsto z \end{aligned}.
$$

It is not hard to verify, that the transition functions of all these charts are indeed holomorphic, for example,

$$
\text{(A.106)} \quad \begin{aligned} \zeta_\pm \circ \zeta_{E_m}^{-1} : U'_{E_m} \cap \zeta_{E_m}(\Pi_\pm) &\to \Pi \cap \zeta_\pm(U_{E_m}), \\ \zeta &\mapsto \zeta^2 + E_m \\ \zeta_{E_m} \circ \zeta_\pm^{-1} : \Pi \cap \zeta_\pm(U_{E_m}) &\to U'_{E_m} \cap \zeta_{E_m}(\Pi_\pm). \\ \zeta &\mapsto \pm\sqrt{\zeta - E_m} \end{aligned}
$$

Compare the following picture ($g = 1$, m odd) and note, that the branch cut of the square root does not belong to its domain!

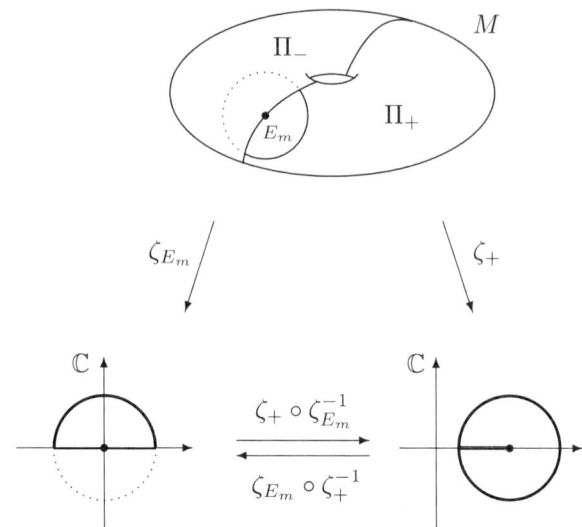

The topology induced by these charts is Hausdorff (take the U's from above and modify C a little if necessary) and second countable (since we only need countably (in particular even finitely) many of these charts to cover M). Consider $\Pi_\pm$, they are connected (since they are homeomorphic to Π). Their closures $\overline{\Pi_\pm}$ are still connected, have points in common and fulfill $M = \overline{\Pi_+} \cup \overline{\Pi_-}$. Thus M is connected and a Riemann surface. But they tell us even more: $\overline{\Pi_\pm}$ is just a sphere with $g+1$ holes, and if we want to get M, we have to identify corresponding points on the boundaries ending up with a sphere having g handles.

We will now define three maps on M. First the sheet exchange map,

(A.107)
$$*: M \to M$$
$$(z, \sigma R_{2g+2}^{1/2}(z)) \mapsto (z, \sigma R_{2g+2}^{1/2}(z))^* = (z, -\sigma R_{2g+2}^{1/2}(z))$$
$$\infty_\pm \mapsto \infty_\mp$$

It is clearly holomorphic, since it is locally just the identity (up to a sign – using the charts of the previous page). And second the projection

(A.108)
$$\pi: M \to \mathbb{C} \cup \{\infty\}$$
$$(z, \sigma R_{2g+2}^{1/2}(z)) \mapsto z$$
$$\infty_\pm \mapsto \infty$$

and the evaluation map

(A.109)
$$R_{2g+2}^{1/2}(.): M \to \mathbb{C} \cup \{\infty\}$$
$$(z, \sigma R_{2g+2}^{1/2}(z)) \mapsto \sigma R_{2g+2}^{1/2}(z)$$
$$\infty_\pm \mapsto \infty$$

Obviously, both are meromorphic functions (use again the local charts and their inversions). π has poles of order 1 at $\infty_\pm$ and two simple zeros at $(0, \pm R_{2g+2}^{1/2}(0))$ if $R_{2g+2}^{1/2}(0) \neq 0$ respectively one double zero at $(0,0)$ if $R_{2g+2}^{1/2}(0) = 0$. $R_{2g+2}^{1/2}(.)$

A.7. Hyperelliptic Riemann surfaces

has poles of order $g+1$ at $\infty_\pm$ and $2g+2$ simple zeros at $(E_n, 0)$. Notice also $\pi(p^*) = \pi(p)$ and $R_{2g+2}^{1/2}(p^*) = -R_{2g+2}^{1/2}(p)$.

Thus we may conclude that M is a two-sheeted, ramified covering of the Riemann sphere ($\cong \mathbb{C} \cup \{\infty\}$), that M is compact (since π is open and $\mathbb{C} \cup \{\infty\}$ is compact), and finally, that M is hyperelliptic (since it admits a function of degree two).

Now consider

$$\frac{d\pi}{R_{2g+2}^{1/2}}. \tag{A.110}$$

Again using local charts we see that $d\pi/R_{2g+2}^{1/2}$ is holomorphic everywhere and has zeros of order $g-1$ at $\infty_\pm$. So we may conclude that

$$\frac{\pi^{j-1} d\pi}{R_{2g+2}^{1/2}}, \qquad 1 \leq j \leq g, \tag{A.111}$$

form a basis for the space of holomorphic differentials. As a consequence we obtain the following result.

Lemma A.20. *Let $\mathcal{D} \geq 0$ and denote by $\tilde{\mathcal{D}}$ the divisor on $\mathbb{C} \cup \{\infty\}$ defined by $\tilde{\mathcal{D}}(z) = \max_{p \in \pi^{-1}(z)} \mathcal{D}(p)$, $z \in \mathbb{C} \cup \{\infty\}$. Then*

$$i(\mathcal{D}) = \max\{0, g - \deg(\tilde{\mathcal{D}})\}. \tag{A.112}$$

Proof. Each $\zeta \in \mathcal{L}^1(\mathcal{D})$ must be of the form

$$\zeta = \frac{(P \circ \pi) d\pi}{R_{2g+2}^{1/2}}, \tag{A.113}$$

where $P(z)$ is a polynomial of degree at most $g - \tilde{\mathcal{D}}(\infty) - 1$. Moreover, $P(z)$ must vanish of order $\tilde{\mathcal{D}}(z)$ at $z \in \mathbb{C}$. Hence we have $\max\{0, g - \deg(\tilde{\mathcal{D}})\}$ free constants for $P(z)$. $\square$

Next, we will introduce the representatives $\{a_\ell, b_\ell\}_{\ell=1}^g$ of a canonical homology basis for M. For a_ℓ we start near $E_{2\ell-1}$ on Π_+, surround $E_{2\ell}$ thereby changing to Π_- and return to our starting point encircling $E_{2\ell-1}$ again changing sheets. For b_ℓ we choose a cycle surrounding $E_0, E_{2\ell-1}$ counter-clock-wise (once) on Π_+. The cycles are chosen so that their intersection matrix reads

$$a_j \circ b_k = \delta_{jk}, \qquad 1 \leq j, k \leq g. \tag{A.114}$$

Visualizing for $g = 2$:

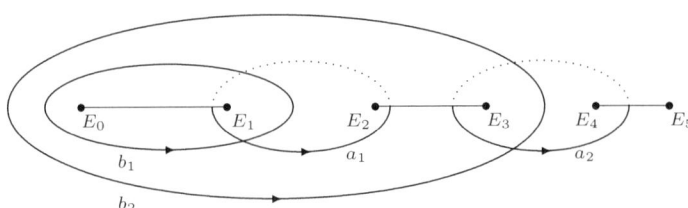

Here the solid lines indicate the parts on Π_+ and the doted ones the parts on Π_-.

Now we can construct a canonical basis $\{\zeta_j\}_{j=1}^g$ for the space of holomorphic differentials in the following way: If we introduce the constants $\underline{c}(.)$ via

(A.115) $\qquad c_j(k) = C_{jk}^{-1}, \qquad C_{jk} = \int_{a_k} \frac{\pi^{j-1} d\pi}{R_{2g+2}^{1/2}} = 2\int_{E_{2k-1}}^{E_{2k}} \frac{z^{j-1} dz}{R_{2g+2}^{1/2}(z)} \in \mathbb{R}.$

(C is invertible since otherwise there would be a nonzero differential with vanishing a-periods.) the differentials

(A.116) $\qquad \underline{\zeta} = \sum_{j=1}^g \underline{c}(j) \frac{\pi^{j-1} d\pi}{R_{2g+2}^{1/2}}$

fulfill

(A.117) $\qquad \int_{a_k} \zeta_j = \delta_{j,k}$

and are thus a basis of the required form. The matrix of b-periods

(A.118) $\qquad \tau_{j,k} = \int_{b_k} \zeta_j = -\sum_{j=1}^g c_j(m) \sum_{\ell=0}^{k-1} 2 \int_{E_{2\ell}}^{E_{2\ell+1}} \frac{z^{m-1} dz}{R_{2g+2}^{1/2}(z)}$

satisfies $\mathrm{Re}(\tau_{j,k}) = 0$ since we have $\mathrm{Im}(c_j(k)) = 0$. In the chart $(U_\pm, w) = (U_{\infty_\pm}, \zeta_{\infty_\pm})$ we have

$$\underline{\zeta} = \pm \sum_{j=1}^g \underline{c}(n) \frac{w^{g-j} dw}{\sqrt{\prod_m (1 - wE_m)}}$$

(A.119) $\qquad = \pm\Big(\underline{c}(g) + \big(\frac{\underline{c}(g)}{2} \sum_{j=1}^{2g+2} E_j + \underline{c}(g-1)\big) w + O(w^2)\Big) dw.$

Finally, we will show how to evaluate some integrals on $\hat{M}$. We fix a base point $p_0 = (E_0, 0)$. Then we have, for example,

(A.120) $\qquad \underline{A}_{p_0}(\infty_+) = -\sum_{j=1}^g \underline{c}(n) \int_{-\infty}^{E_0} \frac{z^{j-1} dz}{R_{2g+2}^{1/2}(z)} - \sum_{j=1}^g \underline{\delta}_j.$

To see this, take the straight line segment from E_0 to $-\infty$ and lift it to Π_+. Unfortunately, we cannot take this lift as our path of integration for it intersects all b cycles, whereas on the other hand we are required to stay in $\hat{M}$. Hence, rather than crossing b_j we will follow the boundary of $\hat{M}$ along b_j, a_j^{-1}, b_j^{-1} until

A.7. Hyperelliptic Riemann surfaces

we hit the corresponding point on b_j^{-1}. The parts on b_j, b_j^{-1} cancel (compare the picture on page 275) but the part on a_j^{-1} contributes. In other words, for each b-cycle we cross, we have to subtract the associated a-period. Similarly, $\underline{\hat{A}}_{p_0}(\infty_-) = -\underline{\hat{A}}_{p_0}(\infty_+) - \sum_{j=1}^{g} \underline{\delta}_j$.

Next, let us try to compute $\hat{\underline{\Xi}}_{p_0}$. As before one shows

(A.121) $$\hat{A}_{p_0,j}((E_{2\ell-1}, 0)) = -\frac{\tau_{j,\ell}}{2}$$

and hence

(A.122) $$\int_{a_\ell} \hat{A}_{p_0,j} \zeta_\ell = -\frac{\tau_{j,\ell}}{2} + \int_{a_\ell} \hat{A}_{(E_{2\ell-1},0),j} \zeta_\ell.$$

The last integral can be split up into two parts, one from $E_{\ell-1}$ to $E_{2\ell}$ on Π_+ and one from E_ℓ to $E_{2\ell-1}$ on Π_-. Since these two parts cancel we infer

(A.123) $$\hat{\Xi}_{p_0,j} = \frac{1}{2} - \frac{1}{2} \sum_{k=1}^{g} \tau_{j,k}.$$

Appendix B

Herglotz functions

The results stated in this appendix are collected from [**16**], [**17**], [**68**], [**73**], [**119**], [**198**], [**199**], [**200**], [**202**], [**203**], [**206**].

In this chapter an important role will be played by **Borel measures** ρ on $\mathbb{R}$. Hence we fix some notation first. For any Borel set $B \subseteq \mathbb{R}$ we will denote by $\rho(B) = \int_B d\rho$ its ρ-measure. Moreover, to each Borel measure ρ on $\mathbb{R}$ corresponds a monotone increasing function (also denoted by ρ for simplicity) which we normalize as

(B.1) $$\rho(\lambda) = \begin{cases} -\rho((\lambda, 0]), & \lambda < 0 \\ 0, & \lambda = 0 \\ \rho((0, \lambda]), & \lambda > 0 \end{cases}.$$

By definition, $\rho(\lambda)$ is right continuous.

Associated to ρ is the (separable) Hilbert space $L^2(\mathbb{R}, d\rho)$ with scalar product

(B.2) $$\langle f, g \rangle = \int_{\mathbb{R}} \overline{f(\lambda)} g(\lambda) d\rho(\lambda).$$

It has the following properties.

Lemma B.1. *Suppose $\rho(\mathbb{R}) < \infty$. The set of all continuous functions is dense in $L^2(\mathbb{R}, d\rho)$. Moreover, if ρ is compactly supported ($\rho(K) = \rho(\mathbb{R})$ for a compact set $K \subseteq \mathbb{R}$), then the set of polynomials is also dense.*

Proof. First of all, bounded functions are dense. In fact, consider

(B.3) $$f_n(\lambda) = \begin{cases} f(\lambda), & |f(\lambda)| \leq n \\ 0, & |f(\lambda)| > n \end{cases}.$$

Then $|f(\lambda) - f_n(\lambda)| \to 0$ a.e. since $|f(\lambda)| < \infty$ a.e.. Since $|f(\lambda) - f_n(\lambda)| \leq |f(\lambda)|$ we obtain $\int |f - f_n|^2 d\rho \to 0$ by the dominated convergence theorem.

So it remains to show that bounded elements can be approximated by continuous ones. By Lusin's theorem, there are continuous functions f_n such that

$\|f_n\|_\infty \leq \|f\|_\infty$ and

(B.4) $$\rho(B_n) \leq \frac{1}{n}, \quad B_n = \{\lambda | f(\lambda) \neq f_n(\lambda)\}.$$

Here $\|f\|_\infty$ denotes the essential supremum of f, that is, the smallest constant M such that $|f(\lambda)| \leq M$ a.e.. Hence

(B.5) $$\|f - f_n\|^2 = \int_{B_n} |f_n(\lambda) - f(\lambda)|^2 d\rho(\lambda) \leq \frac{4\|f\|_\infty^2}{n} \to 0$$

and the first claim follows.

To show the second, it suffices to prove that a continuous element f can be approximated by polynomials. By the Stone-Weierstrass Theorem there is a polynomial P_n approximating f in the sup norm on K such that

(B.6) $$\|f - P_n\|^2 \leq \rho(\mathbb{R})\|f - P_n\|_\infty^2 \leq \frac{\rho(\mathbb{R})}{n} \to 0$$

and we are done. $\square$

If $\rho(\mathbb{R}) < \infty$ we can define the **Borel transform** (also **Cauchy transform**) of ρ by

(B.7) $$F(z) = \int_\mathbb{R} \frac{d\rho(\lambda)}{z - \lambda}, \quad z \in \mathbb{C}_\pm,$$

where $\mathbb{C}_\pm = \{z \in \mathbb{C} | \pm \text{Im}(z) > 0\}$. In addition, we can also define the **Fourier transform**

(B.8) $$T(\lambda) = \int_\mathbb{R} e^{i\lambda t} d\rho(t), \quad \lambda \in \mathbb{R},$$

which is a bounded function and is related to the Borel transform via

(B.9) $$F(z) = \begin{cases} i \int_{-\infty}^0 e^{-iz\lambda} T(\lambda) d\lambda, & \text{Im}(z) > 0 \\ -i \int_0^\infty e^{-iz\lambda} T(\lambda) d\lambda, & \text{Im}(z) < 0 \end{cases}$$

as can be verified using Fubini's theorem.

The Borel transform has the interesting property that it maps the upper half plane $\mathbb{C}_+$ into itself as can be seen from

(B.10) $$\text{Im}(F(z)) = \text{Im}(z) \int_\mathbb{R} \frac{d\rho(\lambda)}{|z - \lambda|^2}.$$

In general, a holomorphic function $F: \mathbb{C}_+ \to \mathbb{C}_+$ is called a **Herglotz function** (sometimes also Pick or Nevanlinna-Pick function). It is no restriction to assume that F is defined on $\mathbb{C}_- \cup \mathbb{C}_+$ satisfying $F(\bar{z}) = \overline{F(z)}$.

The following theorem shows that all Herglotz functions arise as Borel transform (in an extended sense) of some unique measure.

Theorem B.2. *F is a Herglotz function if and only if*

(B.11) $$F(z) = a + bz + \int_\mathbb{R} \left(\frac{1}{\lambda - z} - \frac{\lambda}{1 + \lambda^2} \right) d\rho(\lambda), \quad z \in \mathbb{C}_\pm,$$

where $a, b \in \mathbb{R}$, $b \geq 0$, and ρ is a nonzero measure on $\mathbb{R}$ which satisfies $\int_\mathbb{R} (1 + \lambda^2)^{-1} d\rho(\lambda) < \infty$.

B. Herglotz functions

Moreover, the triple a, b, and ρ is unambiguously determined by F using

(B.12) $$a = \mathrm{Re}\big(F(\mathrm{i})\big), \qquad \lim_{\substack{z \to \infty \\ \mathrm{Im}(z) \geq \varepsilon > 0}} \frac{F(z)}{z} = b \geq 0,$$

and **Stieltjes inversion formula**

(B.13) $$\rho\big((\lambda_0, \lambda_1]\big) = \lim_{\delta \downarrow 0} \lim_{\varepsilon \downarrow 0} \frac{1}{\pi} \int_{\lambda_0 + \delta}^{\lambda_1 + \delta} \mathrm{Im}\big(F(\lambda + \mathrm{i}\varepsilon)\big) d\lambda.$$

The content of Stieltjes inversion formula can be strengthened.

Lemma B.3. *Let F be a Herglotz function with associated measure ρ. Then the measure $\pi^{-1} \mathrm{Im}(F(\lambda + \mathrm{i}\varepsilon)) d\lambda$ converges weakly to $d\rho(\lambda)$ in the sense that*

(B.14) $$\lim_{\varepsilon \downarrow 0} \frac{1}{\pi} \int_{\mathbb{R}} f(\lambda) \mathrm{Im}\big(F(\lambda + \mathrm{i}\varepsilon)\big) d\lambda = \int_{\mathbb{R}} f(\lambda) d\rho(\lambda)$$

for all continuous functions f with $|f| \leq \mathrm{const}(1 + \lambda^2)^{-1}$. In addition,

(B.15) $$\lim_{\varepsilon \downarrow 0} \frac{1}{\pi} \int_{\lambda_0}^{\lambda_1} f(\lambda) \mathrm{Im}\big(F(\lambda + \mathrm{i}\varepsilon)\big) d\lambda = \int_{(\lambda_0, \lambda_1)} f(\lambda) d\rho(\lambda)$$
$$+ \frac{f(\lambda_1)\rho(\{\lambda_1\}) - f(\lambda_0)\rho(\{\lambda_0\})}{2}.$$

The following result shows that a different analytic continuation of F from $\mathbb{C}_+$ to $\mathbb{C}_-$ can exist in some cases.

Lemma B.4. *Let $(\lambda_1, \lambda_2) \subseteq \mathbb{R}$ and suppose $\lim_{\varepsilon \downarrow 0} \mathrm{Re}(F(\lambda + \mathrm{i}\varepsilon)) = 0$ for a.e. $\lambda \in (\lambda_1, \lambda_2)$. Then F can be analytically continued from $\mathbb{C}_+$ into $\mathbb{C}_-$ through the interval (λ_1, λ_2). The resulting function $\tilde{F}(z)$ coincides with $F(z)$ on $\mathbb{C}_+$ and satisfies*

(B.16) $$\overline{\tilde{F}(z)} = -\tilde{F}(\bar{z}).$$

In addition, $\mathrm{Im}(F(\lambda + \mathrm{i}0)) > 0$, $\mathrm{Re}(F(\lambda + \mathrm{i}0)) = 0$ for all $\lambda \in (\lambda_1, \lambda_2)$.

The measure ρ is called spectral measure of F. The set of all **growth points**, that is,

(B.17) $$\sigma(\rho) = \{\lambda \in \mathbb{R} | \rho((\lambda - \varepsilon, \lambda + \varepsilon)) > 0 \text{ for all } \varepsilon > 0\},$$

is called the spectrum of ρ. Invoking Morea's together with Fubini's theorem shows that $F(z)$ is holomorphic for $z \in \mathbb{C} \backslash \sigma(\rho)$. The converse following from Stieltjes inversion formula. Moreover, we have

(B.18) $$\frac{dF(z)}{dz} = -\int_{\mathbb{R}} \frac{d\rho(\lambda)}{(\lambda - z)^2}.$$

In particular, if $\rho\big((\lambda_0, \lambda_1)\big) = 0$, then $F(\lambda)$, $\lambda \in (\lambda_0, \lambda_1)$, is decreasing and hence tends to a limit (in $\mathbb{R} \cup \{\infty\}$) as $\lambda \downarrow \lambda_0$ or $\lambda \uparrow \lambda_1$.

The following result explains why $\sigma(\rho)$ is called the spectrum of ρ.

Lemma B.5. *The set $\sigma(\rho)$ is precisely the spectrum $\sigma(\tilde{H})$ of the multiplication operator $\tilde{H} f(\lambda) = \lambda f(\lambda)$, $\mathfrak{D}(\tilde{H}) = \{f \in L^2(\mathbb{R}, d\rho) | \lambda f(\lambda) \in L^2(\mathbb{R}, d\rho)\}$.*

Proof. If $\lambda \in \sigma(\rho)$, then the sequence $f_n = \rho((\lambda - \frac{1}{n}, \lambda + \frac{1}{n}))^{-1/2}\chi_{(\lambda-\frac{1}{n},\lambda+\frac{1}{n})}$ satisfies $\|f_n\| = 1$, $\|(\tilde{H} - \lambda)f_n\| \to 0$ and hence $\lambda \in \sigma(\tilde{H})$. Conversely, if $z \notin \sigma(\rho)$, then $(\tilde{H} - z)^{-1}f(\lambda) = (\lambda - z)^{-1}f(\lambda)$ exists and is bounded, implying $z \notin \sigma(\tilde{H})$. □

Our main objective is to characterize $\sigma(\rho)$, and various decompositions of $\sigma(\rho)$, in terms of $F(z)$. More precisely, this will be done by investigating the boundary behavior of $F(\lambda + i\varepsilon)$, $\lambda \in \mathbb{R}$, as $\varepsilon \downarrow 0$.

First we recall the unique decomposition of ρ with respect to Lebesgue measure,

(B.19) $$d\rho = d\rho_{ac} + d\rho_s,$$

where ρ_{ac} is **absolutely continuous** with respect to Lebesgue measure (i.e., we have $\rho_{ac}(B) = 0$ for all B with Lebesgue measure zero) and ρ_s is **singular** with respect to Lebesgue measure (i.e., ρ_s is supported, $\rho_s(\mathbb{R}\backslash B) = 0$, on a set B with Lebesgue measure zero). The singular part ρ_s can be further decomposed into a **(singular) continuous** and a **pure point** part,

(B.20) $$d\rho_s = d\rho_{sc} + d\rho_{pp},$$

where ρ_{sc} is continuous on $\mathbb{R}$ and ρ_{pp} is a step function.

By the **Radon-Nikodym Theorem** we have

(B.21) $$d\rho_{ac}(\lambda) = f(\lambda)d\lambda$$

for a locally integrable positive function f which is unique (a.e.). Moreover, we note that the singular part can be characterized by the following lemma.

Lemma B.6. *For any Borel set B we have*

(B.22) $$\rho_s(B) = \lim_{n\to\infty} \sup_{I\in\mathcal{I}, |I|<1/n} \rho(B \cap I),$$

where $\mathcal{I}$ is the (countable) family of finite unions of open intervals, each of which has rational endpoints, and $|I|$ denotes the Lebesgue measure of I.

Proof. Since the supremum decreases as n increases, the limit exists. Denote it by $\tilde{\rho}(B)$. Using $d\rho_{ac}(\lambda) = f(\lambda)d\lambda$ we see

(B.23) $$\rho(B \cap I) \leq \rho_s(B) + R|I| + \rho_{ac}(f^{-1}((R,\infty))).$$

implying $\tilde{\rho}(B) \leq \rho_s(B) + \lim_{R\to\infty} \rho_{ac}(f^{-1}((R,\infty))) = \rho_s(B)$.

Conversely, pick a support B_s for ρ_s of Lebesgue measure $|B_s|$ zero. By $0 = |B_s| = \sup\{|O| \,|\, B_s \subset O, O \text{ open}\}$ there exists a sequence O_n such that $B_s \subset O_n$ and $|O_n| < 1/n$. Moreover, we can find a sequence $I_m \in \mathcal{I}$ such that $I_m \subseteq I_{m+1}$ and $\bigcup I_m = O_n$ (hence $\rho(I_m) \to \rho(O_n)$). Using $\rho(B \cap I_m) \geq \rho(B \cap O_n) - \rho(O_n \backslash I_m)$ concludes the proof

(B.24) $$\sup_{I\in\mathcal{I}, |I|<1/n} \rho(B \cap I) \geq \rho(B \cap O_n) \geq \rho(B \cap B_s) \geq \rho_s(B).$$

□

Now we turn to the concept of a **minimal support** (sometimes also essential support) of a measure. A set M is called a minimal support for ρ if M is a support

B. Herglotz functions

(i.e., $\rho(\mathbb{R}\backslash M) = 0$) and any subset $M_0 \subseteq M$ which does not support M (i.e, $\rho(M_0) = 0$) has Lebesgue measure zero. Let

(B.25)
$$\mathrm{Im}(F(\lambda)) = \limsup_{\varepsilon \downarrow 0} \mathrm{Im}(F(\lambda + i\varepsilon))$$

and let $L(\rho)$ be the set of all $\lambda \in \mathbb{R}$ for which $\lim_{\varepsilon \downarrow 0} \mathrm{Im}(F(\lambda + i\varepsilon))$ exists (finite or infinite). Then we have

Lemma B.7. *Minimal supports M, M_{ac}, M_s for ρ, ρ_{ac}, ρ_s, respectively, are given by*

(B.26)
$$\begin{aligned} M &= \{\lambda \in L(\rho) | 0 < \mathrm{Im}(F(\lambda)) \leq \infty\}, \\ M_{ac} &= \{\lambda \in L(\rho) | 0 < \mathrm{Im}(F(\lambda)) < \infty\}, \\ M_s &= \{\lambda \in L(\rho) | \mathrm{Im}(F(\lambda)) = \infty\}. \end{aligned}$$

In particular, the ρ-measure and the Lebesgue measure of $\mathbb{R}\backslash L(\rho)$ are both zero. This also says that if $\mathrm{Im}(F(\lambda)) < \infty$ for all $\lambda \in (\lambda_1, \lambda_2)$, then ρ is purely absolutely continuous on (λ_1, λ_2). Furthermore, M_s has Lebesgue measure zero and might be nonempty even though $\rho_s = 0$. However, note that if M_s is a countable set, then $\rho_{sc} = 0$.

It is interesting to know when the minimal supports determine the corresponding spectra.

The spectrum of ρ is given by

(B.27)
$$\sigma(\rho) = \overline{M}.$$

To see this observe that F is real holomorphic near $\lambda \notin \sigma(\rho)$ and hence $\mathrm{Im}(F(\lambda)) = 0$ in this case. Thus $M \subseteq \sigma(\rho)$ and since $\sigma(\rho)$ is closed we even have $\overline{M} \subseteq \sigma(\rho)$. Conversely, if $\lambda \in \sigma(\rho)$, then $0 < \rho((\lambda - \varepsilon, \lambda + \varepsilon)) = \rho((\lambda - \varepsilon, \lambda + \varepsilon) \cap M)$ for all $\varepsilon > 0$ and we can find a sequence $\lambda_n \in (\lambda - 1/n, \lambda + 1/n) \cap M$ converging to λ from inside M. This is the remaining part $\sigma(\rho) \subseteq \overline{M}$.

Moreover, $\sigma(\rho_{ac})$ can be recovered from the essential closure of M_{ac}, that is,

(B.28)
$$\sigma(\rho_{ac}) = \overline{M}_{ac}^{ess},$$

where

(B.29)
$$\overline{M}_{ac}^{ess} = \{\lambda \in \mathbb{R} | |(\lambda - \varepsilon, \lambda + \varepsilon) \cap M_{ac}| > 0 \text{ for all } \varepsilon > 0\}.$$

Note that $\overline{M}_{ac}^{ess}$ is closed, whereas we might have $M_{ac} \not\subset \overline{M}_{ac}^{ess}$. To prove (B.28) we use that $0 < \rho_{ac}((\lambda - \varepsilon, \lambda + \varepsilon)) = \rho_{ac}((\lambda - \varepsilon, \lambda + \varepsilon) \cap M_{ac})$ is equivalent to $|(\lambda - \varepsilon, \lambda + \varepsilon) \cap M_{ac}| > 0$. One direction follows from the definition of absolute continuity and the other from minimality of M_{ac}.

Next we define the derivative

(B.30)
$$D\rho(\lambda) = \limsup_{\varepsilon \downarrow 0} \frac{\rho\big((\lambda - \varepsilon, \lambda + \varepsilon)\big)}{2\varepsilon}.$$

Now we can say more about how to extend a Herglotz function F to the real axis.

Theorem B.8. *Let F be a Herglotz function and ρ its spectral measure.*
(i). For all $\lambda \in \mathbb{R}$, $D\rho(\lambda)$ and $\mathrm{Im}(F(\lambda))$ are either both zero, both in $(0, \infty)$, or both infinite. Moreover, $D\rho(\lambda)$ exists as ordinary limit if $\lambda \in L(\rho)$ and

(B.31)
$$D\rho(\lambda) = \frac{1}{\pi}\mathrm{Im}\big(F(\lambda)\big), \qquad \lambda \in L(\rho).$$

Moreover, $d\rho_{ac}(\lambda) = D\rho(\lambda)d\lambda$.

(ii). For almost all $\lambda_0 \in \mathbb{R}$ with respect to ρ and Lebesgue measure, the derivative $D\rho(\lambda_0)$ and

(B.32) $$R_F(\lambda_0) = a + b\lambda_0 + \lim_{\varepsilon \downarrow 0} \int_{\mathbb{R}\setminus(\lambda_0-\varepsilon,\lambda_0+\varepsilon)} \left(\frac{1}{\lambda - \lambda_0} - \frac{\lambda}{1+\lambda^2}\right) d\rho(\lambda)$$

both exist (as ordinary limit) and are finite. For those λ_0 we have

(B.33) $$F(\lambda_0 + i0) = \lim_{\varepsilon \downarrow 0} F(\lambda_0 + i\varepsilon) = R_F(\lambda_0) + i\pi D\rho(\lambda_0).$$

(iii). If $F(z)$ and $G(z)$ are Herglotz functions and the boundary values coincide $F(\lambda+i0) = G(\lambda+i0)$ for λ in a set of positive Lebesgue measure, then $F(z) = G(z)$.

Next, we want to generalize the decomposition of ρ by taking the α-dimensional Hausdorff measure h^α rather than the Lebesgue measure. The splitting $d\rho = d\rho_{ac} + d\rho_s$ will correspond to the case $\alpha = 1$.

For any given Borel set $B \subseteq \mathbb{R}$ and $\alpha \in [0,1]$ we define

(B.34) $$h_\varepsilon^\alpha(B) = \inf\left\{\sum_{j\in\mathbb{N}} |I_j|^\alpha \,\Big|\, |I_j| < \varepsilon, B \subseteq \bigcup_{j\in\mathbb{N}} I_j\right\}, \quad \varepsilon > 0,$$

the infimum over all countable covers by intervals I_j of length at most ε. Since $h_\varepsilon^\alpha(B)$ is increasing with respect to ε (the number of covers decreases) we can define the α-**dimensional Hausdorff measure** of B as

(B.35) $$h^\alpha(B) = \lim_{\varepsilon \downarrow 0} h_\varepsilon^\alpha(B).$$

Note that $h^{\alpha_1}(B) \geq h^{\alpha_2}(B)$ if $\alpha_1 \leq \alpha_2$ and h^0, h^1 correspond to counting, Lebesgue measure, respectively. The **Hausdorff dimension** $\alpha(B)$ of B is the unique number for which $h^\alpha(B) = \infty$, $\alpha < \alpha(B)$ and $h^\alpha(B) = 0$, $\alpha > \alpha(B)$.

Now any measure can be uniquely decomposed with respect to the Hausdorff measure h^α, that is,

(B.36) $$d\rho = d\rho_{\alpha c} + d\rho_{\alpha s},$$

where $\rho_{\alpha c}$ is absolutely continuous with respect to h^α (i.e., $\rho_{\alpha c}(B) = 0$ for all B with $h^\alpha(B) = 0$) and $\rho_{\alpha s}$ is singular with respect to h^α (i.e., $\rho_{\alpha s}$ is supported on a set B with $h^\alpha(B) = 0$).

Next, define the α-**derivative** of ρ by

(B.37) $$D^\alpha \rho(\lambda) = \limsup_{\varepsilon \downarrow 0} \frac{\rho\big((\lambda-\varepsilon,\lambda+\varepsilon)\big)}{(2\varepsilon)^\alpha}.$$

Theorem B.9. *Let F be a Herglotz function with associated measure $d\rho$.*
(i). Let $C^\alpha(\rho) = \{\lambda \in \mathbb{R}|\, D^\alpha\rho(\lambda) < \infty\}$, $\alpha \in [0,1]$, then we have $d\rho_{\alpha c} = \chi_{C^\alpha(\rho)}d\rho$ and $d\rho_{\alpha s} = (1-\chi_{C^\alpha(\rho)})d\rho$.
(ii). Set

$$Q^\alpha \rho(\lambda) = \limsup_{\varepsilon \downarrow 0} \varepsilon^{1-\alpha}\mathrm{Im}F(\lambda + i\varepsilon),$$

(B.38) $$R^\alpha \rho(\lambda) = \limsup_{\varepsilon \downarrow 0} \varepsilon^{1-\alpha}|F(\lambda + i\varepsilon)|.$$

Then $D^\alpha\rho(\lambda)$ and $Q^\alpha\rho(\lambda)$, $\alpha \in [0,1]$, are either both zero, both in $(0,\infty)$, or both infinite.

B. Herglotz functions

If $\alpha \in [0,1)$, $Q^\alpha \rho(\lambda)$ can be replaced by $R^\alpha \rho(\lambda)$ and for any $\alpha \in [0,1]$ we have $2^{\alpha-1} D^\alpha \rho(\lambda) \leq Q^\alpha \rho(\lambda) \leq R^\alpha \rho(\lambda)$.

In particular,

(B.39) $\qquad C^\alpha(\rho) = \{\lambda \in \mathbb{R} \,|\, Q^\alpha \rho(\lambda) < \infty\} \supseteq \{\lambda \in \mathbb{R} \,|\, R^\alpha \rho(\lambda) < \infty\},$

where all three sets are equal if $\alpha \in [0,1)$. For $\alpha = 1$ this is not true in general, as the example $F(\lambda) = \ln(\lambda)$ (see below) shows. For $\alpha = 0$ we have $C^0(\rho) = \mathbb{R}$.

Again there is a value

(B.40) $\qquad\qquad \alpha(\lambda) = \liminf_{\varepsilon \downarrow 0} \frac{\ln \rho\bigl((\lambda - \varepsilon, \lambda + \varepsilon)\bigr)}{\ln \varepsilon}$

such that $D^\alpha \rho(\lambda) = 0$ for $\alpha < \alpha(\lambda)$ and $D^\alpha \rho(\lambda) = \infty$ for $\alpha > \alpha(\lambda)$.

Lemma B.10. *We have*

(B.41) $\qquad\qquad \lim_{\varepsilon \downarrow 0} \frac{\varepsilon}{\mathrm{i}} F(\lambda + \mathrm{i}\varepsilon) = \rho(\{\lambda\})$

and hence $Q^0 \rho(\lambda) = R^0 \rho(\lambda) = \rho(\{\lambda\})$. *Moreover, if* $\rho((\lambda_0, \lambda_1)) = 0$, *we also have*

(B.42) $\qquad \lim_{\varepsilon \downarrow 0} (-\varepsilon) F(\lambda_0 + \varepsilon) = \rho(\{\lambda_0\}) \text{ and } \lim_{\varepsilon \downarrow 0} \varepsilon F(\lambda_1 - \varepsilon) = \rho(\{\lambda_1\}).$

Proof. We only prove the first identity of (B.42), the remaining claims being similar. After splitting the integral in (B.11) into a part over $(\lambda_0 - \delta, \lambda_1)$ and one over $\mathbb{R}\backslash(\lambda_0 - \delta, \lambda_1)$, we see that it suffices to consider only the first one (since the second one is holomorphic near λ_0). By our assumption $\rho((\lambda_0, \lambda_1)) = 0$, the integral is only taken over $(\lambda_0 - \delta, \lambda_0]$ and because of $|\frac{-\varepsilon}{\lambda_0 + \varepsilon - \lambda}| \leq 1$, $\lambda \in (\lambda_0 - \delta, \lambda_0]$, the desired result follows from the dominated convergence theorem. $\square$

Now, we want to consider an alternate integral representation of Herglotz functions connected to the logarithm.

Let $\ln(z)$ be defined such that

(B.43) $\qquad\qquad \ln(z) = \ln|z| + \mathrm{i}\arg(z), \quad -\pi < \arg(z) \leq \pi.$

Then $\ln(z)$ is holomorphic and $\mathrm{Im}\bigl(\ln(z)\bigr) > 0$ for $z \in \mathbb{C}_+$, hence $\ln(z)$ is a Herglotz function. The representation of $\ln(z)$ according to (B.11) reads

(B.44) $\qquad \ln(z) = \int_{\mathbb{R}} \left(\frac{1}{\lambda - z} - \frac{\lambda}{1 + \lambda^2}\right) \chi_{(-\infty, 0)}(\lambda) \, d\lambda, \quad z \in \mathbb{C}_\pm,$

which can be easily verified.

The sum of two Herglotz functions is again a Herglotz function, similarly the composition of two Herglotz functions is Herglotz. In particular, if $F(z)$ is a Herglotz function, the same holds for $\ln\bigl(F(z)\bigr)$ and $-\frac{1}{F(z)}$. Thus, using the representation (B.11) for $\ln\bigl(F(z)\bigr)$, we get another representation for $F(z)$. The main feature of this new representation is that, by Lemma B.7, the corresponding measure is purely absolutely continuous since the imaginary part of $\ln(z)$ is uniformly bounded.

Theorem B.11. *A given function F is Herglotz if and only if it has the representation*

(B.45) $\qquad F(z) = \exp\left\{c + \int_{\mathbb{R}} \left(\frac{1}{\lambda - z} - \frac{\lambda}{1 + \lambda^2}\right) \xi(\lambda) \, d\lambda\right\}, \quad z \in \mathbb{C}_\pm,$

where $c = \ln|F(\mathrm{i})| \in \mathbb{R}$, $\xi \in L^1(\mathbb{R}, (1+\lambda^2)^{-1}d\lambda)$ real-valued and ξ is not identically zero. Moreover,

(B.46) $\qquad \xi(\lambda) = \frac{1}{\pi} \lim_{\varepsilon \downarrow 0} \mathrm{Im}\left(\ln\left(F(\lambda + \mathrm{i}\varepsilon)\right)\right) = \frac{1}{\pi} \lim_{\varepsilon \downarrow 0} \arg\left(F(\lambda + \mathrm{i}\varepsilon)\right)$

for a.e. $\lambda \in \mathbb{R}$, and $0 \leq \xi(\lambda) \leq 1$ for a.e. $\lambda \in \mathbb{R}$. Here $-\pi < \arg(F(\lambda + \mathrm{i}\varepsilon)) \leq \pi$ according to the definition of $\ln(z)$.

Proof. $F(z)$ is Herglotz, therefore $\ln(F(z))$ is Herglotz with

(B.47) $\qquad |\mathrm{Im}(\ln F(z))| = |\arg(F(z))| \leq \pi.$

Hence by Theorem B.2

(B.48) $\qquad \ln(F(z)) = c + \int_{\mathbb{R}} \left(\frac{1}{\lambda - z} - \frac{\lambda}{1+\lambda^2}\right) d\rho(\lambda), \qquad z \in \mathbb{C}_\pm,$

where $\rho(\lambda)$ is (by Lemma B.7) absolutely continuous with respect to Lebesgue measure, that is, $d\rho(\lambda) = \xi(\lambda)d\lambda$, $\xi \in L^1_{loc}(\mathbb{R})$. According to (B.31), ξ is given by (B.46) and taking the Herglotz property of F into account immediately implies $0 \leq \xi(\lambda) \leq 1$ for a.e. $\lambda \in \mathbb{R}$. The converse is easy. $\qquad \square$

Some additional properties are collected in the following lemma.

Lemma B.12. Let F be a Herglotz function with spectral measure ρ and exponential Herglotz measure $\xi(\lambda)d\lambda$.
(i). The set $\{\lambda \in \mathbb{R} | 0 < \xi(\lambda) < 1\}$ is a minimal support for ρ_{ac}. Moreover, if there are constants $0 < c_1 < c_2 < 1$ such that $c_1 \leq \xi(\lambda) \leq c_2$, $\lambda \in (\lambda_1, \lambda_2)$, then ρ is purely absolutely continuous in (λ_1, λ_2).
(ii). Fix $n \in \mathbb{N}$ and set $\xi_+(\lambda) = \xi(\lambda)$, $\xi_-(\lambda) = 1 - \xi(\lambda)$. Then

(B.49) $\qquad \int_{\mathbb{R}} |\lambda|^n \xi_\pm(\lambda) d\lambda < \infty$

if and only if

(B.50) $\qquad \int_{\mathbb{R}} |\lambda|^n d\rho(\lambda) < \infty \quad \text{and} \quad \lim_{z \to \mathrm{i}\infty} \pm F(z) = \pm a \mp \int_{\mathbb{R}} \frac{\lambda d\rho(\lambda)}{1+\lambda^2} > 0.$

(iii). We have

(B.51) $\qquad F(z) = \pm 1 + \int_{\mathbb{R}} \frac{d\rho(\lambda)}{\lambda - z} \quad \text{with} \quad \int_{\mathbb{R}} d\rho(\lambda) < \infty$

if and only if

(B.52) $\qquad F(z) = \pm \exp\left(\pm \int_{\mathbb{R}} \xi_\pm(\lambda) \frac{d\lambda}{\lambda - z}\right) \quad \text{with} \quad \xi_\pm \in L^1(\mathbb{R})$

($\xi_\pm$ from above). In this case

(B.53) $\qquad \int_{\mathbb{R}} d\rho(\lambda) = \int_{\mathbb{R}} \xi_\pm(\lambda) d\lambda.$

Observe that the set

(B.54) $\qquad \tilde{M}_{ac} = \{\lambda \in \mathbb{R} | 0 < \xi(\lambda) < 1\}$

is a minimal support for ρ_{ac}.

B. Herglotz functions

In addition to these results, we will also need some facts on matrix valued measures. Let

$$\text{(B.55)} \qquad d\rho = \begin{pmatrix} d\rho_{0,0} & d\rho_{0,1} \\ d\rho_{1,0} & d\rho_{1,1} \end{pmatrix}, \quad d\rho_{0,1} = d\rho_{1,0},$$

where $d\rho_{i,j}$ are (in general) signed measures. Associated with $d\rho$ is the trace measure $d\rho^{tr} = d\rho_{0,0} + d\rho_{1,1}$.

We require $d\rho$ to be positive, that is, the matrix $(\rho_{i,j}(B))_{0 \leq i,j \leq 1}$ is nonnegative for any Borel set B. Equivalently, $\rho^{tr}(B) \geq 0$ and $\rho_{0,0}(B)\rho_{1,1}(B) - \rho_{0,1}(B)^2 \geq 0$ for any Borel set. This implies that $d\rho_{i,i}$ are positive measures and that

$$\text{(B.56)} \qquad |\rho_{0,1}(B)| \leq \frac{1}{2}\rho^{tr}(B).$$

Hence $d\rho_{i,j}$ is absolutely continuous with respect to $d\rho^{tr}$. Assuming $\rho^{tr}(\mathbb{R}) < \infty$ we can also define the Borel transform

$$\text{(B.57)} \qquad F(z) = \int_{\mathbb{R}} \frac{d\rho(\lambda)}{z - \lambda}.$$

The matrix $F(\lambda)$ satisfies

$$\text{(B.58)} \qquad \pm \text{Im}(F(z)) \geq 0, \quad z \in \mathbb{C}_\pm.$$

Next, consider the sesquilinear form

$$\text{(B.59)} \qquad \langle \underline{f}, \underline{g} \rangle = \int_{\mathbb{R}} \sum_{i,j=0}^{1} \overline{f_i(\lambda)} g_j(\lambda) d\rho_{i,j}(\lambda).$$

Suppose $\underline{f}$ is a simple function, that is, $\underline{f}(\lambda) = \sum_{k=1}^{n} \chi_{B_k}(\lambda)(f_{k,1}, f_{k,2})$, where $(f_{k,1}, f_{k,2}) \in \mathbb{C}^2$ and B_k are disjoint Borel sets. Then our assumptions ensures that $\|\underline{f}\|^2 = \sum_{k=1}^{n} \sum_{i,j=0}^{1} \overline{f_{k,i}} f_{k,j} \rho_{i,j}(B_k) \geq 0$. If $\underline{f}$ is such that $f_i \in L^2(\mathbb{R}, d\rho^{tr})$ we can approximate $\underline{f}$ by simple functions implying $\|\underline{f}\| \geq 0$. As a consequence we get a separable Hilbert space $L^2(\mathbb{R}, \mathbb{C}^2, d\rho)$ with the above scalar product. As before we have

Lemma B.13. *The set $\sigma(\rho^{tr})$ is precisely the spectrum $\sigma(\tilde{H})$ of the multiplication operator $\tilde{H}\underline{f}(\lambda) = \lambda \underline{f}(\lambda)$, $\mathfrak{D}(\tilde{H}) = \{\underline{f} \in L^2(\mathbb{R}, \mathbb{C}^2, d\rho) | \lambda \underline{f}(\lambda) \in L^2(\mathbb{R}, \mathbb{C}^2, d\rho)\}$.*

Proof. The proof is as like the one of Lemma B.5 except that ρ and f_n have to be replaced by ρ^{tr} and $\underline{f}_n = \rho^{tr}((\lambda - \frac{1}{n}, \lambda + \frac{1}{n}))^{-1/2} \chi_{(\lambda - \frac{1}{n}, \lambda + \frac{1}{n})}(1,1)$, respectively. $\square$

The trace measure can even be used to diagonalize $d\rho$ as follows. Since $d\rho_{i,j}$ is absolutely continuous with respect to $d\rho^{tr}$, there is a symmetric (integrable) matrix $R(\lambda)$ such that

$$\text{(B.60)} \qquad d\rho(\lambda) = R(\lambda) d\rho^{tr}(\lambda)$$

by the Radon-Nikodym theorem. Moreover, the matrix $R(\lambda)$ is nonnegative, $R(\lambda) \geq 0$, and given by

$$\text{(B.61)} \qquad R_{i,j}(\lambda) = \lim_{\varepsilon \downarrow 0} \frac{\text{Im}(F_{i,j}(\lambda + i\varepsilon))}{\text{Im}(F_{0,0}(\lambda + i\varepsilon) + F_{1,1}(\lambda + i\varepsilon))}.$$

Note also that we have $\text{tr}(R(\lambda)) = 1$.

Next, there is a measurable unitary matrix $U(\lambda)$ which diagonalizes $R(\lambda)$, that is,

(B.62) $$R(\lambda) = U(\lambda)^* \begin{pmatrix} r_1(\lambda) & 0 \\ 0 & r_2(\lambda) \end{pmatrix} U(\lambda),$$

where $0 \leq r_{1,2}(\lambda) \leq 1$ are the (integrable) eigenvalues of $R(\lambda)$. Note $r_1(\lambda)+r_2(\lambda) = 1$ by $\text{tr}(R(\lambda)) = 1$. The matrix $U(\lambda)$ provides a unitary operator

(B.63)
$$\begin{array}{rcl} L^2(\mathbb{R}, \mathbb{C}^2, d\rho) & \to & L^2(\mathbb{R}, \mathbb{C}^2, \begin{pmatrix} r_1 & 0 \\ 0 & r_2 \end{pmatrix} d\rho^{tr}) = L^2(\mathbb{R}, r_1 d\rho^{tr}) \oplus L^2(\mathbb{R}, r_2 d\rho^{tr}) \\ \underline{f}(\lambda) & \mapsto & U(\lambda)\underline{f}(\lambda) \end{array}$$

which leaves $\tilde{H}$ invariant. This allows us to investigate the spectral multiplicity of $\tilde{H}$.

Lemma B.14. *Define*

(B.64)
$$\begin{aligned} B_1 &= \{\lambda \in \sigma(\rho^{tr})|\det R(\lambda) = r_1(\lambda)r_2(\lambda) = 0\}, \\ B_2 &= \{\lambda \in \sigma(\rho^{tr})|\det R(\lambda) = r_1(\lambda)r_2(\lambda) > 0\}. \end{aligned}$$

Then $\tilde{H} = \chi_{B_1}\tilde{H} \oplus \chi_{B_2}\tilde{H}$ and the spectral multiplicity of $\chi_{B_1}\tilde{H}$ is one and the spectral multiplicity of $\chi_{B_2}\tilde{H}$ is two.

Proof. It is easy to see that $\chi_{B_1}\tilde{H}$ is unitary equivalent to multiplication by λ in $L^2(\mathbb{R}, \chi_{B_1} d\rho^{tr})$. Moreover, since $r_i \chi_{B_2} d\rho^{tr}$ and $\chi_{B_2} d\rho^{tr}$ are mutually absolutely continuous, $\chi_{B_2}\tilde{H}$ is unitary equivalent to multiplication by λ in the Hilbert space $L^2(\mathbb{R}, \mathbb{C}^2, \mathbb{1}_2 \chi_{B_1} d\rho^{tr})$. $\square$

Appendix C

Jacobi Difference Equations with Mathematica

The purpose of this chapter is to show how *Mathematica* can be used to make some calculations with difference equations, in particular, Jacobi difference equations. I assume that you are familiar with *Mathematica*, version 3.0. The calculations require the packages *DiscreteMath'DiffEqs'* and *DiscreteMath'JacOp'* which are available via

- ftp://ftp.mat.univie.ac.at/pub/teschl/book-jac/DiffEqs.m
- ftp://ftp.mat.univie.ac.at/pub/teschl/book-jac/JacDEqs.m

and need to be stored in the AddOns/Applications/DiscreteMath subfolder of your *Mathematica* folder. On multi-user systems, you can install an add-on either in the central *Mathematica* directory (provided you have access to it) or else in your individual user's Mathematica directory (usually ~/.Mathematica/3.0/).

The *Mathematica* notebook used to make the calculations below is also available

- ftp://ftp.mat.univie.ac.at/pub/teschl/book-jac/JacDEqs.nb

C.1. The package *DiffEqs* and first order difference equations

We first load the package DiscreteMath'DiffEqs'.

    In[1]:= Needs["DiscreteMath'DiffEqs'"]

The package *DiscreteMath'DiffEqs'* provides some basic commands for dealing with differences and difference equations.

S[f[n],{n,j}]	shifts the sequence f[n] by j places
FDiff[f[n],n]	(=f[n+1]-f[n]) forward difference operator
BDiff[f[n],n]	(=f[n-1]-f[n]) backward difference operator

SymbSum[f[j],{j,m,n}]	sum for symbolic manipulations
SymbProduct[f[j],{j,m,n}]	product for symbolic manipulations
Casoratian[$f_1,..,f_k$,n]	Casoratian of a list of sequences
SplitSum[expr]	splits sums into smaller parts
SplitProduct[expr]	splits products into smaller parts
SplitAll[expr]	splits sums and products into smaller parts
TestDiff[f[n]==g[n],{n,m}]	simple test for equality based on the fact that f = g if equality holds at one point m and FDiff[f[n] -g[n],n]=0

The main purpose of the built-in Sum command is to search for a closed form. It is, however, not suitable for manipulating sums which cannot be brought to a closed form. For example,

$In[2] := $ Simplify$[\sum_{j=1}^{n+1} f[j] - \sum_{j=1}^{n} f[j]]$

$Out[2] = \sum_{j=1}^{n+1} f[j] - \sum_{j=1}^{n} f[j]$

Moreover, the built-in Sum is zero if the upper limit is smaller than the lower limit. This can lead to, at first sight, surprising results:

$In[3] := S_1[n_] := $ SymbSum$[j, \{j, 1, n\}]$;
$S_2[n_] := $ Sum$[j, \{j, 1, n\}]$;
$S_3[n_] = $ Sum$[j, \{j, 1, n\}]$;
$\{\{S_1[n], S_1[-3]\}, \{S_2[n], S_2[-3]\}, \{S_3[n], S_3[-3]\}\}$//MatrixForm

$Out[3]$//MatrixForm=

$\begin{pmatrix} \sum_{j=1}^{n} {}^* j & 3 \\ \frac{1}{2}n(n+1) & 0 \\ \frac{1}{2}n(n+1) & 3 \end{pmatrix}$

(note that *Mathematica* assumes n to be positive). Observe that SymbSum does not look for a closed form; which can be obtained by switching to the built-in Sum:

$In[4] := S_1[n] $ /. SymbSum$->$ Sum

$Out[4] = \frac{1}{2} n (1+n)$

Observe also the definition for situations where the upper limit is smaller than the lower limit:

$In[5] := \{\sum_{j=0}^{1} {}^* f[j], \sum_{j=0}^{1} f[j], \sum_{j=0}^{-3} {}^* f[j], \sum_{j=0}^{-3} f[j]\}$

$Out[5] = \{f[0] + f[1], f[0] + f[1], -f[-2] - f[-1], 0\}$

Similarly for symbolic products

C.1. The package DiffEqs and first order difference equations

$$In[6] := \{\prod_{j=0}^{1}{}^{*} f[j], \prod_{j=0}^{1} f[j], \prod_{j=0}^{-3}{}^{*} f[j], \prod_{j=0}^{-3} f[j]\}$$

$$Out[6] = \{f[0]f[1], f[0]f[1], \frac{1}{f[-2]\,f[-1]}, 1\}$$

The main purpose of the commands SplitSum and SplitProduct is to simplify expressions involving sums and products (as pointed out earlier, *Mathematica*'s built-in capabilities in this respect are limited). The command SplitSum will try to break sums into parts and pull out constant factors such that the rest can be done with built-in commands. In particular, it will split off extra terms

$$In[7] := \text{SplitSum}[\sum_{j=0}^{n+1}{}^{*} f[j]]$$

$$Out[7] = f[1+n] + \sum_{j=0}^{n}{}^{*} f[j]$$

expand sums and pull out constants

$$In[8] := \text{SplitSum}[\sum_{j=0}^{n}{}^{*} (h\,f[j] + g[j])]$$

$$Out[8] = h \sum_{j=0}^{n}{}^{*} f[j] + \sum_{j=0}^{n}{}^{*} g[j]$$

The command SplitProduct will perform similar operations with products.

$$In[9] := \text{SplitProduct}[\prod_{j=0}^{n-1}{}^{*} \frac{g}{f[j]}]$$

$$Out[9] = \frac{g^n f[n]}{\prod_{j=0}^{n}{}^{*} f[j]}$$

Finally, the command SplitAll also handles nested expressions

$$In[10] := \text{SplitAll}[\sum_{m=0}^{n}{}^{*} \sum_{j=0}^{m}{}^{*} \left(f[j] + \prod_{i=0}^{j}{}^{*} x\,g[i]\right)]$$

$$Out[10] = \sum_{m=0}^{n}{}^{*} \sum_{j=0}^{m}{}^{*} f[j] + \sum_{m=0}^{n}{}^{*} \sum_{j=0}^{m}{}^{*} x^{1+j} \prod_{i=0}^{j}{}^{*} g[i]$$

We can shift a sequence $f(n)$ by j places using S[f[n],{n,j}]. For example,

$In[11]$:=S[f[g[n^2]],{n, 1}]

$Out[11]$= f[g[(1 + n)^2]]

or

$In[12]$:=S[f[n, m],{n, 3}]

$Out[12]$= f[3 + n, m]

Similarly, we can compute forward, backward differences using FDiff[f[n],n], BDiff[f[n],n], respectively. For example,

$In[13]$:={FDiff[f[n], n], BDiff[f[n], n]}

$Out[13]$= {−f[n] + f[1 + n], f[−1 + n] − f[n]}

A simple test for equality of sequences is based on the fact that $f(n) = 0$ if $f(n)$ is constant, that is, $f(n + 1) = f(n)$ and $f(n_0) = 0$ at one fixed point.

$In[14]$:=TestDiff[f[n] == 0, {n, n$_0$}]

$Out[14]$= −f[n] + f[1 + n] == 0 && f[n$_0$] == 0

As a first application we can verify Abel's formula (summation by parts)

$$In[15]:= \sum_{j=m}^{n}{}^{*} \; g[j]\text{FDiff}[f[j], j] ==$$

$$\left(g[n]f[n+1] \; - g[m-1] \; f[m] + \sum_{j=m}^{n}{}^{*} \; \text{BDiff}[g[j], j]f[j] \right) //$$

TestDiff[#, {n, m}]&

$Out[15]$= True

Note also the following product rules

$In[16]$:=FDiff[f[n]g[n], n] == f[n]FDiff[g[n], n] + g[n + 1]FDiff[f[n], n]// Simplify

$Out[16]$= True

$In[17]$:=BDiff[f[n]g[n], n] == f[n]BDiff[g[n], n] + g[n − 1]BDiff[f[n], n]// Simplify

$Out[17]$= True

Next, the product

$$In[18]:=P[n\_] := \prod_{j=1}^{n-1}{}^{*} \; f[j]$$

satisfies the difference equation $P(n + 1) = f(n)P(n)$ plus the initial condition $P(0) = 1$

In[19]:= {P[0], SplitProduct[P[n + 1] − f[n]P[n]]}

Out[19]= {1, 0}

Moreover, the sum

In[20]:= F[n_] := $\sum_{j=1}^{n-1}\!\!\!{}^{*}\ f[j]$

satisfies the difference equation $\partial F(n) = F(n+1) - F(n) = f(n)$ plus the initial condition $F(0) = 0$

In[21]:= {F[0], SplitSum[FDiff[F[n], n] − f[n]]}

Out[21]= {0, 0}

Similarly, the general solution of the difference equation $F(n+1) = f(n)F(n) + g(n)$ with initial condition $F(0) = F_0$ is given by

In[22]:= F[n_] := P[n] $\left(F_0 + \sum_{i=0}^{n-1}\!\!\!{}^{*}\ \frac{g[i]}{P[i+1]} \right)$

In fact, we can easily verify this claim:

In[23]:= {F[0], SplitAll[F[n + 1] − (f[n]F[n] + g[n])]}

Out[23]= {F_0, 0}

C.2. The package *JacDEqs* and Jacobi difference equations

Now we come to Jacobi difference equations. The necessary commands are provided in the package *DiscreteMath'JacDEqs'* which we load first.

In[1]:= Needs["DiscreteMath`JacDEqs`"]

This will also load the package *DiscreteMath'DiffEqs'* if necessary. The following commands are provided in the package *JacDEqs* :

JacobiDE[f[n],n]	computes the Jacobi difference expression, associated with a[n], b[n] of f[n]
JacobiDE[k,f[n],n]	applies the Jacobi difference expression k times.
SolutionJacobi[u]	tells *Mathematica* that u[z,n] solves JacobiDE[u[z,n],n] = z u[z,n]
SolutionJacobi[u,x,y,m]	tells *Mathematica* that u[z,n] solves JacobiDE[u[z,n],n] = z u[z,n] and satisfies the initial conditions u[z,m]=x, u[z,m+1]=y
SolutionJacobi[u,x,y]	tells *Mathematica* that u[z,n,m] solves JacobiDE[u[z,n,m],n] = z u[z,n,m] and

	satisfies the initial conditions u[z,m,m]=x, u[z,m+1,m]=y
SolutionJacobi[f,g]	tells *Mathematica* that f[n] satisfies JacobiDE[f[n],n] = g[n]

The Jacobi difference expression τ applied to a sequence gives the following result

In[2]:= JacobiDE[f[n],n]

Out[2]= $a[-1+n]f[-1+n] + b[n]f[n] + a[n]f[1+n]$

This can also be written as

In[3]:= JacobiDE[f[n],n] == -BDiff[a[n]FDiff[f[n],n],n] + (a[n − 1] + a[n] + b[n])f[n]//Simplify

Out[3]= True

or

In[4]:= JacobiDE[f[n],n] == -FDiff[a[n − 1]BDiff[f[n],n],n] + (a[n − 1] + a[n] + b[n])f[n]//Simplify

Out[4]= True

The command

In[5]:= SolutionJacobi[u]

will tell *Mathematica* that $u(z, n)$ satisfies the Jacobi equation $\tau u(z) = z\, u(z)$. Let's see how this works:

In[6]:= JacobiDE[u[z,n],n] == z u[z,n]

Out[6]= True

We can also define solutions $u(z, n, m)$ satisfying the initial conditions $u(z, m, m) = u_0$ and $u(z, m+1, m) = u_1$ by

In[7]:= Clear[u];
SolutionJacobi[u, u_0, u_1]

Indeed, we obtain

In[8]:= {JacobiDE[u[z,n,m],n] == z u[z,n,m], u[z,m,m], u[z,m + 1,m]}

Out[8]= {True, u_0, u_1}

The solutions $c(z, n, m)$ and $s(z, n, m)$ are predefined by the package

In[9]:= {JacobiDE[c[z,n,m],n] == z c[z,n,m], c[z,m,m], c[z,m + 1,m]}

Out[9]= {True, 1, 0}

In[10]:= {JacobiDE[s[z,n,m],n] == z s[z,n,m], s[z,m,m], s[z,m + 1,m]}

Out[10]= {True, 0, 1}

Note that you can tell *Mathematica* that $u(n)$ satisfies $(\tau - z)u = 0$ using

```
In[11]:=Clear[u];
       SolutionJacobi[u,(z u[#])&]
```

Indeed,

```
In[12]:=JacobiDE[u[n],n] == z u[n]
```

```
Out[12]= True
```

C.3. Simple properties of Jacobi difference equations

Let us first load our package

```
In[1]:= Needs["DiscreteMath`JacDEqs`"]
```

The Wronskian of two solutions $W_n(u(z), v(z))$ is defined as $a(n)$ times the Casoratian of two solutions $C_n(u(z), v(z))$

```
In[2]:= W[u_,v_,n_] := a[n]Casoratian[u,v,n]
```

Please note that the argument n tells *Mathematica* what to consider as index. To compute the value of the Wronskian use

```
In[3]:= W[u[n],v[n],n] /. n-> 0
```

$Out[3]=\quad a[0](-u[1]v[0] + u[0]v[1])$

and not

```
In[4]:= W[u[n],v[n],0]
```

$Out[4]=\quad 0$

Defining two solutions $u(z, n)$ and $v(z, n)$

```
In[5]:= Clear[u,v]; SolutionJacobi[u]; SolutionJacobi[v];
```

we can easily verify that the Wronskian is independent of n

```
In[6]:= FDiff[W[u[z,n],v[z,n],n],n]//Simplify
```

$Out[6]=\quad 0$

Similarly, we can verify Green's formula

$$In[7]:= \sum_{j=m}^{n}{}^{*} (f[j]\text{JacobiDE}[g[j],j] - g[j]\text{JacobiDE}[f[j],j]) == W[f[n],g[n],n] -$$
$$(W[f[n],g[n],n] /. n-> m-1)//\text{TestDiff}[\#,n,m]\&$$

$Out[7]=\quad$ True

We can also define a second solution using the formula

$$In[8]:= \text{Clear}[v]; v[z_,n_] := \sum_{j=0}^{n-1}{}^{*} \frac{u[z,n]}{a[j]u[z,j]u[z,j+1]};$$

Indeed, the sequence $v(z,n)$ defined as above satisfies the Jacobi equation

$In[9]:=$ SplitSum[JacobiDE[v[z,n],n] − zv[z,n]]//Simplify

$Out[9]=$ 0

The Wronskian of $u(z,n)$ and $v(z,n)$ is one:

$In[10]:=$SplitSum[W[u[z,n],v[z,n],n]]//Simplify

$Out[10]=$ 1

Next, recall that the equality $u(n) = v(n)$ holds if equality holds at two consecutive points $n = m, m+1$ and the Jacobi difference equations applied to both sides gives the same result, that is, $\tau u(z) = \tau v(z)$. We can define a simple test using this fact:

$In[11]:=$TestJac[eqn_Equal, {n_, m_}] :=
    Simplify[(JacobiDE[#, n] − z#) & /@ eqn]&&(eqn /. n−> m)&&
    (eqn /. n−> m + 1);

As an application we verify that any solution can be written as a linear combination of the two fundamental solutions $c(z,n,m)$ and $s(z,n,m)$ as follows

$In[12]:=$SolutionJacobi[u, u_0, u_1];
    u[z,n,m] == u_0c[z,n,m] + u_1s[z,n,m]//TestJac[#, {n,m}]&

$Out[12]=$ True

The solutions $c(z,n,m)$ and $s(z,n,m)$ have some interesting properties.

$In[13]:=$s[z,n,m + 1] == $-\dfrac{a[m+1]}{a[m]}$c[z,n,m] // TestJac[#, {n,m}]&

$Out[13]=$ True

$In[14]:=$s[z,n,m − 1] == c[z,n,m] + $\dfrac{z - b[m]}{a[m]}$s[z,n,m] // TestJac[#, {n,m}]&

$Out[14]=$ True

$In[15]:=$c[z,n,m + 1] == $\dfrac{z - b[m+1]}{a[m]}$c[z,n,m] + s[z,n,m] // TestJac[#, {n,m}]&

$Out[15]=$ True

$In[16]:=$c[z,n,m − 1] == $-\dfrac{a[m-1]}{a[m]}$s[z,n,m] // TestJac[#, {n,m}]&

$Out[16]=$ True

Unfortunately, this does not work for the following formula

$In[17]:=$c[z,m,n] == $\dfrac{a[n]}{a[m]}$s[z,n + 1,m] // TestJac[#, {n,m}]&

C.3. Simple properties of Jacobi difference equations

Out[17]= $a[-1+n]c[z,m,-1+n] + (-z+b[n])c[z,m,n] + a[n]\ c[z,m,1+n] ==$
$\frac{1}{a[m]}a[-1+n]^2 s[z,n,m] - a[n](a[n]s[z,n,m] + (-b[n]+b[1+n])s[z,1+n,m])$

and we can only check it for specific values of n:

In[18]:= $\text{Simplify}[c[z,m,n] == \frac{a[n]}{a[m]}s[z,n+1,m]\ /.\ n\text{->}m+3]$

Out[18]= True

Similarly for the following asymptotics $w = \frac{1}{z}$

In[19]:= $\left(\left(\prod_{j=m+1}^{n-1} a[j]\right)\frac{w^{n-m-2}}{-a[m]}c[\frac{1}{w},n,m] - \left(1 - w\sum_{j=m+2}^{n-1}b[j] + O[w]^2\right)\right)\ /.\ n\text{->}m+3$

Out[19]= $O[w]^2$

In[20]:= $\left(\left(\prod_{j=m+1}^{n-1} a[j]\right)w^{n-m-1}s[\frac{1}{w},n,m] - \left(1 - w\sum_{j=m+1}^{n-1}b[j] + O[w]^2\right)\right)\ /.\ n\text{->}m+3$

Out[20]= $O[w]^2$

In[21]:= $\left(\left(\prod_{j=n}^{m-1} a[j]\right)w^{m-n}c[\frac{1}{w},n,m] - \left(1 - w\sum_{j=n+1}^{m}b[j] + O[w]^2\right)\right)\ /.\ n\text{->}m+3$

Out[21]= $O[w]^2$

In[22]:= $\left(\left(\prod_{j=n}^{m-1} a[j]\right)\frac{w^{m-n-1}}{-a[m]}s[\frac{1}{w},n,m] - \left(1 - w\sum_{j=n+1}^{m-1}b[j] + O[w]^2\right)\right)\ /.\ n\text{->}m+3$

Out[22]= $O[w]^2$

Defining the Jacobi matrix

In[23]:= Delta[i_, j_] := If[Simplify[i == j], 1, 0];
J[i_, j_] := a[j]Delta[i − 1, j] + a[i]Delta[i + 1, j]
 + b[j]Delta[i, j];
JacobiMatrix[m_, n_] := Table[J[m + i, m + j], {i, n − m − 1},
{j, n − m − 1}]

```
In[24]:=MatrixForm[JacobiMatrix[m, m + 6]]
```
Out[24]//MatrixForm=
$$\begin{pmatrix} b[1+m] & a[1+m] & 0 & 0 & 0 \\ a[1+m] & b[2+m] & a[2+m] & 0 & 0 \\ 0 & a[2+m] & b[3+m] & a[3+m] & 0 \\ 0 & 0 & a[3+m] & b[4+m] & a[4+m] \\ 0 & 0 & 0 & a[4+m] & b[5+m] \end{pmatrix}$$

we can check

```
In[25]:=(s[z, #, m] ==
         Det[z IdentityMatrix[# − m − 1] − JacobiMatrix[m, #]]
         ─────────────────────────────────────────────────────
                         ∏_{j=m+1}^{#−1} a[j]
        )&[m + 4]
        // Simplify
```
Out[25]= True

or compute traces

```
In[26]:=Tr[M_] := ∑_{i=1}^{Length[M]} M[[i, i]]; (* Not need for Mathematica 4.x *)
```

```
In[27]:=Expand[(Tr[JacobiMatrix[m, #].JacobiMatrix[m, #]]
         − (2 ∑_{j=m+1}^{#−2} a[j]² + ∑_{j=m+1}^{#−1} b[j]²))&[m + 3]]
```
Out[27]= 0

Let us next investigate the inhomogeneous Jacobi operator. The command

```
In[28]:=SolutionJacobi[f, g];
```

tells *Mathematica* that $f(n)$ satisfies the inhomogeneous Jacobi equation $\tau f = g$. In fact,

```
In[29]:=JacobiDE[f[n], n] == g[n]
```
Out[29]= True

Next, let us express $f(n)$ in terms of $s(z, n, m)$. We define the kernel

```
In[30]:=K[z_, n_, j_] := s[z, n, j] / a[j]
```

which satisfies

```
In[31]:=Factor[{K[z, n, n − 2], K[z, n, n − 1], K[z, n, n], K[z, n, n + 1], K[z, n, n + 2]}]
```
Out[31]= $\{\dfrac{z - b[-1 + n]}{a[-2 + n]a[-1 + n]}, \dfrac{1}{a[-1 + n]}, 0, -\dfrac{1}{a[n]}, -\dfrac{z - b[1 + n]}{a[n]a[1 + n]}\}$

and the sequence $f(n)$

C.4. Orthogonal Polynomials

```
In[32]:=Clear[f];
```
$$f[n\_] := \sum_{j=1}^{n}{}^{*} K[z,n,j]g[j]$$

```
In[33]:=SplitSum[JacobiDE[f[n],n] − zf[n]]//Factor
Out[33]= g[n]
```

Finally, we consider the case with $a(n) = 1/2$, $b(n) = 0$.

```
In[34]:=Unprotect[a,b];
```
$$a[n\_] = \frac{1}{2}; b[n\_] = 0;$$

Then we have

```
In[35]:=JacobiDE[(z − (z + √(z² − 1))ⁿ, n] == z(z + √(z² − 1))ⁿ // Simplify
Out[35]= True
```

```
In[36]:=JacobiDE[(z − (z − √(z² − 1))ⁿ, n] == z(z − √(z² − 1))ⁿ // Simplify
Out[36]= True
```

and

$$In[37]:=s[z,n,0] == \frac{(z + \sqrt{z^2-1})^n - (z - \sqrt{z^2-1})^n}{2\sqrt{z^2-1}} \ //\ \mathtt{TestJac}[\#,\{n,0\}]\&$$

```
Out[37]= True
```

```
In[38]:=c[z,n,0] == −s[z,n − 1,0] // TestJac[#,{n,0}]&
Out[38]= True
```

C.4. Orthogonal Polynomials

We first introduce the matrices $C(k)$, $D(k)$, and $P(z,k)$. We use `Cm[k]` for the matrix $C(k)$ and `Cd[k]` for its determinant. Similar for the other three matrices.

```
In[1]:= m[0] := 1;
        Cd[0] = 1; Cd[k_] := det Cm[k];
        Cm[k_?IntegerQ] := Table[m[i + j], {i, 0, k − 1}, {j, 0, k − 1}];
        Dd[0] = 0; Dd[k_] := det Dm[k];
        Dm[k_?IntegerQ] := Table[If[j == k − 1, m[i + j + 1], m[i + j]],
           {i, 0, k − 1}, {j, 0, k − 1}];
        Pd[z_, 0] = 0; Pd[z_, k_] := det Pm[z, k];
        Pm[z_, k_?IntegerQ] := Table[If[i == k − 1, zʲ, m[i + j]],
           {i, 0, k − 1}, {j, 0, k − 1}];
```

Here is how the matrices look like.

In[2]:= Print[MatrixForm[Cm[3]]," ",MatrixForm[Dm[3]]," ",
    MatrixForm[Pm[z,3]]]

$$\begin{pmatrix} 1 & m[1] & m[2] \\ m[1] & m[2] & m[3] \\ m[2] & m[3] & m[4] \end{pmatrix} \begin{pmatrix} 1 & m[1] & m[3] \\ m[1] & m[2] & m[4] \\ m[2] & m[3] & m[5] \end{pmatrix}$$

$$\begin{pmatrix} 1 & m[1] & m[2] \\ m[1] & m[2] & m[3] \\ 1 & z & z^2 \end{pmatrix}$$

The determinants of $C(k)$ and $D(k)$ appear as the first two coefficients in the expansion of the determinant of $P(z,k)$ for large z.

In[3]:= Simplify[(Series[Pd[1/t,k],{t,0,-1}] ==
    Cd[k-1](1/t)^{k-1} - Dd[k-1](1/t)^{k-2})/.k->3]

Out[3]= O[t]^0 == 0

The polynomials $s(z,k)$

In[4]:= s[z_,k_] := $\dfrac{\text{Pd}[z,k]}{\sqrt{\text{Cd}[k]\text{Cd}[k-1]}}$;

are orthogonal with respect to the measure ρ whose moments are $m(k) = \int \lambda^k d\rho(\lambda)$. Moreover, defining

In[5]:= a[k_] := $\dfrac{\sqrt{\text{Cd}[k+1]}\sqrt{\text{Cd}[k-1]}}{\text{Cd}[k]}$; a[0] := 0;

b[k_] := $\dfrac{\text{Dd}[k]}{\text{Cd}[k]} - \dfrac{\text{Dd}[k-1]}{\text{Cd}[k-1]}$;

we have that $s(z,k)$ solves the corresponding Jacobi equation.

In[6]:= Simplify[a[k]s[z,k+1] + a[k-1]s[z,k-1] + (b[k]-z)s[z,k] /.k->3]

Out[6]= 0

Moreover, the polynomials $s(z,k)$ are orthogonal with respect to the scalar product

In[7]:= ScalarProduct[P_,Q_,z_] := $\displaystyle\sum_{i=0}^{\text{Exponent}[P,z]} \sum_{j=0}^{\text{Exponent}[Q,z]} \text{m}[i+j]$
    Coefficient[P,z,i]Coefficient[Q,z,j]/;
    PolynomialQ[P,z]&&PolynomialQ[Q,z]

which can be checked as follows

In[8]:= Simplify[ScalarProduct[s[z,j],s[z,k],z] ==
    If[j==k,1,0] /. {k->3, j->3}]

Out[8]= True

C.5. Recursions

Now we want to look at the matrix elements $g_j(n) = \langle \delta_n, H^j \delta_n \rangle$ and $h_j(n) = 2a(n)\langle \delta_{n+1}, H^j \delta_n \rangle$, which are important when dealing with high energy expansions for the resolvent of our Jacobi operator H.

In[1]:= Needs["DiscreteMath`JacDEqs`"]

For computational purpose it is most convenient to define them using the following recursions.

In[2]:= g[0, n_] = 1; h[0, n_] = 0;
g[j_?IntegerQ, n_] := g[j, n] =
$\frac{1}{2}$(h[j − 1, n] + h[j − 1, n − 1]) + b[n]g[j − 1, n];
h[j_?IntegerQ, n_] := h[j, n] =
$2a[n]^2 \sum_{l=0}^{j-1} g[j-1-1,n]g[1,n+1] - \frac{1}{2}\sum_{l=0}^{j-1} h[j-1-1,n]h[1,n];$

These definitions can be used to compute arbitrary coefficients as needed.

In[3]:= Simplify[g[2, n]]

Out[3]= $a[-1+n]^2 + a[n]^2 + b[n]^2$

In[4]:= Simplify[h[2, n]]

Out[4]= $2a[n]^2(a[-1+n]^2 + a[n]^2 + a[1+n]^2 + b[n]^2 + b[n]b[1+n] + b[1+n]^2)$

In addition, we have the third sequence $\gamma_j^\beta(n) = \langle (\delta_{n+1} + \beta\delta_n), H^j(\delta_{n+1} + \beta\delta_n) \rangle$.

In[5]:= γ[j_, n_] := g[j, n + 1] + $\frac{\beta}{a[n]}$h[j, n] + β^2g[j, n];

But now we want to see that the recursions indeed produce the matrix elements from above. For this purpose we need a delta sequence.

In[6]:= Delta[n_] := If[n == 0, 1, 0];

Using this we have

In[7]:= g[j, n] == JacobiDE[j, Delta[n − m], m] /. j− > 3 /. m− > n//Simplify

Out[7]= True

and

In[8]:= h[j, n] == 2a[n]JacobiDE[j, Delta[n − m], m] /. j− > 3 /. m− > n + 1//Simplify

Out[8]= True

Hence the recursions produce indeed the desired matrix elements.
There is another relation for g_j and h_j which can be verified for given j.

In[9]:= Simplify[h[j + 1, n] − h[j + 1, n − 1] ==
$2(a[n]^2g[j,n+1] - a[n-1]^2g[j,n-1]) + b[n](h[j,n] - h[j,n-1])$ /. j− > 4]

Out[9]= True

Moreover, there is a recursion involving only g_j.

In[10]:= Clear[g];
g[0, n_] = 1;
g[1, n_] = b[n];
g[2, n_] = a[−1 + n]² + a[n]² + b[n]²;
g[j_?IntegerQ, n_] := −a[n]²g[j − 2, n + 1] + a[n − 1]²g[j − 2, n − 1]
$$- b[n]^2 g[j-2,n] + 2b[n]g[j-1,n] - \frac{1}{2}\sum_{l=0}^{j-2} k[j-1-2,n]k[l,n]$$
$$+ 2a[n]^2 \sum_{l=0}^{j-2} g[j-1-2,n+1]g[l,n]$$
$$- 2b[n] \sum_{l=0}^{j-3} g[j-1-3,n]g[l,n+1]$$
$$+ b[n]^2 \sum_{l=0}^{j-4} g[j-1-4,n]g[l,n+1];$$
k[0, n_] = −b[n];
k[j_, n_] := a[n]²g[j − 1, n + 1] − a[n − 1]²g[j − 1, n − 1] + b[n]²g[j − 1, n] −
2b[n]g[j, n] + g[j + 1, n];

Again, it can be checked, that it is equivalent to our other one for given j.

In[11]:= g[j, n] == JacobiDE[j, Delta[n − m], m] /. j−> 4 /. m−> n // Simplify

Out[11]= True

In particular, these relations will be used in the sections on the Toda and Kac van Moerbeke lattice.

C.6. Commutation methods

C.6.1. Single commutation method. In this section we want to consider the single commutation method. As first steps we load our package

In[1]:= Needs["DiscreteMath`JacDEqs`"]

and introduce a few solutions of our original operator

In[2]:= SolutionJacobi[u]; SolutionJacobi[v]; SolutionJacobi[f, g];
W[f_, n_] := a[n]Casoratian[f, n];

Next, we declare the commuted operator H_σ

In[3]:= $a_\sigma[n\_] := \frac{\sqrt{-a[n]}\sqrt{-a[n+1]}\sqrt{u[\lambda,n]}\sqrt{u[\lambda,n+2]}}{u[\lambda,n+1]};$
$b_\sigma[n\_] := \lambda - a[n]\left(\frac{u[\lambda,n]}{u[\lambda,n+1]} + \frac{u[\lambda,n+1]}{u[\lambda,n]}\right);$
JacobiDE$_\sigma$[f_, n_] := a_σ[n] S[f, {n, 1}] + a_σ[n − 1] S[f, {n, −1}] + b_σ[n]f;
W_σ[f_, n_] := a_σ[n]Casoratian[f, n];

and the operator A

C.6. Commutation methods

```
In[4]:= vσ[λ_, n_] := 1/(-a[n]u[λ, n]u[λ, n + 1]);
        A[f_, n_] := vσ[λ, n]W[u[λ, n], f, n];
```

Note that v_σ is a solution of $H_\sigma u = \lambda u$

```
In[5]:= Simplify[JacobiDEσ[vσ[λ, n], n] == λvσ[λ, n]]
Out[5]= True
```

Now we can verify that $Au(z)$ satisfies $H_\sigma(Au(z)) = zAu(z)$

```
In[6]:= Simplify[JacobiDEσ[A[u[z, n], n], n] == zA[u[z, n], n]]
Out[6]= True
```

and that Af satisfies $H_\sigma(Af) = Ag$.

```
In[7]:= Simplify[JacobiDEσ[A[f[n], n], n] == A[g[n], n]]
Out[7]= True
```

Moreover, the following relations for the Wronskians are valid:

```
In[8]:= Simplify[Wσ[A[u[z, n], n], A[v[z, n], n], n] == (z − λ)W[u[z, n], v[z, n], n]]
Out[8]= True
```

```
In[9]:= Simplify[Wσ[A[u[z, n], n], A[v[zz, n], n], n] ==
        -a[n]a[n + 1]Casoratian[u[λ, n], u[z, n], v[zz, n], n] / u[λ, n + 1]]
Out[9]= True
```

```
In[10]:= Simplify[Wσ[A[u[z, n], n], A[v[zz, n], n], n] ==
         (z − λ)W[u[z, n], v[zz, n], n] − (z − zz)v[zz, 1 + n]W[u[z, n], u[λ, n], n] / u[λ, 1 + n]]
Out[10]= True
```

For additional material see the section on the Kac-van Moerbeke lattice below.

C.6.2. Double commutation method.
In this section we want to consider the double commutation method. Again we load our package

```
In[1]:= Needs["DiscreteMath`JacDEqs`"]
```

and introduce two solutions of our original operator.

```
In[2]:= SolutionJacobi[u]; SolutionJacobi[v];
        W[f__, n_] := a[n]Casoratian[f, n];
```

Next, we declare the doubly commuted operator H_γ

$$In[3]:= \ c_\gamma[n\_] = \frac{1}{\gamma} + \sum_{j=-\infty}^{n}{}^* \ u[\lambda, j]^2;$$

$$a_\gamma[n\_] = a[n]\frac{\sqrt{c_\gamma[n-1]}\sqrt{c_\gamma[n+1]}}{c_\gamma[n]};$$

$$b_\gamma[n\_] = b[n] - \text{BDiff}\left[\frac{a[n]u[\lambda, n]u[\lambda, n+1]}{c_\gamma[n]}, n\right];$$

$$\text{JacobiDE}_\gamma[f\_, n\_] := a_\gamma[n]\,S[f, \{n, 1\}] + a_\gamma[n-1]S[f, n, -1] + b_\gamma[n]f;$$

$$W_\gamma[f\_\_, n\_] := a_\gamma[n]\text{Casoratian}[f, n];$$

As a first fact we note that u_γ

$$In[4]:= \ u_\gamma[n\_] = \frac{u[\lambda, n]}{\sqrt{c_\gamma[n-1]}\sqrt{c_\gamma[n]}};$$

satisfies $H_\gamma u_\gamma = \lambda u_\gamma$.

$In[5]:= \ \text{Simplify}[\text{SplitSum}[\text{JacobiDE}_\gamma[u_\gamma[n], n] == \lambda u_\gamma[n]]]$

$Out[5]= \ \text{True}$

To obtain more solutions we introduce the operator A

$$In[6]:= \ A[f\_, \{z\_, n\_\}] := \frac{\sqrt{c_\gamma[n]}}{\sqrt{c_\gamma[n-1]}}f - \frac{1}{z-\lambda}u_\gamma[n]W[u[\lambda, n], f, n];$$

and set

$In[7]:= \ u_\gamma[z\_, n\_] = A[u[z, n], \{z, n\}];$

The sequence $u_\gamma(z)$ satisfies $H_\gamma u_\gamma(z) = z\, u_\gamma(z)$

$In[8]:= \ \text{Simplify}[\text{SplitSum}[\text{JacobiDE}_\gamma[u_\gamma[z, n], n] == zu_\gamma[z, n]]]$

$Out[8]= \ \text{True}$

and for $u_\gamma(z, n)^2$ we have

$$In[9]:= \ \text{Simplify}[\text{SplitSum}[u_\gamma[z, n]^2 ==$$
$$u[z, n]^2 + \frac{1}{(z-\lambda)^2}\text{BDiff}\left[\frac{W[u[\lambda, n], u[z, n], n]^2}{c_\gamma[n]}, n\right]]]$$

$Out[9]= \ \text{True}$

In addition, we note two equivalent forms for $u_\gamma(z)$

$$In[10]:=\text{Simplify}[\text{SplitSum}[u_\gamma[z, n] ==$$
$$\sqrt{\frac{c_\gamma[n-1]}{c_\gamma[n]}}u[z, n] - \frac{1}{z-\lambda}u_\gamma[n]S[W[u[\lambda, n], u[z, n], n], \{n, -1\}]]]$$

$Out[10]= \ \text{True}$

$$In[11]:=\text{AA}[f\_, n\_] := \frac{\sqrt{c_\gamma[n]}}{\sqrt{c_\gamma[n-1]}}f - u_\gamma[n]\sum_{j=-\infty}^{n}{}^* \ u[\lambda, j](f\, /.\ n-> j);$$

C.6. Commutation methods

```
In[12]:=Simplify[AA[u[λ,n],n] == u_γ[n]/γ]
```
Out[12]= True

and the following relations for the Wronskian:

```
In[13]:=Simplify[SplitSum[W_γ[u_γ[n], A[u[z,n],{z,n}],n]]]
```

$$\text{Out[13]}= \frac{\gamma a[n](u[z,1+n]u[\lambda,n]-u[z,n]u[\lambda,1+n])}{1+\gamma \sum_{j=-\infty}^{n}{}^{*}u[\lambda,j]^2}$$

```
In[14]:=Simplify[SplitSum[W_γ[A[u[z,n],{z,n}], A[v[zz,n],{zz,n}],n] ==
        W[u[z,n],v[zz,n],n] + (z-zz)W[u[λ,n],u[z,n],n]W[u[λ,n],v[zz,n],n]/((z-λ)(zz-λ)c_γ[n])]]
```
Out[14]= True

Finally, we turn to the solutions c_β and s_β.

```
In[15]:=u[λ,0] = -Sin[α]; u[λ,1] = Cos[α];
        SolutionJacobi[sb, -Sin[α], Cos[α], 0];
        SolutionJacobi[cb, Cos[α], Sin[α], 0];
```

They are transformed into $c_{\hat\beta,\gamma}$ and $s_{\hat\beta,\gamma}$

```
In[16]:=cn = c_γ[0];
        cp = SplitSum[c_γ[n+1]] /. n- > 0;
        cm = SplitSum[c_γ[n-1]] /. n- > 0;
```
$$\beta = \text{Cot}[\alpha]; \hat\beta = \frac{\sqrt{cm}}{\sqrt{cp}}\beta;$$
$$\delta = \frac{a[0]}{cn(z-\lambda)} - \frac{\beta}{cp(1+\hat\beta^2)};$$
```
        sb_γ[z_, n_] =
```
$$\frac{\sqrt{cm}\sqrt{cp}}{cn}A[sb[z,n],\{z,n\}];$$
```
        cb_γ[z_, n_] =
```
$$\frac{cp}{cn}\frac{1+\hat\beta^2}{1+\beta^2}(A[cb[z,n],\{z,n\}] - \delta A[sb[z,n],\{z,n\}]);$$

as can be seen from

```
In[17]:=Simplify[SplitSum[sb_γ[z,n+1]/sb_γ[z,n]] /. n- > 0]/β̂
```
Out[17]= -1

and

```
In[18]:=Simplify[SplitSum[cb_γ[z,n+1]/cb_γ[z,n]]β̂ /. n- > 0]
```
Out[18]= 1

C.6.3. Dirichlet commutation method.
In this section we want to consider the Dirichlet commutation method. As before we load our package

$In[1]:=$ Needs["DiscreteMath`JacDEqs`"]

and introduce two solutions of our original operator.

$In[2]:=$ SolutionJacobi[u];
SolutionJacobi[v];
u[μ_0, 0] = 0;
W[f__, n_] := a[n]Casoratian[f, n];

Next, we introduce the operator $H_{(\mu,\sigma)}$

$In[3]:=$ $W_{\{\mu,\sigma\}}[n\_] = \dfrac{W[u[\mu_0, n], u[\mu, n], n]}{\mu - \mu_0}$;

$a_{\{\mu,\sigma\}}[n\_] := \dfrac{a[n]\sqrt{W_{\{\mu,\sigma\}}[n-1]}\sqrt{W_{\{\mu,\sigma\}}[n+1]}}{W_{\{\mu,\sigma\}}[n]}$;

$b_{\{\mu,\sigma\}}[n\_] := b[n] - \text{BDiff}[\dfrac{a[n]u[\mu_0,n]u[\mu, n+1]}{W_{\{\mu,\sigma\}}[n]}, n]$;

JacobiDE$_{\{\mu,\sigma\}}$[f_, n_] := $a_{\{\mu,\sigma\}}$[n] S[f, {n, 1}] + $a_{\{\mu,\sigma\}}$[n − 1]S[f, {n, −1}]
       + $b_{\{\mu,\sigma\}}$[n]f;

$W_{\{\mu,\sigma\}}$[f__, n_] := $a_{\{\mu,\sigma\}}$[n]Casoratian[f, n];

We note the alternate expression for $b_{(\mu,\sigma)}$

$In[4]:=$ Simplify[$b_{\{\mu,\sigma\}}$[n] == $\mu_0 - \dfrac{a[n]u[\mu, n+1]W_{\{\mu,\sigma\}}[n-1]}{u[\mu, n]W_{\{\mu,\sigma\}}[n]}$
$- \dfrac{a[n-1]u[\mu, n-1]W_{\{\mu,\sigma\}}[n]}{u[\mu, n]W_{\{\mu,\sigma\}}[n-1]}$]

$Out[4]=$ True

and $W_{(\mu,\sigma)}$

$In[5]:=$ Simplify[SplitSum[FDiff[$\sum_{j=0}^{n}{}^{*} u[\mu_0, j]u[\mu, j] == W_{\{\mu,\sigma\}}[n], n$]]]

$Out[5]=$ True

Defining the quantities

$In[6]:=$ $u_\mu[n\_] := \dfrac{u[\mu_0, n]}{\sqrt{W_{\{\mu,\sigma\}}[n-1]}\sqrt{W_{\{\mu,\sigma\}}[n]}}$;

$u_{\mu_0}[n\_] := \dfrac{u[\mu, n]}{\sqrt{W_{\{\mu,\sigma\}}[n-1]}\sqrt{W_{\{\mu,\sigma\}}[n]}}$;

$A[f\_, z\_, n\_] := \dfrac{\sqrt{W_{\{\mu,\sigma\}}[n]}}{\sqrt{W_{\{\mu,\sigma\}}[n-1]}}f - \dfrac{u_{\mu_0}[n]W[u[\mu_0, n], f, n]}{z - \mu_0}$;

$Out[6]=$ True

C.6. Commutation methods

we get several solutions of our new operator.

$\text{In[7]} := \text{Simplify}[\text{JacobiDE}_{\{\mu,\sigma\}}[u_\mu[n], n] == \mu u_\mu[n]]$

$\text{Out[7]} = \text{True}$

$\text{In[8]} := \text{Simplify}[\text{JacobiDE}_{\{\mu,\sigma\}}[u_{\mu_0}[n], n] == \mu_0 u_{\mu_0}[n]]$

$\text{Out[8]} = \text{True}$

$\text{In[9]} := \text{Simplify}[\text{JacobiDE}_{\{\mu,\sigma\}}[A[u[z,n],z,n],n] == zA[u[z,n],z,n]]$

$\text{Out[9]} = \text{True}$

In addition, we note

$\text{In[10]} := \text{Simplify}[A[u[z,n],z,n]^2 ==$
$\quad u[z,n]^2 + \dfrac{u[\mu,n]}{(z-\mu_0)^2} u[\mu_0,n] \text{BDiff}\Big[\dfrac{W[u[\mu_0,n],u[z,n],n]^2}{W_{\{\mu,\sigma\}}[n]}, n\Big]]$

$\text{Out[10]} = \text{True}$

and the following formulas for Wronskians

$\text{In[11]} := \text{Simplify}[W_{\{\mu,\sigma\}}[u_{\mu_0}[n], A[u[z,n],z,n], n] == \dfrac{W[u[\mu,n],u[z,n],n]}{W_{\{\mu,\sigma\}}[n]}]$

$\text{Out[11]} = \text{True}$

$\text{In[12]} := \text{Simplify}[W_{\{\mu,\sigma\}}[u_\mu[n], A[u[z,n],z,n], n] == \dfrac{z-\mu}{z-\mu_0} \dfrac{W[u[\mu_0,n],u[z,n],n]}{W_{\{\mu,\sigma\}}[n]}]$

$\text{Out[12]} = \text{True}$

$\text{In[13]} := \text{Simplify}[W_{\{\mu,\sigma\}}[A[u[z,n],z,n], A[v[zz,n],zz,n], n] ==$
$\quad \dfrac{z-\mu}{z-\mu_0} W[u[z,n],v[zz,n],n] +$
$\quad \dfrac{z-zz}{(z-\mu_0)(zz-\mu_0)} \dfrac{W[u[\mu,n],u[z,n],n]W[u[\mu_0,n],v[zz,n],n]}{W_{\{\mu,\sigma\}}[n]}]$

$\text{Out[13]} = \text{True}$

Finally, we turn to the solutions $c(z,n)$ and $s(z,n)$ which transform into

$\text{In[14]} := s_{\{\mu,\sigma\}}[\text{z\_},\text{n\_}] = \dfrac{\sqrt{W_{\{\mu,\sigma\}}[m+1]}}{\sqrt{W_{\{\mu,\sigma\}}[m]}} A[s[z,n,0],z,n] \;/.\; \text{m\_} -> 0;$

$\quad c_{\{\mu,\sigma\}}[\text{z\_},\text{n\_}] = \dfrac{z-\mu_0}{z-\mu} \Big(A[c[z,n,0],z,n]$
$\quad + \dfrac{(\mu-\mu_0)u[\mu,m+1]A[s[z,n,0],z,n]}{(z-\mu_0)} u[\mu,m] \Big) \;/.\; \text{m\_} -> 0;$

$\text{Out[14]} = \text{True}$

as can be seen from

$In[15]:= s_{\{\mu,\sigma\}}[z, 0]$

$Out[15]= 0$

$In[16]:= \text{Simplify}[s_{\{\mu,\sigma\}}[z, 1]]$

$Out[16]= 1$

$In[17]:= \text{Simplify}[c_{\{\mu,\sigma\}}[z, 0]]$

$Out[17]= 1$

$In[18]:= \text{Simplify}[c_{\{\mu,\sigma\}}[z, 1]]$

$Out[18]= 0$

Note also

$In[19]:= \text{Simplify}[\frac{W_{\{\mu,\sigma\}}[n+1]}{W_{\{\mu,\sigma\}}[n]} \,/.\, n->0]$

$Out[19]= \dfrac{a[0]u[\mu, 0] + (-\mu + \mu_0)u[\mu, 1]}{a[0]u[\mu, 0]}$

C.7. Toda lattice

Now we come to the Toda lattice. Our package is needed again.

$In[1]:= \text{Needs}[\text{"DiscreteMath`JacDEqs`"}]$

To define the Toda hierarchy, we recall the recursions for g_j and h_j found before. We only consider the homogeneous case for simplicity.

$In[2]:= g[0, n\_] = 1;$
$\quad h[0, n\_] = 0;$
$\quad g[j\_?\text{IntegerQ}, n\_] := g[j, n] =$
$\quad\quad \frac{1}{2}(h[j-1, n] + h[j-1, n-1]) + b[n]g[j-1, n];$
$\quad h[j\_?\text{IntegerQ}, n\_] := h[j, n] =$
$\quad\quad 2a[n]^2 \sum_{l=0}^{j-1} g[j-1-l, n]g[l, n+1] - \frac{1}{2}\sum_{l=0}^{j-1} h[j-1-l, n]h[l, n];$

The following command will compute the r'th (homogeneous) Toda equation for us

$In[3]:= TL[r\_] := \text{Simplify}[$
$\quad \{\partial_t a[n, t] == a[n](g[r+1, n+1] - g[r+1, n]),$
$\quad \partial_t b[n, t] == h[r+1, n] - h[r+1, n-1]\}\,/.\,$
$\quad \{a[n\_] -> a[n, t], b[n\_] -> b[n, t]\}$

C.7. Toda lattice

For example

$In[4] := \text{TL}[1]//\text{TableForm}$

$Out[4]//\text{TableForm}=$
$$a^{(0,1)}[n,t] == -a[n,t](a[-1+n,t]^2 - a[1+n,t]^2 + b[n,t]^2 - b[1+n,t]^2)$$
$$b^{(0,1)}[n,t] == -2a[-1+n,t]^2(b[-1+n,t] + b[n,t])$$
$$\qquad + 2a[n,t]^2(b[n,t] + b[1+n,t])$$

Next, we introduce the second operator in the Lax pair which will be denoted by TodaP.

$In[5] := \text{TodaP}[r\_,f\_,n\_] := -\text{JacobiDE}[r+1,f,n] +$
$$\sum_{j=0}^{r} 2(a[n]g[j,n]S[\text{JacobiDE}[r-j,f,n],\{n,1\}]$$
$$- h[j,n]\text{JacobiDE}[r-j,f,n]) + g[r+1,n]f;$$

Here is the first one applied to a sequence $f(n)$

$In[6] := \text{TodaP}[0,f[n],n]$

$Out[6] = -a[-1+n]f[-1+n] + a[n]f[1+n]$

or, alternatively, in matrix form (cut off after 2 terms in both directions).

$In[7] := \text{m} = 2;$
$\qquad \text{Table}[\text{Coefficient}[\text{TodaP}[0,f[n],n],f[n+i-j]],$
$\qquad \{i,-m,m\},\{j,-m,m\}]//\text{MatrixForm}$
$\qquad \text{m} =.;$

$Out[7]//\text{MatrixForm}=$
$$\begin{pmatrix} 0 & -a[-1+n] & 0 & 0 & 0 \\ a[n] & 0 & -a[-1+n] & 0 & 0 \\ 0 & a[n] & 0 & -a[-1+n] & 0 \\ 0 & 0 & a[n] & 0 & -a[-1+n] \\ 0 & 0 & 0 & a[n] & 0 \end{pmatrix}$$

Similarly, the second one applied to a sequence $f(n)$

$In[9] := \text{TodaP}[1,f[n],n]//\text{Simplify}$

$Out[9] = -a[-2+n]a[-1+n]f[-2+n] - a[-1+n](b[-1+n] + b[n])f[-1+n] +$
$\qquad a[n](b[n]f[1+n] + b[1+n]f[1+n] + a[1+n]f[2+n])$

The following command will display the coefficients in a handy format.

$In[10] := \text{CoeffTodaP}[k\_] := \text{Block}[\{\text{dummy},i,n\},$
$\qquad \text{dummy} = \text{TodaP}[k,f[n],n];$
$\qquad \text{TableForm}[$
$\qquad\qquad \text{Table}[\{f[n+i]," : ",\text{Simplify}[\text{Coefficient}[\text{dummy},f[n+i]]]\},$
$\qquad\qquad \{i,-k-2,k+2\}]]$
$\qquad];$

For example we obtain for $r = 0$

$In[11] := \text{CoeffTodaP}[0]$

Out[11]//TableForm=
$$\begin{array}{rcl} f[-2+n] & : & 0 \\ f[-1+n] & : & -a[-1+n] \\ f[n] & : & 0 \\ f[1+n] & : & a[n] \\ f[2+n] & : & 0 \end{array}$$

and $r = 1$

```
In[12]:=CoeffTodaP[1]
```

Out[12]//TableForm=
$$\begin{array}{rcl} f[-3+n] & : & 0 \\ f[-2+n] & : & -a[-2+n]a[-1+n] \\ f[-1+n] & : & -a[-1+n](b[-1+n]+b[n]) \\ f[n] & : & 0 \\ f[1+n] & : & a[n](b[n]+b[1+n]) \\ f[2+n] & : & a[n]a[1+n] \\ f[3+n] & : & 0 \end{array}$$

We can also compute the commutator of H and P_{2r+2}

```
In[13]:=Commutator[r_,f_,n_] :=
          JacobiDE[TodaP[r,f,n],n] - TodaP[r,JacobiDE[f,n],n];
```

```
In[14]:=CoeffCommutator[r_] := Block[{dummy,i,n},
          dummy = Commutator[r,f[n],n];
          TableForm[Table[{f[n+i]," : ",
            Simplify[
              Coefficient[dummy,f[n+i]]]},{i,-2,2}]]   ];
```

and display its matrix coefficients

```
In[15]:=CoeffCommutator[0]
```

Out[15]//TableForm=
$$\begin{array}{rcl} f[-2+n] & : & 0 \\ f[-1+n] & : & a[-1+n](b[-1+n]-b[n]) \\ f[n] & : & 2(a[-1+n]^2 - a[n]^2) \\ f[1+n] & : & a[n](b[n]-b[1+n]) \\ f[2+n] & : & 0 \end{array}$$

C.8. Kac-van Moerbeke lattice

Finally we come to the Kac van Moerbeke lattice.

```
In[1]:= Needs["DiscreteMath`JacDEqs`"]
```

Again we start with the recursions for G_j and H_j and only consider the homogeneous case for simplicity.

C.8. Kac-van Moerbeke lattice

$In[2]:=$ $G[0, n_{-}] = 1; H[0, n_{-}] = 0;$
$G[j_{-}?IntegerQ, n_{-}] := G[j, n] =$
$\frac{1}{2}(H[j-1, n] + H[j-1, n-2]) + (\rho[n]^2 + \rho[n-1]^2)G[j-1, n];$
$H[j_{-}?IntegerQ, n_{-}] := H[j, n] =$
$2(\rho[n]\rho[n+1])^2 \sum_{l=0}^{j-1} G[j-l-1, n]G[l, n+1] - \frac{1}{2}\sum_{l=0}^{j-1} H[j-l-1, n]H[l, n];$

We can easily compute the first few

$In[3]:=$ $\text{Table}[\text{Simplify}[\{G[j, n], H[j, n]\}], \{j, 0, 1\}]//\text{TableForm}$

$Out[3]//TableForm=$
$\begin{array}{ll} 1 & 0 \\ \rho[-1+n]^2 + \rho[n]^2 & 2\rho[n]^2\rho[1+n]^2 \end{array}$

This command will display the r'th Kac-van Moerbeke equation

$In[4]:=$ $KM[r_{-}] := \text{Simplify}[$
$\partial_t\rho[n, t] == \rho[n](G[r+1, n+1] - G[r+1, n])] /. \rho[n_{-}] -> \rho[n, t];$

For example

$In[5]:=$ $KM[1]$

$Out[5]=$ $\rho^{(0,1)}[n, t] == -\rho[n, t](\rho[-2+n, t]^2\rho[-1+n, t]^2 + \rho[-1+n, t]^4$
$+ \rho[-1+n, t]^2\rho[n, t]^2 - \rho[1+n, t]^2(\rho[n, t]^2 + \rho[1+n, t]^2 + \rho[2+n, t]^2)$

Next, we need two pairs of sequences a_1, b_1 and a_2, b_2

$In[6]:=$ $a_1[n_{-}] := \rho_e[n]\rho_o[n]; b_1[n_{-}] := \rho_e[n]^2 + \rho_o[n-1]^2;$
$a_2[n_{-}] := \rho_e[n+1]\rho_o[n]; b_2[n_{-}] := \rho_e[n]^2 + \rho_o[n]^2;$

plus the corresponding quantities $g_{k,j}$, $h_{k,j}$, $k = 1, 2$.

$In[7]:=$ $g_{k_{-}}[0, n_{-}] = 1; h_{k_{-}}[0, n_{-}] = 0;$
$g_{k_{-}}[j_{-}?IntegerQ, n_{-}] := g_k[j, n] =$
$\frac{1}{2}(h_k[j-1, n] + h_k[j-1, n-1]) + b_k[n]g_k[j-1, n];$
$h_{k_{-}}[j_{-}?IntegerQ, n_{-}] := h_k[j, n] =$
$2a_k[n]^2 \sum_{l=0} j - 1g_k[j-l-1, n]g_k[l, n+1] -$
$\frac{1}{2}\sum_{l=0} j - 1h_k[j-l-1, n]h_k[l, n];$

They satisfy several relations which can be checked for given j:

$In[8]:=$ $\text{Simplify}[(g_2[j+1, n] ==$
$\rho_o[n]^2g_1[j, n+1] + \rho_e[n]^2g_1[j, n] + h_1[j, n]) /. j -> 5]$

$Out[8]=$ True

$In[9]:=$ $\text{Simplify}[(h_2[j+1, n] - h_1[j+1, n] == \rho_o[n]^2(2(\rho_e[n+1]^2g_1[j, n+1] -$
$\rho_e[n]^2g_1[j, n]) + (h_1[j, n+1] - h_1[j, n]))) /. j -> 5]$

$Out[9]=$ True

```
In[10]:=Simplify[(g₁[j + 1, n] == ρₑ[n]²g₂[j, n] + ρₒ[n − 1]²g₂[j, n − 1] + h₂[j, n − 1]) /. j− > 5]
Out[10]= True
```

$$\text{In[10]}:=\text{Simplify}[(g_1[j+1,n] == \rho_e[n]^2 g_2[j,n] + \rho_o[n-1]^2 g_2[j,n-1] + h_2[j,n-1]) \,/.\, j->5]$$

Out[10]= True

$$\text{In[11]}:=\text{Simplify}[(h_2[j+1,n-1] - h_1[j+1,n] == \rho_e[n]^2(-2(\rho_o[n]^2 g_2[j,n] - \rho_o[n-1]^2 g_2[j,n-1]) - (h_2[j,n] - h_2[j,n-1]))) \,/.\, j->5]$$

Out[11]= True

$$\text{In[12]}:=\text{Simplify}[(h_2[j,n-1] - h_1[j,n] == 2\rho_e[n]^2(g_1[j,n] - g_2[j,n])) \,/.\, j->5]$$

Out[12]= True

$$\text{In[13]}:=\text{Simplify}[(h_1[j+1,n] - h_2[j+1,n-1] == 2(a_1[n]^2 g_1[j,n] - a_2[n-1]^2 g_2[j,n-1]) - \rho_o[n]^2(h_2[j,n-1] - h_1[j,n]) + \rho_e[n]^2(h_2[j,n] - h_2[j,n-1])) \,/.\, j->5]$$

Out[13]= True

Now we introduce the corresponding operators A, A^*, H_k, $P_{k,2r+2}$, D, Q_{2r+2}, $k = 1, 2$.

$$\text{In[14]}:=\text{A}[f_-,n_-] := \rho_o[n]S[f,\{n,1\}] + \rho_e[n]f;$$
$$\text{A}^*[f_-,n_-] := \rho_o[n-1]S[f,\{n,-1\}] + \rho_e[n]f;$$
$$\text{JacobiDE}_k[f_-,n_-] := a_k[n]S[f,\{n,1\}] + S[a_k[n]f,\{n,-1\}] + b_k[n]f;$$
$$\text{JacobiDE}_k[r_-?\text{IntegerQ}, f_-, n_-] := \text{Nest}[\text{JacobiDE}_k[\#,n]\&, f, r];$$
$$\text{DiracDE}[\{f_-,g_-\}, n_-] := A[g,n], A^*[f,n];$$
$$\text{TodaP}_k[r_-, f_-, n_-] := -\text{JacobiDE}_k[r+1,f,n]$$
$$+ \sum_{j=0} r(2a_k[n]g_k[j,n]S[\text{JacobiDE}_k[r-j,f,n],n,1]$$
$$- h_k[j,n]\text{JacobiDE}_k[r-j,f,n]) + g_k[r+1,n]f;$$
$$\text{KMQ}[r_-, \{f_-,g_-\}, n_-] := \{\text{TodaP}_1[r,f,n], \text{TodaP}_2[r,g,n]\};$$

Again we can use *Mathematica* to check some simple formulas. For example $A^*A = H_1$

$$\text{In[15]}:=\text{Simplify}[A^*[A[f[n],n],n] == \text{JacobiDE}_1[f[n],n]]$$

Out[15]= True

or $AA^* = H_2$

$$\text{In[16]}:=\text{Simplify}[A[A^*[f[n],n],n] == \text{JacobiDE}_2[f[n],n]]$$

Out[16]= True

Moreover, we have $H_2^r A = A H_1^r$

$$\text{In[17]}:=\text{Simplify}[(\text{JacobiDE}_2[r, A[f[n],n], n] == A[\text{JacobiDE}_1[r,f[n],n],n]) \,/.\, r->5]$$

Out[17]= True

respectively $A^*H_2^r = H_1^r A^*$

In[18]:=Simplify[(A*[JacobiDE$_2$[r, f[n], n], n] == JacobiDE$_1$[r, A*[f[n], n], n])
   /. r− > 5]

Out[18]= True

and $P_{1,2r+2}A - AP_{2,2r+2} = \rho_o(g_{1,r+1}^+ - g_{2,r+1})S^+ + \rho_e(g_{2,r+1} - g_{1,r+1})$

In[19]:=Simplify[(TodaP$_2$[r, A[f[n], n], n] − A[TodaP$_1$[r, f[n], n], n] ==
   ρ_o[n](g$_1$[r + 1, n + 1] − g$_2$[r + 1, n])f[n + 1] + ρ_e[n](g$_2$[r + 1, n] − g$_1$[r + 1, n])f[n]) /. r− > 5]

Out[19]= True

respectively $P_{1,2r+2}A^* - A^*P_{2,2r+2} = -\rho_o^-(g_{2,r+1}^- - g_{1,r+1})S^- + \rho_e(g_{2,r+1} - g_{1,r+1})$

In[20]:=Simplify[(TodaP$_1$[r, A*[f[n], n], n] − A*[TodaP$_2$[r, f[n], n], n] ==
   − ρ_o[n − 1](g$_2$[r + 1, n − 1] − g$_1$[r + 1, n])f[n − 1] + ρ_e[n](g$_2$[r + 1, n] − g$_1$[r + 1, n])f[n]) /. r− > 5]

Out[20]= True

Bibliography

[1] R. Abraham, J.E. Marsden, and T. Ratiu, *Manifolds, Tensor Analysis, and Applications*, 2nd edition, Springer, New York, 1983.

[2] M. Adler, *On the Bäcklund transformation for the Gel'fand-Dickey equations*, Commun. Math. Phys. **80**, 517–527 (1981).

[3] M. Adler, L. Haine, and P. van Moerbeke, *Limit matrices for the Toda flow and periodic flags for loop groups*, Math. Ann. **296**, 1–33 (1993).

[4] R. P. Agarwal, *Difference Equations and Inequalities*, Marcel Dekker, New York, 1992.

[5] C. D. Ahlbrandt, *Continued fraction representation of maximal and minimal solutions of a discrete matrix Riccati equation*, Siam J. Math. Anal. **24**, 1597–1621 (1993).

[6] C. D. Ahlbrandt, *Dominant and recessive solutions of symmetric three term recurrences*, J. Diff. Eqs. **107**, 238–258 (1994).

[7] C. D. Ahlbrandt and W. T. Patula, *Recessive solutions of block tridiagonal nonhomogenous systems*, J. Difference Eqs. Appl. **1**, 1–15 (1995).

[8] N. I. Akhiezer, *A continuous analogue of orthogonal polynomials on a system of intervals*, Sov. Math. Dokl. **2**, 1409–1412 (1961).

[9] N. Akhiezer, *The Classical Moment Problem*, Oliver and Boyd, London, 1965.

[10] N. Akhiezer, *Elements of the Theory of Elliptic Functions*, Transl. Math. Monographs, vol. 79, Amer. Math. Soc., Providence, R. I., 1990.

[11] S. J. Al'ber, *Associated integrable systems*, J. Math. Phys. **32**, 916–922 (1991).

[12] A. J. Antony and M. Krishna, *Nature of some random Jacobi matrices*, preprint (1993).

[13] A. J. Antony and M. Krishna, *Inverse spectral theory for Jacobi matrices and their almost periodicity*, Proc. Indian Acad. Sci. (Math. Sci.) **104**, 777–818 (1994).

[14] A. J. Antony and M. Krishna, *Almost periodicity of some Jacobi matrices*, Proc. Indian Acad. Sci. (Math. Sci.) **102**, 175–188 (1992).

[15] A. I. Aptekarev, A. Branquinho, and F. Marcellan, *Toda-type differential equations for the recurrence coefficients of orthogonal polynomials and Freud transformation*, J. Comput. Appl. Math. **78**, 139–160 (1997).

[16] N. Aronszajn, *On a problem of Weyl in the theory of singular Sturm–Liouville equations*, Am. J. of Math. **79**, 597–610 (1957).

[17] N. Aronszajn and W. Donoghue, *On the exponential representation of analytic functions in the upper half-plane with positive imaginary part*, J. d'Analyse Math. **5**, 321–388 (1956-57).

[18] T. Asahi, *Spectral theory of the difference equations*, Prog. Theor. Phys. Suppl., **36**, 55–96 (1966).

[19] F. Atkinson, *Discrete and Continuous Boundary Problems*, Academic Press, New York, 1964.

[20] M. Audin, *Vecteurs propres de matrices de Jacobi*, Ann. Inst. Fourier **44**, 1505–1517 (1994).

[21] J. Avron and B. Simon *Singular continuous spectrum of a class of almost periodic Jacobi operators*, Bull. Amer. Math. Soc. **6**, 81–85 (1982).

[22] J. Avron and B. Simon *Almost periodic Schrödinger operators II. The integrated density of states*, Duke Math. J. **50**, 369–391 (1983).

[23] H. F. Baker, *Note on the foregoing paper, "Commutative ordinary differential operators", by J. L. Burchnall and T. W. Chaundy*, Proc. Roy. Soc. London **A118**, 584–593 (1928).

[24] B. Beckermann and V. Kaliaguine, *The diagonal of the Pade table and the approximation of the Weyl function of second-order difference operators*, Constructive Approximation **13**, 481–510 (1997).

[25] E. D. Belokolos, A. I. Bobenko, V. Z. Enol'skii, A. R. Its, and V. B. Matveev, *Algebro-geometric Approach to Nonlinear Integrable Equations*, Springer, Berlin, 1994.

[26] M. Benammar and W.D. Evans, *On the Friedrichs extension of semi-bounded difference operators*, Math. Proc. Camb. Phil. Soc. **116**, 167–177 (1994).

[27] Yu. M. Berezanskiĭ, *Expansions in Eigenfunctions of Self-adjoint Operators*, Transl. Math. Monographs, vol. 17, Amer. Math. Soc., Providence, R. I., 1968.

[28] Yu. M. Berezanskiĭ, *Integration of nonlinear difference equations by the inverse spectral problem method*, Soviet Math. Dokl., **31** No. 2, 264–267 (1985).

[29] Yu. M. Berezanski, *The integration of semi-infinite Toda chain by means of inverse spectral problem*, Rep. Math. Phys., **24** No. 1, 21–47 (1985).

[30] Yu. M. Berezansky, *Integration of nonlinear nonisospectral difference-differential equations by means of the inverse spectral problem*, in "Nonlinear Physics. Theory and experiment", (eds E. Alfinito, M. Boiti, L. Martina, F. Pempinelli), World Scientific, 11–20 (1996).

[31] Yu. M. Berezansky and M. I. Gekhtman, *Inverse problem of the spectral analysis and non-Abelian chains of nonlinear equations*, Ukrain. Math. J., **42**, 645–658 (1990).

[32] Yu. Berezansky and M. Shmoish, *Nonisospectral flows on semi-infinite Jacobi matrices*, Nonl. Math. Phys., **1** No. 2, 116–146 (1994).

[33] D. Bättig, B. Grébert, J. C. Guillot, and T. Kappeler, *Fibration of the phase space of the periodic Toda lattice*, J. Math. Pures Appl. **72**, 553–565 (1993).

[34] O. I. Bogoyavlenskii, *Algebraic constructions of integrable dynamical systems–extensions of the Volterra system*, Russian Math. Surv. **46:3**, 1–64 (1991).

[35] W. Bulla, F. Gesztesy, H. Holden, and G. Teschl *Algebro-Geometric Quasi-Periodic Finite-Gap Solutions of the Toda and Kac-van Moerbeke Hierarchies*, Memoirs of the Amer. Math. Soc. **135/641**, 1998.

[36] J. L. Burchnall and T. W. Chaundy, *Commutative ordinary differential operators*, Proc. London Math. Soc. Ser. 2, **21**, 420–440 (1923).

[37] J. L. Burchnall and T. W. Chaundy, *Commutative ordinary differential operators*, Proc. Roy. Soc. London **A118**, 557–583 (1928).

[38] P. Byrd and M. Friedman *Handbook of Elliptic Integrals for Engineers and Physists*, Springer, Berlin, 1954.

[39] L. Breiman, *Probability*, Addison-Wesley, London, 1968.

[40] R. Carmona and J. Lacroix, *Spectral Theory of Random Schrödinger Operators*, Birkhäuser, Boston, 1990.

[41] R. Carmona and S. Kotani, Inverse spectral theory for random Jacobi matrices, J. Stat. Phys. **46**, 1091–1114 (1987).

[42] K. M. Case, *Orthogonal polynomials from the viewpoint of scattering theory*, J. Math. Phys. **14**, 2166–2175 (1973).

[43] K. M. Case, *The discrete inverse scattering problem in one dimension*, J. Math. Phys. **15**, 143–146 (1974).

[44] K. M. Case, *Orthogonal polynomials II*, J. Math. Phys. **16**, 1435–1440 (1975).

[45] K. M. Case, *On discrete inverse scattering problems. II*, J. Math. Phys. **14**, 916–920 (1973).

[46] K. M. Case and S. C. Chiu *The discrete version of the Marchenko equations in the inverse scattering problem*, J. Math. Phys. **14**, 1643–1647 (1973).

[47] K. M. Case and M. Kac, *A discrete version of the inverse scattering problem*, J. Math. Phys. **14**, 594–603 (1973).

[48] S. Clark and D. Hinton, *Strong nonsubordinacy and absolutely continuous spectra for Sturm-Liouville equations*, Diff. Int. Eqs. **6**, 573–586 (1993).

[49] W. Craig and B. Simon, *Subharmonicity of the Lyaponov exponent*, Duke Math. J. **50**, 551–560 (1983).

[50] H. L. Cyclon, R. G. Froese, W. Kirsch and B. Simon, *Schrödinger Operators*, Springer, Berlin, 1987.

[51] D. Damanik and D. Lenz, *Uniform spectral properties of one-dimensional quasicrystals, I. Absence of eigenvalues*, Comm. Math. Phys. (to appear).

[52] D. Damanik and D. Lenz, *Uniform spectral properties of one-dimensional quasicrystals, II. The Lyapunov exponent*, preprint (1999).

[53] E. Date and S. Tanaka *Analogue of inverse scattering theory for the discrete Hill's equation and exact solutions for the periodic Toda lattice*, Prog. Theoret. Phys. **56**, 457–465 (1976).

[54] P. Deift, *Applications of a commutation formula*, Duke Math. J. **45**, 267–310 (1978).

[55] P. A. Deift and L. C. Li, *Generalized affine Lie algebras and the solution of a class of flows associated with the QR eigenvalue algorithm*, Commun. Pure Appl. Math. **42**, 963–991 (1989).

[56] P. Deift and L. C. Li, *Poisson geometry of the analog of the Miura maps and Bäcklund-Darboux transformations for equations of Toda type and periodic Toda flows*, Commun. Math. Phys. **143**, 201–214 (1991).

[57] P. Deift, *Orthogonal Polynomials and Random Matrices: A Riemann-Hilbert Approach*, Courant Lecture Notes in Mathematics 3, Courant Institute, New York, 1999.

[58] P. Deift and R. Killip, *On the absolutely continuous spectrum of one dimensional Schrödinger operators with square summable potentials*, preprint (1999).

[59] P. Deift and K. T-R. McLaughlin, *A Continuum Limit of the Toda Lattice*, Memoirs of the Amer. Math. Soc. **131/624**, 1998.

[60] P. Deift and E. Trubowitz, *Inverse scattering on the line*, Comm. Pure Appl. Math. **32**, 121–251 (1979).

[61] P. Deift and X. Zhou, *Long-time asymptotics for integrable systems. Higher order theory.*, Commun. Math. Phys. **165**, 175–191 (1994).

[62] P. Deift and X. Zhou, *A steepest descent method for oscillatory Riemann-Hilbert problems. Asymptotics for the MKdV equation*, Ann. of Math. **137**, 295–368 (1993).

[63] P. Deift, T. Kriecherbauer, and S. Venakides, *Forced lattice vibrations. I*, Commun. Pure Appl. Math. **48**, 1187–1250 (1995).

[64] P. Deift, T. Kriecherbauer, and S. Venakides, *Forced lattice vibrations. II*, Commun. Pure Appl. Math. **48**, 1251–1298 (1995).

[65] P. Deift, L.C. Li, and C. Tomei, *Toda flows with infinitely many variables*, J. Func. Anal. **64**, 358–402 (1985).

[66] P. Deift, L. C. Li, and C. Tomei, *Matrix factorizations and integrable systems*, Commun. Pure Appl. Math. **42**, 443–521 (1989).

[67] P. Deift, S. Kamvissis, T. Kriecherbauer, and X. Zhou, *The Toda rarefaction problem*, Commun. Pure Appl. Math. **49**, 35–83 (1996).

[68] R. del Rio, S. Jitomirskaya, Y. Last, and B. Simon, *Operators with Singular Continuous Spectrum, IV. Hausdorff dimensions, rank one perturbations, and localization*, J. Anal. Math. **69**, 153–200 (1996).

[69] F. Delyon and B. Souillard, *The rotation number for finite difference operators and its properies*, Comm. Math. Phys. **89**, 415–426 (1983).

[70] F. Delyon and B. Souillard, *Remark on the continuity of the density of states of ergodic finite difference operators*, Comm. Math. Phys. **94**, 289–291 (1984).

[71] J. Dombrowski, *Absolutely continuous measures for systems of orthogonal polynomials with unbounded recurrence coefficients*, Constructive Approximation **8**, 161–167 (1992).

[72] J. Dombrowski and S. Pedersen, *Orthogonal polynomials, spectral measures, and absolute continuity*, J. Comput. Appl. Math. **65**, 115–124 (1995).

[73] W. Donoghue, *Monotone mMatrix Functions and Analytic Continuation*, Springer, Berlin, 1974.

[74] B. Dubrovin, *Theta functions and nonlinear equations*, Russian Math. Surveys. **362**, 11–92 (1981).

[75] B. Dubrovin, I. Krichever, and S. Novikov *Integrable systems I*, in "*Dynamical Systems IV*", (eds. V. Arnold, S. Novikov), 173–280 (1990).

[76] B. A. Dubrovin, V. B. Matveev, and S. P. Novikov, *Non-linear equations of Korteweg-de Vries type, finite-zone linear operators, and Abelian varieties*, Russian Math. Surv. **31:1**, 59–146 (1976).

[77] J. Edwards, *Spectra of Jacobi matrices, differential equations on the circle, and the $su(1,1)$ Lie algebra*, SIAM J. Math. Anal. **24**, No. 3, 824–831 (1993).

[78] S. N. Eilenberger, *An Introduction to Difference Equations*, Springer, New York, 1996.

[79] G. Elaydi, *Solitons*, Springer, Berlin, 1981.

[80] L. Faddeev and L. Takhtajan, *Hamiltonian Methods in the Theory of Solitons*, Springer, Berlin, 1987.

[81] H. Farkas and I. Kra, *Riemann Surfaces*, 2nd edition, GTM 71, Springer, New York, 1992.

[82] J. Fay, *Theta Functions on Riemann Surfaces*, LN in Math. 352, Springer, Berlin, 1973.

[83] A. Finkel, E. Isaacson, and E. Trubowitz, *An explicit solution of the inverse periodic problem for Hill's equation*, SIAM J. Math. Anal. **18**, 46–53 (1987).

[84] H. Flaschka, *On the Toda lattice. I*, Phys. Rev. B**9**, 1924–1925 (1974).

[85] H. Flaschka, *On the Toda lattice. II*, Progr. Theoret. Phys. **51**, 703–716 (1974).

[86] O. Forster, *Lectures on Riemann Surfaces*, GTM 81, Springer, New York, 1991.

[87] T. Fort, *Finite Differences and Difference Equations in the Real Domain*, Oxford University Press, London, 1948.

[88] L. Fu, *An Inverse theorem for Jacobi matrices*, Linear Algebra Appl. **12**, 55–61 (1975).

[89] L. Fu and H. Hochstadt, *Inverse theorems for Jacobi matrices*, J. Math. Anal. Appl. **47**, 162–168 (1974).

[90] H. Furstenberg and H. Kesten, *Products of random matrices*, Ann. Math. Stat. **31**, 457–469 (1960).

[91] F. R. Gantmacher, *The Theory of Matrices*, Vol. 1, Chelsa, New York, 1990.

[92] M. G. Gasymov and G. Sh. Gusejnov, *On the inverse problems of spectral analysis for infinite Jacobi matrices in the limit-circle*, Sov. Math. Dokl. **40**, 627–630 (1990).

[93] I. M. Gel'fand and B. M. Levitan, *On a simple identity for eigenvalues of a second order differential operator* Dokl. Akad. Nauk SSSR **88**, 593–596 (1953); English translation in Izrail M. Gelfand, *Collected papers*, Vol. I (S. G. Gindikin et al, eds), pp 457–461, Springer, Berlin, 1987.

[94] J. S. Geronimo, *A relation between the coefficients in the recurrence formula and the spectral function for orthogonal polynomials*, Trans. Amer. Math. Soc. **260**, 65–82 (1980).

[95] J. S. Geronimo, *On the spectra of infinite–dimensional Jacobi matrices*, J. App. Th. **53**, 251–265 (1988).

[96] J. S. Geronimo, *An upper bound on the number of eigenvalues of an infinite–dimensional Jacobi matrix*, J. Math. Phys. **23**, 917–921 (1982).

[97] J. S. Geronimo, *Scattering theory, orthogonal polynomials, and q-series*, Siam J. Anal. **25**, 392–419 (1994).

[98] J. S. Geronimo and W. Van Assche, *Orthogonal polynomials with asymptotically periodic recurrence coefficients*, J. App. Th. **46**, 251–283 (1986).

[99] K. M. Case and J. S. Geronimo, *Scattering theory and polynomials orthogonal on the unit circle*, J. Math. Phys. **20**, 299–310 (1979).

[100] J. S. Geronimo and R. A. Johnson, *Rotation number associated with difference equations satisfied by polynomials orthogonal on hte unit circle*, J. Diff. Eqs., **132**, 140–178 (1996).

[101] F. Gesztesy, *A complete spectral characterization of the double commutation method*, J. Funct. Anal. **117**, 401–446 (1993).

[102] F. Gesztesy and H. Holden, *Trace formulas and conservation laws for nonlinear evolution equations*, Rev. Math. Phys. **6**, 51–95 (1994).

[103] F. Gesztesy and H. Holden, *Hierarchies of Soliton Equations and their Algebro-Geometric Solutions*, monograph in preparation.

[104] F. Gesztesy and B. Simon, *m-functions and inverse spectral analysis for finite and semi-infinite Jacobi matrices*, J. Anal. Math. **73**, 267–297 (1997).

[105] F. Gesztesy and B. Simon, *The xi function*, Acta Math. **176**, 49–71 (1996).

[106] F. Gesztesy and B. Simon, *Uniqueness theorems in inverse spectral theory for one-dimensional Schrödinger operators*, Trans. Amer. Math. Soc. **348**, 349–373 (1996).

[107] F. Gesztesy and B. Simon, *Rank one perturbations at infinite coupling*, J. Funct. Anal. **128**, 245–252 (1995).

[108] F. Gesztesy and R. Svirsky, *(m)KdV-Solitons on the Background of Quasi-Periodic Finite-Gap Solutions*, Memoirs of the Amer. Math. Soc. **118**, 1995.

[109] F. Gesztesy and G. Teschl, *On the double commutation method*, Proc. Amer. Math. Soc. **124**, 1831–1840 (1996).

[110] F. Gesztesy and G. Teschl, *Commutation methods for Jacobi operators*, J. Diff. Eqs. **128**, 252–299 (1996).

[111] F. Gesztesy and Z. Zhao, *Critical and subcritical Jacobi operators defined as Friedrichs extensions*, J. Diff. Eq. **103**, 68–93 (1993).

[112] F. Gesztesy, H. Holden, and B. Simon, *Absolute summability of the trace relation for certain Schrödinger operators*, Com. Math. Phys. **168**, 137–161 (1995).

[113] F. Gesztesy, M. Krishna, and G. Teschl, *On isospectral sets of Jacobi operators*, Com. Math. Phys. **181**, 631–645 (1996).

[114] F. Gesztesy, R. Ratnaseelan, and G. Teschl, *The KdV hierarchy and associated trace formulas*, in "*Proceedings of the International Conference on Applications of Operator Theory*", (eds. I. Gohberg, P. Lancaster, and P. N. Shivakumar), Oper. Theory Adv. Appl., 87, Birkhäuser, Basel, 125–163 (1996).

[115] F. Gesztesy ,R. Nowell, and W. Pötz, *One-dimensional scattering for quantum systems with nontrivial spatial asymptotics*, Diff. Integral Eqs. **10**, 521–546 (1997).

[116] F. Gesztesy, B. Simon, and G. Teschl, *Zeroes of the Wronskian and renormalized oscillation Theory*, Am. J. Math. **118** 571–594 (1996).

[117] F. Gesztesy, H. Holden, B. Simon, and Z. Zhao, *On the Toda and Kac-van Moerbeke systems*, Trans. Amer. Math. Soc. **339**, 849–868 (1993).

[118] F. Gesztesy, H. Holden, B. Simon, and Z. Zhao, *Higher order trace relations for Schrödinger operators*, Rev. Math. Phys. **7**, 893–922 (1995).

[119] D. J. Gilbert, *On subordinacy and analysis of the spectrum of Schrödinger operators with two singular endpoints*, Proc. Roy. Soc. Edinburgh. **112A**, 213–229 (1989).

[120] D. J. Gilbert and D. B. Pearson, *On subordinacy and analysis of the spectrum of Schrödinger operators*, J. Math. Anal. Appl. **128**, 30–56 (1987).

[121] I. M. Glazman, *Direct Methods of Qualitative Spectral Analysis of Singular Differential Operators*, I.P.S.T., Jerusalem 1965.

[122] A. Ya. Gordon, *On the point spectrum of one-dimensional Schrödinger operators (in Russian)* Usp. Mat. Nauk **31**, 257-258 (1976).

[123] E. van Groesen and E.M. de Jager, *Mathematical Structures in Continuous Dynamical Systems*, North–Holland, Amsterdam 1994.

[124] G. S. Guseinov, *The inverse problem of scattering theory for a second-order difference equation on the whole axis*, Soviet Math. Dokl., **17**, 1684–1688 (1976).

[125] G. S. Guseinov, *The determination of an infinite Jacobi matrix from the scattering data*, Soviet Math. Dokl., **17**, 596–600 (1976).

[126] G. S. Guseinov, *Scattering problem for the infinite Jacobi matrix*, Izv. Akad. Nauk Arm. SSR, Mat. **12**, 365–379 (1977).

[127] H. Hancock, *Theory of Elliptic Functions*, Dover Publications, New York, 1958.

[128] P. Hartman, *Difference equations: Disconjugacy, principal solutions, Green's functions, complete monotonicity*, Trans. Amer. Math. Soc., **246**, 1–30 (1978).

[129] R. Hartshorne, *Algebraic Geometry*, Springer, Berlin, 1977.

[130] A. J. Heeger, S. Kivelson, J. R. Schrieffer, and W. P. Su, *Solitons in conducting polymers*, Revs. Mod. Phys. **60**, 781–850 (1988).

[131] E. Hellinger, *Zur Stieltjeschen Kettenbruch Theorie*, Math. Ann. **86**, 18–29 (1922).

[132] D. B. Hinton and R. T. Lewis, *Spectral analysis of second order difference equations*, J. Math. Anal. Appl. **63**, 421–438 (1978).

[133] J. W. Hooker and W. T. Patula, *Riccati type transformations for second-order linear difference equations*, J. Math. Anal. Appl. **82**, 451–462 (1981).

[134] J. W. Hooker, M. K. Kwong and W. T. Patula, *Oscillatory second-order linear difference equations and Riccati equations*, SIAM J. Math. Anal., **18**, 54–63 (1987).

[135] E. K. Ifantis, *A criterion for the nonuniqueness of the measure of orthogonality*, J. Approximation Theory **89**, 207–218, (1997).

[136] A. R. Its and V. B. Matveev, *Schrödinger operators with finite-gap spectrum and N-soliton solutions of the Korteweg-de Vries equation*, Theoret. Math. Phys. **23**, 343–355 (1975).

[137] C. G. T. Jacobi, *Über eine neue Methode zur Integration der hyperelliptischen Differentialgleichung und über die rationale Form ihrer vollständigen algebraischen Integralgleichungen*, J. Reine Angew. Math. **32**, 220–226 (1846).

[138] S. Jitomirskaya, *Almost everything about the almost Mathieu operator, II*, in "Proceedings of XI International Congress of Mathematical Physics, Paris 1994", Int. Press, 373–382 (1995).

[139] S. Jitomirskaya and Y. Last, *Dimensional Hausdorff properties of singular continuous spectra*, Phys. Rev. Lett. **76**, 1765–1769 (1996).

[140] S. Jitomirskaya and B. Simon, *Operators with singular continuous spectrum, III. Almost periodic Schrödinger operators*, Commun. Math. Phys. **165**, 201–205 1994.

[141] C. Jordan, *Calculus of Finite Difference Equations*, Chelsea, New York, 1979.

[142] M. Kac and P. van Moerbeke, *On an explicitly soluble system of nonlinear differential equations, related to certain Toda lattices*, Adv. Math. **16**, 160–169 (1975).

[143] M. Kac and P. van Moerbeke, *On some periodic Toda lattices*, Proc. Nat. Acad. Sci. USA **72**, 2879–2880 (1975).

[144] M. Kac and P. van Moerbeke, *A complete solution of the periodic Toda problem*, Proc. Nat. Acad. Sci. USA **72**, 1627–1629 (1975).

[145] S. Kahn and D.B. Pearson, *Subordinacy and spectral theory for infinite matrices*, Hel. Phys. Acta **65**, 505–527 (1992).

[146] Y. Kato, *On the spectral density of periodic Jacobi matrices*, in "Non-linear Integrable Systems–Classical Theory and Quantum Theory", (eds., M. Jimbo and T. Miwa), World Scientific, 153–181, (1983).

[147] W. Kelley and A. Peterson, *Difference Equations: An Introduction with Applications*, Academic Press, San Diego, 1991.

[148] J. Kingman, *Subadditive ergodic theory*, Ann. Prob. **1**, 883–909 (1973).

[149] A. Kiselev, *Absolutely continuous spectrum of one-dimensional Schrödinger operators and Jacobi matrices with slowly decreasing potentials*, Comm. Math. Phys. **179**, 377-400 (1996).

[150] M. Klaus, *On bound states of the infinite harmonic crystal*, Hel. Phys. Acta **51**, 793–803 (1978).

[151] O.Knill, *Factorization of random Jacobivoperators and Bäcklund transformations*, Commun. Math. Phys. **151** , 589–605 (1993).

[152] O. Knill, *Renormalization of random Jacobi operators*, Commun. Math. Phys. **164**, 195–215 (1993).

[153] S. Kotani and M. Krishna, Almost periodicity of some random potentials, J. Funct. Anal. **78**, 390–405 (1988).

[154] A. Krazer, *Lehrbuch der Thetafunktionen*, Chelsea Publ. Comp., New York, 1970.

[155] M. G. Krein, *Perturbation determinants and a formula for the traces of unitary and self-adjoint operators*, Soviet Math. Dokl. **3**, 707–710 (1962).

[156] I. M. Krichever, *Algebraic curves and nonlinear difference equations*, Russian Math. Surveys. **334**, 255–256 (1978).

[157] I. M. Krichever, *Nonlinear equations and elliptic curves*, Rev. of Science and Technology **23**, 51–90 (1983).

[158] I. M. Krichever, *Algebro-geometric spectral theory of the Schrödinger difference operator and the Peierls model*, Soviet Math. Dokl. **26**, 194–198 (1982).

[159] I. M. Krichever, *The Peierls model*, Funct. Anal. Appl. **16**, 248–263 (1982).

[160] I. Krichever, *Algebraic-geometrical methods in the theory of integrable equations and their perturbations*, Acta Appl. Math. **39**, 93–125 (1995).

[161] B. A. Kupershmidt, *Discrete Lax equations and differential-difference calculus*, Astérisque **123**, 1985.

[162] Y. Last, *Almost everything about the almost Mathieu operator, I*, in "Proceedings of XI International Congress of Mathematical Physics, Paris 1994", Int. Press, 366–372 (1995).

[163] Y. Last and B. Simon *Eigenfunctions, transfer matrices, and absolutely continuous spectrum of one-dimensional Schrödinger operators*, Invent. Math., **135**, 329–367 (1999).

[164] P. D. Lax *Integrals of nonlinear equations of evolution and solitary waves*, Comm. Pure and Appl. Math. **21**, 467–490 (1968).

[165] F. G. Maksudov, Eh. M. Bajramov, and R. U. Orudzheva, *The inverse scattering problem for an infinite Jacobi matrix with operator elements*, Russ. Acad. Sci., Dokl., Math. **45**, No.2, 366–370 (1992).

[166] F. Mantlik and A. Schneider, *Note on the absolutely continuous spectrum of Sturm-Liouville operators*, Math. Z. **205**, 491–498 (1990).

[167] V. A. Marchenko, *Sturm–Liouville Operators and Applications*, Birkhäuser, Basel, 1986.

[168] H. P. McKean, *Integrable systems and algebraic curves*, in "Global Analysis", (eds., M. Grmela and J. E. Marsden), Lecture Notes in Mathematics 755, Springer, Berlin, 83–200 (1979).

[169] H. McKean, *Variation on a theme of Jacobi*, Comm. Pure Appl. Math. **38**, 669–678 (1985).

[170] H. P. McKean and P. van Moerbeke, *Hill and Toda curves*, Commun. Pure Appl. Math. **33**, 23–42 (1980).

[171] R. E. Mickens, *Difference Equations. Theory and Applications*, 2nd ed., Van Nostrand Reinhold, New York, 1990.

[172] R. M. Miura, *Korteweg-de Vries equation and generalizations. I. A remarkable explicit nonlinear transformation*, J. Math. Phys. **9**, 1202–1204 (1968).

[173] P. van Moerbeke, *The spectrum of Jacobi Matrices*, Inv. Math. **37**, 45–81 (1976).

[174] P. van Moerbeke, *About isospectral deformations of discrete Laplacians*, in "*Global Analysis*", (eds., M. Grmela and J. E. Marsden), Lecture Notes in Mathematics 755, Springer, Berlin, 313–370 (1979).

[175] P. van Moerbeke and D. Mumford *The spectrum of difference operators and algebraic curves*, Acta Math. **143**, 97–154 (1979).

[176] F. Sh. Mukhtarov, *Reconstruction of an infinite nonselfadjoint Jacobi matrix by scattering data*, Dokl., Akad. Nauk Az. SSR **40**, No.10, 10–12 (1984).

[177] D. Mumford, *An algebro-geometric construction of commuting operators and of solutions to the Toda lattice equation, Korteweg de Vries equation and related non-linear equations*, Intl. Symp. Algebraic Geometry, 115–153, Kyoto, 1977.

[178] D. Mumford, *Tata Lectures on Theta II*, Birkhäuser, Boston, 1984.

[179] S. N. Naboko and S. I. Yakovlev *The discrete Schrödinger operator. The point spectrum lying on the continuous spectrum*, St. Petersbg. Math. J. **4**, No.3, 559–568 (1993).

[180] S. N. Naboko and S. I. Yakovlev *On the point spectrum of discrete Schrödinger operators*, Funct. Anal. Appl. **26**, No.2, 145-147 (1992).

[181] R. Narasimhan, *Compact Riemann Surfaces*, Birkhäuser, Basel, 1992.

[182] R. Nevanlinna *Asymptotische Entwicklung beschränkter Funktionen und das Stieltjesche Momentenproblem*, Ann. Acad. Sci. Fenn. A **18**, 5, 52pp. (1922).

[183] A. J. Niemi and G. W. Semenoff, *Fermion number fractionization in quantum field theory*, Phys. Rep. **135**, 99–193 (1986).

[184] S. Novikov, S. V. Manakov, L. P. Pitaevskii, and V. E. Zakharov, *Theory of Solitons*, Consultants Bureau, New York, 1984.

[185] F. W. J. Olver, *Asymptotics and Special Functions*, Academic Press, New York, (1974).

[186] B. P. Osilenker, *M. G. Krein's ideas in the theory of orthogonal polynomials* (English translation) Ukr. Math. J. **46**, 75–86 (1994).

[187] L. Pastur, *Spectral properties of disordered systems in one-body approximation*, Comm. Math. Phys. **75**, 179–196 (1980).

[188] W. T. Patula, *Growth and oscillation properties of second order linear difference equations*, SIAM J. Math. Anal. **6**, 55–61 (1979).

[189] W. T. Patula, *Growth, oscillation and comparison theorems for second order linear difference equations*, SIAM J. Math. Anal. **6**, 1272–1279 (1979).

[190] A. Perelomov, *Integrable Systems of Classical Mechanics and Lie Algebras*, Birkhäuser, Basel, (1990).

[191] CH. G. Philos and Y. G. Sficas, *Positive solutions of difference equations*, Proc. of the Amer. Math. Soc. **108**, 107–115 (1990).

[192] M. Reed and B. Simon, *Methods of Modern Mathematical Physics I. Functional Analysis*, rev. and enl. edition, Academic Press, San Diego, 1980.

[193] M. Reed and B. Simon, *Methods of Modern Mathematical Physics II. Fourier Analysis, Self-Adjointness*, Academic Press, San Diego, 1975.

[194] M. Reed and B. Simon, *Methods of Modern Mathematical Physics III. Scattering Theory*, Academic Press, San Diego, 1979.

[195] M. Reed and B. Simon, *Methods of Modern Mathematical Physics IV. Analysis of Operators*, Academic Press, San Diego, 1978.

[196] W. Renger, *Toda soliton limits on general background*, in Differential equations, asymptotic analysis, and mathematical physics, (ed. M. Demuth), 287-291, Math. Res. 100, Akademie Verlag, (1997).

[197] F. S. Rofe-Beketov, *A test for the finiteness of the number of discrete levels introduced into gaps of a continuous spectrum by perturbations of a periodic potential*, Soviet Math. Dokl. **5**, 689–692 (1964).

[198] C. A. Rodgers, *Hausdorff Measures*, Cambridge Univ. Press, London, 1970.

[199] C. A. Rodgers and S. J. Taylor, *The analysis of additive set functions in Euclidean space*, Acta Math., Stock., **101**, 273–302 (1959).

[200] C. A. Rodgers and S. J. Taylor, *The analysis of additive set functions in Euclidean space*, Acta Math., Stock., **109**, 207–240 (1963).

[201] D. Ruelle, *Ergodic theory of differentiable dynamical systems.*, Publ. Math., Inst. Hautes Etud. Sci. **50**, 27–58 (1979).

[202] W. Rudin *Real and Complex Analysis*, 2nd ed., McGraw Hill, New York, 1974.

[203] B. Simon, *Trace Ideals and their Applications*, Cambridge Univ. Press, Cambridge, 1979.

[204] B. Simon, *Schrödinger semigroups*, Bulletin Amer. Math. Soc. **7**, 447-526 (1982).

[205] B. Simon, *Kotani theory for one dimensional Jacobi matrices*, Com. Math. Phys. **89**, 227–234 (1983).

[206] B. Simon, *Spectral Analysis of Rank One Perturbations and Applications*, proceedings, "Mathematical Quantum Theory II: Schrödinger Operators", (eds. J. Feldman, R. Froese, and L. M. Rosen), CRM Proc. Lecture Notes 8, 109–149 (1995).

[207] B. Simon, *Bounded eigenfunctions and absolutely continuous spectra for one-dimensional Schrödinger operators*, Proc. Amer. Math. Soc. **124**, 3361–3369 (1996).

[208] B. Simon, *Operators with singular continuous spectrum. I: General operators*, Ann. Math.,**141**, 131–145 (1995).

[209] B. Simon, *The classical moment problem as a self-adjoint finite difference operator*, Advances in Math. **137**, 82–203 (1998).

[210] Smirnov, *Finite-gap solutions of Abelian Toda chain of genus 4 and 5 in elliptic functions*, Teoret. Mat, Fiz., **78**, No.1, 11–21 (1989).

[211] A. Sinap and W. Van Assche, *Orthogonal matrix polynomials and applications*, J. Comput. Appl. Math. **66**, 27–52 (1996).

[212] M.L. Sodin and P.M. Yuditskiĭ, *Infinite-zone Jacobi matrices with pseudo-extendible Weyl functions and homogeneous spectrum*, Russian Acad. Sci. Dokl. Math. **49**, 364–368 (1994).

[213] M. Sodin and P. Yuditskii, *Almost periodic Jacobi matrices with homogeneous spectrum, infinite dimensional Jacobi inversion, and Hardy spaces of character-automorphic functions*, preprint, (1994).

[214] J. M. Steele, *Kingman's subadditive ergodic theorem* Ann. Inst. Henri Poincare, Probab. Stat. **25**, No.1, 93–98 (1989)

[215] G. Stolz, *Spectral Theory for slowly oscillating potentials I. Jacobi Matrices*, Man. Math. **84**, 245–260 (1994).

[216] J.C.F. Sturm *Mémoire sur les équations différentielles linéaires du second ordre*, J. Math. Pures Appl., **1**, 106–186 (1836).

[217] G. Szegö, *Orthogonal Polynomials*, Amer. Math. Soc., Providence, R.I., 1978.

[218] H. Tanabe, *Equations of Evolution*, Pitman, London, 1979.

[219] G. Teschl, *Inverse scattering transform for the Toda hierarchy*, Math. Nach. (to appear).

[220] G. Teschl, *On the Toda and Kac-van Moerbeke hierarchies*, Math. Z. (to appear).

[221] G. Teschl, *Spectral Deformations of Jacobi Operators*, J. Reine. Angew. Math. **491**, 1–15 (1997).

[222] G. Teschl, *Trace Formulas and Inverse Spectral Theory for Jacobi Operators*, Comm. Math. Phys. **196**, 175–202 (1998).

[223] G. Teschl, *Oscillation theory and renormalized oscillation theory for Jacobi operators*, J. Diff. Eqs. **129**, 532–558 (1996).

[224] G. Teschl, *On the initial value problem for the Toda and Kac-van Moerbeke hierarchies*, in "Proceedings of the International Conference on Differential Equations and Mathematical Physics, Birmingham 1999" (to appear).

[225] B. Thaller, *The Dirac Equation*, Springer, Berlin, 1991.

[226] B. Thaller, *Normal forms of an abstract Dirac operator and applications to scattering theory*, J. Math. Phys. **29**, 249–257 (1988).

[227] E. Titchmarsh, *Eigenfunction Expansions Associated with Second-Order Differential Equations, Part I*, Oxford University Press, Oxford, 1946.

[228] E. Titchmarsh, *Eigenfunction Expansions Associated with Second-Order Differential Equations, Part II*, Oxford University Press, Oxford, 1958.

[229] E. C. Titchmarsh, *The Theory of Functions*, 2nd edition, Oxford Univ. Press, Oxford, 1985.

[230] M. Toda, *Theory of Nonlinear Lattices*, 2nd enl. ed., Springer, Berlin, 1989.

[231] M. Toda, *Theory of Nonlinear Waves and Solitons*, Kluwer, Dordrecht, 1989.

[232] M. Toda, *Selected pPapers of Morikazu Toda*, ed. by M. Wadati, World Scientific, Singapore, 1993.

[233] M. Toda, *Vibration of a chain with nonlinaer interactions*, J. Phys. Soc. Jpn **22**, 431–436 (1967).

[234] M. Toda and M. Wadati, *A canonical transformation for the exponential lattice*, J. Phys. Soc. Jpn **39**, 1204–1211 (1975).

[235] S. Venakides and P. Deift, R. Oba *The Toda shock problem*, Comm. Pure Appl. Math. **44**, 1171–1242 (1991).

[236] K. Ueno and K. Takasaki, *Toda lattice hierarchy*, in "*Advanced Studies in Pure Mathematics 4*", (ed. K. Okamoto), North-Holland, Amsterdam, 1–95 (1984).

[237] K. Ueno and K. Takasaki, *Toda lattice hierarchy. I'*, Proc. Japan Acad., Ser. A **59**, 167-170 (1983).

[238] K. Ueno and K. Takasaki, *Toda lattice hierarchy. II'*, Proc. Japan Acad., Ser. A **59**, 215-218 (1983).

[239] M. Wadati and M. Toda, *Bäcklund transformation for the exponential lattice*, J. Phys. Soc. Jpn **39**, 1196–1203 (1975).

[240] H. Wall, *Analytic Theory of Continued Fractions*, Chelsa, New York, 1973.

[241] J. Weidmann, *Linear Operators in Hilbert Spaces*, Springer, New York, 1980.

[242] S. T. Welstead, *Selfadjoint extensions of Jacobi matrices of limit-circle type*, J. Math. Anal. Appl. **89**, 315–326 (1982).

[243] S. T. Welstead, *Boundary conditions at infinity for difference equations of limit-circle type*, J. Math. Anal. Appl. **89**, 442–461 (1982).

[244] H. Weyl, *Über gewöhnliche Differentialgleichungen mit Singularitäten und die zugehörige Entwicklung willkürlicher Funktionen*, Math. Ann. **68**, 220–269 (1910).

[245] Y. Xu, *Block Jacobi matrices and zeros of multivariate orthogonal polynomials*, Trans. Am. Math. Soc. **342**, 855–866 (1994).

[246] V. E. Zakharov, S. L. Musher, and A. M. Rubenchik, *Nonlinear stage of parametric wave excitation in a plasma*, JETP Lett. **19**, 151–152 (1974).

Glossary of notations

$a_R(n)$	$= a(2n_0 - n - 1)$ reflected coefficient, 15
$\underline{A}_{p_0}$	...Abel map, 281
$\underline{\alpha}_{p_0}$	...Abel map for divisors, 282
$\arg(.)$	...argument of a complex number, $\arg(z) \in (-\pi, \pi]$
$b_R(n)$	$= b(2n_0 - n)$ reflected coefficient, 15
$b^{(\ell)}(n)$	...expansion coefficient of $\ln g(z, n)$, 112
$b^{\beta,(\ell)}(n)$	...expansion coefficient of $\ln \gamma^\beta(z, n)$, 113
$B_\pm(\tau)$	...67
$B_{\pm,f}(\tau)$	...67
$c(z, n)$	$= c(z, n, 0)$
$c(z, n, n_0)$	...fundamental solution of $\tau u = zu$ vanishing at $n_0 + 1$, 6
$c_\beta(z, n)$	$= c_\beta(z, n, 0)$
$c_\beta(z, n, n_0)$	...fundamental solution of $\tau u = zu$ corresponding to general boundary conditions at n_0, 37
$\mathbb{C}$	...the set of complex numbers
$\mathbb{C}_\pm$	$= \{z \in \mathbb{C}|\ \pm \operatorname{Im}(z) > 0\}$
$C^\alpha(\rho)$	...support of the α-continuous part of ρ, 302
$C(M_1, M_2)$	...continuous functions from M_1 to M_2
$C^\infty(M_1, M_2)$	...smooth functions from M_1 to M_2
$\chi_\Lambda(.)$	...characteristic function of the set Λ
deg	...degree of a divisor (or of a polynomial), 277
δ_n	...$\delta_n(m) = \delta_{n,m}$
$\delta_{n,m}$	...Kronecker delta $\delta_{n,m} = 0$, if $n \neq m$ and $\delta_{n,n} = 1$
dim	...dimension of a linear space
$\operatorname{dist}(A, B)$	...distance between two sets A, B
D^α	...α-derivative, 302
$\mathcal{D}$	...divisor on a Riemann surface, 277
$\operatorname{Div}(M)$	...ring of divisors on M, 277
$\mathfrak{D}(.)$	...domain of an operator

e	... exponential function, $e^z = \exp(z)$
$\mathbb{E}(.)$	... expectation of a random variable, 87
$f_\pm(k,n)$	... Jost solutions, 167
$g(z,n)$	$= G(z,n,n)$
$G_r(z,n)$	... polynomial, 144
$G(z,n,m)$	... Green function of H, 16
$\gamma^\pm(z)$	... Lyapunov exponent, 7
$\gamma^\beta(z,n)$	$= g(z,n+1) + \frac{\beta}{a(n)}h(z,n) + \beta^2 g(z,n)$, 18
h^α	... Hausdorff measure, 302
$h(z,n)$	$= 2a(n)G(z,n,n+1) - 1$
H	... Jacobi operator, 13
$H^\beta_{\pm,n_0}$	... Jacobi operator on $\ell^2(n_0,\pm\infty)$, 16
$H_{\pm,n_0}$	$= H^\infty_{\pm,n_0}$
$H_\pm$	$= H_{\pm,0}$
$H^{\beta_1,\beta_2}_{n_1,n_2}$	... Jacobi operator on $\ell^2(n_1,n_2)$, 16
H_{n_1,n_2}	$= H^{\infty,\infty}_{n_1,n_2}$
H_R	... reflected Jacobi operator, 15
$\hat{H}$	... Helmholtz operator, 20
$\mathfrak{H}$	... (separable) Hilbert space
$H_{r+1}(z,n)$	... polynomial, 145
hull(c)	... hull of a sequence, 100
$\text{Hyp}_\pm(\Phi)$	... set of hyperbolic numbers, 7
i	... complex unity, $i^2 = -1$
$i(\mathcal{D})$	$= \dim \mathcal{L}^1(\mathcal{D})$ index of specialty, 279
Im$(.)$	... imaginary part of a complex number
$\text{Iso}_R(\Sigma)$	... set of reflectionless Jacobi operators with spectrum Σ, 141
J_{n_1,n_2}	... Jacobi matrix, 12
$J(M)$	... Jacobian variety of the Riemann surface M, 281
$\text{KM}_r(\rho)$	... Kac-van Moerbeke hierarchy, 258
$\widetilde{\text{KM}}_r(\rho)$	... homogeneous Kac-van Moerbeke hierarchy, 259
Ker	... kernel of an operator
$K^\beta_{r+1}(z,n)$	... polynomial, 150
$\ell(I,M)$	... set of M-valued sequences with index set $I \subseteq \mathbb{Z}$, 3
$\ell(I)$	$= \ell(I,\mathbb{C})$
$\ell^p(I,M)$	... sequences in $\ell(I,M)$ being p summable, 3
$\ell^p(I)$	$= \ell^p(I,\mathbb{C})$
$\ell^\infty(I,M)$	... sequences in $\ell(I,M)$ being bounded, 3
$\ell^\infty(I)$	$= \ell^\infty(I,\mathbb{C})$
$\ell_0(I,M)$	... sequences in $\ell(I,M)$ with only finitely many nonzero values, 3
$\ell_0(I)$	$= \ell_0(I,\mathbb{C})$
$\ell^p_\pm(\mathbb{Z},M)$	... sequences in $\ell(I,M)$ being ℓ^p near $\pm\infty$, 3
$\ell_\pm(\mathbb{Z})$	$= \ell_\pm(I,\mathbb{C})$
$\ell^2(\mathbb{Z};w)$	$=$ weighted ℓ^2 space, 20

Glossary of notations

$L^1_{loc}(x_0, x_1)$	...space of (equivalence classes of) locally integrable functions over the interval (x_0, x_1)
$L^2(x_0, x_1)$	...Hilbert space of (equivalence classes of) square integrable functions over the interval (x_0, x_1)
$\mathcal{L}(\mathcal{D})$	$= \{f \in \mathcal{M}(M) \mid f = 0 \text{ or } (f) \geq \mathcal{D}\}$, 279
$\mathcal{L}^1(\mathcal{D})$	$= \{\omega \in \mathcal{M}^1(M) \mid \omega = 0 \text{ or } (\omega) \geq \mathcal{D}\}$, 279
λ	...a real number
$\lambda_j^\beta(n)$	...eigenvalues associated with general boundary conditions, 139
$m^\pm(z)$	...Floquet multipliers, 116
$m_\pm(z, n)$	...Weyl m-functions, 27
$m_\pm(z)$	$= m_\pm(z, 0)$
$\tilde{m}_\pm(z, n)$	...Weyl $\tilde{m}$-functions, 29
$\tilde{m}_\pm(z)$	$= \tilde{m}_\pm(z, 0)$
$m_\pm^\beta(z)$	...Weyl m-functions associated with general boundary conditions, 29
$\tilde{m}_\pm^\beta(z)$	...Weyl $\tilde{m}$-functions associated with general boundary conditions, 39
M	...manifold or Riemann surface
$M(z)$	...Weyl matrix, 46
$M^\beta(z)$	...Weyl matrix corresponding to general boundary conditions, 46
$\mathcal{M}(M)$	...meromorphic functions on M
$\mathcal{M}^1(M)$	...meromorphic forms on M
$\mu_j(n)$	...Dirichlet eigenvalue, 135
$\mathbb{N}$	...the set of positive integers
$\mathbb{N}_0$	$= \mathbb{N} \cup \{0\}$
$N_\pm(\tau)$	...set of λ for which no subordinate solution exists, 66
p	...point on M
$P_\Lambda(.)$	...family of spectral projections of an operator
$\Phi(z, n, n_0)$	...fundamental matrix, 7
$\Phi(z, n)$	$= \Phi(z, n, 0)$
$\mathbb{Q}$	...the set of rational numbers
$q(z)$	...Floquet momentum, 116
Q^α	...302
$r(\mathcal{D})$	$= \dim \mathcal{L}(\mathcal{D})$, 279
$\operatorname{Re}(.)$	...real part of a complex number
$\mathbb{R}$	...the set of real numbers
$R_{2r+2}^{1/2}(z)$	...143
$R_j(n)$	...residue associated with the Dirichlet eigenvalue $\mu_j(n)$, 136
$R_\pm(k)$	...reflection coefficient(constant background), 173
$R_\pm(\lambda)$	...reflection coefficient, 129
Ran	...range of an operator
R^α	...302
ρ	...a Borel measure

$s(z,n)$	$= s(z,n,0)$
$s(z,n,n_0)$	...fundamental solution of $\tau u = zu$ vanishing at n_0, 6
$s_\beta(z,n)$	$= s_\beta(z,n,0)$
$s_\beta(z,n,n_0)$	...fundamental solution of $\tau u = zu$ corresponding to general boundary conditions at n_0, 37
$S(k)$	...scattering matrix (constant background), 173
$S(\lambda)$	...scattering matrix, 129
$S^\pm$	...shift expressions, 4
$S_\pm(H)$	...scattering data of H, 174
$S(\tau)$	...59
$S_f(\tau)$	...59
$S_\pm(\tau)$	...58
$S_{f,\pm}(\tau)$	...58
$\sigma(H)$	...spectrum of an operator H, 15
$\sigma_{ac}(H)$	...absolutely continuous spectrum of H, 41, 45
$\sigma_{sc}(H)$	...singular continuous spectrum of H, 41, 45
$\sigma_{pp}(H)$	...pure point spectrum of H, 41, 45
$\sigma_s(H)$	$= \sigma_{sc}(H) \cup \sigma_{pp}(H)$
$\sigma_p(H)$	...point spectrum (set of eigenvalues) of H, 41
$\sigma_{ess}(H)$	...essential spectrum of H, 60
$\sigma_{\alpha c}(H)$	...α-continuous spectrum of H, 41
$\sigma_{\alpha s}(H)$	...α-singular spectrum of H, 41
$\sigma_j(n)$	...sign associated with the Dirichlet eigenvalue $\mu_j(n)$, 136
$\sigma_j^\beta(n)$	...sign associated with the eigenvalue $\lambda_j^\beta(n)$, 151
$\mathrm{sgn}(.)$	...sign of a real number, $\mathrm{sgn}(\lambda) = -1$ if $\lambda < 0$, $\mathrm{sgn}(0) = 0$, and $\mathrm{sgn}(\lambda) = 1$ if $\lambda > 0$
$T(k)$	...transmission coefficient(constant background), 173
$T(\lambda)$	...transmission coefficient, 129
τ	...second order, symmetric difference expression, 5
$\underline{\tau}$	$= (\tau_{i,j})_{i,j=1}^g$, matrix of b periods of ζ_j, 276
$\theta(\underline{z})$	...Riemann theta function, 284
$\mathrm{TL}_r(a,b)$	...Toda hierarchy, 227
$\widetilde{\mathrm{TL}}_r(a,b)$	...homogeneous Toda hierarchy, 228
tr	...trace of an operator
$u(z,n)$	...solution of $\tau u = zu$
$u_\pm(z,.)$	...solution of $\tau u = zu$ being square summable near $\pm\infty$, 30
$U(t,s)$	...unitary propagator, 228
$W_n(.,..)$	...Wronskian of two sequences, 6
$\xi(\lambda,n)$	...xi function for Dirichlet boundary conditions, 112
$\xi^\beta(\lambda,n)$	...xi function for general boundary conditions, 113
$\underline{\Xi}_{p_0}$	...vector of Riemann constants, 285
$\mathbb{Z}$	...the set of integers
z	...a complex number

Glossary of notations

$\mathbb{1}$	...identity operator
$\sqrt{z}$	...standard branch of the square root, 19
$\overline{z}$	...complex conjugation
$\|\cdot\|_p$	...norm in the Banach space ℓ^p, 3
$\|\cdot\|$	...norm in the Hilbert space ℓ^2, 13
$\langle .,..\rangle_{\mathfrak{H}}$	...scalar product in $\mathfrak{H}$, 4
$\langle .,..\rangle$	...scalar product in $\ell^2(I)$, 13
$\oplus$	...direct sum of linear spaces or operators
∂	...forward difference, 5
∂^*	...backward difference, 5
$\sum^*$	...indefinite sum, 9
$\prod^*$	...indefinite product, 10
$\top$	...transpose of a matrix
$\#(u)$	...total number of nodes of u, 75
$\#_{(m,n)}(u)$	...number of nodes of u between m and n, 75
$\#W(u_1,u_2)$	...total number of nodes of $W_.(u_1,u_2)$, 81
$\#_{(m,n)}W(u_1,u_2)$	...number of nodes of $W_.(u_1,u_2)$ between m and n, 81
$[.]_n$	...Weyl bracket, 31
$[.]_\pm$	...upper/lower triangular part of a matrix, 5
$[.]_0$	...diagonal part of a matrix, 5
$[\![x]\!]$	$= \sup\{n \in \mathbb{Z} \,\|\, n < x\}$, 74
$\overline{M}^{ess}$	...essential closure of a set, 301
(λ_1, λ_2)	$= \{\lambda \in \mathbb{R} \,\|\, \lambda_1 < \lambda < \lambda_2\}$, open interval
$[\lambda_1, \lambda_2]$	$= \{\lambda \in \mathbb{R} \,\|\, \lambda_1 \le \lambda \le \lambda_2\}$, closed interval

Index

Abel transform, 5
Abel's map, 281
Abel's theorem, 282
Abelian differentials, 275
Additive process, 91
Almost Mathieu operator, 101
Almost periodic sequence, 100
Anderson model, 88
Antisymmetric part, 15
Asymptotic expansion, 106

Bäcklund transformation, 259
Baker-Akhiezer function, 148
Banach space, 3
Bessel function, 25
Birkhoff's ergodic theorem, 91
Birman-Schwinger type bound, 70
Borel
　measure, 297
　set, 297
　transform, 298
Boundary condition, 16, 18
Branch number, 274
Burchnall-Chaundy polynomial, 234

Canonical divisor, see divisor
Canonical form, 40
Canonical homology basis, 274
Casoratian, 189
Cauchy
　transform, 298
Characteristic polynomial, 12
Closure
　essential, 301
　operator, 47
Commutation method, 185
　Dirichlet deformation method, 207
　double, 198
　single, 187
Commutator, 224
Conservation law, 252
Continued fraction, 10
Crystal
　exponential interaction, 221
　linear interaction (harmonic), 22
C^*-algebra, 227
Cyclic vector, 40

Deficiency indices, 48
Degree, 274
Degree of a divisor, see divisor
Density of states, 95
Deterministic sequences, 100
Difference expression, 4
　commuting, 231
　matrix representation, 4
　order, 4
　second order, symmetric, 5
　skew-symmetric, 4
　symmetric, 4
　triangular parts, 5
Dirac operator, 262
Dirichlet boundary condition, 16
Dirichlet eigenvalues, 134
Discriminant, 116
Dispersion, 25
　relation, 25
Divisor, 277
　associated with eigenvalues, 154
　canonical, 278
　degree, 277
　index of specialty, 279
　nonspecial, 279
　positive, 279
　principal, 278
　relatively prime, 279
Divisor class group, 278
Divisor of poles, 279
Divisor of zeros, 279
Double commutation method, 198
Dynamical system, 87
　ergodic, 88

Eigenfunction expansion, 40, 45
Elliptic functions, 162
Elliptic integral, 163
Ergodic, 88
Essential spectrum, 60

349

Finite lattices, 231
Floquet
  discriminant, 116
  functions, 118
  momentum, 117
  monodromy matrix, 115
  multipliers, 117
  solutions, 117
  theory, 115
Fourier transform, 20, 298
Fundamental matrix, 7
Fundamental polygon, 273
Fundamental solutions, 6

Gel'fand-Levitan-Marchenko equation, 176
Genus, 273
Green function, 16
Green's formula, 6
Gronwall-type inequality, 177

Haar measure, 100
Hamburger moment problem, 42
Hamiltonian system, 22, 222
Hausdorff dimension, 302
Hausdorff measure, 302
Helmholtz operator, 20
Herglotz function, 298
Hilbert space, 4, 13
  weighted, 20
Hilbert-Schmidt operator, 51
Holomorphic differentials, 274
Holomorphic structure, 273
Homology basis, 274
Hull, 100
Hyperbolic numbers, set of, 8
Hyperelliptic Riemann surface, 281
Hypergeometric function (generalized), 25

Index of specialty, see divisor
Initial conditions, 6
Initial value problem
  Kac-van Moerbeke system, 258
  Toda system, 230
Intersection matrix, 293
Inverse scattering transform, 251
Isospectral operators, 140

Jacobi equation, 6
  inhomogeneous, 9
Jacobi inversion problem, 288
Jacobi matrix, 12, 119
  trace, 12
Jacobi operator, 13
  almost periodic, 100
  critical, 36
  free, 19
  maximal, 47
  minimal, 47
  periodic, 115
  quasi-periodic, finite-gap, 153
  random, 87
  reflectionless, 133
    finite-gap, 142
  subcritical, 36
  supercritical, 36
  unbounded, 47
Jacobian variety, 281
Jensen's inequality, 93, 102
Jordan canonical form, 116
Jost solutions, 167

Kac-van Moerbeke hierarchy, 258
  homogeneous, 259
Krein spectral shift, 109

Langmuir lattice, 262
Lax equation, 224
Lax operator, 227
  homogeneous, 225
Lebesgue decomposition, 41, 45
Limit circle, 39, 47
Limit point, 39, 47
Liouville number, 103
Lusin's theorem, 297
Lyapunov exponent, 7

Matrix valued measure, 44
Measure
  Baire, 100
  Borel, 297
  ergodic, 88
  growth point, 299
  invariant, 88
  Jacobi, 52
  probability, 87
  stationary, 88
Meromorphic differentials, 274
Meromorphic functions, 274
m-function, 27, 50
$\tilde{m}$-function, 29, 50
Minimal positive solution, 35
Miura-type transformation, 259
Möbius transformation, 38
Moment problem, 42, 52
  monic, 230
Monodromy matrix, 115

Neumann expansion, 105
Nevanlinna-Pick function, 298
Newton's interpolation formula, 123
Node
  sequence, 75
  Wronskian, 81
Noether gap, 281
Nonspecial divisor, see divisor
Norming constants, 174

oliton solutions, 265
Order
  difference expression, 4
  function, 274
Orthogonal polynomials, 40
Oscillatory, 76

Periodic coefficients, 115
Perturbation determinant, 111, 172
Phase space, 23
Picard group, 278
Pick function, 298
Plücker identity, 50
Poisson-Jensen formula, 174
Positive divisor, see divisor
Principal divisor, see divisor
Principal solutions, 35
Probability space, 87
Propagator, 228
Prüfer variables, 73

Quantum mechanics
  supersymmetric, 15
Quasi-periodic, 153

Index

Radon-Nikodym Theorem, 300
RAGE Theorem, 90
Random
    Jacobi operator, 87
    variable, 87
Recessive solutions, 35
Reflection, 14
Reflection coefficient, 174
Reflectionless, see Jacobi operator
Relatively prime, see divisor
Residue, 274
Resolvent, 16
    Neumann expansion, 105
    set, 16
Riccati equation (discrete), 10
Riemann inequality, 280
Riemann surface, 273
Riemann theta function, 284
Riemann vanishing theorem, 288
Riemann-Hilbert problem, 272
Riemann-Hurwitz relation, 274
Riemann-Roch Theorem, 280
Rotation number, 95

Scalar product, 4, 13
Scattering data, 174
Scattering matrix, 173
Scattering theory, 129, 167
Shift expressions, 4
Single commutation method, 187
Solution
    weak, 235
Spectral measure
    Herglotz function, 299
    Jacobi operator, 28, 44
Spectral projections, 28
Spectral resolution, 28
Spectral shift function, 111
Spectrum, 15
    absolutely continuous, 41, 45
    point, 41, 59
    pure point, 41, 45
    singular continuous, 41, 45
Stable manifold, 8
Stieltjes inversion formula, 299
Stone-Weierstrass Theorem, 298
Sturm's separation theorem, 75
Sturm-Liouville expression, 5
Subadditive ergodic theorem, 91
Subadditive process, 91
Subharmonic, 92
Submean, 92
Subordinacy, 65
Subordinacy, principle of, 68
Subordinate solution, 65
Summation by parts, 5
Supersymmetric, 15, 262
Support, 300
    essential, 300
    minimal, 300
Symmetric part, 15
Symmetric power of a manifold, 286
Symplectic form, 23
Symplectic transform, 23

Theorem
    Birkhoff's ergodic, 91
    Lusin, 297
    Radon-Nikodym, 300
    RAGE, 90
    Riemann-Roch, 280

    Stone-Weierstrass, 298
    Sturm, 75
    subadditive ergodic, 91
    Torelli, 283
Thouless formula, 97
Tight binding approximation, 87
Toda hierarchy, 227
    homogeneous, 228
    stationary, 231
Topological group, 100
Torelli's theorem, 283
Total branching number, 274
Trace measure, 45
Trace relations, 120
Transfer matrix, 7
Transformation operator, 169, 188, 212
Transmission coefficient, 174

Unstable manifold, 8
Uppersemicontinuous, 92

Variation of constants formula, 9
Vector of Riemann constants, 286
Vector valued polynomial, 44
Volterra sum equation, 126

Weak solution, 235
Weierstrass gap, 281
Weierstrass point, 281
Weyl
    bracket, 31
    circle, 38
    matrix, 46
    m-function, 27, 50
    $\bar{m}$-function, 29, 50
Wronskian, 6

Xi function, 112

Young's inequality, 179

Zeros of the Riemann theta function, 286

Selected Titles in This Series

(Continued from the front of this publication)

- 40.4 **Daniel Gorenstein, Richard Lyons, and Ronald Solomon,** The classification of the finite simple groups, number 4, 1999
- 40.3 **Daniel Gorenstein, Richard Lyons, and Ronald Solomon,** The classification of the finite simple groups, number 3, 1998
- 40.2 **Daniel Gorenstein, Richard Lyons, and Ronald Solomon,** The classification of the finite simple groups, number 2, 1995
- 40.1 **Daniel Gorenstein, Richard Lyons, and Ronald Solomon,** The classification of the finite simple groups, number 1, 1994
- 39 **Sigurdur Helgason,** Geometric analysis on symmetric spaces, 1994
- 38 **Guy David and Stephen Semmes,** Analysis of and on uniformly rectifiable sets, 1993
- 37 **Leonard Lewin, Editor,** Structural properties of polylogarithms, 1991
- 36 **John B. Conway,** The theory of subnormal operators, 1991
- 35 **Shreeram S. Abhyankar,** Algebraic geometry for scientists and engineers, 1990
- 34 **Victor Isakov,** Inverse source problems, 1990
- 33 **Vladimir G. Berkovich,** Spectral theory and analytic geometry over non-Archimedean fields, 1990
- 32 **Howard Jacobowitz,** An introduction to CR structures, 1990
- 31 **Paul J. Sally, Jr. and David A. Vogan, Jr., Editors,** Representation theory and harmonic analysis on semisimple Lie groups, 1989
- 30 **Thomas W. Cusick and Mary E. Flahive,** The Markoff and Lagrange spectra, 1989
- 29 **Alan L. T. Paterson,** Amenability, 1988
- 28 **Richard Beals, Percy Deift, and Carlos Tomei,** Direct and inverse scattering on the line, 1988
- 27 **Nathan J. Fine,** Basic hypergeometric series and applications, 1988
- 26 **Hari Bercovici,** Operator theory and arithmetic in H^∞, 1988
- 25 **Jack K. Hale,** Asymptotic behavior of dissipative systems, 1988
- 24 **Lance W. Small, Editor,** Noetherian rings and their applications, 1987
- 23 **E. H. Rothe,** Introduction to various aspects of degree theory in Banach spaces, 1986
- 22 **Michael E. Taylor,** Noncommutative harmonic analysis, 1986
- 21 **Albert Baernstein, David Drasin, Peter Duren, and Albert Marden, Editors,** The Bieberbach conjecture: Proceedings of the symposium on the occasion of the proof, 1986
- 20 **Kenneth R. Goodearl,** Partially ordered abelian groups with interpolation, 1986
- 19 **Gregory V. Chudnovsky,** Contributions to the theory of transcendental numbers, 1984
- 18 **Frank B. Knight,** Essentials of Brownian motion and diffusion, 1981
- 17 **Le Baron O. Ferguson,** Approximation by polynomials with integral coefficients, 1980
- 16 **O. Timothy O'Meara,** Symplectic groups, 1978
- 15 **J. Diestel and J. J. Uhl, Jr.,** Vector measures, 1977
- 14 **V. Guillemin and S. Sternberg,** Geometric asymptotics, 1977
- 13 **C. Pearcy, Editor,** Topics in operator theory, 1974
- 12 **J. R. Isbell,** Uniform spaces, 1964
- 11 **J. Cronin,** Fixed points and topological degree in nonlinear analysis, 1964
- 10 **R. Ayoub,** An introduction to the analytic theory of numbers, 1963
- 9 **Arthur Sard,** Linear approximation, 1963
- 8 **J. Lehner,** Discontinuous groups and automorphic functions, 1964

For a complete list of titles in this series, visit the
AMS Bookstore at **www.ams.org/bookstore/**.